Freshwater Mussels of Central America

This book describes a fauna of global conservation concern and provides a framework for researchers to begin to test hypotheses regarding the evolution of freshwater mussels in Central America. Placing the Central American fauna in context with that of North and South America, the authors fill a gap in our knowledge of this endangered and largely endemic fauna. This book:

- Provides a full-color treatment of all 81 native and introduced freshwater mussel species found from the Río Grande in Texas, USA, and México to Colombia, South America
- Includes 110 distribution maps of all species
- Presents complete synonymies in every systematic species account
- Offers historical information and bios of previous workers in the field
- Cites complete literature and references on freshwater mussels in the region

Through photographs, distribution maps, and a complete list of species described and reported from Central America, *Freshwater Mussels of Central America* will inspire researchers to begin filling in the gaps regarding the systematics and ecology of freshwater mussels in Central America. It will be an essential source for conservationists, aquatic biologists, and those interested in the natural history of temperate and tropical freshwater ecosystems.

Freshwater Mussels of Central America

Kevin S. Cummings
Daniel L. Graf
John M. Pfeiffer
Jeremy S. Tiemann

CRC Press
Taylor & Francis Group
Boca Raton London New York

CRC Press is an imprint of the
Taylor & Francis Group, an **informa** business

First edition published 2025
by CRC Press
2385 NW Executive Center Drive, Suite 320, Boca Raton FL 33431

and by CRC Press
4 Park Square, Milton Park, Abingdon, Oxon, OX14 4RN

CRC Press is an imprint of Taylor & Francis Group, LLC

ISBN: 9781032323176 (hbk)
ISBN: 9781032323183 (pbk)
ISBN: 9781003314448 (ebk)

DOI: 10.1201/9781003314448

Typeset in Times
by codeMantra

Contents

PART 3 Freshwater Mussel Studies in Central America

PART 4 Zoogeography of Central American Freshwater Mussels

PART 5 Species Accounts

About the Authors

Kevin S. Cummings was born and raised in Chicago and is a lifelong resident of Illinois. He obtained B.S. and M.A. degrees in Zoology from Southern Illinois University in Carbondale and was employed as a Research Scientist and Curator of Mollusks at the Illinois Natural History Survey, University of Illinois, Urbana-Champaign for 38 years. He is also a Research Associate at the Field Museum in Chicago, the Carnegie Museum of Natural History in Pittsburgh, and the Academy of Natural Sciences of Drexel University, Philadelphia.

He has conducted research in the conservation, systematics, and ecology of freshwater mollusks, as well as the protection of freshwater habitats, primarily streams. He has conducted fieldwork on mollusks around the world including the Amazon, Orinoco, and Xingu rivers of South America; the Zambezi and Congo rivers in Africa; and the Mighty Mississippi River in North America. His recent research has centered around the global diversity of freshwater mussels, particularly in Central and South America.

He was a Past President of the Freshwater Mollusk Conservation Society and received the Society's Lifetime Achievement Award in March 2015. He also received the Illinois Chapter of the American Fisheries Society, the Stephen A. Forbes Excellence in Fisheries Award in 2008, and the Outstanding Public Service Award from Prairie Rivers Network in 2013. He has served on various state, federal, and professional society committees on the conservation of freshwater mollusks including the Illinois Endangered Species Technical Advisory Committee on Invertebrates, the American Fisheries Society, Endangered Species Committee, Endangered Freshwater Mussels Subcommittee, the American Malacological Society Conservation Committee, the Upper Mississippi River Conservation Committee, and the International Union for the Conservation of Nature (IUCN).

Daniel L. Graf, after completing his B.S. from the College of Biological Sciences at the University of Minnesota (Twin Cities), Minnesota, studied Invertebrate Zoology while earning M.S. and Ph.D. degrees from Northeastern University in Boston and the University of Michigan, respectively. He was hired as Assistant Curator of Malacology at the Academy of Natural Sciences in Philadelphia and then as Assistant Professor of Biology at the University of Alabama (Tuscaloosa). He currently holds the title of Professor of Aquatic Invertebrate Zoology at the University of Wisconsin-Stevens Point.

The common throughlines of Prof. Graf's research have been using natural history collections and applying the methods of biodiversity informatics to address questions about the origin and maintenance of freshwater molluscan diversity. He is a member of the American Malacological Society and the Freshwater Mollusk Conservation Society.

John M. Pfeiffer was born in Detroit and grew up in Oakland County, Michigan. He earned a bachelor's degree from Northern Michigan University, a master's degree from the University of Alabama, and a Ph.D. from the University of Florida. Currently, he serves as a Research Zoologist and Curator of Bivalves at the Smithsonian's National Museum of Natural History. His research uses genomics, biodiversity informatics, and natural history collections to explore the evolution, ecology, and conservation of freshwater mussels. His fieldwork specializes in collecting elusive and enigmatic freshwater mussels, and he has been chasing "Unicorns" in Thailand, Cambodia, Borneo, Sumatra, Costa Rica, and Panama. He lives in Maryland with his partner, Cassi, and their three children—Cypress, Clover, and Brightly. Outside of work, he enjoys camping, fly fishing, and mayonnaise.

Jeremy S. Tiemann is a Kansas native. He received his Bachelor of Science from the University of Kansas and his Master of Science from Emporia State University in Emporia, Kansas. He has been employed as an Associate Aquatic Ecologist for the Illinois Natural History Survey at the University of Illinois in Urbana-Champaign since 2002. His research interests include documenting basic natural history information of non-game fishes and freshwater mollusks, addressing the effects of anthropogenic disturbances on stream ecosystems, and restoring stream habitats and aquatic faunas. He has collaborated on research projects in North America, Central America, South America, and Africa.

Jeremy has served on various state, federal, and professional society committees on the conservation of fishes and freshwater mollusks, including the Illinois Endangered Species Technical Advisory Committee on both fishes and aquatic invertebrates, the U.S. Geological Survey and U.S. Fish & Wildlife Service Black Carp work group, the American Fisheries Society's Endangered Species Committee—Aquatic Gastropods Subcommittee, and the American Fisheries Society's Aquatic Invasive Species Policy work group. He is also a Past President of both the Freshwater Mollusk Conservation Society and the Illinois Chapter of the American Fisheries Society.

Acknowledgments

We would like to thank the current and former curators and collection managers of the museums that we visited for allowing us access to examine and photograph specimens under their care: Paula Mikkelsen and Marla Coppolino (AMNH); Winston Ponder (AMS); Gary Rosenberg and Paul Callomon (ANSP); Dawn Roberts (CHAS); Tim Pearce (CM); Dr. Edna Naranjo-Garcia (CNMO); Deneb Ortigosa (COMA); Liz Shea (DMNH); John Slapcinsky and the late Fred Thompson (FLMNH); Rüdiger Bieler, Jochen Gerber, and Marty Pryzdia (FMNH); Rachel Vinsel and Alison Stodola (INHS), Adam Baldinger, the late Richard Johnson, Gonzalo Giribet, and Jennifer Trimble (MCZ); Philippe Bouchet and Virginie Héros (MNHN); Art Bogan and Jamie Smith (NCSM); Kathie Way and Jon Ablett (NHMUK); the late Tom Watters (OSUM); Ronald Janssen and Karl Otto-Nagel (SMF); Diarmaid O'Foighil and Taehwan Lee (UMMZ); Robert Hershler, Ellen Strong, Tim Coffer, Kathryn Ahlfeld, and Paul Greenhall (USNM); and Thomas von Rintelen, Matthias Glaubrecht, Frank Köhler, and Christine Zorn (ZMB).

We thank Michael Hart, Anna Pieri, Charles Randklev (Texas A&M), Mike Larimore (KY), Amy Burden (UK), Adriana Reyes-Gomez (Universidad de Guadalajara, México), José DeJesus de la Rosa Bolón (Universidad Autónoma de Chiapas (UNACH), Alejandro Arreola Reza (Queretaro, México), Nate Johnson (FLMNH), Kentaro Inoue (Shedd Aquarium, Chicago, Illinois), Eduardo Gonzalez, Enrique Corona, and Pablo Gonzalez (Universidad Autónoma de Tamaulipas, México) for assistance in the field. Dr. Edna Naranjo-Garcia assisted with permit acquisition in México. Thanks to John Lyons for sharing his knowledge on fishes in México and pointing us to the relevant piscine literature. We would like to thank Ashley Sharpe for help with the zooarchaeology section of the manuscript. Partial funding for field work in Mexico was provided by the U.S. Fish and Wildlife Service and the Mohamed bin Zayed Species Conservation Fund. Base maps from Natural Earth. Free vector and raster map data @ naturalearthdata.com.

We would also like to thank the following for providing us with photographs: Topiltzin Contreras-MacBeath, José DeJesus de la Rosa Bolón, Adrián González-Guillén (Cuba), Kristen Grace (University of Florida), Kentaro Inoue, John Karges (Nature Conservancy) Aaron Koch, Charley Morones, Kevin J. Roe (Iowa State University), Mark H. Sabaj (ANSP), and Rebecca Young.

Chris Harper, Gary Pandolfi, and Susan Oetker (U.S. Fish and Wildlife Service) provided funding (F17AC01008) for some of our work in the ríos Conchos and Pánuco. Samples were collected under TE79165C-0 (USFWS permit for *P. popeii*) and SGPA/DGVS/002175/18 (México *Popenaias*).

We especially thank Aaron Koch and Heather Warman (SUP Kentucky), Enrique Aguado, and Charly Morones (Kayak Huasteca) and J. Alfredo Rodríguez-Pineda (Instituto de Biología, Departamento de Zoología, Universidad Nacional Autónoma de México) for logistical support and taking very good care of us while on various trips to México.

Kevin would like to thank his mentors Larry Page and Brooks Burr and INHS lab members Alice Stodola, Rachel Vinsel, and Sarah Douglas for their encouragement and help. I would also like to thank my friends Larry Gates, Anthony Gravino, Ian Shepherd, Brian Wilm, Rob Krumm, Peter Blackmon, Candy Foster, the Markland's, the Gallo's, The Kane's, the Zooter Brothers, Martin, Charlie and the rest of my Rose Bowl family. Thanks to my neighbor Janna Pflugmacher APRN, Reinaldo J Sanchez-Torrez, MD, and Lindsay Roberts, APRN, and all my other friends in the Urbana-Champaign area for their encouragement and support in my time of need. Thanks Hector Romero, Sue Davis, and Lori Simon for the food! I would also like to offer my gratitude to lifelong friends Mariann Tsilis, Jamyn Fiorito, and Lynda Bihl Jaros and friends and family back home in the Chicago area for helping me across the finish line when I needed it the most. I especially want to give very special thanks to my beautiful and talented wife Kathy Harden for her unwavering love and support and for keeping the home fires burning while I was in the field and on the mend.

Introduction

The Recent freshwater mussels of the order Unionoida are represented by six families: Unionidae Rafinesque, 1820; Margaritiferidae Haas, 1940; Hyriidae Swainson, 1840; Iridinidae Swainson, 1840; Mycetopodidae Gray, 1840; and Etheriidae Deshayes, 1832. Only the families Unionidae and Mycetopodidae are represented in Central America, but the fauna is poorly known and is badly in need of revision. Many species were described based upon one or two specimens, and the type localities of many taxa are vague: some to the level of country only. The last comprehensive treatments of the fauna were made by Fischer and Crosse (1870–1878; 1878–1902) and von Martens (1890–1901). Species lists for Central America with general comments on distribution were included in Simpson (1900; 1914), Frierson (1927), Haas (1969), and Graf and Cummings (2007; 2021). A few regional or basin surveys have been conducted in Central America, most notably those in México by Anson A. Hinkley in the Pánuco basin (Pilsbry 1910; Pilsbry and Hinkley 1910), H.B. Baker in the Papaloapán drainage in Veracruz (Baker 1922), Calvin Goodrich and Henry van der Schalie in the Usumacinta system in Guatemala (Goodrich and van der Schalie 1937), Herbert Athearn in the 1960s (unpublished), and D. Bereza et al. in the 1980s (unpublished) collected samples from the Pánuco and Papaloapán basins. Czaja et al. (2023:85–88) reported that the native Mexican freshwater bivalve fauna totaled 97 species in five families (Unionidae, Sphaeriidae, Cyrenoididae, Cyrenidae, and Dreissenidae) and 32 genera. They stated that twenty-six species and one genus (*Actinonaias*) were endemic to México and noted that the Unionidae were by far the richest family with 76 species. The geographic ranges they presented were limited to the level of Méxican states. A summary of the major works by previous researchers can be found in the Historical Surveys & Biographies of Malacologists section below.

We define Central America as those basins in México south of the Colorado River and Río Grande del Norte, through Central America to the Río Tuira in Panamá and the Caribbean Island of Cuba following Graf and Cummings (2021:4). We consider the Río Grande del Norte (Gulf of Mexico) and Colorado River (Pacific Ocean) drainages in the North America Region following Graf and Cummings (2021:4). We have included species lists and accounts for the freshwater mussels found in the Colorado River and Río Grande del Norte basins (Table 16) to provide a complete list of the freshwater mussels that occur in México.

The Central American mussel fauna is entirely distinct from that of North America: there are no native species shared between North and Central America, but further data are needed (particularly on the genera *Megalonaias* and *Pyganodon*) to confirm that. Our designation of various species as endemic is a working hypothesis. In some cases, we are confident that we have enough comprehensive species-level data to make that determination, but insufficient collection makes it far from certain. Molecular phylogenetic analyses based on multiple mitochondrial and nuclear genes coupled with morphological and life history studies will result in some widespread taxa being broken up as cryptic species are discovered, and others lumped.

In their world checklist, Graf and Cummings (2007) stated that "Our objective here is to review and summarize the global and regional diversity of the Unionoida. How many species of freshwater mussels are there and how are they distributed?" Additional work needs to be done to further document the biodiversity of the freshwater mussel fauna of Central America, and much more fieldwork will be necessary to achieve that goal. We hope that providing a complete list of species described, distribution maps, photographs of type specimens, and georeferenced type locality data for all of the names applied to Central American freshwater mussels will encourage additional studies to better document species distributions and test hypotheses regarding the evolution and systematics of the rich and interesting Central America freshwater mussel fauna.

DOI: 10.1201/9781003314448-1

Part 1

Phylogeny, Anatomy, Life History, and Conservation

Phylogeny

Unlike North America where the intergeneric relationships among unionids have been relatively well studied, little information regarding the distribution and evolutionary history is available on the mussel fauna of Central America, and the taxonomy and estimates of species richness are based on woefully outdated literature. Most of the genera were poorly described and those of Frierson (1927) were presented without diagnosis or justification whatsoever (e.g., *Martensnaias*, *Nephritica*, *Popenaias*, and *Reticulatus*) and that Frierson's classification was followed at all by subsequent workers (Haas 1929) is a mystery. We have reproduced, and in some cases translated, the descriptions and diagnoses of the freshwater mussel genera from the original literature in the species accounts section below. Shell morphology alone is insufficient to accurately diagnose mussel genera, and new analyses using soft anatomy and genetic markers will be needed to identify natural phylogenetic groups and test them for monophyly. Graf and Cummings (2007:294) stated, "No strictly Mesoamerican species have been included in published molecular phylogenetic studies to date, although certain genera can be assigned with confidence to tribes well diagnosed by morphological characters." Little has changed since then with only a handful of systematic studies published to date (e.g., Pfeiffer et al. 2018; 2019a,b; Inoue et al. 2018; 2020a; Smith et al. 2020). A cladogram of the phylogenetic relationships of the tribes of the Unionoida (simplified from Graf and Cummings 2006, Lopes-Lima et al. 2017; Pfeiffer et al. 2019a,b; Huang et al. 2019; Zieritz et al. 2024) is given in Figure 1.

ORDER UNIONOIDA

Family Unionidae Rafinesque, 1820

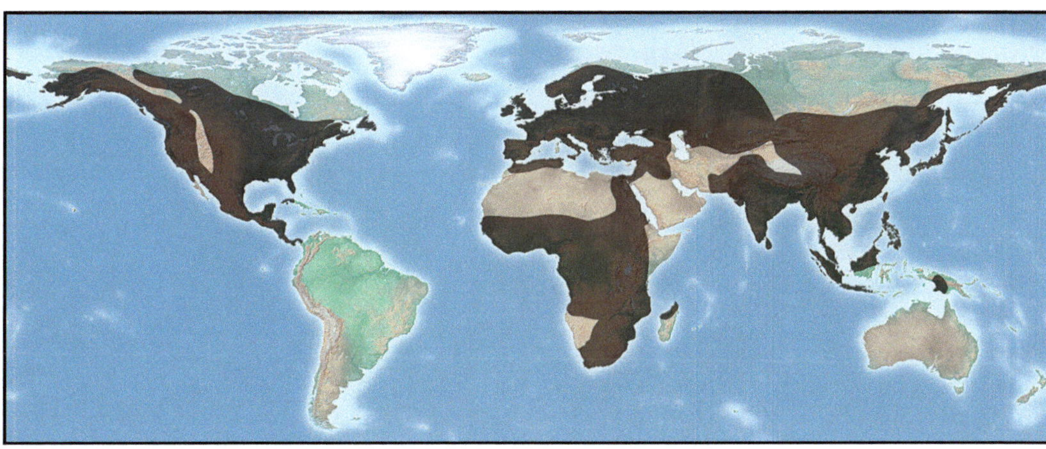

MAP 1 Distribution of the Unionidae (Graf and Cummings 2019a:202).

The bivalve family Unionidae is monophyletic and has been classified with the Margaritiferidae in the superfamily Unionoidea Rafinesque, 1820 (Bieler et al., 2010). However, that classification is tenuous (Pfeiffer et al. 2019b:124). The Unionidae Rafinesque, 1820 is the most species-rich of the six extant families of the order Unionoida. The family is composed of 792 Recent species in 167 genera, classified among at least five subfamilies: Unioninae (s.s.), Ambleminae Rafinesque, 1820, Gonideinae Ortmann, 1916, Modellnaiinae Brandt, 1974, and Parreysiinae Henderson, 1935 (Graf and Cummings 2021:6). The Unionidae is the most geographically widespread family,

DOI: 10.1201/9781003314448-3

occurring in all continents except South America and Antarctica, with only two species represented in the Australian region on the island of New Guinea (Map 1). The Unionidae reach their greatest species richness in East Asia (333 spp.) and the Nearctic (302 spp.) (Graf and Cummings 2021:11).

Family Unionidae, subfamily Unioninae. The Unioninae contains three tribes (Figure 1) and is represented in Central America by the Anodontini characterized by a set of distinctive morphological (large and ovate thin shells, and large triangular and hooked glochidia), anatomical (demibranchs with perforated septa, secondary water tubes in the outer demibranchs, and marsupium in the external demibranchs that are distend laterally when gravid, Plate 9), and ecological characters (most species seem to be generalists concerning habitat and host fish). The subfamily Anodontinae Rafinesque in 1820 was used in many classifications in the 20th century (e.g., Haas 1969; Heard and Guckert 1970; Modell 1964; Ortmann 1910). Subsequent studies sunk Anodontinae to a tribe within Unioninae due to the shared subtriangular hooked glochidia (Bieler et al. 2010; Graf 2002; Graf and Cummings 2021). However, the rank change of Anodontinae into Anodontini has been recently disputed based on glochidia morphology (Huang et al. 2013:8).

Family Unionidae, subfamily Ambleminae. As recognized today, the subfamily Ambleminae contains 5 tribes, three of which (Lampsilini, Popenaiadini, and Quadrulini) are represented in Central America (Figure 1). The Lampsilini are characterized by a suite of anatomical and life history characters, including long-term brooding and a well-developed marsupium restricted to a limited portion of the outer demibranchs capable of great expansion (Graf and Ó Foighil 2000). Based on geography and phylogenetic analyses performed to date, the subfamily Ambleminae is a monophyletic lineage endemic to North and Central America, including Cuba in the Caribbean (Graf and Cummings 2007:295; Pfeiffer et al. 2019a:111). The Lampsilini is sister to the Popenaiadini, which in turn are sister to the Amblemini. Those three tribes are sister to the Pleurobemini and then to the Quadrulini (Lampsilini+Popenaiadini)+Amblemini)+Pleurobemini)+Quadrulini) (Graf and Cummings 2006; Lopes-Lima et al. 2017; Pfeiffer et al. 2019a,b; Huang et al. 2019; Zieritz et al. 2024; Figure 1). The affinities of these taxa will be resolved as molecular analyses progress and the taxa are better distinguished with morphological characters.

Family Mycetopodidae Gray, 1840

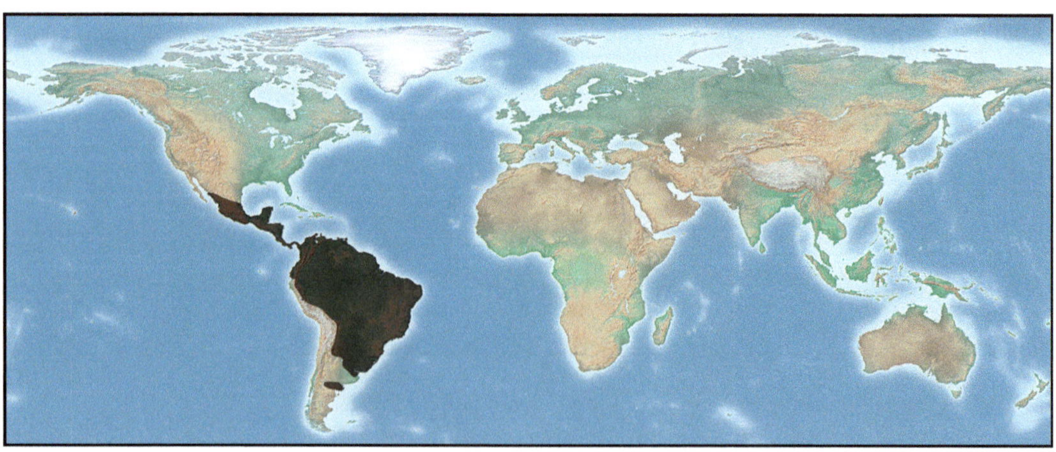

MAP 2 Distribution of the Mycetopodidae (Graf and Cummings 2019b:222).

Species in the family Mycetopodidae Gray, 1840 are found throughout the Neotropics, from México south to Argentina (Map 2). The family as recognized by Graf and Cummings (2021:11) comprised 53 species (49 in South America, 7 in Central America) in 11 genera, divided among four subfamilies: Mycetopodinae s.s., Anodontitinae Modell, 1942, Leilinae de Morretes, 1949, and Monocondylaeinae Modell, 1942 (Graf and Cummings 2019b:222).

Graf and Cummings (2021:28) recognized 11 genera in four subfamilies in the Mycetopodidae: *Mycetopoda*, *Mycetopodella*, *Anodontites*, *Lamproscapha*, *Leila*, *Monocondylaea*, *Diplodontites*, *Fossula*, *Haasica*, *Iheringella*, and *Tamsiella*, largely following Simone (2006) and Pereira et al. (2014). *Mycetopoda* d'Orbigny, 1835 (4 spp.) and the monotypic *Mycetopodella* Marshall, 1927, comprise the Mycetopodinae Gray, 1840. *Mycetopodella falcata* (Higgins, 1868) is endemic to the Upper Amazon, but *Mycetopoda* extends from Central America south to Argentina (Paraná–La Plata Dr.). The Anodontitinae Modell, 1942 is the most species-rich subfamily with 28 species (*Anodontites* 26 spp.; *Lamproscapha* 2 spp.). *Leila* Gray, 1840 (2 spp.) is the sole genus of the Leilinae de Morretes, 1949, widespread in the Amazon and Paraná–La Plata basins. The Monocondylaeinae Modell, 1942 has the highest generic richness of the family: *Monocondylaea* d'Orbigny, 1835 (8 spp.), *Fossula* Lea, 1870 (1 sp.), *Haasica* Strand, 1932 (1 sp.), *Iheringella* Pilsbry, 1893 (2 spp.), *Diplodontites* Marshall, 1922 (3 spp.), and *Tamsiella* Haas, 1931 (3 spp.). The first four genera are restricted to the Paraná–La Plata Basin and, with one exception, southern tributaries of the Amazon, whereas *Tamsiella* and *Diplodontites* are distributed from the Amazon, Orinoco, and Magdalena drainages.

Eight species of mycetopodids were reported as part of the Central American fauna by Graf and Cummings in 2007: *Mycetopoda siliquosa* (Spix in Wagner, 1827), *Anodontites cylindracea* (Lea, 1838), *A. depexa* (von Martens, 1900), *A. inaequivalva* (Lea, 1868), *A. tortilis* (Lea, 1852), *A. trapesialis* (Lamarck, 1819), and *A. trigona* (Spix in Wagner, 1827). In their 2021 checklist, Graf and Cummings listed seven species in the family for Central America: *Mycetopoda siliquosa* (Spix in Wagner, 1827), *Anodontites cylindracea* (Lea, 1838), *A. depexa* (von Martens, 1900), *A. montezuma* (Lea, 1841), *A. tehuantepecensis* (Crosse & Fischer, 1893), *A. tortilis* (Lea, 1852), and *A. trapesialis* (Lamarck, 1819). In this work, we recognize eight species in the family Mycetopodidae (seven in the genus *Anodontites* and one in *Mycetopoda*), and the changes from the 2021 checklist are discussed below. There are four species of freshwater mussels shared between Central and South America, all in the family Mycetopodidae: *Anodontites tehuantepecensis* (Crosse & Fischer, 1893), *Anodontites tortilis* (Lea, 1852), *Anodontites trapesialis* (Lamarck, 1819), and *Mycetopoda subsinuata* (Sowerby, 1868).

There is little evidence against the monophyly of the Mycetopodidae, but taxon sampling in phylogenetic studies to date has been too sparse for a robust test. Morphological (Graf 2000) and molecular (Bogan and Hoeh 2000; Walker et al. 2006; Whelan et al. 2011; Guerra et al. 2019; Huang et al. 2019) analyses have included representatives of only two subfamilies at a time (three subfamilies now). Graf and Cummings (2006:370) identified only a single synapomorphy to distinguish the family from the other families of the Etherioidea Deshayes, 1832: pedal elevator scars that are "inconspicuous." Otherwise, mycetopodids have lasidium-type parasitic larvae and mantle fusion between their incurrent and excurrent apertures (homologous with the conditions seen in Etheriidae Deshayes, 1832). Mycetopodid shells are largely edentulous except for the genera of the Monocondylaeinae in which the hinge teeth are poorly developed.

Graf and Cummings (2007:295) identified an assemblage of 102 species of freshwater mussels (94 Unionidae and 8 Mycetopodidae) in Central America. At that time, little current information was available on the mussel fauna of Central America, and their 2007 checklist relied largely upon the woefully outdated opinions of Frierson (1927) and Haas (1969) for their estimates of freshwater mussel species richness in the region. Graf and Cummings updated their checklist, redefined the regions and subregions (2021:11, fig. 1, Tables 1, 3–4), and recognized 87 species of Unionidae in Central America (Subregion CA1), 82 of which they considered endemic.

In this work, we recognize 56 native species (all endemic) in the family Unionidae in Central America (Tables 3 & 4). Twelve native freshwater mussel genera are endemic to Central America: *Brachyanodon* (Unioninae: Anodontini), *Actinonaias*, *Arotonaias*, *Delphinonaias*, *Friersonia*, *Mesonaias*, *Pachynaias* (Ambleminae: Lampsilini), *Micronaias*, *Nephritica*, *Nephronaias*, *Psoronaias*, and *Sphenonaias* (Ambleminae: Popenaiadini). Only three native genera extend into North America: *Pyganodon* (Unioninae), *Popenaias*, and *Megalonaias* (Ambleminae) (but see the species account for *Disconaias conchos* p. 419). We have reorganized the fauna with respect to geography and grouped species into malacological provinces, as defined below. Nomenclatural changes from previously published studies and the Graf and Cummings 2021 checklist are discussed in the

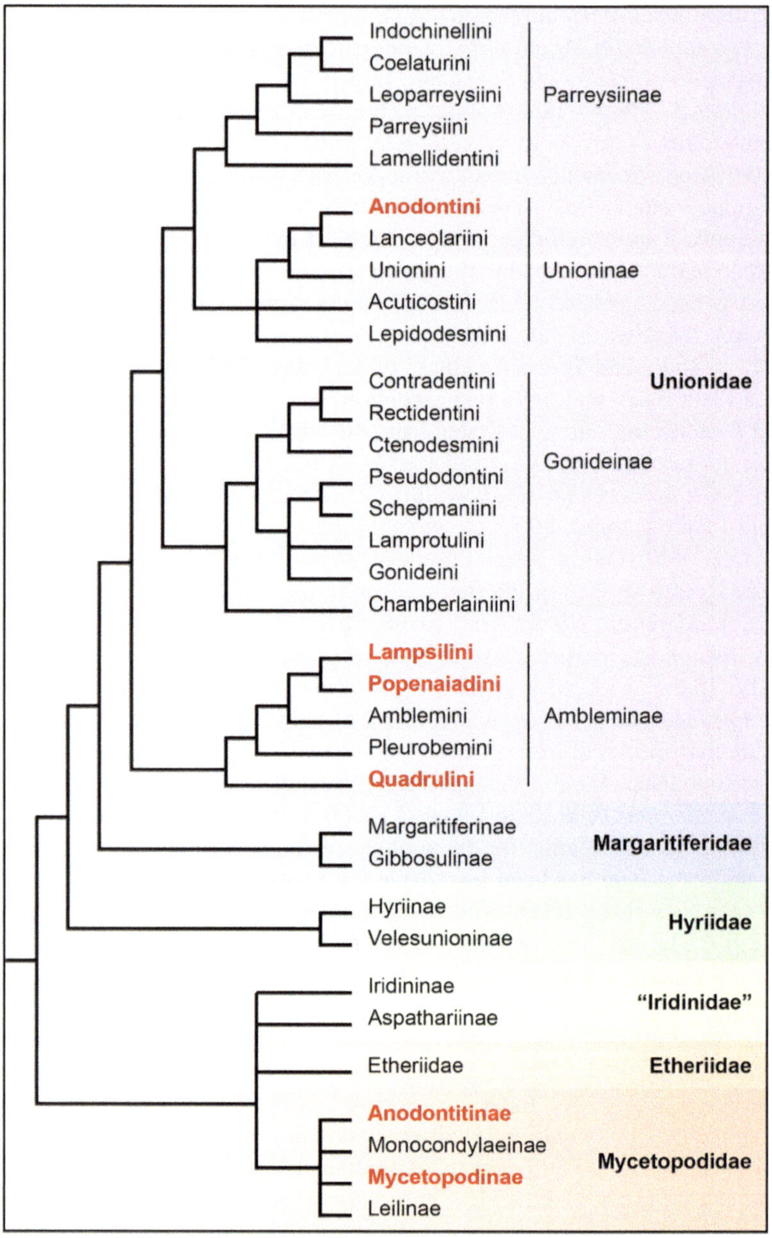

FIGURE 1 Schematic cladogram of family-group level taxa of the order Unionoida. Taxa occurring in Central America are highlighted in red. Phylogenetic relationships simplified from Graf and Cummings (2006), Lopes-Lima et al. (2017), Pfeiffer et al. (2019a,b), Huang et al. (2019), and Zieritz et al. (2024).

accounts below. We hope that the information compiled on the Central American mussel fauna will provide a framework for testing and identifying the synapomorphies that diagnose clades will follow and improve on the classification presented here.

Detailed information and a summary to date on the anatomy, life history, and other aspects of biology of the order Unionoida and other freshwater bivalves can be found in Cummings and Graf (2010). Some general comments on anatomy, life history, and conservation status follow. A glossary of terms is provided in Appendix 1.

Anatomy

To identify freshwater bivalves, knowledge about their basic anatomy is essential. Cummings and Graf (2010:310–318, 370) described in detail the internal and external anatomy of members of the order Unionoida. In general, freshwater mussel shells vary in lateral outline from elongate to ovate to rhomboid to trigonal. The beaks or umbos represent the oldest parts of the shell, and they are frequently eroded in older specimens of larger individuals. The shell of a mussel is composed largely of calcium carbonate ($CaCO_3$). As the mollusk grows, the shell increases in size with the secretion of new shell material (organic and inorganic) along the leading edge of the mantle.

The outer surface of the shell of many species of mussels is completely smooth, while others possess knobs, pustules, spines, wrinkles, or folds. Often, the size, combination, or arrangement of these surface features is diagnostic of a particular species. Another important character used in identification is beak sculpture: ridges or raised lines found on the beak or umbo. Beak sculpture can vary from concentric rings and V-shaped ridges to a series of wavy double-looped ridges. It is often the case that the periostracum becomes thin or is completely eroded in spots (often on the beak or umbo), thus obscuring the beak sculpture and making the identification of some species difficult. This is common in older individuals or mussels living in poorly buffered streams. The shape, size, thickness, and color of the shell can vary enormously within or among species of freshwater mussels. The color and patterns can be important characters for identification. Variations in the color of the periostracum range from light yellow to green, brown, or black, and the shells of many species are patterned with various chevron markings or rays. The color of the nacre in freshwater mussels varies from a pure silvery white through shades of pink, salmon, to dark purple.

The mantle is bound to the inner shell surface through various muscle attachments, leaving a number of conspicuous scars. Running concentrically around the inner surface, parallel with the valve margin, is the pallial line, the site of attachment for the pallial muscles. Dorsally, the anterior and posterior adductor muscles connect the two valves, providing the contractile force to close the valves. Closing the valves stretches the dorsal ligament, and relaxing the adductors allows the ligament to pull the valves back to gaping. The anterior and posterior pedal retractors (distally attached to the foot) are anchored on the shell. These muscles are often closely associated with the adductors and often leave inseparable scars. Immediately posterior to the anterior adductor, and often in close association, is the pedal protractor muscle, and the pedal elevator muscles attach beneath the umbos leaving a series of minute pits in the beak cavity. The soft tissues also include a large muscular foot used for locomotion, labial palps that move food particles to the mouth, and two pairs of gills. The gills have three functions: (1) respiration, (2) filter-feeding, and (3) reproduction; in females, the gills incubate larval mussels until they are mature and ready to be released (Figure 2).

Internal organ systems include an open circulatory system powered by a heart; a digestive system that consists of mouth, stomach, gut, and anus; a decentralized nervous system that controls the movement of the foot and adductor muscles; and reproductive organs that usually occur separately in males and females.

DOI: 10.1201/9781003314448-4

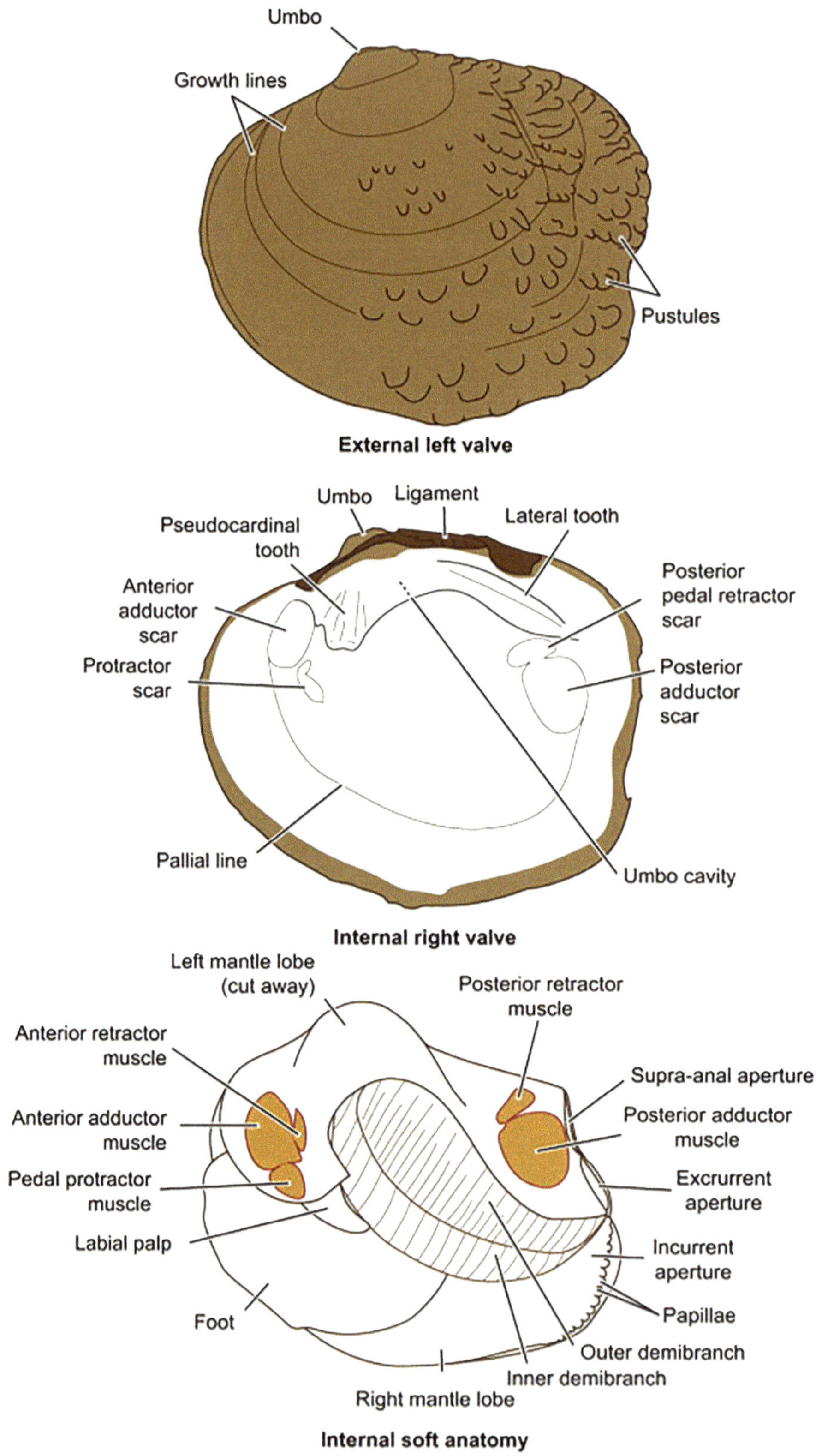

FIGURE 2 Freshwater mussel anatomy. (From Cummings and Graf 2010:314).

The anterior end of a freshwater mussel can be determined by the position of the umbo relative to the pseudocardinal teeth, which are always located anteriorly. The typical freshwater mussel hinge bears schizodont dentition in which the teeth are large and possess parallel ridges at right angles to the axis of the teeth. The left valve typically bears a single tooth. The hinge teeth, or the lack thereof, are an important feature in identifying bivalves because the teeth are generally similar within the major taxonomic groups. Historically, the hinge teeth have provided a convenient means by which to construct classification schemes and attempt to indicate the phylogenetic relationships within bivalves.

Members of the Mycetopodidae and Unionidae (tribe Anodontini Rafinesque, 182) have secondarily reduced or lost their hinge teeth. Edentate mussels can be identified to family by the presence (Mycetopodidae) or absence (Unionidae) of ligamental sinus: a brown triangular-shaped blotch at the posterior end of the hinge (Figure 3). Among those bearing hinge dentition, the lateral teeth run posteriorly from beneath the umbos, parallel to the ligament. Typically, two elongate laterals are found in the right valve, with a single interlocking lateral tooth in the left. However, large individuals may present additional teeth to these combinations. The function of the hinge dentition is to resist twisting of the valves along the plane of articulation.

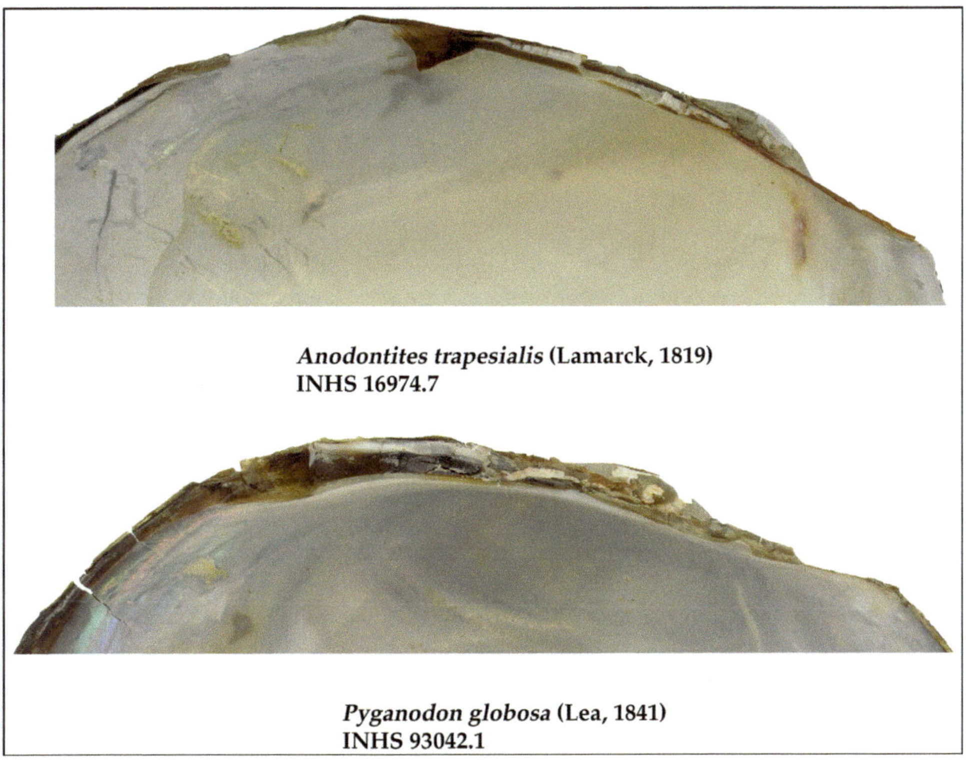

FIGURE 3 Freshwater mussel hinge lines showing the ligamental sinus present in members of the genus *Anodontites* and absent in *Pyganodon*. **Top:** *Anodontites trapesialis* (Lamarck, 1819) INHS 16974.7. Small creek (Orinoco Dr.), Mapuey, Distrito San Carlos, Cojedes, Venezuela [9.6333 -68.6317]. **Bottom:** *Anodontites globosa* (Lea, 1841) INHS 93042.1. trib. Rio San Juan (Papaloapán Dr.), Nopalapan, Veracruz, Mexico [18.1136, -95.3264].

Life History

FOOD AND FEEDING

Freshwater bivalves filter phytoplankton, zooplankton, bacteria, and particulate organic matter from the water column and from the sediment by deposit or pedal feeding. Bivalve feeding is based upon the general model of active suspension feeding seen in other invertebrate groups. A water current through the mantle cavity is generated by ciliary action of the ctenidia. Environmental water is drawn in via the incurrent aperture to the infrabranchial chamber. The current passes through pores into the interlamellar spaces of the demibranchs, then to the suprabranchial space, and out the excurrent aperture. Food and other suspended particles are captured by the ctenidia, partially sorted, and passed anteriorly for further sorting and ingestion. Much of this process is similar among the suspension-feeding bivalves, but in many freshwater lineages, the mechanics are somewhat disrupted during the brooding period in females when some or all their demibranchs may be gravid with developing larvae (Allen 1914).

AGE AND GROWTH

Freshwater mussels are among the most long-lived animals in the world, with many species living from 10 years to over 200 years. The oldest known freshwater mussel was found in the Görjeån River Norrbotten, Sweden and was estimated to be approximately 280 years old (Degerman et al. 2009). Freshwater mussel shell growth occurs rapidly in the early years of their lives and slows down considerably with advanced age. In contrast, somatic growth remains constant or increases with the onset of sexual maturity. In general, thin-shelled species grow faster and have shorter life spans than thick-shelled species.

It has been a long-held assumption that freshwater mussels lay down annual rings like fish scales, otoliths, or tree rings. The most used method to age mussels is the counting of external or internal growth rings or annuli. Although external annuli are unambiguous and easy to identify in young mussels, they are not easy to discern or count in older individuals. Other aging methods include thin sectioning, acetate peels, and shell ashing. Thin sectioning involves cutting the shell with a diamond saw from the umbo to the ventral margin, following the vector of maximum growth. External growth rings appear as dark lines on the external surface of the shell, presumably representing pauses in growth during winter months. Internal annuli appear as dark layers in a cross section of a shell. It has been assumed for decades that the thickest, most opaque bands on the surface of the shell are annually generated, while thinner, less opaque bands are produced by some nonannual event. It has also been assumed that these annuli are indeed annual; that is, they are formed each year. Recent studies based on direct measurement of marked individuals have concluded that counting external or internal annuli to estimate the age of freshwater mussels is insufficient and may substantially underestimate their true age (Downing et al. 1992; Rypel et al. 2008).

Freshwater mussel life spans may surpass, by an order of magnitude, most of the ages estimated by traditional annulus counting methods. Ages predicted from actual growth rates have suggested that some individuals frequently reach ages in excess of 100 years (Kessler and Downing 1997). If the recent estimates are correct, the ancient life spans of some populations have significant implications on the management, recovery, and commercial harvest of mussels, which would be among the most long-lived animals on Earth. However, not all authors have come to the same conclusion. Confirmation of annulus formation across multiple species, rivers, and years has suggested that

DOI: 10.1201/9781003314448-5

the laying down of annual rings is a pervasive phenomenon in temperate North America (Haag and Commons-Carson 2008). Age at maturity has not been established for most species of freshwater mussels. Age at maturity estimates for *Pyganodon grandis* from Central Alberta, Canada, ranged from 4 to 5 years, whereas European *Margaritifera margaritifera* ranged from 6 to 12 years (Cosgrove et al. 2000; Hanson et al. 1989).

The growth rates of the thin-shelled *Utterbackia imbecillis* were studied for three years from five ponds in West Virginia, and the maximum age was estimated to be about 4 years (Harmon and Joy 1990). The maximum age estimated for *Pyganodon grandis* in Canada was 12 years (Hanson et al. 1989). Using annual rings, a shell of *Megalonaias nervosa* from the Ohio River, Illinois, was estimated to be 54 years old (Chamberlain 1933). The maximum observed life span of the European freshwater pearl mussel *Margaritifera margaritifera* has been reported to range from 30–132 years to over 200 years (San Miguel et al. 2004).

PREDATION

A wide diversity of animals, including fishes, turtles, birds, and mammals (particularly otters and raccoons), are known to prey upon freshwater mussels and their young. The rhabdocoel turbellarian flatworm (*Macrostomum* sp.) feeds upon juvenile mussels (Zimmerman et al. 2003). Exotic fishes, particularly the black carp (*Mylopharyngodon pisceus*), pose a serious threat to native freshwater mussels (Nico et al. 2005). Adult and juvenile black carp feed almost exclusively on freshwater mollusks. They possess large pharyngeal teeth, which they use to crush snails and small- to medium-sized bivalves. The prey they consume is limited by their gape size, but gape size is related to body size, and black carp can grow up to 1.5 m in total length and 70 kg or more in weight. A black carp 1 m long would have a gape size of approximately 57 mm, and a fish 1.5 m in length would have a gape size of 84 mm, which would allow them to consume all but the largest native mussels. Reptiles, particularly turtles, are known to consume small freshwater mussels (Shively and Vidrine 1984), and there is at least one report of a mussel catching a turtle (Plummer and Goy 1997:88, fig. 1). Charles Darwin (1878:120–121) reported that *Elliptio complanata* (Lightfoot, 1786), a large freshwater mussel, was found attached to the toe of a duck shot in Danversport, Massachusetts. Large piles of freshly cleaned mussels, called middens, can be found along the banks of a river or lake where otters or raccoons have been actively foraging. These middens often contain a wide variety of species and can be one of the best places to find shells. There are no published studies on mussel predation in Central America.

REPRODUCTION

Freshwater mussels have a parasitic life cycle that is unique among bivalves. In most species of freshwater mussels, the sexes are separate. Males release sperm into the water, and eggs are fertilized internally by sperm that enters the female via the incurrent aperture. The larval stage in the mussel family Unionidae is known as a glochidium. The portion of the demibranchs that serve for brooding larvae is referred to as the marsupium. Glochidia are released from the female via the excurrent aperture into the water column and must encounter an appropriate host, usually a fish, to which they attach and form a cyst. The mussel larvae undergo metamorphosis transforming from larvae to juveniles (Figure 4).

North and Central American freshwater mussels in the family Unionidae brood ova and developing larvae in either all four demibranchs (tetragenous) or only the outer pair (or some portion of the outer demibranchs—exogenous). The portion of the demibranchs that serve for brooding is referred to as the marsupium. Characters associated with brooding morphology and duration have long been considered of taxonomic importance, and much of the substance of those hypotheses have been supported by recent phylogenetic analyses. Ova and developing larvae are kept in the interlamellar space. Among unionids, the interlamellar space of the demibranch is divided by vertical septa into

FIGURE 4 Generalized life cycle of a freshwater mussel. (Figure by Kyle Luckenbill).

a series of water tubes. Gravid females are evident, as the water tubes of the marsupium are visibly charged with offspring. Further, among the ectobranchous freshwater mussels, two lineages, the Anodontini and Lampsilini, have made significant modifications to their marsupia that can accommodate long-term brooding, and these derived conditions provide a useful contrast with the plesiomorphic morphology.

Two basic brooding periods exist among the Unionidae: short term and long term. These patterns were first characterized as summer and winter brooding, respectively. Ortmann introduced the terms tachytictic and bradytictic for short- and long-term brooders, respectively (nouns: tachytictia and bradytictia). Short-term (tachytictic) brooders tend to spawn gametes in the early spring; the embryos develop in the marsupial demibranchs and then are released to infect their hosts by summer. Bradytictia has apparently evolved twice among the Unionidae, in the Anodontini and the Lampsilini. Both of those lineages have independently evolved modifications to their marsupial demibranchs that may ameliorate the effects of brooding on other functions of the ctenidia (Cummings and Graf 2010:335).

Species of the Anodontini (subfamily Unioninae) employ the whole of the outer (lateral) pair of demibranchs for brooding, with two major modifications. The first is that the ventral margins of the marsupial demibranchs are augmented by additional tissue, allowing significant lateral expansion. Among anodontines, the margin is rounded, and the marsupium appears inflated; it resembles an air mattress, with the interlamellar septa dividing the turgid chambers. This modification for greater expansion was perhaps an adaptation to maintain reasonably high fecundity with the relatively larger, anodontine hooked-type glochidia. See Cummings and Graf (2010:333–335) for addition information.

The species of the Lampsilini (subfamily Ambleminae) have also reduced the effect of brooding on feeding and gas exchange by limiting the marsupium to only a restricted portion of the outer demibranch. The ventral margin of the marsupial portion of the outer demibranch expands both laterally and ventrally. This is usually some posterior fraction, although some taxa constrain the marsupium to the central section or most of the marsupium (see Plates 29–31).

Few species in the Popenaiadini (subfamily Ambleminae) have been examined with respect to gill morphology and marsupium position. *Popenaias popeii* (Lea, 1857) has been classified as both tachytictic and bradytictic based on limited data. Histological analysis of gonadal tissue from specimens collected in New Mexico indicated gonochoric individuals; hermaphroditism was not observed in any specimens (Smith et al. 2003). Gravid females were observed from March through August, with all stages of embryos and larvae present on every sampling date during that period. Contrary to previous reports, *P. popeii* is considered an asynchronous short-term brooder with an extended period (late winter to mid-summer) of oviposition. A small percentage of gravid females used posterior portions of inner demibranchs as marsupia. This character suggests that inner demibranchs might function in a facultative capacity as an overflow reservoir for larvae unable to fit into full marsupia of the specialized outer demibranchs. Smith et al. (2003:337, fig. 4) described and figured the glochidia of *P. popeii*. The glochidial shells were semi-elliptical in shape with pitted valves. Their figured specimen measured 219 mm in length and 208 mm in height. The hinge was long (165 mm) with a slight convexity and a well-developed ligament. The lateral margins were unequal; the anterior margin was flatly convex, whereas the posterior margin was somewhat truncated and more flattened. The ventral margin was semicircular. The inner part of the occlusal surface of each valve possessed rows of simple, conical denticles on a platform just inside the lip. The largest denticles were in the outer most row and were approximately 2 mm long.

In *Popenaias berezai* Inoue et al., 2020, the gills are creamy white to tan and may show streaks of dark pigmentation posteriorly in non-gravid and partially gravid individuals (Plate 60). The female outer gill is marsupial, and the entire gill is utilized to brood glochidia except for the extreme posterior portion and is distended or pad-like when gravid. Partially gravid females of *Psoronaias semigranosus* (von dem Busch in Philippi, 1843) (developing eggs, no glochidia) were collected from the Río Huichihuayán (Pánuco–Moctezuma–Axtla Dr.), 8 km NE Xilitla, San Luis Potosí, México, on 10 December 2018 (INHS 90189), and it appears that the larvae occupy most of the outer demibranch (Plate 69).

In North America, female members of the tribe Lampsilini in the genera *Lampsilis* and *Toxolasma* have modified mantle margins that mimic prey to attract a potential host fish and elicit an attack on the gravid female. The mantle flap modifications mimic fishes (complete with lateral stripes and eyespots), crayfish, and other potential prey items (Haag 2012) and can be variable within a population (Zanatta et al. 2007). An attack on the mussel by the potential fish host ruptures the marsupium positioned adjacent to the lure and protruding between the valves and extracts glochidia in the process. In *Lampsilis*, the lamellae of the protruding marsupial demibranchs can separate at the distal edge and are easily ruptured upon attack. These specializations increase the likelihood of glochidia contacting a potential host by minimizing the diffusion of the brood prior to contact. It is unknown if any Central American lampsiline mussels have fish-attracting lures.

Among the Unionoida are two distinct types of parasitic larvae: glochidia—calcified bivalved larvae with a single adductor muscle, and lasidia—uncalcified trilobed larvae with a long larval thread (Graf and Cummings 2006, fig. 4). Molecular and morphological phylogenetic analyses sampling representatives of each unionoid family have consistently recovered the glochidia-bearing mussels as paraphyletic; either the Hyriidae is recovered as sister to the remaining five families (Whelan et al. 2011; Pfeiffer and Graf 2015; Lopes-Lima et al. 2017; Huang et al. 2018) or the Hyriidae is recovered as sister to the lasidia-bearing mussels (Graf 2000; Graf and Cummings 2006).

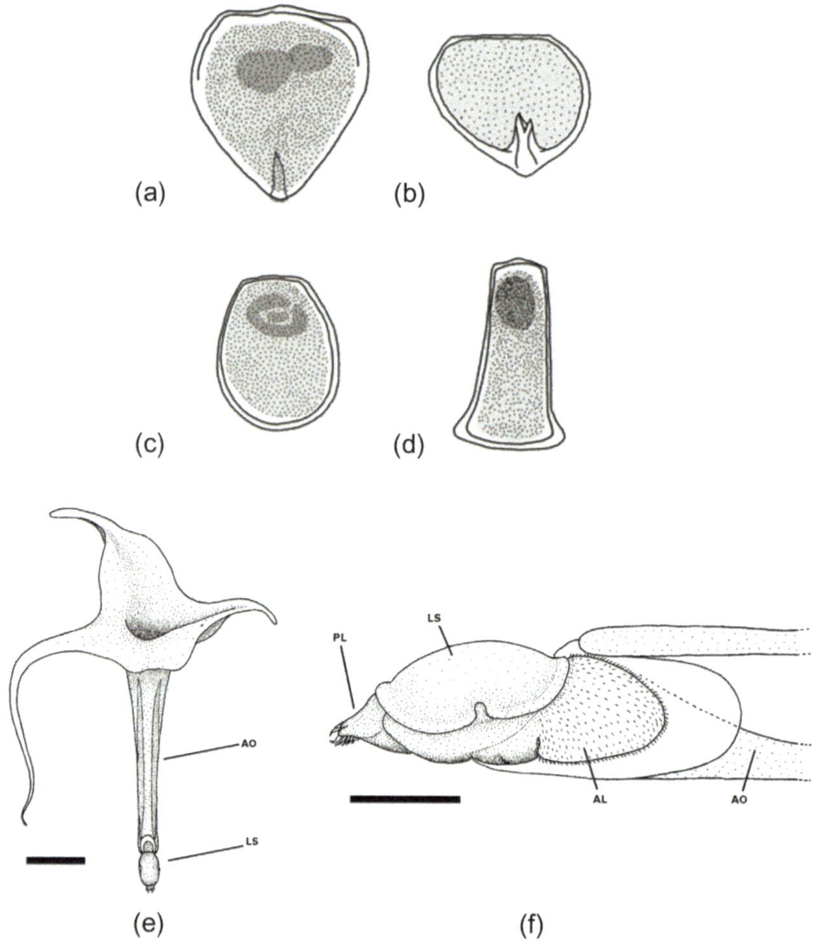

FIGURE 5 Freshwater mussel larval morphology: (a, b) hooked glochidium, (c) unhooked glochidium, (d) axe-head glochidium, (e) lasidium—dorsal view, and (f) lasidium—lateral view. Abbreviations: AL, anterior lobe; AO, adhesive organ; LS, larval shell; PL, posterior lobe. Scale bar: (A) 150 µm and (B) 50 µm. (a–d from Graf and Cummings 2006:361; e, f from Ituarte and Mansur 2020:388).

In Central America, members of the Unionidae possess glochidial larvae and those of the Mycetopodidae lasidia. Glochidia are small (50–450 µm) with calcified, bivalved larval shells and a single adductor muscle (Figure 5, A–D). They attach to either gill or body/fin tissue of their host fish by clamping on with their valves and becoming enclosed in a cyst or capsule. Lasidia are also small (85–150 µm, not including the "larval thread"), but with an uncalcified, univalved larval shell and a trilobed body (Figure 5, E). They attach by either forming a cyst. Lasidia are found only in the largely southern tropical families of the Etherioidea. At least one mussel, *Mutela bourguignati* (Bourguignat, 1885), from Africa has a modified lasidium known as a "haustorium" (Figure 5, F) (Graf and Cummings 2006).

Among the North American species of the Unionidae and Margaritiferidae, there are three principle morphological types of glochidia: (1) subtriangular, hooked-type (Figure 5, A); (2) semi-elliptical, unhooked-type (Figure 5, C); and (3) axe-head-type glochidia (Figure 5, D). Hooked glochidia are known only from the Unioninae (e.g., *Pyganodon*), although a similar, convergent morphology is found among the Hyriidae of South America (Figure 5, B) and Australasia. Axe-head glochidia are found only in the genus *Potamilus* (Lampsilini). The smaller, hookless glochidia found in the Ambleminae are typically attached to the gills of the host, whereas the larger, triangular, hooked-type glochidia found in Anodontini are more commonly found attached to the skin or fins.

The glochidia are typically found on the gills or fins of fishes. Some species are host-specific, while others can use a wide variety of fishes as hosts. Amphibians are also rarely used as hosts. Freshwater mussels use fishes from across the taxonomic spectrum, and an excellent summary of fish-host relationships can be found in Watters (1994). The degree of host specificity varies widely among species, from generalists to single-species specialists. There are also rare instances where some mussel species have been reported to transform without a host, but this is an exception to the rule. After metamorphosis, the young mussel breaks free from the cyst and drops to the bottom to begin an independent life. The nomenclature of fishes mentioned in this work has been updated using FishBase, a global biodiversity information system on fishes (fishbase.mnhn.fr/home.htm). In North America, the Lampsilini are for the most part hosted by sight-feeding perciform fishes (Centrarchidae, Percidae, and Cottidae). Cichlid fishes are a likely host for Central American lampsiline mussels. There are no studies on the host fish of native freshwater mussels that have been conducted in Central America. All of the data on hosts were from studies conducted in North or South America. Additional information on all aspects of mussel reproduction can be found in Cummings and Graf (2010:330–341).

Mussels have evolved a wide variety of morphological and behavioral adaptations to facilitate the transmission of larvae to their eventual host, many of which have only recently been discovered and described. Released glochidia can survive a few days to a week before they must find and attach to a suitable host, and there is considerable variability between species in the long-term viability of larvae, much of which is temperature dependent (Zimmerman and Neves 2002). The period of attachment varies from weeks to months depending on the host, location of attachment, and temperature.

Conservation Status

Freshwater mussels are one of the most imperiled groups of animals in the world (Williams et al. 1993; Lydeard et al. 2004; Strayer et al. 2004; Böhm et al. 2020) and discovering the origins of species richness and biogeographic patterns is essential for understanding evolutionary processes and developing effective conservation strategies.

Invasive species, habitat degradation, and overexploitation have led to worldwide freshwater mussel population reductions, although the global magnitude of these declines is poorly documented in most regions (Graf and Cummings 2021). The International Union for the Conservation of Nature (IUCN) Red List of Threatened Species was established in 1964 to provide information on the global conservation status of animal, fungi, and plant species. The IUCN Red List Categories divide species into nine categories: NE—Not Evaluated, DD—Data Deficient, LC—Least Concern, NT—Near Threatened, VU—Vulnerable, EN—Endangered, CR—Critically Endangered, EW—Extinct in the Wild, EX—Extinct, and NA—Not Applicable (IUCN 2024). When available, the IUCN status is reported below.

There are 549 species and 18 subspecies and varieties of freshwater mussels (order Unionoida) that have been assessed by the IUCN for their Red List (IUCN, June 2024). The categorical breakdown of the 567 taxa are as follows: EX (35), CR (78), EN (70), VU (43), NT (53), LC (204), and DD (84). However, that tally is an underestimate, as only 567 taxa have been of the currently recognized 996 species of freshwater mussels worldwide have been assessed. Except for North America and Africa, the majority of freshwater mussel species in most regions have not been evaluated or lack sufficient data to make an assessment (Graf and Cummings 2019a:206). That is particularly true for Central America, where only nine species have been assessed. However, three of the species listed do not occur in Central America: *Anodonta nuttalliana* Lea, 1838 (VU), *Fusconaia mitchelli* (Simpson in Dall, 1895) (CR), and *Anodontites trigona* (Spix in Wagner, 1827) (DD); five are strictly North American: *Pyganodon grandis* (Say, 1829), *Megalonaias nervosa* (Rafinesque, 1820), *Uniomerus tetralasmus* (Say, 1831), *Truncilla donaciformis* (Lea, 1828), and *Lampsilis teres* (Rafinesque, 1820) (all as LC), leaving only one Central American species, *Brachyanodon impura* (Say, 1829) (as *Anodonta impura*), which has been assessed and is classified as DD—Data Deficient. Much work needs to be done to provide a clear picture of the conservation status of freshwater mussels in Central America. The Red List and Czaja et al. (2023:86) both listed *Anodonta nuttalliana* Lea, 1838 as occurring in Chihuahua and Sonora, México. *Anodonta nuttalliana* is found only in the Pacific Northwest, and these records are most likely mis-identified *Anodonta californiensis* Lea, 1852.

The conservation status of freshwater fish is more well understood than freshwater mussels. Contreras-MacBeath et al. (2014:5429–5431) identified three hotspots of fish species richness in México: The Central Southeastern (Coatzacoalcos, Papaloapán, and Grijalva–Usumacinta) (208 species), the Mesa Central (111 species), and lower Río Grande (90 species). Burkhead (2012:799, citing Minckley and Douglas 1991; Burr and Mayden 1992) reported that fishes in the western United States and México exhibit relatively lower species richness per drainage unit, but they generally have higher endemism and disproportionate numbers of extinct species. Burkhead also noted that México had many fishes that were vulnerable to extinction in the future, primarily because of the high levels of endemism. Of the 280 North American fishes he identified as endangered from North America, 65 were exclusively from México.

Lyons et al. (2020:10) assessed 536 freshwater fish species in México for the IUCN for their conservation status. They found that at the global level, 165 species were threatened with extinction, representing 39.9% of all species assessed—assuming that all Data Deficient species were

DOI: 10.1201/9781003314448-6

threatened in the same proportion as those for which enough information was available. Threatened species were divided into the categories: Vulnerable, that included 9.3% (50), Endangered, that comprised 13.2% (71), and Critically Endangered, 8.2% (44). A further 3.4% (18) of species were classified as Near Threatened. An additional 2.2% (12) species of México freshwater fishes were considered Globally Extinct, and 1.5% (8) species were Extinct in the Wild. Of these, all were endemic to México, and all but one belonged to the order Cyprinodontiformes. They noted that those numbers were likely an underestimate of the true extinction rate, given that several Critically Endangered species had not been recorded from Méxican freshwaters in the recent past (Miller et al. 2005; Burkhead 2012).

El Salvador and México are the only countries in Central America that have assessed the conservation status of freshwater mussel species nationally. El Salvador has *Mycetopoda subsinuata* (Sowerby, 1868), *Nephronaias goascoranensis* (Lea, 1858) (as *Nephronaias lempensis* Marshall, 1926), and *N. rowellii* (Lea, 1859) listed as endangered (Ministério de Medio Ambiente y Recursos Naturales 2023). The Mexican government placed 204 species under some category of risk, 13 are considered as probably extinct, 81 threatened, 80 endangered, and 30 were subject to special protection, but only two freshwater mussels [*Cyrtonaias tampicoensis tecomatensis* (Lea, 1841) and *Megalonaias nickliniana* (Lea,1834)] were listed (Semarnat 2019), and no comprehensive country-wide conservation status has been conducted.

Use by Indigenous Peoples

The lowland Maya culture of México, Guatemala, and Belize utilized a wide variety of animal resources as food and ritual items, including mammals, fishes, birds, reptiles, crustaceans, and several mollusk species: both freshwater and marine. Powis (2004) described the archaeological use and distribution of "*Nephronaias* spp." by the Maya from the beginning of permanent settlements around 900 B.C. until the Spanish Conquest in the mid-1500s—a period of more than 2,500 years from numerous archeological sites in Belize (Barton Ramie, Blackman Eddy, Cahal Pech, and Caracol, Cayo District; Colha, Cuello, and Quiroz Cave, Orange Walk District; Altun Ha, Belize District) and Guatemala (Altar de Sacrificios, Sayaxché Department; Tikal, Seibal, and Uaxactún, Peten Department) (Figure 6). Interestingly, mussels have been recovered at three archeological sites in Yucatán, México (Chichén Itzá, Dzibilchaltún, and Uxmal) although no records of native mussels exist in that region. Powis (2004:138) suggested that although mussels were important as a supplemental food source, they were not a staple food item during either the Preclassic (1800 B.C.–250 A.D.), Classic (250–1000 A.D.), or Postclassic (1000–1500 A.D.) periods.

Richards and Boekelman (1937:169, pl. 6, fig. 9) reported on *Nephronaias dysonii* (Lea, 1859) (as *N. goascoranensis* and *Nephronaias* cf. *yzabalensis*) collected from a Mayan burial site at San Jose, Orange Walk District, Belize. Coe (1959:55, pl. 64, fig. 19) reported on two whole, unworked *Psoronaias crocodilorum* (Morelet, 1849) (as *Quadrula quadrata* Simpson, 1914) from a Late Classic burial site at Piedras Negras along the Río Usumacinta in Petén Department, Guatemala. Andrews (1969:32) reported on three unworked and four perforated specimens of *Nephronaias dysonii* (Lea, 1859) (as *N. ortmanni*) from a burial site at Barton Ramie Archeological Site, Cayo District, Belize, citing Willey et al. (1965, fig. 309, l–n) and *Nephronaias calamitarum* (Morelet, 1849) (as *Psoronaias semigranosus* on page 33 and as *Nephronaias* aff. *calamitarum* Morelet, page 104, pl. 21, fig. g) from Dzibilchaltún and Mayapán, Yucatán.

Emery (2008:622) conducted zooarchaeological analyses of animal remains from Mayan Petexbatun sites in the Usumacinta basin of Guatemala and reported that they consumed various freshwater mollusks that included freshwater mussels (*Psoronaias* spp.) and gastropods [*Pachychilus* spp. and *Pomacea flagellata* (Say, 1827)]. Sharpe et al. (2020:16, 26) recorded at least 1,032 individuals in the family Unionidae (identified as primarily *Psoronaias semigranosa*) at the large site of Seibal (Ceibal) in the Petexbatun, where they were most common in the Preclassic period. They also noted that the people living around Sayaxche, including many of the Guatemalan excavators working on the project still collect and consume these mussels, which are usually not cooked before consumption. Further down the Río Usumacinta in Tabasco, a 1200 B.C. cache of 140 intentionally stacked mussel valves have recently been found under the central monumental platform of Agauda Fénix, one of the largest in pre-Hispanic Mesoamerica (Inomata et al. 2021, Supplementary Discussion 4:25).

In the Classic period, there was a clear emphasis on the use of freshwater mussel shells for both ritual and ornamental purposes. These shell artifacts, cut into various forms such as beads, pendants, inlays, and rings, were now being deposited primarily in burials and dedicatory caches (Figure 6; Powis 2004:139). Mussel shells were also used as paint pots, shell pendants, shell discs sewn into clothing like iridescent sequins, and ground shell powder that may have been an early form of glitter (Figure 7).

Freshwater shells were also used in ritual contexts, and they, like marine shells, were strongly tied to water symbolism in Maya cosmology. A ritual deposit containing thousands of freshwater mussels and snails was found at the base of Structure B1 at Blackman Eddy in Belize. The basin-shaped deposit contained alternating layers of freshwater mussels with small gastropods

DOI: 10.1201/9781003314448-7

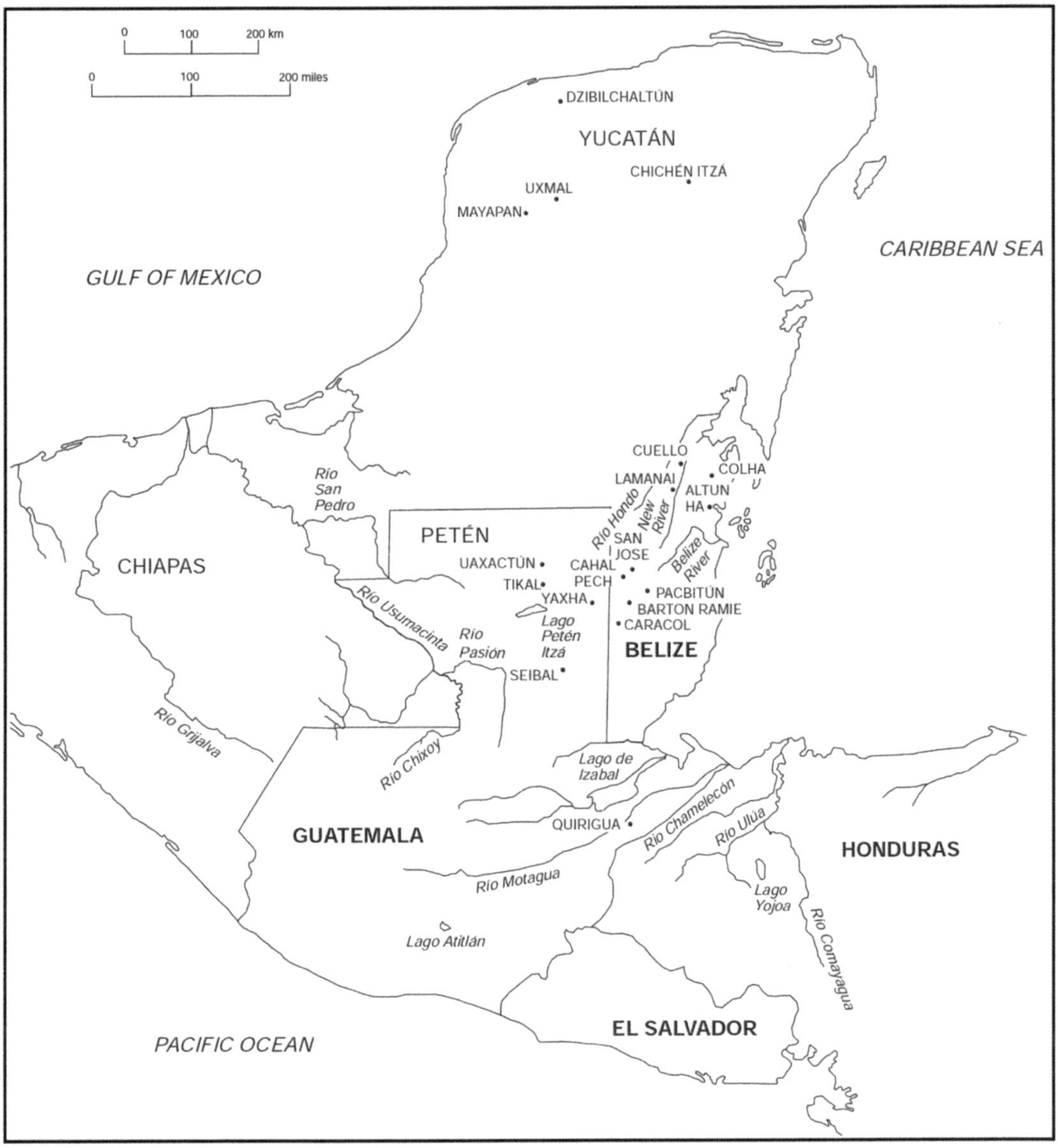

FIGURE 6 Archaeological sites in the Yucatán. (From Powis 2004:126).

(*Pachychilus*) placed predominantly at the base, mussel shells placed in the next layer, and large *Pachychilus* and *Pomacea* snails layered on top (Brown 2001:14).

In Nuevo León, México, unionid artifacts in the form of pendants have been found in caves (Cueva de La Candelaria and Cueva de La Paila) in Coahuila, México and Boca de Potrerillos, an archaeological site located about 14 km from the municipal seat of the municipality of Mina in the intermountain valleys of the Sierra Madre Oriental [26.0443, -100.6413], and La Morita II: a cave located on the northwest slope of Mesa El Cedral. In the La Paila cave, two complete earrings made with freshwater mussels were found. Other unionid pieces were found in Cueva Ahumada, an archaeological site located at the foot of Cerro La Mota Grande, near the town of Rinconada, Nuevo León, Mexicó [25.7064, -100.7203]. At the Coyotito site at La Lata, a fragment of the middle part of a valve of *Popenaias popeii* (Lea, 1857) and, at the Los Agaves site, two cut fragments of *Cyrtonaias*

FIGURE 7 Freshwater mussel pendant found at a Maya burial site in Guatemala. (Photo courtesy of Kristen Grace, University of Florida).

berlandierii (Lea, 1857) were found (Zepeda Quintero and Valadez Moreno *In* Velázquez Castro et al., 2011:50). The use of freshwater species probably owes to the fact that they were also used as a food resource and as receptacles (especially in the case of the large valves), and some others as raw material for the production of ornamental pieces.

Habitats and Collection

Freshwater mussels vary considerably with respect to the habitats they occupy. Some are restricted to a specific stream type (e.g., small caños or large rivers), whereas others can be found in almost any permanent body of water. Mussels can be found in a wide variety of in-stream including pools, stream runs, or riffle areas where they live in a mixture of sand, mud, and/or gravel. Shells are occasionally encountered along stream banks when the water levels are at their lowest. Although a few species (particularly members of the family Mycetopodidae and the genus *Uniomerus*) can withstand some desiccation and estivate (Haas 1939), most are found in permanently flowing streams or lakes that contain water year-round.

A wide variety of techniques are used to collect freshwater bivalves, and the method selected largely depends on the objective of the study. The method chosen will influence the way data are collected and interpreted. Qualitative surveys (often in the form of timed searches) are those that document the presence/absence and the relative abundance of species at a particular site. Quantitative methods (typically in the form of linear transects or quadrat samples) are used to gather data on community and population structure but increase survey costs considerably. Timed searches tended to overestimate large, sculptured species and underestimate small, smooth-shelled species (Vaughn et al. 1997:159).

The most common method for collecting freshwater mussels is by hand-picking in the substrate or picking up shells along the shore or stream bottom. With practice, one can learn to see mussels in the shallows of streams with clear water. Usually, the only part visible is the posterior end with the incurrent and excurrent apertures protruding from the bottom of the stream. Often, a mussel can be seen at the end of a meandering trail left in the stream bottom while it moves from place to place. Rakes or dredges can be used in shallow water with sandy or fine substrates to bring mussels to the surface. Quadrats are usually placed on the bottom and all substrate is removed to a depth of a few centimeters and run through a series of sieves. This method is especially effective in recovering juveniles that are easily missed by hand-picking. Diving with either scuba tanks or surface-supplied air is often used for sampling freshwater bivalves in deep water. These tools allow for quantitative work in otherwise inaccessible areas and provide the greatest ability to assess density and diversity in deep-water habitats. Both of these methods require diver certification and should not be used without adequate logistical support. A summary of collecting methods can be found in Strayer and Smith (2003) and Cummings et al. (2016).

The best time to collect freshwater bivalves is when streams are at their lowest levels. When collecting, all available habitat types (pools, runs, riffles, backwater areas, etc.) should be searched so that species restricted to specific habitats are not overlooked. Most of the field gear needed to collect freshwater mussels is inexpensive. Mesh bags to gather mussels in the field can be purchased from a commercial supplier or made at home. Coolers are essential to transport specimens to the lab alive, without appreciable mortality. A viewing bucket (constructed by cutting out the bottom of a 5-gallon plastic pail and replacing it with a piece of clear plexiglass adhered with silicon) can be used to locate and observe freshwater mussels in clear water.

Specimens are often collected, photographed, and returned, but voucher specimens (live or dead) should always be retained to document the occurrence. Recently, citizen science-sourced monitoring networks such as iNaturalist (www.iNaturalist.org) have helped tremendously to document species occurrence and biodiversity. Before collecting mussels, it is always advisable to contact the local natural resource agency to find out if there are any restrictions and, if required, obtain the necessary permits. Once collected, specimens can be cleaned with warm water and a small brush or scrub pad to bring out the colors and other markings needed for identification. Foreign substances

DOI: 10.1201/9781003314448-8

like varnish, petroleum jelly, or other substances used to coat shells should never be used. Labels should accompany a collection and include the following information at a minimum: the name of the body of water, road or bridge crossing, distance and direction from the nearest town, the county or administrative division, state, date, the name of the collector(s), and most importantly, latitudinal and longitudinal coordinates. Other information such as water temperature, depth, current velocity, bottom type, and collecting effort can be recorded in a field notebook. Locality data should be written in pencil or permanent ink on quality label paper.

Arrangements should be made with a curator for deposition of voucher specimens in a natural history museum. Museums play essential roles as repositories for specimens and associated data to document changes in biodiversity. Carefully collected and documented collections are essential to these efforts. All research (site or regional survey, taxonomic, genetic, ecological, or anatomical) should be documented with vouchers placed in an established museum to verify the identification of specimens used in the study.

Part 2

Area of Coverage, Data Sources, Maps, and Photographs

Area of Coverage, Data Sources, Maps, and Photographs

In their updated global checklist of the Unionoida Graf and Cummings (2021), each nominal species was assigned to a valid species (or treated as a *nomen dubium*). In the checklist, they reviewed the modern literature beginning with Simpson (1900; 1914) and Haas (1969) and then proceeded through the major works listed in Graf and Cummings (2007:292) and numerous minor works. For each work, the taxonomic history was captured by associating a "taxonomic opinion" with each nominal species. For each nomen, there is a related set of taxonomic opinions indicating the author(s), date of publication, whether the nomen was considered valid, which nominal species was treated as the senior synonym (if invalid), and the combination of genus and species applied. Thus, each valid species was related by synonymy to a comprehensive history of how the taxonomic components of that species were variously treated by different authorities. This allowed them to form a framework to address the question of how many species of mussels are in the families Unionidae and Mycetopodidae in Central America and how they are distributed.

AREA OF COVERAGE AND DATA SOURCES

The area of coverage in this book is the Río Grande del Norte and Colorado River drainages in the USA and México south to Colombia and western Venezuela in South America (Maps 3–5). We compiled the species lists and distribution maps by a thorough review of the published geographic and taxonomic data from the last 200 plus years and examination of over 3,300 specimen lots including over 300 types housed in 21 collections in the United States, México, Europe, and Australia during the period from December 2002 to June 2023 (Table 1). Another 1,600 records from the Río Grande and Colorado basins and over 10,000 records from North America north of the Río Grande were used to complete the total range maps for those species whose range extends beyond Central America. The supplemental records were obtained from InvertEBase (invertebase.org), iDigBio (www.idigbio.org), iNaturalist (www.inaturalist.org), Mussels of Texas, (mussels.nri.tamu.edu), or other online collection resources (Table 2). A plot of all records is shown in Map 3.

DOI: 10.1201/9781003314448-10

TABLE 1
Museum Collections Examined (3,300 Plus Records)

AMNH	American Museum of Natural History, New York, New York
AMS	Australian Museum, Sydney, New South Wales, Australia
ANSP	Academy of Natural Sciences of Drexel University, Philadelphia, Pennsylvania
CHAS	Chicago Academy of Science, Illinois
CM	Carnegie Museum, Pittsburgh, Pennsylvania
CNMO	Universidad Nacional Autónoma de México (UNAM), México City
COMA	Malacological Collection Dr. Antonio García-Cubas, México City
DMNH	Delaware Museum of Natural History, Wilmington
FLMNH	Florida Museum of Natural History, Gainesville
FMNH	Field Museum Natural History, Chicago, Illinois
INHS	Illinois Natural History Survey, Champaign
IRSNB	Institut royal des Sciences naturelles de Belgique, Brussels
MCZ	Museum of Comparative Zoology, Harvard University, Cambridge, Massachusetts
MNHN	Muséum national d'Histoire naturelle, Paris, France
NCSM	North Carolina Museum of Natural Sciences, Raleigh
NHMUK	British Museum of Natural History, London, United Kingdom
OSUM	Ohio State Museum of Biological Diversity, Columbus
SMF	Senckenberg Forschungsinstitut und Naturmuseum, Frankfurt, Germany
UMMZ	University of Michigan Museum of Zoology, Ann Arbor
USNM	National Museum of Natural History, Smithsonian, Washington D.C.
ZMB	Museum für Naturkunde der Humboldt–Universität, Berlin, Germany

TABLE 2
Additional Specimen Records Obtained from Online Databases or Publications (12,000 Plus Records)

AUM	Auburn University Museum of Natural History Mollusk Collection, Alabama
CCMSH	Corpus Christie Museum of Science and History, Corpus Christie, Texas
CMN	Canadian Museum of Nature, Ottawa, Canada
JFBM	J. Ford Bell Museum of Natural History, Minneapolis, Minnesota
LACM	Natural History Museum of Los Angeles County, Los Angeles, California
MHNG	Muséum d'histoire naturelle, Genève, Switzerland
MNHNS	Museo Nacional de Historia Natural, Santiago, Chile
MALA	Muséum d'histoire naturelle de Rouen, France
NMMNH	New Mexico Museum of Natural History, Albuquerque, New Mexico
NRI	Joseph Britton Freshwater Mollusk Collection, Texas A&M, Natural Resources Institute, Dallas, Texas
OMNH	Sam Noble Museum of Natural History, University of Oklahoma, Norman, Oklahoma
PRI	Paleontological Research Institution, Ithaca, New York
RMNH	Rijksmuseum Van Natuurlijke Historie Leiden—Naturalis Biodiversity Center
SBMNH	Santa Barbara Museum of Natural History, Santa Barbara, California
SMBU	Strecker Museum, Baylor University, Waco, Texas
TM	Terry Myers Collection
UCM	University of Colorado Museum of Natural History, Boulder, Colorado
UT	University of Texas, Non-vertebrate Paleontology Collection (NPL), Austin, Texas
UTEP	University of Texas El Paso, Texas
ZMA	Zoologisch Museum Amsterdam—Naturalis Biodiversity Center

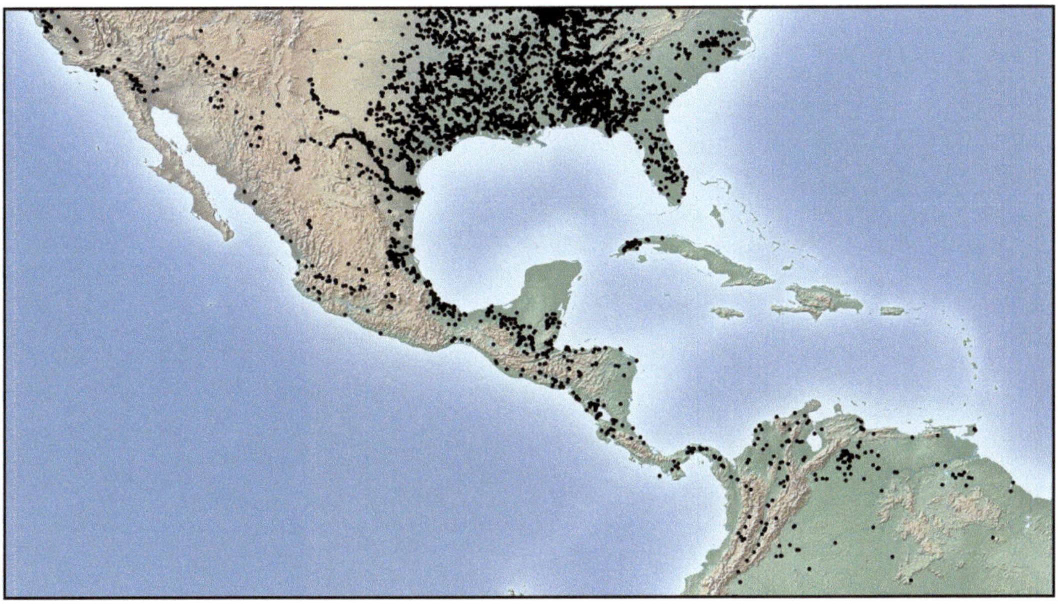

MAP 3 Plot of all specimen records used in this work.

MAP 4 Principal rivers of the USA, México, and Guatemala.

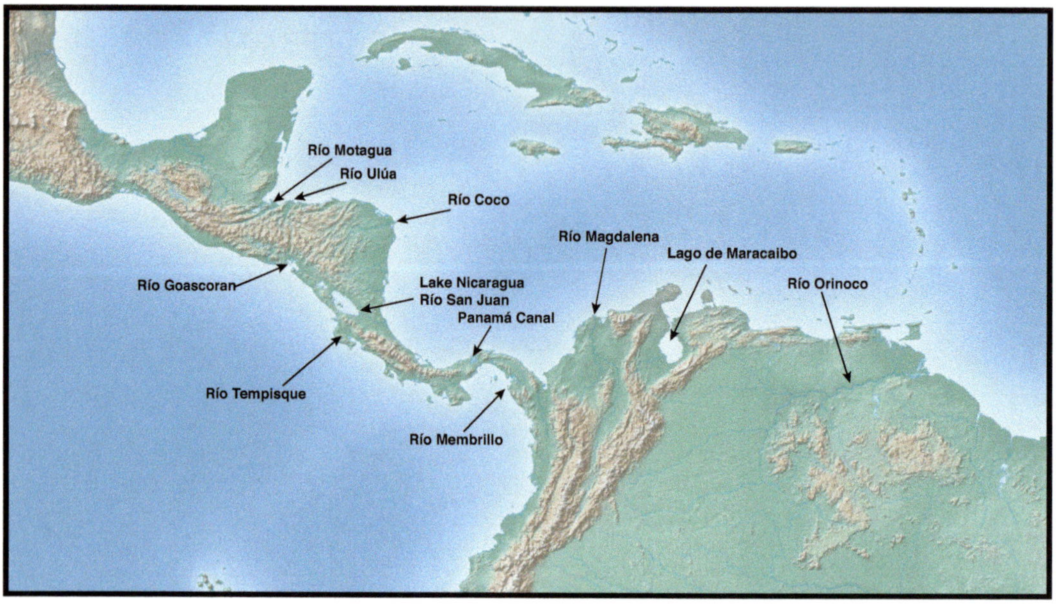

MAP 5 Principal rivers of Guatemala, Belize, Honduras, El Salvador, Nicaragua, Costa Rica, Panamá, and northern South America.

PHOTOGRAPHS

Each museum lot we examined was digitally photographed to document shell morphology and original label information. Textual data (catalogue number, previous identifications, collection locality, etc.) were captured subsequently from the images. Locality data were geo-referenced (if possible) from a variety of sources, including online geographic name servers, published lists, the original source references, specimen labels, and GPS in the field. Images were also databased along with taxonomic information. Depending upon the geographic extent of a locality, geocodes may represent an approximate position (on a continental scale) or may not be possible at all. Details of our data model and all specimen records are available on the MUSSEL Project Website (www.mussel-project.net/).

Part 3

Freshwater Mussel Studies in Central America

Historical Surveys and Biographies of Malacologists

There have been few comprehensive surveys or collection efforts in Central America, and the coverage has been extremely uneven. Most of the early work on the freshwater mussels of Central America was conducted by European and North American scientists. The vast majority of collections have been made in México (more than 1,900 out of over 3,100 records) and just four of the 31 Mexican States (Veracruz, San Luis Potosí, Tamaulipas, and Tabasco) account for nearly 70% of the collections made in México. Twenty-one of the 31 states in México have fewer than 10 records, with five of those (Aguascalientes, Baja California Sur, Puebla, Tlaxcala, and Zacatecas) having no records whatsoever. Most of the surveys have concentrated on the east coast in rivers that drain to the Gulf of Mexico or the Caribbean, whereas the interior and Pacific basins have been largely ignored. Some large river systems have never been investigated. Efforts focused on data-poor regions will certainly result in new and significant findings.

For the 356 lots for which we have plottable data for Guatemala, just over 40% are from the Usumacinta (181), followed by the Río Dulce (116), Lago de Izabal (73), Río Motagua (39), and Río Michatoya (16). Other countries lack sufficient data to make definitive statements about the richness of their mussel fauna. The number of plottable records by country is as follows: Belize (62), Costa Rica (29), El Salvador (39), Honduras (51), Nicaragua (188, 148 from Lake Nicaragua and its tributaries), and Panamá (49).

The following short biographies include personal information and the scientific achievements of malacologists who conducted surveys or described mussels from Central America. The biographies were compiled from Abbott and Young (1973), Coan and Kabat (2024b), and several sources cited in the accounts. Additional information on other early workers (Deppe, Uhde, Liebmann, Berendt, Strebel, Stoll, and others) and the expeditions they conducted collecting land and freshwater mollusks can be found in von Martens (1890–1901:vi–xii).

ISAAC LEA (1792–1886)

The first comprehensive treatment of the freshwater mussels that included freshwater mussels found in Central America was that of Isaac Lea. Isaac Lea was born on 4 March 1792 in Wilmington, Delaware, and died on 8 December 1886 in Philadelphia, Pennsylvania. Lea inherited a prosperous publishing business from his father-in-law that became one of the most successful in America. Lea was granted an honorary LL.D. from Harvard in 1852. He was elected as a member of the Academy of Natural Sciences, Philadelphia in 1815 and served as President from 1858 to 1863 [or 1853–1858 (Anon. 1887)]. Lea served as President of the American Association for the Advancement of Science (AAAS) in 1860. He had an early interest in minerals, geology, and paleontology, but later switched to conchology. Lea began publishing on unionids in 1828 and his last paper on the subject was in 1876. He traveled to Europe in 1832, where he met several prominent scientists and examined the unionids in the British Museum. Lea also traveled to Paris, where he met Baron Férussac and helped organize the mollusk collection at the Jardin des Plantes. He retired from the publishing business in 1851 and returned to Europe in 1852. He also met Alexander Humbolt in Berlin and Charles Lyell in London.

DOI: 10.1201/9781003314448-12

Lea did not collect much, but he described many species sent to him by others. Lea published most of his descriptions in the Proceedings of the Academy of Natural Sciences of Philadelphia. He later redescribed and figured them in the Journal of the Academy of Natural Sciences of Philadelphia or the Transactions of the Philosophical Society. He later compiled those in his monumental *Observations on the Genus Unio* in 13 volumes from 1834 to 1874. Geologist Richard Owen, acknowledging the receipt of one of the volumes, said, "They represent a kind or class of labors the most genuine and important and lasting, in the hard endeavor to gain a knowledge of Nature." According to his Popular Science Monthly obituary (Anon. 1887), he described 872 species of mollusks, including 21 of the 66 species of freshwater mussels currently recognized as valid in Central America. Lea's extensive mollusk collection has been deposited at the USNM and ANSP.

RUDOLPH AMANDUS PHILIPPI (1808–1904)

Rudolph Amandus Philippi was born on 14 September 1808 in Charlottenburg (now part of Berlin). He studied at the Königliche Gymnasium zum Grauen Kloster (equivalent to a high school). In 1826, Philippi studied at Friedrich-Wilhelms-Universität Berlin (now known as the Humboldt-Universität zu Berlin), where he graduated in 1830 at the age of 21. Although his degree was nominally in medicine and surgery, his coursework included botany, entomology, comparative anatomy, vertebrate zoology, and, perhaps most importantly, "Physische Geographie," taught by Alexander von Humboldt (1769–1859). In February 1835, he was hired as a teacher in natural history at the Höhere Gewerb-Schule in Cassel, a small university town in northern Germany, where he remained for many years. On 20 July 1851, Philippi left Germany, never to return, sailing from Hamburg to Chile. Philippi became the director of the Lyceum (high school) in Valdivia in 1853. In October 1853, Manuel Montt, the President of Chile, appointed Philippi to be the Professor of Zoology and Botany at the Universidad de Chile (Santiago), and that same month also appointed him to serve as Director of the Museo de Historia Natural, which was then located at the university.

His scientific work began in Germany in the 1830s and continued unabated until his death in Chile in 1904. Philippi contributed significantly to malacology: he described over 2,500 new taxa of Recent and fossil mollusks from around the world (2,528 species, 40 genera, and three families), particularly from Italy and Chile. Philippi initially published primarily on Recent and fossil mollusks from Europe in the 1830s, then expanded to marine mollusks from around the world by the 1840s. Philippi arrived in Chile in December 1856 and lived near Valdivia, where he spent his summers while on vacation from teaching in Santiago. Unfortunately, in November 1863, a fire destroyed his Valdivia residence, resulting in the loss of part of his personal archive, research notes, and manuscripts. In 2007, a fire at the Universidad Austral de Chile, Valdivia, destroyed some of his specimens and publications along with copies of Darwin's and Humboldt's work.

In 1890, the University in Berlin awarded Philippi an honorary doctorate degree, on the sixtieth anniversary of his 1830 thesis. Ten years later, Berlin awarded Philippi a second honorary degree, on the seventieth anniversary of his thesis. In early July 1904, Philippi was diagnosed with pneumonia and died on 23 July 1904. The next day, his body was laid in state in the grand hall of the Universidad, where it was viewed by over 10,000 people. Rudolph Amandus Philippi's first name is sometimes misspelled in the subsequent literature as the German variant "Rudolf." Philippi described five species from Central America that are still considered valid today: *Arotonaias cyrenoides* (Philippi, 1847), *Delphinonaias largillierti* (Philippi, 1847), *Nephronaias aztecorum* (Philippi, 1847), *Psoronaias semigranosa* (von dem Busch in Philippi, 1843), and *Sphenonaias liebmanni* (Philippi, 1847). (Coan and Kabat 2017; Kabat and Coan 2017).

PIERRE MARIE ARTHUR MORELET (1809–1892)

Arthur Morelet was born on 26 August 1809 at Château de Lays (Saône-et-Loire) and died on 9 October 1892 at Château de Velars near Dijon, France. Arthur Morelet received his education at the 'École de Droit' close to his home in Dijon. Morelet traveled throughout Europe and northern Africa including Italy, Algeria, Corsica, Sardinia, Spain, and Portugal. From 1846 to 1848, he undertook an expedition to Central America visiting Cuba, the states of Yucatán, Tabasco, and Veracruz México and the Departments of Petén and Alta Verapaz in Guatemala. According to von Martens (1890–1901:ix), he traveled mostly on foot. At that time, Central America was as unexplored as the interior of Africa, and the interior appeared as blank areas on maps. In 1849, he published the first part of the results of his voyage, describing 85 new species, followed by the second part in 1851 in which Morelet described another 65 species. A narrative of Morelet's expedition can be found in *Voyage dans l'Amérique Centrale, l'Ile de Cuba et le Yucatan* (Morelet 1857a,b). He described nine currently valid freshwater mussel species from Cuba, Guatemala, and México (Morelet 1849, 1851). Many of his type specimens are in the British Museum of Natural History, London or the Muséum national d'Histoire naturelle, Paris (Drouët 1892, 1893; Crosse 1893). An excellent and extraordinarily detailed account of his life and travels and catalogue of his taxa can be found in Breure et al. (2018).

PAUL HENRI FISCHER (1835–1893)

French Naturalist Henri Fischer was born on 7 July 1835 in Paris and died on 29 November 1893. He studied science and medicine, obtaining doctorates in both, and was appointed as Assistant in Paleontology and Assistant Naturalist at the Muséum national d'Histoire naturelle, Paris in 1872. He was joint editor of the *Journal de Conchyliologie* and served several terms as president of the Société Géologique de France and the Société Zoologique de France. With Hypolyte Crosse, he published a large monograph—Études sur les *mollusques terrestres et fluviatiles du Mexique et du Guatemala, In* M.H.–Milne Edwards (ed.). *Mission Scientifique au Mexique et dans l'Amérique Centrale. Recherches Zoologiques*—in three volumes from 1870 to 1902. Together they erected many of the Central American freshwater mussel genera that are in current use today.

JOSEPH CHARLES HIPPOLYTE CROSSE (1826–1898)

French Conchologist Joseph C. Crosse was born on 1 October 1826 in Paris and died on 7 August 1898 at his residence in Château d'Argeville at Vernou-la-Celle-sur-Seine, near Paris. Crosse studied at the Collège royal de Bourbon in Paris (also known as Lycée Condorcet). Early travels included the French coast of the Mediterranean, Corsica and Sicily in 1840. He published his first paper in 1851 in the *Journal de Conchyliologie*. He was co-editor of the same journal (with Paul-Henri Fischer) beginning in 1861 and would continue for the next 37 years. He was a Corresponding Member of the Zoological Society of London and Member of the Malacological Society of London. Crosse co-authored Études sur les *mollusques terrestres et fluviatiles du Mexique et du Guatemala, In* M.H.–Milne Edwards (ed.). *Mission Scientifique au Mexique et dans l'Amérique Centrale. Recherches Zoologiques*, with Henri Fischer in three volumes from 1870 to 1902. Crosse began the work in 1869 and it continued for 25 years. Over the course of his career, Crosse authored over 300 works on marine, land, and freshwater mollusks. Crosse and Fischer also published several shorter papers in which they described five Central American mussels we recognize as valid today (Cooke 1899; Nördlinger 1899; Poyard 1899).

KARL EDUARD VON MARTENS (1831–1904)

Eduard von Martens was born on 18 April 1831 in Stuttgart, Germany, and died on 14 August 1904. von Martens entered the University of Tübingen in 1849 to study medicine and obtained his M.D., but never entered practice. There are conflicting accounts regarding the date of his doctorate. Günther stated that he obtained his Ph.D. *honoris causa* from the University of Rostock in 1872. Kabat and Boss (1997:3) wrote that he received his doctorate from Tübingen on 6 June 1855 with a thesis on the distribution of the land and freshwater gastropods of Europe, which, though issued privately, also appeared in the Jahreshefte des Vereins für vaterländische Naturkunde in Württemberg (von Martens 1855). In 1855, he moved to Berlin and was commissioned to arrange and curate the mollusk collection at the Museum für Naturkunde der Humboldt–Universität (ZMB), which he carried on for nearly 50 years until his death on 14 August 1904 in Berlin. He was conferred the academic title of Privatdozent in 1873 and Extraordinary Professor in 1874. von Martens was made the Second Director of the ZMB in 1877 and Director from 1883 to 1887, but willingly relinquished his duties as an administrator to return to life as a curator. As the curator, he acquired the important collections of Johann Christian Albers, Wilhelm Dunker, and Friedrich Paetel. He was the naturalist on a Prussian expedition to East Asia departing on board the SMS Thetis on 20 February 1860. The journey took him through the Atlantic and Indian oceans to Japan, China, Sulawesi, Java, and Thailand. On 16 March 1862, he departed the Thetis and traveled to Sumatra, Java, Sulawesi, the Moluccas, Timor, and Borneo returning to Berlin in 1864. He always published alone and contributed some classic monographs on mollusks around the world. von Martens lavishly illustrated Land and Freshwater Mollusca, *In* F.D. Godman and O. Salvin (eds.), *Biologia Centrali-Americana*. Vol. 9, xxviii+706 pp., 44 plates, published in parts from 1890 to 1901, is one of the primary references on the mussels of Central America and a must-have for any student of the fauna. von Martens described 155 new genera (150 of them mollusks) and almost 1800 species (including around 1680 mollusks, 39 crustaceans, and 50 echinoderms) (Günther 1905). von Martens described three species: *Nephronaias chapalanus* (von Martens, 1900), *Nephronaias copanensis* (von Martens, 1900), and *Psoronaias rubicunda* (von Martens, 1900) that we recognize as valid today. An extensive volume on his life and work (400+ pages) and a comprehensive bibliography was published by Kabat and Boss (1997).

Isaac Lea (1792–1886)

Rudolph A. Philippi (1808–1904)

Pierre Marie Arthur Morelet (1809–1892)

Paul Henri Fischer (1835–1893)

PLATE 1 Isaac Lea (1792–1886), Rudolph A. Philippi (1808–1904), Pierre Marie Arthur Morelet (1809–1892), and Paul Henri Fischer (1835–1893).

PLATE 2 Illustration of *Unio nicklinianus* Lea, 1834. (From Plate 1 in Lea, I. 1834. Observations on the naiades; and descriptions of new species of that and other families. *Transactions of the American Philosophical Society* 5(N.S.):23–119 + 19 plates).

PLATE 3 Illustrations of 1. *Unio digitatus* Morelet, 1851, 2. *Unio morini* Morelet, 1851, 3. *Unio crocodilorum* Morelet, 1849, 4. *Unio crocodilorum* var. *β*, *semipustulata* Fischer & Crosse 1894, and 5. *Unio crocodilorum* var. γ, *praestricta* Fischer & Crosse 1894. (From Plate 60 in Fischer, P., and H. Crosse. 1894. Études sur les mollusques terrestres et fluviatiles du Mexique et du Guatemala. In M.H.–Milne Edwards (ed.). *Mission Scientifique au Mexique et dans l'Amérique Centrale. Recherches Zoologiques pour servir á l'Histoire de la Faune de l'Amérique Centrale et du Mexique.* Septiéme Partie. 15th Livraison. Feuilles 62–72, pp. 489–576, plates 59–62, Imprimerie Nationale, Paris Volume 2. 731 pp).

PLATE 4 Joseph Charles Hippolyte Crosse (1826–1898), Karl Eduard von Martens (1831–1904), Anson A. Hinkley (1857–1920), and Lorraine Screven Frierson (1861–1936).

PLATE 5 Illustrations of *Unio semigranosus* von dem Busch in Philippi, 1843. (From Plate 30 in Martens, E., von. 1890–1901. Land and freshwater Mollusca. *In* F.D. Godman & O. Salvin (eds.). *Biologia Centrali-Americana*. Vol. 9, xxviii+706 pp.+44 plates).

CHARLES TORREY SIMPSON (1846–1932)

Charles Torrey Simpson was born on 3 June 1846 in Tiskilwa, Illinois, and died on 17 December 1932 in Lemon City, Florida. Sc.D., University of Miami, Florida, 1927. Simpson was an American botanist, malacologist, and conservationist. From 1889 to 1902, he was an Aid in the Division of Mollusks at the United States National Museum, now the National Museum of Natural History. Simpson published the first synthesis of world-wide freshwater mussels, *Synopsis of the Naiades, or Pearly Fresh-Water Mussels* in 1900. Simpson retired to Florida in 1902, where he wrote two books in the natural history of the state. In 1914, Simpson's work on unionids, augmented by descriptions, expanded with revisions, and brought up to date, was privately published by Bryant Walker of Detroit, Michigan, under the title, *A Descriptive Catalogue of the Naiades or Pearly Fresh-water Mussels*.

ANSON A. HINKLEY (1857–1920)

Anson A. Hinkley was born on 26 November 1857 in Farmersville, Indiana, USA, but lived most of his life in Rockford and DuBois, Illinois, where he died on 23 July 1920. Hinkley published 27 papers on the land and freshwater mollusks of North and Central America. He was a prodigious collector, and the results of his collecting endeavors can be found in museums throughout the world. Of particular importance are his turn-of-the-century collections from the southeastern United States, which helped to document the incredibly diverse molluscan fauna of that region.

Hinkley collaborated with Henry A. Pilsbry of the Academy of Natural Sciences, Philadelphia and together or separately published over 50 papers on the freshwater and terrestrial mollusks of Central America. Of importance here are the papers they published on the Pánuco Basin, México (Hinkley 1907a,b; Pilsbry 1909, 1910; Pilsbry and Hinkley 1910) and Guatemala (Hinkley 1920). Most of the historical data we have on the Pánuco basin in México is from the collecting efforts of Hinkley, who sampled at 41 sites in the drainage between 1907 and 1909 (Map 6). In 1912, Hinkley made his first expedition to Guatemala, a second trip in 1913 and a third in 1917. These expeditions resulted in a large series of specimens from that region and the subsequent addition of many land and freshwater species new to science. Hinkley is credited with discovering approximately 115 new species (17 of which were named after him), a record of accomplishment that secures his place in the study of North and Central American land and freshwater mollusks. We examined over 450 specimen lots collected by Hinkley in Central America, which accounted for nearly 15% of all the specimens we examined.

LORRAINE SCREVEN FRIERSON (1861–1936)

Frierson's family settled in South Carolina and later moved to Louisiana, where he was born in Lakeville, DeSoto Parrish, Louisiana, on 7 August 1861. After the Civil War, his family emigrated to Honduras, where they remained for several years. They returned to the United States and settled in Florida until 1876. They then returned to Lakeville, Louisiana, and formed the town of Frierson's Mill, afterward called Frierson, where they lived for many years. In 1919, Frierson moved to Gayle, South Caddo Parrish, south of Shreveport, where he raised cotton. In 1888, he became interested in freshwater mussels and began to collect them locally in nearby Bayou Pierre. He collected primarily in Louisiana and Texas with Dr. William S. Strode of Illinois and Texans Henry Garrison Askew and Joseph Daniel Mitchell. He described *Lampsilis fimbriata* Frierson, 1907 and *Lampsilis iridella* Pilsbry & Frierson, 1908 from specimens collected by Anson A. Hinkley in México. He also described *Unio* (*Nephronaias*) *ortmanni* Frierson, 1913, *Elliptio* (*Nephritica*) *haricotti* Frierson, 1927, *Elliptio* (*Nephronaias*) *sajensis* Frierson, 1927, *Psoronaias kuxensis* Frierson, 1917, and *Psoronaias martensi* Frierson, 1927 largely from Hinkley's Guatemala collections. A list of all the

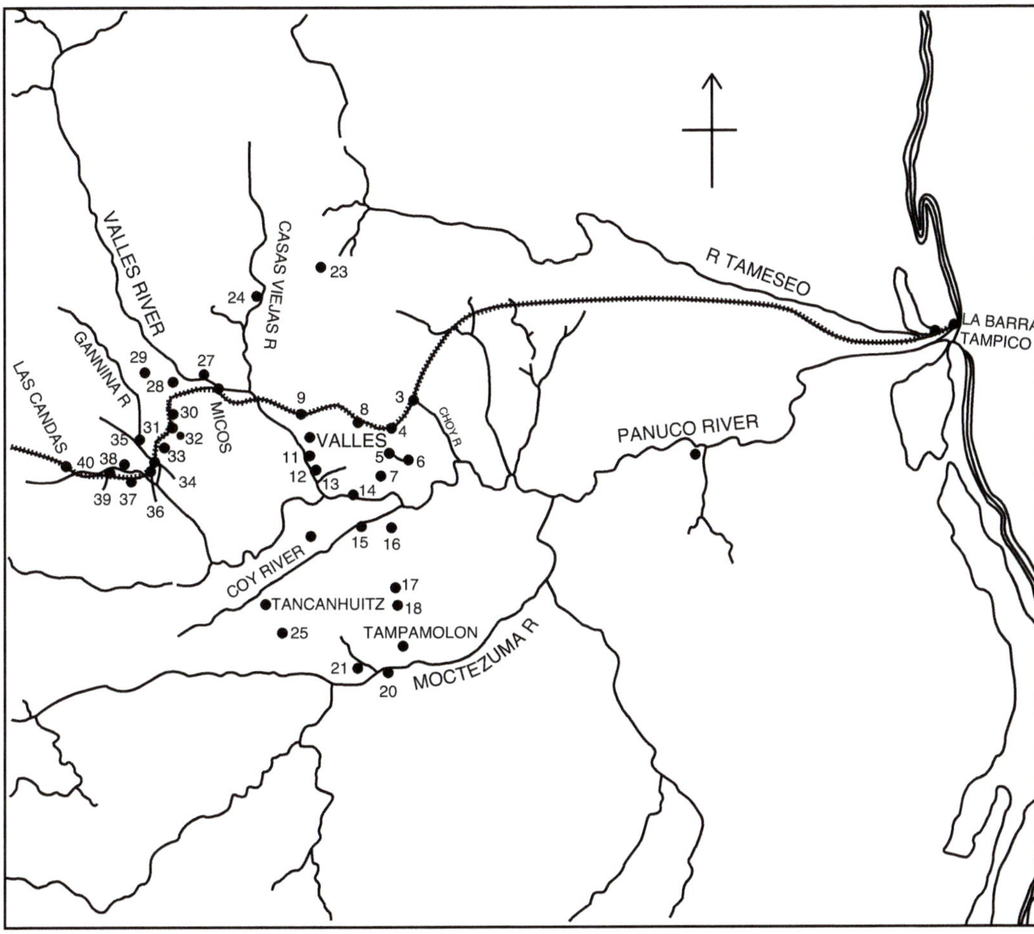

MAP 6 Hinkley's hand-drawn map of his collection sites in the Pánuco basin, México, where he collected in 1907 and 1909. (From Pilsbry, H.A., and A.A. Hinkley. 1910. Melaniidae of the Panuco River system, Mexico. *Proceedings of the Academy of Natural Sciences of Philadelphia* 61(3):519–531 + 2 plates).

taxa described by Frierson, including the five considered valid from Central America, can be found in Johnson (1972a). Frierson's collection (including the Truman H. Aldrich collection and half of the William A. Marsh collection) was purchased by Bryant Walker and now resides primarily in the UMMZ, USNM, and the ANSP (Strecker 1929; Vidrine and Schwartz 1978). Frierson died in his home on 11 November 1936 at the age of 75 and is buried in a family plot in Evergreen Cemetery, Frierson, DeSoto Parish, Louisiana, USA.

HENRY AUGUSTUS PILSBRY (1862–1957)

Henry Pilsbry was born on 7 December 1862 near Iowa City, Iowa; died on 26 October 1957 in Lantana, Florida; and was buried at Bala, Pennsylvania. Pilsbry obtained his B.S. from the University of Iowa in 1882 and his Honorary D. Sci. degrees from the University of Iowa in 1899, University of Pennsylvania in 1940, and Temple University in 1941, respectively. He was a newspaper reporter in Iowa City and Davenport from 1882 to 1887. He was the Assistant Curator of Mollusks at the ANSP under G.W. Tryon in 1887, Conservator of Conchology at the ANSP from 1888 to 1895, and Curator at the ANSP from 1895 to 1957. He was the editor of the *Manual of Conchology* from 1889 to 1932 and *The Nautilus* from 1889 to 1957. He was an honorary member of the Conchological Society of Great Britain, California Academy of Science, Senckenberg Gesellschaft für Naturforschung, and the Society of Malacology "Carlos de la Torre." He was the first President of the American Malacological Society in 1907. He was also the first President of the American Malacological Union (AMU) in 1931 and Honorary President from 1936 to 1957. He was the Honorary Life President of the Philadelphia Shell Club from 1955 to 1957. He traveled to Florida, Cuba, the southwestern USA (with James H. Ferriss and Lorenzo E. Daniels), Hawaii, Japan, Australia, West Indies, Galapagos, and various places in South and Central America. He published over 3,000 papers on mollusks and barnacles (Baker 1957). As noted above, Henry A. Pilsbry collaborated with A.A. Hinkley and together or separately published over 50 papers on both the freshwater and terrestrial mollusks of Central America. Six of the 64 native species of freshwater mussels found in Central America were described by Pilsbry.

FRITZ HAAS (1886–1969)

Fritz Haas was born on 4 January 1886 in Frankfurt am Main, Germany, and died on 26 December 1969 in Hollywood, Florida, just short of his 84th birthday. Haas was Keeper of Invertebrate Zoology at the Senckenberg Forschungsinstitut und Naturmuseum (SMF) in Frankfurt am Main, Germany from 1911 to 1936. The Nazis forcibly removed him from his position on 30 June 1936. He fled Nazi Germany to Spain and then to the Field Museum of Natural History, Chicago, where he was Curator of Lower Invertebrates from 1938 to 1959. When he began his tenure at the Field Museum, the only invertebrates were leftover exhibits from the Columbian Exposition of 1893, and only minimal literature on mollusks was available. By 1954, 54,000 entries comprised the Field Museum's mollusk collection. Essentially, all of these had been labeled, catalogued, and reidentified by Haas.

Haas received his Ph.D. from Heidelberg in 1910. He was trained in biology by herpetologist and malacologist Oscar Böttger (1844–1910) and malacologist Wilhelm Kobelt (1840–1916). Early in 1920, he became editor of the Archiv für Molluskenkunde, a position he held for many years. In 1954, the Senckenberg Natural Research Society honored his scientific achievements by awarding him their Silver Cretzschmar Medal. He spoke German, English, French, Spanish, and Catalonian, and could read and understand Portuguese, Italian, Dutch, Swedish, Danish, Latin, and Greek. In the 60 years he worked in Frankfurt and Chicago, Haas introduced 385 new names to science (77 genera, 308 species and subspecies). Haas used the humorous pseudonym "Lep." (= *Lepus*, the rabbit) that he occasionally signed to his essays if they did not seem particularly noteworthy. Haas was pleased with the dedication of the species *Pisidium lepus* Kuiper, 1957 to him using the pseudonym he coined 50 years earlier. Helene Ganz was assigned to help Fritz Haas at the Senckenberg Museum. She proved indispensable, and on 30 March 1922, shortly after Fritz had been promoted to Keeper of Invertebrate Zoology, she became Mrs. Fritz Haas.

Haas' research interests were the Unionoida and, to a lesser degree, freshwater and terrestrial gastropods. He traveled to Norway in 1910; Pyrenees, Spain, France in 1914–1919 (in exile); southern Africa in 1931–1932 (as a member of the Schomburgk Expedition); Brazil in 1937 (aided by R. von Ihering, nephew of the Hermann von Ihering); Bermuda, Cuba, and Canada. He was President of the AMU in 1950. Haas published the second comprehensive world review of freshwater mussels *Superfamilia Unionacea* in 1969, a required reference for any student of freshwater mussels. Haas' Central American specimens are in the Field Museum Natural History, Chicago, and the Senckenberg Forschungsinstitut und Naturmuseum, Frankfurt (Solem 1967a,b; Zilch 1970).

HORACE BURRINGTON BAKER (1889–1971)

H.B. Baker was born on 25 January 1889 in Sioux City, Iowa, and died on 11 March 1971 in Havertown, Pennsylvania. Baker attended High School in Flint, Michigan and enrolled at the

Henry Augustus Pilsbry (1862–1957)

Fritz Haas (1886–1969)

Horace Burrington Baker (1889–1971)

Daniel J. Bereza (1950–2007)

PLATE 6 Henry Augustus Pilsbry (1862–1957), Fritz Haas (1886–1969), Horace Burrington Baker (1889–1971), and Daniel J. Bereza (1950–2007).

University of Michigan in 1906. In his senior year (1910), he accompanied A.G. Ruthven in collecting mollusks in México (H.B. Baker 1922). He was an instructor at the Douglas Lake Field Station in Michigan and in 1913 appointed as instructor at Colorado College, Colorado Springs. He fought in World War I, was commissioned a 2nd Lt., Field Artillery, in 1917, and spent eighteen months in France and Germany with the 18th Field Artillery, Third Division. He returned to the University of Michigan after the war and participated in fieldwork on mollusks in Venezuela in 1919. He received his Ph.D. from the University of Michigan in 1920. The research for his first doctoral dissertation on midwestern grasshoppers was completed, but before it was submitted to his graduate committee, it was lost in a fire at the University of Michigan and he had to start all over again. He was appointed as an Instructor at the University of Pennsylvania in 1920, Assistant Professor in 1926, Associate Professor in 1928, Full Professor in 1939, and finally Emeritus Professor in 1959. Baker collected in México (1910 and 1926), Dutch Leeward Islands (1922), southern Appalachians (1928), Idaho (1929 and 1931), Jamaica (1933), Hawaii (1935), Puerto Rico (1939), and made several trips to Florida. Baker was the Business Manager of *The Nautilus* from 1932 to 1956 and editor from 1957 to 1970. He was a Charter Member and 9th president (1940) of the American Malacological Union (AMU) (Abbott and Wurtz 1971). Baker's Méxican specimens are in the mollusk collections of the UMMZ, ANSP, MCZ, and CM. Photo taken on 27 April 1968 in Philadelphia, Pennsylvania (courtesy of the MCZ). Baker was part of the University of Michigan-Walker Expedition to Veracruz, México, which collected terrestrial and freshwater mollusks there for six weeks (10 July–20 August) in 1910 (Baker 1922, 1930). Much of the historical knowledge on the mussels of the Papaloapán basin was from that effort.

DANIEL J. BEREZA (1950–2007)

Dan Bereza was born Danylo Jaroslav Illyich Bereza on 22 October 1950 in Philadelphia, Pennsylvania. He attended St. Nicholas parochial school and then Central High School in Philadelphia, Pennsylvania. Dan was a member of the Philadelphia Shell Club since about the age of 10. He volunteered for Tucker Abbott at the Department of Malacology at the ANSP. From 1973 to 1974, Dan was a grant-supported staff member in the Department of Malacology and a first-year graduate student at Jefferson Medical University. Dan gave up graduate school and attended Temple University, School of Pharmacy, where he was awarded a B.S. in 1973. He was a practicing pharmacist for nearly all his adult career, but he was active in field work and maintained an extensive collection (nearly all self-collected) of preserved unionids and an outstanding malacological library. He was the sole collector or part of a collecting team for more than 1,200 lots at ANSP, totaling more than 60,000 specimens from 1968 to 1979. He was a close friend of Samuel Liberty Harvey Fuller (1942–2001) and Malcom Vidrine (1949–), and they conducted a lot of fieldwork together.

Dan Bereza died on 4 June 2007 at the age of 56. His brother and sister donated his collection of unionids (including soft parts) to the Smithsonian's National Museum of Natural History, in Washington, DC, where they are available for study (Luebke and Graf 2013; Gary Rosenberg pers. comm. 2019; Jerry Harasewych pers. comm. 2019; Inoue et al. 2020a). In 1978 and 1982, Bereza and colleagues collected at over 100 sites in central México in the states of Coahuila, Queretaro, San Luis Potosí, and Tamaulipas. Extensive collections were made in the Pánuco, Carrizal, Actopan, Cazones, Pantapec, Jamapa, and Papaloapán basins.

FRED THOMPSON (1934–2016)

Fred Thompson was born in 1934 in Cleveland, Ohio. Herpetology was his initial interest, but during his travels, he was struck by the diversity of non-marine mollusks and became intrigued by the lack of published research on the group. His career was a 65-year-long odyssey to nearly every corner of the planet to study and publish on terrestrial and freshwater mollusks. Fred's first papers on the systematics of land snails in México and Central America were published while he was an undergraduate at the University of Michigan, from where he graduated in 1958. He received his Master of Science degree at Wayne State University, Michigan in 1961. After graduating, Fred moved to Florida, where he received his Ph.D. at the University of Miami in 1964. His dissertation led to his seminal work on the hydrobiid snails of Florida's springs (Thompson 1968).

In 1965, Fred moved to the University of Florida to work as a Research Associate, and in September 1966, he was hired as the Florida Museum of Natural History's first mollusk curator. Through his fieldwork and acquisition of private and orphaned collections, he took a small collection of a few hundred local mollusks and built it into one of the largest mollusk collections in the country, with over half a million lots comprising approximately three million specimens (Sierwald et al. 2018). The Florida Museum of Natural History collection is one of the largest non-marine snail collections in the world. It has especially large holdings from the southeastern USA, West Indies, Andean South America, Southeast Asia, and Oceania. Throughout his career, he continued his fieldwork on non-marine snails of México and Central America, describing many new species and resulting in the only modern compilation of the freshwater and terrestrial gastropod fauna of the area (Thompson 2011), an indispensable resource for students interested in Central American gastropods. Although his interests were primarily snails, he sampled in the Río Pánuco drainage in 1965 and again in 1989, 1990, 1993, 1998, and 2001, and he collected numerous freshwater mussels during his fieldwork there. Fred retired from the Florida Museum of Natural History, becoming Curator Emeritus, in October 2012, and continued to work until his death on 27 December 2016 at the age of 82 (Slapcinsky 2017).

HERBERT D. ATHEARN (1923–2011)

Herbert David Athearn was born on 3 July 1923 in Fall River, Bristol County, Massachusetts, and began collecting mollusks at the age of 16. After serving in World War II, he lived in Boston where he volunteered at the MCZ. An ardent collector and student of freshwater mollusks, he moved to Cleveland, Tennessee, in 1955, having determined that this was the center of molluscan diversity in North America. Upon his arrival, he began to collect aquatic mollusks throughout the southeastern USA until about 2000, when his health no longer permitted long field trips. He maintained a personal collection, the Museum of Fluviatile Mollusks, in his home until 2007, when it was donated to the North Carolina Museum of Natural Sciences, Raleigh (Smith et al., 2007). In 2001, the Freshwater Mollusk Conservation Society presented Athearn with its William J. Clench Memorial Award for extraordinary contributions to the field of malacology. (Bogan and Johnson 2011). Athearn made collections in Oaxaca, San Luis Potosí, Tamaulipas, and Veracruz México in 1966 and 1969.

EDNA NARANJO-GARCIA (1950–)

Edna Naranjo-Garcia got her M.S. (1984) and Ph.D. (1988) degrees, both from the Department of Ecology and Evolutionary Biology, University of Arizona, Tucson, USA and the Licenciatura en Biología, Facultad de Ciencias, Universidad Nacional Autónoma de México (UNAM), México, 1979. She first became interested in freshwater mollusks as an undergrad from a class on invertebrates. Her undergraduate essay was on aquatic mollusks as intermediate host of parasites. Her focus as a graduate student became terrestrial gastropods, but she has also published papers on freshwater snails and mussels (Hoeh et al. 1995, 1998; Naranjo-García 2003a,b; Tiemann et al. 2020; Ramírez et al. 2022). Some non-marine fossils had been also the subject of her research (Perrilliat et al. 2008; Vega et al. 2019). She has mentored many bachelor and graduate students, and she is currently the curator of the mollusk collection and teaches at UNAM.

CUMMINGS, TIEMANN, PFEIFFER, AND INOUE (2017–2024)

Various field surveys over the past seven years in México, Costa Rica, El Salvador, Guatemala, and Panamá. Cummings, Tiemann, and colleagues sampled for mussels in the four major basins that drain to the Gulf of Mexico—from north to south: the Río Conchos, a Río Grande tributary in Chihuahua (2018); the Río Pánuco in San Luis Potosí (2017, 2018); the Río Papaloapán in Veracruz (2022); and the Río Usumacinta in Chiapas and Tabasco along the Guatemalan border (2019). Pfeiffer and colleagues sampled in the Chagres (Caribbean Dr.), Sabanas, Santa Maria, and Turia basins (Pacific Dr.) Panamá in 2023 and Costa Rica in 2024. Kentaro Inoue (Shedd Aquarium, Chicago) and colleagues have ongoing studies in Costa Rica, El Salvador, and Guatemala.

Part 4

Zoogeography of Central American Freshwater Mussels

Zoogeography of Central American Freshwater Mussels

We treat 77 native species of freshwater mussels (Unionidae—69; Mycetopodidae—8) as valid in this work. An additional four species have been introduced into the region for a total of 81 species. Central America proper has 66 species, 64 that are native (60 of which are endemic) and two that have been introduced (Table 3). The Río Grande del Norte basin has (or had) 12 native (5 endemics) and three introduced species (Tables 6 and 16), and the Colorado River has one native and two introduced species (Tables 7 and 16). *Megalonaias*, *Pyganodon*, and *Popenaias* are the only freshwater mussel genera native to both Central and North America. Nomenclatorial changes made since the 2021 checklist of Graf and Cummings are given in the genus and species accounts below.

The Gulf coastal, Caribbean, and Pacific drainages of Central America harbor unique assemblages of aquatic species, and the area is a transition zone between the Nearctic and Neotropic biogeographic regions. From the Río Soto la Marina to the Río Papaloapán on the Gulf coastal plain of central México, the composition of the freshwater fish fauna shifts from 75% North American to 95% Central American species (Miller and Smith 1986). This faunal transition is largely derived from a series of past geological and climatic changes, such as the uplift of the Sierra Madre Oriental and Sierra Madre Occidental and the formation of the Mesa Central during the Paleogene to Neogene periods. Miller and Smith (1986) hypothesized that the current distribution of freshwater fishes in central México is described by six major biogeographic patterns: (1) endemic species that have undergone adaptive radiation within the area, (2) species that have originated from Central or South American faunal groups, (3) species that have originated from groups of the Río Grande and/or eastern North America, (4) species that are associated with groups of the Pacific fauna, (5) species that have originated from groups of anadromous species, and (6) species that have invaded from the sea. While the biogeography and evolutionary histories of freshwater fishes in México have been relatively well studied, information on other aquatic groups, including freshwater mussels, in México and the rest of Central America is scarce. The vast majority of freshwater mussel species in Central America are found in basins that drain to the Gulf of Mexico or the Caribbean Sea. Five rivers draining into the Gulf of Mexico: Pánuco, Papaloapán, Coatzacoalcos, Grijalva, and Usumacinta (north to south), account for over 50% of México's average annual flow discharged into ocean basins (Hudson et al. 2005:1036). The principal rivers of Central America and surrounding countries are given in Maps 4 and 5.

FRESHWATER MUSSEL REGIONS AND PROVINCES OF CENTRAL AMERICA

We have grouped the freshwater mussel fauna of Central America into six Regions and 10 Provinces based upon their species richness, endemism, shared species, and drainage hydrology (Table 5, Map 8). The criteria we used were similar to those applied in other assessments of aquatic biodiversity (Miller 1966; Bussing 1976; Olson et al. 1998; Abell et al. 2000; 2008; Lyons et al. 2020; Hulsey and López-Fernandez 2011). The large drainages were consolidated as a single basin unit, with adjacent coastal drainages historically connected, whereas smaller coastal basins were grouped together based on proximity and geography. As mentioned above, although not strictly part of Central America, we included the Río Grande del Norte and Colorado basins in this work to account for all species that occur in México. The province, the number of native species contained in each, those that are endemic, and the number of species that have been introduced are listed in Table 5.

DOI: 10.1201/9781003314448-14

TABLE 3

Freshwater Mussels (Unionidae and Mycetopodidae) of Central America.
Unionidae: 58 species (56 native; all endemic, 2 introduced). Mycetopodidae: 8 native species (4 endemic). Total: 66 species (64 native, 60 endemic, 2 introduced). Major Basin Codes: Panucó (PA), Papaloapán (PP), Usumacinta (US), Mesa Central and Balsas (MC), Oaxaca to Colombia (OC), Gulf of Honduras and Caribbean (GH), San Juan (SJ), Cuba (CU)

Basin	PA	PP	US	MC	OC	GH	SJ	CU
Family Unionidae Rafinesque, 1820								
Subfamily Unioninae Rafinesque, 1820								
Tribe Anodontini Rafinesque, 1820								
Brachyanodon impura (Say, 1829)	X			X				
Pyganodon globosa (Lea, 1841)		X	X					
Sinanodonta woodiana (Lea, 1834)[1]					X	X		
Utterbackia imbecillis (Say, 1829)[1]	X	X						
Subfamily Ambleminae Rafinesque, 1820								
Tribe Lampsilini von Ihering, 1901								
Actinonaias coyensis (Pilsbry, 1910)	X							
Actinonaias medellina (Lea, 1838)	X	X						
Actinonaias sapotalensis (Lea, 1841)		X						
Actinonaias signata (Pilsbry, 1910)	X							
Arotonaias cyrenoides (Philippi, 1847)							X	
Arotonaias encarpa (Lea, 1868)							X	
Arotonaias gabbiana (Lea, 1868)							X	
Arotonaias reticulata (Simpson, 1900)						X		
Delphinonaias largillierti (Philippi, 1847)			X			X		
Delphinonaias scutulata (Morelet, 1849)			X					
Disconaias disca (Lea, 1838)	X							
Disconaias fimbriata (Frierson, 1907)	X	X						
Friersonia iridella (Pilsbry & Frierson, 1908)	X							
Mesonaias explicata (Morelet, 1849)			X					
Mesonaias tampicoensis (Lea, 1838)	X	X						
Mesonaias umbrosa (Lea, 1856)		X						
Pachynaias spheniopsis (Morelet, 1849)			X					
Tribe Popenaiadini Heard & Guckert, 1970								
Micronaias arata (Lea, 1843)							X	
Micronaias fallaciosa (Simpson, 1914)							X	
Micronaias granadensis (Lea, 1868)							X	
Nephritica haricotti (Frierson, 1927)								
Nephritica poeyana (Lea, 1857)				X				
Nephronaias aeruginosa (Morelet, 1849)			X					
Nephronaias aztecorum (Philippi, 1847)	X	X				X		
Nephronaias calamitarum (Morelet, 1849)			X		X			
Nephronaias caldwellii (Lea, 1858)				X				
Nephronaias chapalanus (von Martens, 1900)				X				
Nephronaias copanensis (von Martens, 1900)						X		
Nephronaias dysonii (Lea, 1859)						X		

(*Continued*)

TABLE 3 *(Continued)*

Freshwater Mussels (Unionidae and Mycetopodidae) of Central America. Unionidae: 58 species (56 native; all endemic, 2 introduced). Mycetopodidae: 8 native species (4 endemic). Total: 66 species (64 native, 60 endemic, 2 introduced). Major Basin Codes: Panucó (PA), Papaloapán (PP), Usumacinta (US), Mesa Central and Balsas (MC), Oaxaca to Colombia (OC), Gulf of Honduras and Caribbean (GH), San Juan (SJ), Cuba (CU)

Basin	PA	PP	US	MC	OC	GH	SJ	CU
Nephronaias flucki Bartsch, 1906						X		
Nephronaias goascoranensis (Lea, 1858)					X		X	
Nephronaias gundlachi (Dunker, 1858)								X
Nephronaias hjalmarsoni (Dunker, 1858)						X		
Nephronaias macnielii (Lea, 1869)					X		X	
Nephronaias ortmanni (Frierson, 1913)						X		
Nephronaias ravistella (Morelet, 1849)						X		
Nephronaias rowellii (Lea, 1859)					X			
Nephronaias sallei (Crosse & Fischer, 1894)			X					
Nephronaias scamnata (Morelet, 1849)								X
Nephronaias tehuantepecensis (Crosse & Fischer, 1894)					X			
Nephronaias tempisquensis Pilsbry, 1920					X			
Nephronaias yzabalensis (Crosse & Fischer, 1893)						X		
Popenaias berezai Inoue et al., 2020	X	X						
Popenaias metallica (Say, 1831)				X				
Popenaias semirasa (Pilsbry, 1910)	X							
Psoronaias crocodilorum (Morelet, 1849)			X					
Psoronaias distincta (Crosse & Fischer, 1893)		X						
Psoronaias kuxensis Frierson, 1917			X					
Psoronaias rubicunda (von Martens, 1900)						X		
Psoronaias semigranosa (von dem Busch in Philippi, 1843)	X							
Sphenonaias liebmanni (Philippi, 1847)		X						
Sphenonaias opacata (Crosse & Fischer, 1894)		X						
Sphenonaias plexus (Conrad, 1838)		X						
Tribe Quadrulini von Ihering, 1901								
Megalonaias nickliniana (Lea, 1834)	X	X	X					
Family Mycetopodidae Gray, 1840								
Subfamily Mycetopodinae Gray, 1840								
Mycetopoda subsinuata (Sowerby, 1868)					X	X	X	
Subfamily Anodontitinae Modell, 1942								
Anodontites cylindracea (Lea, 1838) [E]	X	X	X					
Anodontites jaliscoensis Pilsbry, 1920 [E]				X				
Anodontites lurulenta (Morelet, 1849) [E]			X					
Anodontites montezuma (Lea, 1841) [E]							X	
Anodontites tehuantepecensis (Crosse & Fischer, 1893)					X		X	
Anodontites tortilis (Lea, 1852)					X	X	X	
Anodontites trapesialis (Lamarck, 1819)				X	X	X	X	

[E]=Endemic Mycetopodidae; [I]=Introduced.

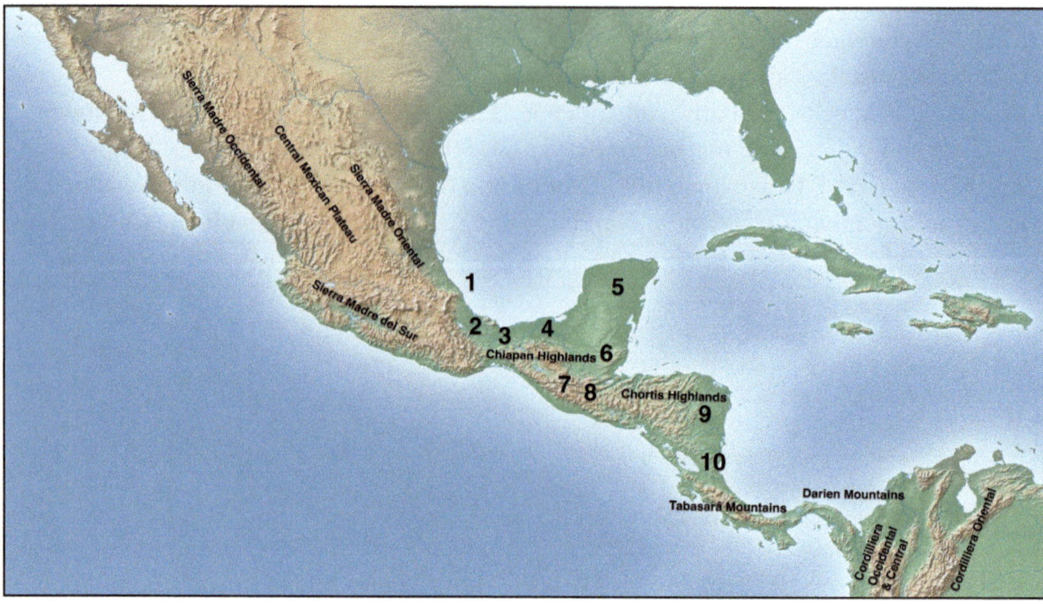

MAP 7 Major geological features of Central America and northwestern South America. The key is given in Table 4.

TABLE 4
Major Geological Features of Central America and Northwestern South America

1. Punta del Morro
2. Veracruz Lowlands
3. Isthmus of Tehuantepec
4. Tabasco Lowlands
5. Yucatán
6. Maya Mountains
7. Polochic-Motagua Fault
8. Central American Volcanic Front
9. Mosquitia
10. Nicaraguan Depression

Miller (1966, fig. 2) identified four Provinces in Central America based upon fishes: (1) the Usumacinta, (2) the San Juan, (3) the Chiapas–Nicaraguan, and (4) the Isthmian. Miller defined Usumacinta Province as the area from the Rio Papaloapán of southeastern México eastward and southward to the Río Polochic of eastern Guatemala and that it was the best-marked fish province in Middle America. He reported that there were 108 species, 29 genera, and 9 families of primary and secondary fishes in Usumacinta Province. Miller (1966:780) noted that the Usumacinta–Grijalva had 38 endemics out of a total of 67 species of primary and secondary freshwater fishes (57%) and that it contained more than half (53%) of the cichlid fauna of Central America and nearly 60% of the species assignable to the genus *Cichlasoma*. Miller (1966, following Regan 1906–08) created the San Juan Province to include only the San Juan watershed (including Lake Managua [Lago de Xolotlán] and Lake Nicaragua [Lago de Nicaragua]) that is shared by Nicaragua and Costa Rica. The clupeid genus *Dorosoma* reaches its southern limit in the Nicaraguan lakes, where it

MAP 8 The freshwater mussel provinces of Central America. The key is given in Table 5.

TABLE 5
The Freshwater Mussel Regions and Provinces of Central America

North American Region	
• Río Grande del Norte (Black)	15 species (12 native, 5 endemics, 3 introduced)
• Colorado (Blue)	3 species (1 native, 2 introduced)
Gulf of Mexico Region (North to South)	
• Pánuco (Dark Red)	15 species (14 native, 6 endemics, 1 introduced)
• Papaloapán (Purple)	15 species (14 native, 6 endemics, 1 introduced)
• Usumacinta (Yellow)	13 species (7 endemics)
Northern Pacific + Endorheic Region	
• Mesa Central and Balsas (Green)	6 species (4 endemics)
Southern Pacific Region	
• Oaxaca, México to Colombia (Dark Blue)	11 species (10 native, 3 endemics, 1 introduced)
Caribbean Region	
• Gulf of Honduras and Northern Caribbean (Bright Red)	15 species (14 native, 5 endemics, 1 introduced)
• San Juan (Orange)	13 species (6 endemics)
Cuban Region	
• Cuba (Maroon)	2 species (2 endemics)

is represented by the endemic species *D. chavesi* Meek, 1907. Miller stated that there was insufficient collectionng of the Caribbean fish fauna of Honduras and Nicaragua from the Río Motagua, Guatemala south to the Río San Juan, Nicaragua to delimit a province at that time.

Miller (1966:781–783) identified two provinces with streams that drain primarily to the Pacific Coast: (1) the Chiapas–Nicaraguan Province extended along the Pacific slope from the Rio Tehuantepec in Oaxaca, México to the Nicoya Peninsula in Guanacaste Province, western Costa Rica, with 30 species of primary and secondary fishes, and (2) the Isthmian Province that ran from

Lake Nicaragua south through Panamá to Colombia with a rich and diversified fish fauna of 136 species, 55 genera, and 12 families of primary and secondary fishes. We have combined the two provinces of Miller as the Oaxaca, México to Panamá Province (see below).

Bussing (1976:157–158, fig. 1) used approximately the same provincial divisions as Miller (1966), but extended Usumacinta Province west by including the Río Motagua basin in Guatemala and then south along the Atlantic coast to the Río Coco, Nicaragua. He reported that 130 species in 34 genera and 10 families of freshwater fishes occur in the province. The northern Atlantic boundary Bussing's Isthmian Province lied between Península Valiente (forming part of Laguna de Chiriqui) and Rio Coclé del Norte, Panamá. The Pacific boundary ran from Punta Mala, Costa Rica, southeast to the Panamá border with Colombia. Bussing (1976) expanded Miller's San Juan Province to include the Atlantic slope watersheds from the Río Prinzapolka, Nicaragua to the Golfo de los Mosquitos in western Panamá. Bussing (1976:173) reported that the San Juan fish assemblage consisted of eight families, 25 genera, and 54 species of freshwater fishes and noted that it was less diverse than the Usumacinta and Isthmian assemblages. Only the Chiapas–Nicaraguan assemblage had fewer genera and species of fishes.

In their discussion of neotropical fishes, Hulsey and López-Fernandez (2011) identified an area from approximately 7° to 19° N latitude they termed *Nuclear Central America* (NCA) that spanned numerous political boundaries and including present-day northern Costa Rica, Nicaragua, El Salvador, Honduras, Belize, Guatemala, and southeastern México. They also provided an excellent, detailed geological history of the region and how fishes evolved and diversified in the basins created by the tectonic plates and the long history of mountain building along the fault zones. A list and map showing some of the major geological features of Central America taken from Hulsey and López-Fernandez (2011) is given in Map 7, Table 4.

The NCA of Hulsey and López-Fernandez (2011) contained numerous creeks, rivers, and lakes, and it was located between the Trans-Mexican Volcanic Belt (TMVB) and the Nicaraguan Depression. They noted the high endemism in the area and that the freshwater fish fauna of NCA differed substantially from the South American fauna in that many species do not co-occur with congeners. They sub-divided NCA into four provinces, based largely on the scheme proposed by Bussing (1976). Three of these extend along the Atlantic coast: (1) the Usumacinta, (2) the Honduran, and (3) the San Juan. They grouped all of the Pacific drainages (Río Tehuantepec, Río de los Perros, Río Lempa, Río Goascorán, and Río Grande de Choluteca) into Chiapas–Nicaraguan Province (4).

Several groups of fishes present south of the TMVB have phylogenetic affinities with North American taxa and no relatives in South America or the Antilles. The catfish *Ictalurus meridionalis* (Günther, 1864) reaches as far south as the Belize River (Miller 1966:786). In the family Catostomidae, two species of *Ictiobus*, *Ictiobus bubalus* (Rafinesque, 1818) and *Ictiobus meridionalis* (Günther, 1868), reach to the Papaloapán and Usumacinta basins, respectively (Miller 1966:786). Gars in Central America also have close affiliations with extant groups in North America (Wiley 1976). A monophyletic group of cypriniform fishes closely allied to the genus *Notropis* also has undergone restricted diversification south of the TMVB, representing the southernmost extension of the family Leuciscidae in the Western Hemisphere (Schönhuth and Doadrio 2003:333).

Lyons et al. (2020:16. fig. 3.4.3) noted that in México the highest regions of endemism for fishes occurred in the Lerma–Santiago basin (Pacific drainage) in Mesa Central Province and the Pánuco and Grijalva–Usumacinta River drainages on Gulf. They also found that the Lerma–Santiago basin had the highest concentrations of threatened freshwater fishes, including species that occur within lakes Pátzcuaro, Cuitzeo, and Chapala. Other basins that had high concentrations were the Yucatán, endorheic watersheds in northern México (particularly Cuatro Ciénegas in Coahuila), and the headwaters of the Pánuco in San Luis Potosí. They also identified three major centers of species richness: (1) Southeastern México, particularly the Usumacinta, Coatzacoalcos, and Papaloapán drainages,

(2) The Pacific Slope of Central México, primarily the Lerma–Santiago basin, and (3) The Río Grande del Norte–Conchos drainage.

Beltran-López et al. (2023:123) identified the same three regions of highest species richness for fishes in México with slightly different numbers: Usumacinta (249 species), Mesa Central (103 species), and Río Grande del Norte (119 species). Both Lyons et al. (2020) and Beltran-López et al. (2023) included the Coatzacoalcos, Papaloapán in the Grijalva–Usumacinta. We place the Coatzacoalcos in Papaloapán Province and the Grijalva in Usumacinta Province.

The mussel provinces we have erected here approximate those created for fishes by Minckley et al. (2005:24–47, Map 3.3) and mussels by Graf and Cummings (2021:11, Table 5), both of which are similar to the Freshwater Ecoregions of the World (FEOW) proposed by Abell et al. (2008:408–409)—also based largely on fishes. Minckley et al. (2005) identified eight ichthyological provinces in México: Río Bravo [Río Grande del Norte], Northwest Pacific Slope, Baja California Peninsula, Tamesí–Pánuco, Mesa Central, Balsas, Chiapas–Nicaraguan, and Usumacinta. Minckley et al. (2005:28) also identified regions that represented zones of transition between provinces. The provinces south of México are similar to those of Miller (1966), Bussing (1976), and Hulsey and López-Fernandez (2011). A short description of each province, the number of species contained in each, those that are endemic, and those introduced follow below.

North American Region

The freshwater mussel fauna of the Río Grande del Norte (also called the Río Grande or Río Bravo del Norte) and Colorado River basin group with the North America fauna belong to the Gulf– Florida and Pacific Regions (NA4 and NA5, respectively) of Graf and Cummings (2021:4). These drainage basins were also included in the freshwater biodiversity conservation assessment of Latin America and the Caribbean by Olson et al. (1998). Although not strictly part of Central America, we include them here to provide a complete list of freshwater mussels found in México.

RÍO GRANDE DEL NORTE PROVINCE

MAP 9 Collection sites in Río Grande del Norte Province.

The Río Grande del Norte is the largest basin (in terms of drainage area) in México. The Río Grande forms the natural border between the U.S. state of Texas and the Mexican states of Chihuahua, Coahuila, Nuevo León, and Tamaulipas (Map 9). This province corresponds to the Upper Río Grande–Bravo (132), Pecos (133), Río Conchos (134), Lower Río Grande–Bravo (135), Río Salado (137), and Río San Juan (138) Ecoregions of Abell et al. (2008). The Río Grande del Norte originates in the San Juan Mountains of southern Colorado and follows a 2,830 km course before it flows into the Gulf of Mexico near Brownsville, Texas. It is the 5th longest river system in North America and the 24th longest river in the world (Dahm et al. 2005:186). The Río Grande's watershed covers about 870,000 km^2; however, a large proportion of the river's sub-basins are endorheic and only about half of the total area, or about 450,000 km^2, contributes to the river's overall flow (Dahm et al. 2005:186).

The Río Grande is one of the most impaired rivers in the world (Dahm et al. 2005:192). Both water quantity and water quality are major issues. The central portion of the Río Grande from Elephant Butte and Caballo dams to the confluence of the Río Conchos operates largely as a ditch

DOI: 10.1201/9781003314448-15

for water delivery for agriculture and a rapidly growing population. The Río Grande itself is at times dry in the portion between El Paso, Texas and the mouth of the Río Conchos, Chihuahua (Chernoff et al. 1982:29–30). Amistad and Falcon dams, in the lower border reach, further regulate flow and divert water for irrigation. The Río Grande failed to reach the Gulf of Mexico in much of 2002 and 2003 (Dahm et al. 2005:189). Water-quality problems include elevated salinity, nutrients, bacteria, metals, pesticides, herbicides, organic solvents, and a decline in riparian vegetation, with non-native species dominating in many reaches (Dahm et al. 2005:192).

By volume, the Río Conchos is the largest tributary of the Río Grande. The Río Conchos originates in the Sierra Madre Occidental and then flows eastward through the Chihuahuan Desert (Gutíerrez and Borrego 1999; Johnson 1999a, Figures 8 and 9). Most of the basin lies within Chihuahua, México, but the Conchos also drains a portion of northern Durango. The Río Conchos is nearly 750 km in length and drains an area of about 67,000 km^2 (Barrios et al. 2009; Montero-Martínez and Ibáñez-Hernández 2017) (or 77,090 km^2 according to Miller et al. 2005:17) that accounts for 14% of the Río Grande del Norte drainage. However, because the Río Grande now has very little flow upstream of the confluence, the Conchos now contributes a large proportion to total discharge below the confluence. The Río Conchos joins the Río Grande at the town of Ojinaga, México, opposite Presidio, Texas, and the basin comprises 14% of the binational (México–United States) Río Grande drainage (Barrios et al. 2009). In fact, the Río Conchos supplies >75% of the flow through the Big Bend Region of Texas (Gutíerrez and Borrego 1999).

Like many rivers throughout North America, the fauna of the Río Grande has been irrevocably altered. Historically, about 121 primary freshwater fish species inhabited the basin (Smith and Miller 1986:477, Table 13.1). A survey in the Río Grande proper found that three species are now extinct, others are threatened or endangered, and most of the remaining native fishes are declining in both range and numbers. The most recent data suggests that 17 of the 40 primary native freshwater fishes have been either extirpated in part or throughout the Río Grande mainstem (Calamusso et al. 2005:205).

The mussel species in Río Grande del Norte Province group with the North American fauna as pointed out by Graf and Cummings (2021:4) and no species in the Río Grande are naturally found

FIGURE 8 Río Conchos (Rio Grande del Norte Dr.), Peguis Canyon, Chihuahua, México. 29.5042, -104.7624. 7 May 2019.

FIGURE 9 Río Nonoava (Río Conchos Dr.), Nonoava, Chihuahua, México. 27.47294, -106.7397. 8 May 2019.

south of the basin. This is somewhat similar for fishes, but unlike mussels, the Río Grande shares many fish species with the San Fernando and Soto la Marina drainages to the southeast and a few with the Mesa Central (Smith and Miller 1986:458, Tables 13.1 and 13.2; Minckley et al. 2005). About 57% of the Río Grande fish fauna is thought to be endemic (Smith and Miller 1986:477).

The Río Grande del Norte has been well sampled for mussels in the United States. We have compiled over 2,000 records for the basin from museums and literature sources (e.g., Neck and Metcalf 1988; Burlakova et al. 2011; Karatayev et al. 2012). The Río Grande basin on the México side has not been well sampled with most of the records in the drainage coming from the Conchos (43 records) and Salado de Nadadores (77 records) basins. We have compiled only 172 records from the Méxican side. The Río Conchos was sampled at 11 sites in May 2018 (Tiemann et al. 2020:182), and of the three species known from the Conchos basin, only one live individual of *Utterbackia imbecillis* (Say, 1829) was encountered.

Fifteen species of mussels are known from Río Grande del Norte Province, 12 of which are native and three are thought to have been introduced (Tables 6 and 16). Five of the 12 native species known from the basin are endemic and include *Disconaias conchos* Taylor, 1997, *Potamilus metnecktayi* Johnson, 1998, *Truncilla cognata* (Lea, 1860), *Quadrula couchiana* (Lea, 1860), and *Popenaias popeii* (Lea, 1857). Of the remaining seven species, six [*Lampsilis teres* (Rafinesque, 1820), *Megalonaias nervosa* (Rafinesque, 1820), *Toxolasma parvum* (Barnes, 1823), *Toxolasma texasiense* (Lea, 1857), *Uniomerus tetralasmus* (Say, 1831), and *Utterbackia imbecillis* (Say, 1829)] are wide-ranging in North America and reach their natural southern limits in the Río Grande basin. *Cyrtonaias berlandierii* (Lea, 1857) is widely distributed from the Brazos River, Texas south to the Río Grande del Norte in Texas and in Tamaulipas, Nuevo León, and Coahuila, México (Johnson 1999a:27; Howells 2003:60). Populations in the Trinity River of Texas may be introduced (Howells et al. 1996:48). The three species that have been theorized to have been introduced in the basin are *Pyganodon grandis* (Say, 1829), *Potamilus purpuratus* (Lamarck, 1819), and *Quadrula quadrula* (Rafinesque, 1820) [as *Quadrula apiculata* (Say, 1829)] (Metcalf and Smartt 1972:144; Howells

TABLE 6

Freshwater Mussels of Río Grande del Norte Province.
Total: 15 species (12 native, 5 endemic [E], 1 extinct [EX], 3 introduced [I])

Order Unionoida
- **Superfamily Unionoidea**
  - **Family Unionidae**
    - **Subfamily Unioninae**
      - **Tribe Anodontini**
        - *Pyganodon grandis* (Say, 1829) [I]
        - *Utterbackia imbecillis* (Say, 1829)
    - **Subfamily Ambleminae**
      - **Tribe Lampsilini**
        - *Cyrtonaias berlandierii* (Lea, 1857)
        - *Disconaias conchos* Taylor, 1997 [E, EX]
        - *Lampsilis teres* (Rafinesque, 1820)
        - *Potamilus metnecktayi* Johnson, 1998 [E]
        - *Potamilus purpuratus* (Lamarck, 1819) [I]
        - *Toxolasma parvum* (Barnes, 1823)
        - *Toxolasma texasiense* (Lea, 1857)
        - *Truncilla cognata* (Lea, 1860) [E]
      - **Tribe Popenaiadini**
        - *Popenaias popeii* (Lea, 1857) [E]
      - **Tribe Quadrulini**
        - *Megalonaias nervosa* (Rafinesque, 1820)
        - *Quadrula couchiana* (Lea, 1860) [E, EX]
        - *Quadrula quadrula* (Rafinesque, 1820) [I]
        - *Uniomerus tetralasmus* (Say, 1831)

et al. 1996:102; Johnson 1999a:8,13,17; Karatayev et al. 2012:209). *Quadrula quadrula* (Rafinesque, 1820) appears to be expanding its range in the drainage (Karatayev et al. 2012:213).

Karatayev et al. (2012:208) surveyed for mussels at 162 sites in the Río Grande drainage in Texas from 2001 to 2011 and noted a considerable reduction in the fauna. They collected ten species alive. The most common mussels in the basin calculated as the percentage of sites where the species was found were *Cyrtonaias berlandierii* and *Quadrula quadrula*, which were found at 28.6% of the sites sampled (Karatayev et al. 2012:210). The rarest species were *P. metnecktayi*, which was found alive at only one site in the mainstem, and *Toxolasma texasiense*, which was found at only one tributary site. *Potamilus metnecktayi* had the lowest percentage of sites where live mussels were found, compared to the total number of sites where shells were present (13%).

The mussel fauna of the Río Grande del Norte has experienced a decline in species richness, range fragmentation, local extirpations, and introduction of widespread common species. The Texas Hornshell, *Popenaias popeii*, was added to the U.S. Federal Endangered Species list in 2018 (Federal Register 2018). *Quadrula couchiana* was considered extirpated from the Texas part of the Río Grande basin (Karatayev et al. 2012:212). The last live record we encountered was collected in 1981 from the Rio Salado de Nadadores in Progresso, Coahuila, México (USNM uncatalogued). Although the 2018 survey of the Conchos was limited to 11 sites, no live or fresh-dead shells of *Quadrula couchiana* were encountered, and given the water quality issues in the basin, it is likely extirpated from the Conchos drainage (if it ever occurred there) and likely extinct (Williams et al. 2017:47; Tiemann et al. 2020:182). Several tributary streams in the drainage in Texas and New Mexico have lost all or a significant number of species including Las Moras Creek, Devils River

(Figure 10), and the Pecos River. Regarding *Fusconaia mitchelli* (Simpson in Dall, 1895) and its supposed presence in the Río Grande, see comments under species erroneously attributed to the region below.

Czaja et al. (2023:90) identified five "hot spots" of species richness for freshwater mollusks in México. They defined a hot spot as a "concentration of native species, especially endemics, in a specifically defined region." In this province, they identified two: the Conchos–Guzman basin and the Cuatro Ciénegas–Río Sabinas drainage. However, their designation was based on gastropods only.

FIGURE 10 Devils River (Río Grande Dr.), Texas, USA. Lower Canyon reach. Habitat for *Popenaias popeii* (Lea, 1857) and the type locality of *Unio bairdianus* Lea, 1857 [=*Toxolasma texasiense* (Lea, 1857)]. (Photo by John Karges, Nature Conservancy).

COLORADO PROVINCE

MAP 10 Collection sites in Colorado Province, México and USA.

Colorado Province includes the Colorado River and its tributaries and a few adjacent endorheic basins. This province corresponds to the Colorado (130), Gila (131), Southern California Coastal–Baja California (159), Sonora (160), and Guzmán–Samalayuca (161) Ecoregions of Abell et al. (2008), approximates the Northwest Pacific Slope and Baja California Provinces of Minckley et al. (2005:31–32), and largely corresponds to the Colorado River Complex of Olson et al. (1998:24). However, ours includes the endorheic Guzmán system that Olson et al. placed in the Río Bravo [Río Grande del Norte] Complex. We placed the Guzmán in the Colorado because there are records of *Anodonta californiensis* Lea, 1852 there—a species not found in the Río Grande. Colorado Province is part of the Pacific Region of Graf and Cummings (2021:4, NA5) that runs from the Aleutian Islands and the Yukon River south to the Colorado River, México.

The Colorado River Basin lies within the Intermontane Plateaus of the American Southwest. The basin encompasses a large geographic area ranging from the Rocky Mountains in Wyoming and Colorado to the Colorado River delta in México (Map 10). The Colorado basin drains seven states including parts of Colorado, Wyoming, Utah, New Mexico, Nevada, and California, and 95% of Arizona's land mass. The entire basin drains an area of about 640,000 km^2. A series of creeks, rivers, and arroyos drain the region, with water mostly flowing in a northwesterly or south-southwesterly direction. The major rivers originate in the Wind River Range of the Wyoming Rockies, the Colorado Rockies, and the Méxican Highlands of west-central New Mexico and east-central Arizona. Major tributaries include (North to South) the Green, Yampa, Gunnison, San Juan, Little Colorado Virgin, and Gila.

The Colorado River is one of the most physically developed and controlled rivers in North America, even though it passes through some of the most arid and remote regions of the continent. Historically, the Colorado River drained to the Gulf of California, forming a large estuarine delta at the mouth of the river in México. Today, with the construction of numerous diversions and reservoirs, the channel is nearly dry by the time it reaches the Gulf (Blinn and Poff 2005).

The Guzmán Basin is an endorheic basin located in northern México and the southwestern United States. It occupies the northwestern portion of Chihuahua in México and extends into southwestern New Mexico in the United States. Notable rivers of the Guzmán Basin are the Casas Grandes River, which empties into Lake Guzmán, the Santa Maria River, which empties into Lake Santa Maria, the Carmen River, and the Mimbres River of New Mexico.

The Salton Sea is a shallow, endorheic, highly saline lake in Riverside and Imperial counties in southern California. It lies on the San Andreas Fault within the Salton Trough, which stretches to the Gulf of California in México. *Anodonta californiensis* Lea, 1852 has been found in both the Guzmán and Salton Sea basins.

Only three freshwater mussel species are known from the province, two of which are likely introduced (Table 7). The larger basins that contain mussels are the Colorado (two species) and the Yaqui (Figure 11) (one species). The fish fauna of the Yaqui basin is relatively depauperate with only 13 native primary freshwater fishes (Hendrickson et al. 1980:84). Baja California has only six native fish species (Minckley et al. 2005:32) and only one mussel species, *Anodonta californiensis*, which reaches its southern limit in Colorado Province.

TABLE 7

Freshwater Mussels of Colorado Province.
Total: 3 Species (1 Native, 2 Introduced [1])

Order Unionoida

Superfamily Unionoidea

Family Unionidae

Subfamily Unioninae

Tribe Anodontini

Anodonta californiensis Lea, 1852

Pyganodon grandis (Say, 1829)[1]

Utterbackia imbecillis (Say, 1829)[1]

FIGURE 11 Río Yaqui (Pacific–Gulf of California Dr.), Sonora, México. 27.9610, -109.8190. Habitat for *Anodonta californiensis* Lea, 1852. (Photo by Tomas Castelazo, Wikipedia).

Gulf of Mexico Region

Rivers draining to the Gulf of Mexico harbor some of the highest species richness of freshwater mussels in Central America. Three Provinces have large river basins that drain into the Gulf of Mexico: the Pánuco, Papaloapán, and Usumacinta (North to South). The mussel species richness in all the major Gulf of Mexico basins is similar, and each has about the same number of endemic species. The Río Pánuco has 14 native and six endemics, the Papaloapán has 14 native and six endemics, and the Usumacinta has 14 species, seven of which are endemic.

The species richness of the tribe Lampsilini in the major Gulf of Mexico drainages decreases as you go from the North to the South. The Pánuco has seven species, five of which are endemic: three in *Actinonaias*, two in *Disconaias*, and the monotypic *Friersonia iridella* (Pilsbry & Frierson, 1908). *Actinonaias medellina* (Lea, 1838) and *Mesonaias tampicoensis* (Lea, 1838) are shared between the Pánuco and Papaloapán. The Papaloapán has five lampsilines: *Actinonaias medellina* (Lea, 1838) and *Actinonaias sapotalensis* (Lea, 1841) (endemic), *Disconaias fimbriata* (Frierson, 1907), *Mesonaias tampicoensis* (Lea, 1838), and *Mesonaias umbrosa* (Lea, 1856). All four members of the tribe Lampsilini found in the Usumacinta basin, *Delphinonaias largillierti* (Philippi, 1847), *Delphinonaias scutulata* (Morelet, 1849), *Mesonaias explicata* (Morelet, 1849), and *Pachynaias spheniopsis* (Morelet, 1849), are endemic.

Gulf of Mexico members of the Unionidae, the subfamily Ambleminae, tribe Popenaiadini include *Popenaias semirasa* (Pilsbry, 1910) and *Psoronaias semigranosa* (von dem Busch in Philippi, 1843), which are endemic to the Pánuco and *Popenaias berezai* Inoue et al., 2020—found primarily in the Pánuco. However, there are two records from Papaloapán Province: one from the Río Jamapa collected in 1927, and three specimens from Arroyo de Hueyapan, a tributary to the Río San Juan collected in 1949. *Nephronaias aztecorum* (Philippi, 1847) is found in both Pánuco and Papaloapán Provinces. *Psoronaias distincta* (Crosse & Fischer, 1893), *Sphenonaias liebmanni* (Philippi, 1847), *Sphenonaias opacata* (Crosse & Fischer, 1894), and *Sphenonaias plexus* (Conrad, 1838) are all endemic to Papaloapán Province. Endemic Popenaiadini in Usumacinta include *Nephronaias aeruginosa* (Morelet, 1849), *Nephronaias sallei* (Crosse & Fischer, 1894), *Psoronaias crocodilorum* (Morelet, 1849), and *Psoronaias kuxensis* Frierson, 1917. *Nephronaias calamitarum* (Morelet, 1849) is found in rivers draining to the Gulf of Mexico (Usumacinta) and Caribbean–Gulf of Honduras (Río Dulce) in the Gulf of Honduras and Northern Caribbean Province.

DOI: 10.1201/9781003314448-16

PÁNUCO PROVINCE

MAP 11 Collection sites in Pánuco Province.

Our Pánuco Province is basically the same as the Pánuco Ecoregion (167) of Abell et al. (2008), with the exception of the Río San Fernando. Because of the similarity of the mussel fauna, we added the Gulf Coastal drainages south of the Río Grande del Norte (ríos Carrizal and San Fernando) and those basins from the Pánuco south to, but not including the Río Jamapa (Pantepec–Tuxpan, Cazones, Tecolutla, Nautla, Misantla, Colipa, Actopan, and La Antigua) (Map 11, Figures 12–17). Czaja et al. (2023:92) identified this area as a hot spot (Río Verde–Río Gauyalejo System) based only upon gastropods with no mention of the mussel species richness found there.

The Río Pánuco is the third largest (98,227 km^2) watershed in México draining to the Gulf of México. It and its lowermost tributary, the Tamesí, drain the Tamaulipas–Veracruz ecoregion and join near the port city of Tampico. The Pánuco basin (not including the Río Tamesí) drains 79,100 km^2. The headwaters of the Pánuco begin in the arid Mexican Altiplano and the mesic Sierra Madre Oriental and flow across the Coastal Plain, draining parts of the states of Tamaulipas, San Luis Potosí, Hidalgo, Querétaro, México, Guanajuato, and Veracruz. The Río Pánuco proper is formed by the confluence of the ríos Tamuin (33,260 km^2) and Moctezuma (42,726 km^2) after they cross the Sierra Madre Oriental (Hudson et al. 2005:1037).

The Tampaón and Moctezuma are the two largest tributaries of the Pánuco. The Río Coy forms from one of the largest springs in the world and joins the Río Tampaón upstream of Ciudad Tamuin to become the Río Tamuin (Fish 1977; Hudson et al. 2005:1040). The Río Tamuin meanders a short distance before joining the Río Moctezuma to form the Río Pánuco in the western Gulf Coastal Plain, 185 km upstream of the Gulf of Mexico. The Río Tamuin is often not labeled on online maps and simply considered to be the Río Tampaón. The Cascada de Tamul (Figure 12) is a 102 m waterfall at the confluence of the ríos Gallinas and Santa María, forming the Río Tampaón. Discharge and runoff are uniformly low from December through early May, but increase rapidly with the onset of summer trade winds in late May (Hudson et al. 2005:1041). Although information on water chemistry of the Pánuco basin is lacking, the river and its tributaries are threatened by pollution from

FIGURE 12 Río Gallinas flowing into the Río Tampaón (Pánuco Dr.), Cascada de Tamul, and San Luis Potosí. 21.8011, -99.1782. December 2017. Río Gallinas a few km upstream of the falls is habitat for *Actinonaias coyensis* (Pilsbry, 1910), *Friersonia iridella* (Pilsbry & Frierson, 1908), and is the type locality of *Popenaias semirasa* (Pilsbry, 1910). (Photo by Charly Morones).

agrochemicals, municipal discharge, and salinization (Abell et al. 2000). Although México City is not within the Pánuco basin, a network of canals deliver raw sewage from the metropolitan area to the Río Tula, a tributary of the Río Moctezuma (Hudson et al. 2005:1043).

The Pánuco basin has been well-studied for freshwater mollusks, beginning with the surveys by Anson A. Hinkley in 1907 and 1909 (and reported on by Pilsbry 1910) and more recently in 2017 and 2018 (Inoue et al. 2020a; Kiser et al. 2022). Kiser et al. (2022) used species distribution models to predict the distribution of 11 freshwater mussel species in the Pánuco Basin, determine habitat factors shaping freshwater mussel occurrence, and predict occupancy to identify mussel biodiversity hotspots. A total of 621 species-specific observations at 87 sites were used to create species-specific ensembles. These predictive species ensembles were then combined to create local diversity hotspot maps to guide current conservation efforts and prioritize areas for future conservation and study.

We have compiled records from over 800 lots collected in the province. Fifteen species of mussels are known from Pánuco Province, 14 of which are thought to be native, and one introduced (Table 8). Six of the 14 native species are endemic and include *Actinonaias coyensis* (Pilsbry, 1910), *Actinonaias signata* (Pilsbry, 1910), *Disconaias disca* (Lea, 1838), *Friersonia iridella* (Pilsbry & Frierson, 1908), *Popenaias semirasa* (Pilsbry, 1910), and *Psoronaias semigranosa* (von dem Busch in Philippi, 1843). Six of the remaining seven species known from the Pánuco are wide-ranging in Central America: *Anodontites cylindracea* (Lea, 1838), *Actinonaias medellina* (Lea, 1838), *Brachyanodon impura* (Say, 1829), *Megalonaias nickliniana* (Lea, 1834), *Mesonaias tampicoensis* (Lea, 1838), and *Nephronaias aztecorum* (Philippi, 1847).

FIGURE 13 Río Gallinas (Pánuco–Tampaón Dr.), Hacienda Maria Bonita, 25.7 km WSW Valles, San Luis Potosí. 21.8961, -99.2453. 4 December 2017. Habitat for *Actinonaias coyensis* (Pilsbry, 1910), *Friersonia iridella* (Pilsbry & Frierson, 1908), *Popenaias semirasa* (Pilsbry, 1910), and *Nephronaias aztecorum* (Philippi, 1847). (Photo by Charly Morones).

FIGURE 14 Río Valles (Pánuco–Tampaón Dr.), Micos, NW of Valles, San Luis Potosí. 22.1035, -99.1565. December 2017. Habitat for numerous mussel species including *Actinonaias coyensis* (Pilsbry, 1910), *Actinonaias medellina* (Lea, 1838), *Actinonaias signata* (Pilsbry, 1910), *Disconaias fimbriata* (Frierson, 1907), *Friersonia iridella* (Pilsbry & Frierson, 1908), and *Popenaias berezai* (Inoue et al., 2020).

Disconaias fimbriata (Frierson, 1907) is known primarily from the Pánuco, but there is a population in the Actopan basin near José Cardel (MCZ 288416, MCZ 288430, and MCZ 288432) near Punta del Morro: a lava flow from the Trans-Mexican Volcanic Belt that extends almost to the coastal line in the Méxican state of Veracruz, found at the northern border of the Neotropical zone where

FIGURE 15 Río Moctezuma (Pánuco Dr.), Chiquinteco, San Luis Potosí. 21.5020, -98.7903. 7 December 2017. Habitat for *Actinonaias coyensis* (Pilsbry, 1910), *Actinonaias signata* (Pilsbry, 1910), *Disconaias disca* (Lea, 1838), and *Friersonia iridella* (Pilsbry & Frierson, 1908). (Photo by Charly Morones).

FIGURE 16 Río Huichihuayán (Pánuco–Moctezuma–Axtla Dr.), Río Huichihuayán, 8 km NE Xilitla, San Luis Potosí, México [21.4363, -98.9334]. 10 December 2018. Habitat for numerous mussel species including *Actinonaias medellina* (Lea, 1838), *Disconaias fimbriata* (Frierson, 1907), *Friersonia iridella* (Pilsbry & Frierson, 1908), and *Popenaias berezai* Inoue et al., 2020. (Photo by Aaron Koch).

FIGURE 17 Río Tampaón (Pánuco Dr.), Puente de Dios, 22.1 km SSW of Ciudad Valles, San Luis Potosí. 21.8069, -99.0745. 14 December 2018. Habitat for *Disconaias disca* (Lea, 1838), *Disconaias fimbriata* (Frierson, 1907), and *Friersonia iridella* (Pilsbry & Frierson, 1908). (Photo by Charly Morones).

TABLE 8
Freshwater Mussels of Pánuco Province.
Total 15 Species (14 Native, 6 Endemic [E], 1 Introduced [I])

Order Unionoida
Superfamily Unionoidea
Family Unionidae
Subfamily Unioninae
Tribe Anodontini
Brachyanodon impura (Say, 1829) ?
Utterbackia imbecillis (Say, 1829) [I]
Subfamily Ambleminae
Tribe Lampsilini
Actinonaias coyensis (Pilsbry, 1910) [E]
Actinonaias medellina (Lea, 1838)
Actinonaias signata (Pilsbry, 1910) [E]
Disconaias disca (Lea, 1838) [E]
Disconaias fimbriata (Frierson, 1907)
Friersonia iridella (Pilsbry & Frierson, 1908) [E]
Mesonaias tampicoensis (Lea, 1838)
Tribe Popenaiadini
Nephronaias aztecorum (Philippi, 1847)
Popenaias berezai Inoue et al., 2020
Popenaias semirasa (Pilsbry, 1910) [E]
Psoronaias semigranosa (von dem Busch in Philippi, 1843) [E]
Tribe Quadrulini
Megalonaias nickliniana (Lea, 1834)

Superfamily Etherioidea
Family Mycetopodidae
Subfamily Anodontitinae
Anodontites cylindracea (Lea, 1838)

it terminates along a transition zone in México (Pérez-Miranda et al. 2020:3). Punta del Morro has been demonstrated to act as an effective barrier for freshwater species (Contreras-Balderas et al. 1996). The fish fauna drastically changes from being 75% of Nearctic origin north of Punta del Morro to 95% Neotropical origin south (Obregón-Barboza et al. 1994; Miller et al. 2005). Punta del Morro was not found to be a barrier to *Dorosoma petenense* (Günther, 1867) (family Clupeidae) (Elías et al. 2022:102).

The U.S. Federally Endangered *Popenaias popeii* (Lea, 1857) was thought to have a large population in the Pánuco, but genetic analysis confirmed that the Mexican populations south of the Río Grande del Norte represent a distinct phylogenetic lineage and were recently described as a distinct species (Inoue et al. 2020a). All records of *Popenaias berezai* Inoue et al., 2020 are from Pánuco Province, with the exception being one specimen from the Río Jamapa in Veracruz (ANSP 142600).

There are two records of *Brachyanodon impura* (Say, 1829) from the Pánuco—one from the Río Tula north of México City collected in 1927 (USNM 134837) and a lot collected at Valles by Seth E. Meek in 1903 (FMNH 11631). Meek (1902; 1903) does not mention collecting at Valles. Whether or not these are valid records is unknown. Minckley et al. (2005:33) identified 10 species of fishes that originated on Mesa Central and are now found in the Pánuco, originating from headwater stream capture, so it is possible that they are native. *Utterbackia imbecillis* (Say, 1829) has been introduced in the province and was first collected in 1982 (USNM 1282737, $n=6$).

PAPALOAPÁN PROVINCE

MAP 12 Collection sites in Papaloapán Province.

Our Papaloapán Province is comparable to the Papaloapán (171) and Coatzacoalcos (172) Ecoregions of Abell et al. (2008). The Río Papaloapán is the largest in the province with an area of around 37,380 km^2 and drains the central state of Veracruz (Map 12). The Río Coatzacoalcos, to the southeast, has a drainage area of 21,120 km^2 and drains the northern slope of the Isthmus of Tehuantepec. The Papaloapán and Coatzacoalcos basins were considered the northernmost Division of Usumacinta Province by Minckley et al. (2005:46), but it is considered here as a distinct province.

Primary rivers include the Actopan, Cotaxtla (also called the Atoyac in the headwaters), Jamapa, San Juan, Tesechoacan, and the Río Grande de Catemaco (including Laguna de Catemaco) (Figures 18–20). Eyipantla Falls effectively isolates Laguna de Catemaco from the rest of the Papaloapán basin (Figure 18). Laguna de Catemaco has been biogeographically isolated since it was formed in a caldera along the Gulf of Mexico coast less than two million years ago (Miller and Van Conner 1997; Hulsey and López-Fernandez 2011). Laguna de Catemaco has a highly endemic ichthyofauna including *Poeciliopsis catemaco* Miller, 1975, *Poecilia catemaconis* Miller, 1975, and *Xiphophorus milleri* Rose, 1960, and most of the known records for the freshwater mussel, *Sphenonaias opacata* (Crosse & Fischer, 1894), are from the lake.

We consider it a full province because of its high endemicity. One-third of the fish species are endemics. Streams in the province have been periodically sampled for mussels, first in 1910 (Baker 1922), then the 1960s (Fred Thompson in 1965 and Herbert Athearn in 1969, both unpublished), 1982 (Bereza et al. unpublished), and December 2022 (Cummings et al., herein). Fourteen native mussel species are known from the province, six of which are endemic: *Actinonaias sapotalensis* (Lea, 1841), *Mesonaias umbrosa* (Lea, 1856), *Psoronaias distincta* (Crosse & Fischer, 1893), and *Sphenonaias liebmanni* (Philippi, 1847), *Sphenonaias opacata* (Crosse & Fischer, 1894), and

FIGURE 18 Rio Grande de Catemaco (Papaloapán Dr.), Salto de Eyipantla, Veracruz, México. 18.3848, -95.2076. 5 December 2022.

Sphenonaias plexus (Conrad, 1838). *Utterbackia imbecillis* (Say, 1829) has been introduced in the province (Table 9). There is one questionable record of *Popenaias berezai* Inoue et al., 2020 known from the province; ANSP 142600 (*n* = 1) from the Río Jamapa collected in 1927. It is unknown if it is still extant in the basin. Czaja et al. (2023:93) identified this area as a hot spot (D. Río Papaloapán–Río Blanco System) based on gastropods and the large number of bivalves that occur there.

FIGURE 19 Laguna de Catemaco, (Papaloapán–Rio Grande de Catemaco Dr.), Veracruz, México. 18.4174, -95.1000. Collecting *Sphenonaias opacata* (Crosse & Fischer, 1894) from the lake on 5 December 2022.

FIGURE 20 Laguna de Catemaco (Papaloapán–Rio Grande de Catemaco Dr.), Veracruz, México. 18.4174, -95.1000. 5 December 2022.

TABLE 9
Freshwater Mussels of Papaloapán Province.
Total: 15 Species (14 Native, 6 Endemic [E], 1 Introduced [I])

Order Unionoida

Superfamily Unionoidea

Family Unionidae

Subfamily Unioninae

Tribe Anodontini

Pyganodon globosa (Lea, 1841)

Utterbackia imbecillis (Say, 1829) [I]

Subfamily Ambleminae

Tribe Lampsilini

Actinonaias medellina (Lea, 1838)

Actinonaias sapotalensis (Lea, 1841) [E]

Disconaias fimbriata (Frierson, 1907)

Mesonaias tampicoensis (Lea, 1838)

Mesonaias umbrosa (Lea, 1856) [E]

Tribe Popenaiadini

Nephronaias aztecorum (Philippi, 1847)

Popenaias berezai Inoue et al., 2020 ?

Psoronaias distincta (Crosse & Fischer, 1893) [E]

Sphenonaias liebmanni (Philippi, 1847) [E]

Sphenonaias opacata (Crosse & Fischer, 1894) [E]

Sphenonaias plexus (Conrad, 1838) [E]

Tribe Quadrulini

Megalonaias nickliniana (Lea, 1834)

Superfamily Etherioidea

Family Mycetopodidae

Subfamily Anodontitinae

Anodontites cylindracea (Lea, 1838)

USUMACINTA PROVINCE

MAP 13 Collection sites in Usumacinta Province, México and Guatemala.

Usumacinta Province as defined here includes the Usumacinta–Grijalva basin and nearby direct tributaries to the Gulf of México, plus the Yucatán peninsula. It roughly corresponds to the Grijalva–Usumacinta (173), Upper Usumacinta (174), and Yucatán (175) Ecoregions of Abell et al. (2008). Usumacinta Province as defined by Minckley et al. (2005:45) was much larger and included the Papaloapán–Coatzacoalcos, Usumacinta–Grijalva, and Yucatán as Divisions of Usumacinta Province. We treat the Usumacinta here as a separate province (Map 13).

The Río Usumacinta serves as the international border between México and Guatemala. The Usumacinta–Grijalva is the longest and largest river system in México, with a combined total of 1,521 km of main-stream channel in México draining 112,550 km^2 (Figures 21–26). It originates from the tropical rain forests Guatemala and the Méxican states of Chiapas and Tabasco emptying into the southern Gulf of Mexico near Villahermosa, Tabasco. The main rivers within this system are the Usumacinta and Grijalva, both with headwaters in Guatemala (35% of entire basin) but with 45% of the basin in Chiapas, México's southernmost state (Hudson et al. 2005:1046).

FIGURE 21 Río Usumacinta (Grijalva Dr.), Emiliano Zapata, Tabasco, México. 17.7154, -91.7129. 3 December 2019. Shifting sand substrate with no mussels found.

FIGURE 22 Laguna de Catazajá (Grijalva–Usumacinta Dr.), Catazajá, Chiapas, México. 17.7297, -92.0183. 2 December 2019. Habitat for *Delphinonaias largillierti* (Philippi, 1847).

FIGURE 23 Río Saguila (Grijalva–Usumacinta–Chacamax Dr.), SE Emiliano Zapata, off Rt. 10, Tabasco, México. 17.7115, -91.7408. 3 December 2019. Habitat for *Mesonaias explicata* (Morelet, 1849).

FIGURE 24 Río Chacamax (Grijalva–Usumacinta Dr.), La Libertad, Chiapas, México. 17.6917, -91.7173. 3 December 2019. Habitat for *Mesonaias explicata* (Morelet, 1849) and *Delphinonaias largillierti* (Philippi, 1847).

FIGURE 25 Río Michol (Grijalva–Usumacinta–Tulija Dr.), 11 km WSW Palenque, Chiapas, México. 17.4946, -92.0454. 1 December 2019. Habitat for *Nephronaias calamitarum* (Morelet, 1849) and *Anodontites cylindracea* (Lea, 1838).

The Río Grijalva begins as the Río San Miguel in Guatemala becomes the Grijalva in México and is joined by several large tributaries, including the Río Suchiapa, which drains the Sierra Madre de Chiapas of western Chiapas. As the river flows toward the Coastal Plain, the Grijalva also receives drainage from the eastern margins of the states of Oaxaca and Veracruz. The Grijalva and Usumacinta join, but only partially, near Frontera, Tabasco, about 15 km above their mouth in the Gulf of Mexico (Rodiles-Hernández 2004; Hudson et al. 2005:1044).

In contrast to the Pánuco system, which has its headwaters in the arid and semi-arid Central Mesa, the upper reaches of the Usumacinta–Grijalva system receive abundant rainfall. The Usumacinta drains the extensive Cretaceous limestone of the Sierra de San Cristobal. As is common in limestone areas with high precipitation, there is considerable subsurface dissolution that creates numerous karst landforms, particularly in the eastern portions of the basin (West et al. 1969; Hudson et al. 2005:1045). The much smaller Río Candelaria (also in the Usumacinta–Grijalva) to the northeast is one of the larger rivers draining the highly karstic Yucatán Peninsula (Hudson et al. 2005:1036).

The most pristine waters, containing the greatest biodiversity and the most intact ecosystems, are found in the mountainous regions and tropical rain forest of southern México. However, these rain forest rivers are facing extensive exploitation, with plans for multiple hydropower dams that would flood vast natural areas as well as important archaeological sites. The Usumacinta–Grijalva system is not only México's largest river system, but it also has more natural areas than any other drainage in Central America. It contains part of the largest remaining tropical rain forest north of the Amazon. In addition, it has immense archeological value as the center of the ancient Maya culture. Although the Usumacinta is presently not impounded, the construction of a series of dams on the main stem has been discussed since at least 1987 and may be forthcoming. The Río Grijalva has two large dams (Presa Nezahualcóyotl or Malpaso and Presa de la Angostura or Belisario Domínguez) constructed in the 1960s and 1970s, and so now serves as a major source for hydro-electric power. The biosphere reserve sites in the Usumacinta basin are in considerable peril, both from impoundment and forest exploitation (Hudson et al. 2005:1036).

FIGURE 26 Río Seco (Grijalva–Usumacinta Dr.), 1 km SW La Cascada, Cascadas Welib Ha, Chiapas, México. 17.3755, -91.7968. 6 December 2019. Habitat for *Psoronaias kuxensis* Frierson, 1917.

Usumacinta Province harbors the largest and most diverse fish fauna in México with over 200 described species (Minckley et al. 2005:45). The Usumacinta–Grijalva Division has about 115 species, 36% of which are endemics (Minckley et al. 2005:46). Most of the fish species are largely of Central or South American derivation. It is also one of the least known to science as illustrated by the fact that a newly described family of catfishes (Lacantuniidae) was discovered in the basin in 1996 (Rodiles-Hernández et al. 2005).

The Yucatán Division of Minckley et al. (2005:46) encompasses the rivers that drain the Yucatán Peninsula north of 18°30' N. This Division corresponds to the Yucatán (175) Ecoregion of Abell et al. (2008). It is a relatively young area geologically, and it was only at the end of the Pliocene that the peninsula was above water. There is evidence that most of the peninsula dried out during the Holocene (about 11,650 YPB) so there has been little time for mussels to colonize the area. For freshwater fishes, Minckley et al. (2005) pointed out that the young age and scarcity of surface water has resulted in a depauperate fish fauna and unsurprisingly that is also reflected in the freshwater mussel fauna. Although there are six lots of mussel specimens labeled as "Yucatán" in the collections we examined, there is only one record of any freshwater mussel from Campeche (COMA 1092)

TABLE 10
Freshwater Mussels of Usumacinta Province. Total: 13 Species (All Native, 7 Endemic [E])

Order Unionoida
- **Superfamily Unionoidea**
  - **Family Unionidae**
    - **Subfamily Unioninae**
      - **Tribe Anodontini**
        - *Pyganodon globosa* (Lea, 1841)
    - **Subfamily Ambleminae**
      - **Tribe Lampsilini**
        - *Delphinonaias largillierti* (Philippi, 1847)
        - *Delphinonaias scutulata* (Morelet, 1849) [E]
        - *Mesonaias explicata* (Morelet, 1849) [E]
        - *Pachynaias spheniopsis* (Morelet, 1849) [E]
      - **Tribe Popenaiadini**
        - *Nephronaias aeruginosa* (Morelet, 1849)
        - *Nephronaias calamitarum* (Morelet, 1849)
        - *Nephronaias sallei* (Crosse & Fischer, 1894) [E]
        - *Psoronaias crocodilorum* (Morelet, 1849) [E]
        - *Psoronaias kuxensis* Frierson, 1917 [E]
      - **Tribe Quadrulini**
        - *Megalonaias nickliniana* (Lea, 1834)
- **Superfamily Etherioidea**
  - **Family Mycetopodidae**
    - **Subfamily Anodontitinae**
      - *Anodontites cylindracea* (Lea, 1838)
      - *Anodontites lurulenta* (Morelet, 1849) [E]

and no records from the Yucatán peninsula with specific locality data (Lozada-Flores 2011a,b). A survey of the land and freshwater mollusks was conducted at fifteen sites in the Yucatán in 1890 by Angelo Heilprin, Witmer Stone, J.E. Ives, F.C. Baker, and Roberts Le Boutillier and published by Pilsbry (1891). They collected numerous freshwater gastropods in the families Lymnaeidae Rafinesque, 1815, Ancylidae Rafinesque, 1815, Planorbidae Rafinesque, 1815, Physidae Fitzinger, 1833, Ampullariidae J.E. Gray, 1824, Valvatidae Gray, 1840, and Amnicolidae Tryon, 1863, but no freshwater bivalves. The Yucatán is an area that warrants further investigation.

The Río Usumacinta drainage has been well-studied, and we have over 400 records from the basin with the earliest collections dating from the 1840s. In 1935, Goodrich and van der Schalie (1937) sampled the headwaters of the Río San Pedro and the lower reaches of the Río de la Pasión in the Department of Petén, Guatemala, as well as the upper parts of the Rio de la Pasión where it leaves the Department of Alta Vera Paz and flows through Petén. K.S. Cummings and colleagues sampled for mussels at 10 sites in the basin in Chiapas and Tabasco, México, in December 2019.

Usumacinta Province (including the Grijalva) harbors a diverse mussel fauna with 13 native species (Table 10). It is also an area of high endemism. Over half of the species known from the province are endemic: *Anodontites lurulenta* (Morelet, 1849), *Delphinonaias scutulata* (Morelet, 1849), *Mesonaias explicata* (Morelet, 1849), *Nephronaias sallei* (Crosse & Fischer, 1894), *Pachynaias spheniopsis* (Morelet, 1849), *Psoronaias crocodilorum* (Morelet, 1849), and *Psoronaias kuxensis* Frierson, 1917. Czaja et al. (2023:93) identified this area as a hot spot (E. Río Grijalva–Río Usumacinta System) based primarily upon the number of endemic freshwater gastropods in the genus *Pachychilus*.

PACIFIC BASINS

The Pacific Coast drainages of México were divided into three ichthyofaunal provinces by Minckley et al. (2005). They identified a large transition zone between the Northwest Pacific Slope and Chiapas–Nicaraguan Provinces for fishes. The streams along the Pacific coast of Central America are mostly short and high-gradient rivers and generally unsuitable for mussels. Fischer and Crosse (1902:672) stated that the genus *Unio* is wanting ("fait défáut") within México in the streams running to the Pacific. Of the over 3,000 specimen lots we examined from Central America, just over 350 are from Pacific drainages. We have included the rivers that flow to the Pacific in three different Regions and Provinces: (1) the North American Region; Colorado Province (treated above), (2) the Northern Pacific and Endorheic Region; Mesa Central and Balsas Province, and (3) the Southern Pacific Region; Oaxaca México to Colombia Province. The Pacific drainages of Central America are rather depauperate. No members of the Lampsilini and only five members of the Popenaiadini are found in any Pacific basin outside of the Central Mexican Plateau: *Nephronaias caldwellii* (Lea, 1858), *Nephronaias goascoranensis* (Lea, 1858), and *Nephronaias macnielii* (Lea, 1869), *Nephronaias tehuantepecensis* (Crosse & Fischer, 1894), and *Nephronaias tempisquensis* Pilsbry, 1920.

In the Mycetopodidae, *Anodontites tortilis* is known in Pacific drainages in the Isthmus of Darien, Canal Zone, and Rio Membrillo, both in Darién Province, Panamá, the Río Michatoya, Escuintla Department, Guatemala, Rio Grande de Choluteca, Honduras, and ríos Tempisque and Bebedero in Guanacaste, Costa Rica. *Anodontites trapesialis* (Lamarck, 1819) is known in Central America primarily along the Pacific coast from Sinaloa, Jalisco, and Guerrero, México south to Nicaragua and Panamá and in a few rivers draining to the Caribbean. The primarily South American genus *Anodontites* reaches its northern limit in Mazatlan, Sinaloa, México that is also the northern limit of the family Cichlidae on the Pacific side (Meek 1903:783; Chakrabarty 2004:99, fig. 1). *Mycetopoda subsinuata* (Sowerby, 1868) is found in two Caribbean drainages and a few Pacific basins: the Río Michatoya in Guatemala and the Río Grande de San Miguel and Lempa drainages in El Salvador.

Northern Pacific and Endorheic Region

MESA CENTRAL AND BALSAS PROVINCE

MAP 14 Collection sites in Mesa Central and Balsas Province, México.

Mesa Central and Balsas Province occupies much of Central México. The province lies south of the Colorado and west of the Pánuco provinces (Map 14). This province corresponds to the Sinaloa (162), Mayran–Viesca (163), Río Santiago (164), Lerma–Chapala (165), Ameca–Manantlan (168), Río Balsas (also called the Atoyac) (169), and Sierra Madre del Sur (170) ecoregions of Abell et al. (2008). There are also numerous endorheic basins in this province (e.g., the Llanos El Salado (166) ecoregion of Abell et al. 2008). This province corresponds to the Sinaloan Coastal (4), Lerma/Santiago (15–19), and Balsas (21) complexes of Olson et al. (1998:24).

The province is largely drained by the Lerma–Santiago system that flows to the Pacific. The Río Lerma mainstem is more than 700 km long and empties into the Lago de Chapala, México's largest lake (Figure 27) and then into the Río Grande de Santiago, which drains into the Pacific (Miller and Smith 1986:490; Meek 1903:778; Contreras-Balderas and Hales 2019a). The Río Grande de Santiago is one of the longest rivers in México at about 500 km in length.

The Río Ameca has a total length of 260 km and drainage basin of about 17,000 km^2 (López-López and Paulo-Maya 2001:180). It begins in the Bosque de la Primavera in Jalisco, about 20 km west of Guadalajara; flows through the city of Ameca; and then forms the boundary between Jalisco and Nayarit, México, on its way to the Pacific. Although it has a relatively diverse fish fauna for a basin its size (20 species), there are only two mussels known from the Ameca drainage: *Anodontites trapesialis* (Lamarck, 1819) and *Brachyanodon impura* (Say, 1829).

DOI: 10.1201/9781003314448-17

FIGURE 27 Lago de Chapala (Pacific–Rio Grande de Santiago Dr.), Jalisco, México. Habitat for *Nephronaias chapalanus* (von Martens, 1900). (Photo by Gil Garza, 2010, CC0, via Wikimedia Commons).

FIGURE 28 Río Balsas, Coyuca, Guerrero, México. 18.3289, -100.6948. 18 March 2008. Habitat for *Nephritica poeyana* (Lea, 1857) (Photo by Sealight, Wikipedia).

FIGURE 29 Río Amacuzac, a tributary of the upper Río Balsas, about 20 km above the confluence, Morelos, México. Habitat for *Nephritica poeyana* (Lea, 1857). (Photo by Topiltzin Contreras-MacBeath).

The Río Balsas (also locally known as the Río Mezcala or Río Atoyac; Figure 28) has its headwaters in the western slope of the Continental Divide southeast of the city of Puebla and flows 770 km to the southwest where it empties into the Pacific Ocean at the border of Guerrero and Michoacan. The Río Balsas drainage basin has an area of about 117,000 km^2 (Pease et al. 2023:1021). This is a highly mountainous province, with the Sierra Madre Occidental reaching its southern limit here and the Río Balsas cutting through the Sierra Madre del Sur and Eje Volcánico Transversal. The Río Amacuzac (Figure 29) is a tributary of the Balsas in Morelos, México, and is significant in that five of the nine records of *Nephritica poeyana* (Lea, 1857) have been reported from that basin.

The Balsas is a depauperate system with only 36 species of fishes, 16 of which are native and seven of which are endemic (Contreras-Balderas and Hales 2019b; Pease et al. 2023:1021). The fish fauna of the Central Mexican Plateau is composed of species with sister groups located in the Río Grande to the north and Río Lerma to the south (Smith and Miller 1986:475). There is a single cyprinid fish, *Notropis moralesi* de-Buen, 1956, shared between the Balsas of the Mesa Central and the Papaloapán basin (Miller and Smith 1986:513; Schönhuth and Doadrio 2003:325).

Mesa Central and Balsas Province has not been extensively sampled for mussels (156 records), and it is lacking in species richness, with only six species, four of which are endemic: *Anodontites jaliscoensis* Pilsbry, 1920, *Nephritica poeyana* (Lea, 1857), *Nephronaias chapalanus* (von Martens, 1900), and *Popenaias metallica* (Say, 1831) (Table 11). *Popenaias metallica* (Say, 1831) was found only in the Endorheic Lago de Chalco, but it is likely extinct since the lake is heavily polluted and has been periodically drained. Although not strictly an endemic, all but two records of *Brachyanodon impura* (Say, 1829) (both supposedly from the Pánuco basin) are from the province. *Anodontites trapesialis* (Lamarck, 1819) reaches its northern limit in this province in the Río Fuerte along the Pacific coast in Sinaloa, México (Map 14).

TABLE 11

Freshwater Mussels of Mesa Central and Balsas Province.
Total: 6 Species (All Native, 4 endemic [E])

Order Unionoida

Superfamily Unionoidea

Family Unionidae

Subfamily Unioninae

Tribe Anodontini

Brachyanodon impura (Say, 1829)

Subfamily Ambleminae

Tribe Popenaiadini

Nephritica poeyana (Lea, 1857) [E]

Nephronaias chapalanus (von Martens, 1900) [E]

Popenaias metallica (Say, 1831) [E]

Superfamily Etherioidea

Family Mycetopodidae

Subfamily Anodontitinae

Anodontites jaliscoensis Pilsbry, 1920 [E]

Anodontites trapesialis (Lamarck, 1819)

Southern Pacific Region

OAXACA MÉXICO TO COLOMBIA PROVINCE

MAP 15 Collection sites in Oaxaca México to Colombia Province.

Oaxaca México to Colombia Province extends along the Pacific slope from the Isthmus of Tehuantepec in Oaxaca and Chiapas, México south to Panamá on the border with Colombia, South America (Map 15). It largely corresponds to the Chiapas Province of Minckley et al. (2005:44) and the Río Chiapas–Fonseca (201), Estero Real–Tempisque (204), Chiriqui (206), Isthmus Caribbean (207), Santa Maria (208), Chagres (209), and Río Tulra (210) ecoregions of Abell et al. (2008). This province corresponds to the Pacific Central Complex (22), parts of the Atlantic Central Complex [Central America Caribbean Lowlands (25), Talamanca Highlands (26), and the Isthmus Atlantic and Pacific Complexes (35–36)] of Olson et al. (1998:24). All of the mussel records in Oaxaca México to Colombia Province are from Pacific drainages except for the Chagres (Caribbean Dr.) in Panamá. Some of the important rivers in this province with freshwater mussels include the Florida in Chiapas, México, the Michatoya of Guatemala, the ríos Lempa and Rio Grande de San Miguel in El Salvador, the Goascorán of Honduras, the Estero Real in Nicaragua, the Tempisque of Costa Rica, and the Chagres basin in Panamá (Figures 30–34).

Many of the drainages in this province are short and have little flow due to limited precipitation. The fish fauna of this region in México is depauperate, with only 30 primary or secondary fish. However, there is relatively high endemism in the fishes; seven of the ten cichlids in the province from México are endemics (Minckley et al. 2005:44).

The Isthmus of Tehuantepec has been an important avenue for fish movement from Atlantic to Pacific drainages. This low-altitude, narrow (<200 km) passage is nearly traversed by the Río

DOI: 10.1201/9781003314448-18

FIGURE 30 Laguna El Jocotal (Pacific–Rio Grande de San Miguel Dr.), 11.5 km ESE El Transito, San Miguel Department, El Salvador. 13.3249, -88.2467. Habitat for *Mycetopoda subsinuata* (Sowerby, 1868) and *Nephronaias goascoranensis* (Lea, 1858). (Photo by Kentaro Inoue).

FIGURE 31 Río Goascorán (Pacific Dr.), Concepcion de Oriente, 27.1 km NE Santa Rosa de Lima, La Unión Department, El Salvador. 13.7914, -87.7100. Habitat for *Nephronaias goascoranensis* (Lea, 1858). (Photo by Kentaro Inoue).

FIGURE 32 Río Membrillo, ca. 5 km upstream Sinai, Darien, Panamá. 8.6328, -77.8061. Habitat for *Nephronaias caldwellii* (Lea, 1858) and *Anodontites tortilis* (Lea, 1852). March 2023.

FIGURE 33 Río Chagres (Caribbean Dr.), Rt. 9 bridge, near Guayabalito, Colon Province, Panamá. 9.1935, -79.6530. Habitat for *Sinanodonta woodiana* (Lea, 1834). 20 March 2023.

FIGURE 34 Río Florido (Pacific–Tapachula Dr.), 11.2km SW of Tapachula, Chiapas, México. 14.8474, -92.3448. Habitat for *Nephronaias* cf. *caldwellii* (Lea, 1858), possibly an undescribed species. (Photo by José DeJesus de la Rosa Bolón).

TABLE 12
Freshwater Mussels of Oaxaca México to Colombia Province.
Total: 11 species (10 native, 3 endemic [E], 1 introduced [I])

Order Unionoida
Superfamily Unionoidea
Family Unionidae
Subfamily Unioninae
Tribe Anodontini
Sinanodonta woodiana (Lea, 1834) [I]
Subfamily Ambleminae
Tribe Popenaiadini
Nephronaias caldwellii (Lea, 1858) [E]
Nephronaias goascoranensis (Lea, 1858)
Nephronaias macnielii (Lea, 1869)
Nephronaias rowellii (Lea, 1859)
Nephronaias tehuantepecensis (Crosse & Fischer, 1894) [E]
Nephronaias tempisquensis Pilsbry, 1920 [E]

Superfamily Etherioidea
Family Mycetopodidae
Subfamily Mycetopodinae
Mycetopoda subsinuata (Sowerby, 1868)
Subfamily Anodontitinae
Anodontites tehuantepecensis (Crosse & Fischer, 1893)
Anodontites tortilis (Lea, 1852)
Anodontites trapesialis (Lamarck, 1819)

Coatzacoalcos and is the only region in México where multiple groups of aquatic and riparian animals appear to have spread between the Gulf of Mexico and Pacific drainages (Hulsey and López-Fernandez 2011:289).

The Province has not been extensively sampled for mussels (just over 230), but from what we know is that it is relatively diverse with 10 native species, three of which are endemic: *Nephronaias caldwellii* (Lea, 1858), *Nephronaias tehuantepecensis* (Crosse & Fischer, 1894), and *Nephronaias tempisquensis* Pilsbry, 1920 (Table 12). The Asian unionid, *Sinanodonta woodiana*, has been introduced into this province, and it has been found in Lago de Guija (Río Lempa basin) in Santa Anna Department, El Salvador, Laguna de Arenal (Río San Juan drainage), fish culture ponds at Estacion Experimental Enrique Jimenez Nunez in Guanacaste Province, Costa Rica, Gatun Lake (Río Chagres basin) in Colon Province, Panamá, and the Río Chagres near Guayabalito by Pfeiffer et al. in 2023 (Figure 33).

Kentaro Inoue and Yasmín Quintana of the Shedd Aquarium, Chicago sampled at sites in the Lempa, Goascorán, and Rio Grande de San Miguel basins El Salvador in March 2023 (Figures 30 and 31), as well as a few basins in Guanacaste and Alajuela provinces in Costa Rica. They found live individuals of *Nephronaias goascoranensis* (Lea, 1858), *Mycetopoda subsinuata* (Sowerby, 1868), and the introduced *Sinanodonta woodiana* (Lea, 1834). Their preliminary results indicated the presence of three native species each in El Salvador and Costa Rica, including the endemic *Nephronaias tempisquensis* Pilsbry, 1920 in Costa Rica. Additionally, the invasive *Sinanodonta woodiana* (Lea, 1834) was found in both countries (Quintana and Inoue 2024). A survey of 38 sites in Panamá was conducted in March 2023 by John Pfeiffer and colleagues (Figures 32 and 33), and they found freshwater mussels at six sites including *Nephronaias caldwellii* (Lea, 1858), *Anodontites tortilis* (Lea, 1852), and the aforementioned *Sinanodonta woodiana*.

Caribbean Region

Freshwater mussels found in rivers draining to the Caribbean can be found in three different Provinces: Gulf of Honduras and Northern Caribbean Province, San Juan, and Oaxaca, México to Colombia. In the Unionidae, the genera *Arotonaias* (four species) and *Micronaias* (three species) are found only in Caribbean basins (but see the species account for *Arotonaias cyrenoides*). *Delphinonaias largillierti* (Philippi, 1847) is found in the Usumacinta basin (Gulf of Mexico drainage) in México and Guatemala and in a few streams that drain into the Caribbean Sea in Quintana Roo, México, Belize, and Honduras. Ten of the 20 species of *Nephronaias* (*N. copanensis*, *N. dysonii*, *N. flucki*, *N. ortmanni*, *N. ravistella*, and *N. yzabalensis*) are found only in Caribbean basins. *Nephronaias goascoranensis*, *N. hjalmarsoni*, *N. macnielii*, and *N. rowellii* are found in both Pacific and Caribbean drainages. As noted above, *Nephronaias aztecorum* (Philippi, 1847) and *Nephronaias aeruginosa* (Morelet, 1849) occur in Gulf of Mexico drainages, and *Nephronaias calamitarum* (Morelet, 1849) is the only species of *Nephronaias* that occurs in a Gulf of Mexico basin that also extends south into a Caribbean basin (Río Dulce) of Guatemala.

In the Mycetopodidae, four species of the genus *Anodontites* are found in Caribbean basins. *Anodontites montezuma* (Lea, 1841) is endemic to Lake Nicaragua and the Río Malacatoya in the Río San Juan basin, Nicaragua. *Anodontites tortilis* (Lea, 1852) and *Anodontites tehuantepecensis* (Crosse & Fischer, 1893) are South American species that extend north in both Caribbean and Pacific basins. In the Caribbean, *Anodontites tortilis* (Lea, 1852) is known from the Río Sixaola in Costa Rica and the Río Motagua, Guatemala.

Anodontites tehuantepecensis has two Caribbean records: one from Juigalpa, Departamento de Chontales, Nicaragua, and one from Limon Province in Costa Rica. *Anodontites tehuantepecensis* is also found in Pacific basins at Cacoprieto, Isthmus of Tehuantepec, and the Río Novillero in Oaxaca, México, and the Río Michatoya, Escuintla Department, Guatemala. *Anodontites trapesialis* (Lamarck, 1819) is known in Central America, primarily along the Pacific coast from Sinaloa, México south to Nicaragua and Panamá, but it also occurs in a few rivers draining to the Caribbean: the Río Patuca in Honduras, Río Sixaola in Costa Rica, Río Coco in Nicaragua, and Lake Nicaragua and adjacent Lake Managua, Nicaragua.

Mycetopoda subsinuata (Sowerby, 1868) is found in two Caribbean drainages: the Río San Juan in Nicaragua and the Río Motagua in Guatemala in addition to a few Pacific basins in Guatemala and El Salvador. Although no records exist from Costa Rica or Panamá, it may eventually be found there.

DOI: 10.1201/9781003314448-19

GULF OF HONDURAS AND NORTHERN CARIBBEAN PROVINCE

MAP 16 Collection sites in the Gulf of Honduras and Northern Caribbean Province.

The Gulf of Honduras and Northern Caribbean Province extends along the Caribbean coast from the southeastern part of the Yucatán Peninsula south to northern Nicaragua (Map 16, Figures 35 and 36). It largely corresponds to the Quintana Roo–Motagua (202) and Río Mosquitia (213) ecoregions of Abell et al. (2008). Larger river systems that include freshwater mussels are the ríos Dulce, Hondo (border between Quintana Roo, México, and Belize), and the Polochic and Motagua rivers of Guatemala. Lago de Izabal in Guatemala is the largest lake in the region with a surface area of 717 km^2 (Hulsey and López-Fernandez 2011:284). Lago de Izabal drains into the Gulf of Honduras via Amatique Bay (Contreras-Balderas and Bussing 2019). The province includes all the Caribbean basins of Honduras including the ríos Humuya, Ulúa, Patuca, the Río Coco on the Honduras-Nicaragua border, the Belize, Bravo, and New basins in Belize, the Kukalaya–Wawa drainage of Nicaragua, and Lago de Yojoa, the largest lake in Honduras (Bussing and Hales 2019).

Miller (1966:780) noted that the Polochic–Izabal–Dulce basin had a total of 28 species of freshwater fishes with six endemics belonging to the Poeciliidae and Cichlidae. Some drainages in this province have been relatively well-sampled for mussels, particularly in the Río Dulce basin in Guatemala that were sampled by Anson A. Hinkley in the early 1900s, but new collections are needed to improve coverage and check the conservation status of species previously collected. We have just over 300 records from the province. The Gulf of Honduras and Northern Caribbean Province has high species richness with 14 native and five endemic species: *Arotonaias reticulata* (Simpson, 1900), *Nephronaias copanensis* (von Martens, 1900), *Nephronaias flucki* Bartsch, 1906, *Nephronaias ortmanni* (Frierson, 1913), and *Psoronaias rubicunda* (von Martens, 1900) (Table 13). *Psoronaias rubicunda* (von Martens, 1900) is the only member of the genus in the Province and its generic placement is uncertain (see its species account below). The genus *Mycetopoda* reaches its northern limit here in the Río Motagua, Izabal Department, Guatemala. *Sinanodonta woodiana* (Lea, 1834) has been introduced and was found in the Río Polochic, Izabal Department, Guatemala in 2014 (OSUM 84881).

FIGURE 35 Swasey Branch (Caribbean–Gulf of Honduras–Monkey River Dr.), San Pablo, 2.5 km W Red Bank Village, Toledo District, Belize. 16.6158, -88.5817. Habitat for *Nephronaias dysonii* (Lea, 1859). (Photo by Rebecca Young).

FIGURE 36 Río Motagua (Caribbean–Gulf of Honduras Dr.), Zacapa, Guatemala (Kymacpherson, CC BY-SA 3.0 Wikimedia Commons).

TABLE 13

Freshwater Mussels of Gulf of Honduras and Northern Caribbean Province. Total: 15 Species (14 Native, 5 endemic [E], 1 introduced [I])

Order Unionoida

Superfamily Unionoidea

Family Unionidae

Subfamily Unioninae

Tribe Anodontini

Sinanodonta woodiana (Lea, 1834) [I]

Subfamily Ambleminae

Tribe Lampsilini

Arotonaias reticulata (Simpson, 1900) [E]

Delphinonaias largillierti (Philippi, 1847)

Tribe Popenaiadini

Nephronaias calamitarum (Morelet, 1849)

Nephronaias copanensis (von Martens, 1900) [E]

Nephronaias dysonii (Lea, 1859)

Nephronaias flucki Bartsch, 1906 [E]

Nephronaias hjalmarsoni (Dunker, 1858)

Nephronaias ortmanni (Frierson, 1913) [E]

Nephronaias ravistella (Morelet, 1849)

Nephronaias yzabalensis (Crosse & Fischer, 1893)

Psoronaias rubicunda (von Martens, 1900) [E]

Superfamily Etherioidea

Family Mycetopodidae

Subfamily Mycetopodinae

Mycetopoda subsinuata (Sowerby, 1868)

Subfamily Anodontitinae

Anodontites tortilis (Lea, 1852)

Anodontites trapesialis (Lamarck, 1819)

SAN JUAN PROVINCE

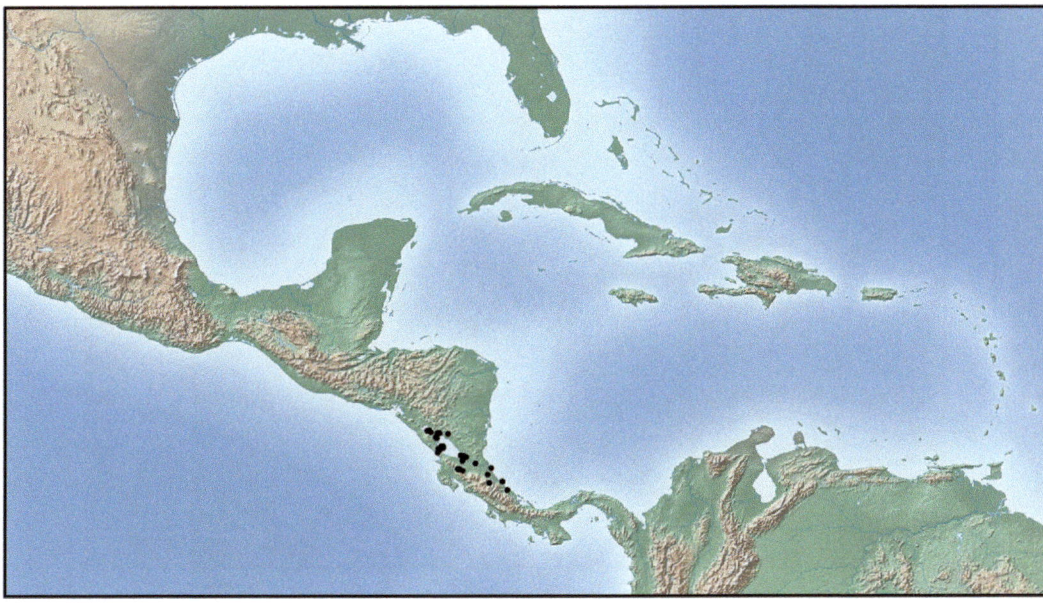

MAP 17 Collection sites in San Juan Province.

San Juan Province corresponds largely to the San Juan (Nicaragua/Costa Rica) (205) ecoregion of Abell et al. (2008) (Map 17, Figure 37). The main river in the province is the San Juan, which is shared by Costa Rica and Nicaragua. Although Lake Nicaragua (also called Lago Cocibolca) is only 16 km from the Pacific and about 100 km from the Atlantic, it drains to the Caribbean Sea via the Río San Juan. The Río San Juan extends about 180 km in a northwest direction from its outlet at San Carlos, Nicaragua. to the coast. Lake Nicaragua is the largest lake in Central America, with a surface area of approximately 8,000 km^2 (Bussing and Smith 2019), and it is connected to adjacent Lake Managua by the Río Tipitapa. Lake Managua is the second largest lake in Central America with a surface area of around 1,000 km^2 (Hulsey and López-Fernandez 2011:284). The Lake Nicaragua basin drains an area of approximately 41,600 km^2 (Gill and Bransford 1877:175; Montenegro-Guillén 2003:1).

San Juan Province has high species richness for an area of its size, with 13 native and six endemic freshwater mussel species: *Anodontites montezuma* (Lea, 1841), *Arotonaias gabbiana* (Lea, 1868), *Arotonaias encarpa* (Lea, 1868), *Micronaias arata* (Lea, 1843), *Micronaias fallaciosa* (Simpson, 1914), and *Micronaias granadensis* (Lea, 1868) (Table 14). The genera *Arotonaias* and *Micronaias* are largely found in the San Juan basin, with the exception of *Arotonaias reticulata* (Simpson, 1900), which is found in the Coco and Patuca basins in the northern Caribbean, and one questionable record (UF 93197) of *Arotonaias cyrenoides* (Philippi, 1847) consisting of two half valves from Río Escondido, San Juan del Sur on the Pacific coast of Nicaragua (see below).

FIGURE 37 Lake Nicaragua, Nicaragua. Habitat for 11 species of mussels. (Photo by Kevin J. Roe).

TABLE 14
Freshwater Mussels of San Juan Province.
Total: 13 Species (All Native, 6 endemic [E])

Order Unionoida

Superfamily Unionoidea

Family Unionidae

Subfamily Ambleminae

Tribe Lampsilini

Arotonaias cyrenoides (Philippi, 1847)
Arotonaias encarpa (Lea, 1868) [E]
Arotonaias gabbiana (Lea, 1868) [E]

Tribe Popenaiadini

Micronaias arata (Lea, 1843) [E]
Micronaias fallaciosa (Simpson, 1914) [E]
Micronaias granadensis (Lea, 1868) [E]
Nephronaias goascoranensis (Lea, 1858)
Nephronaias macnielii (Lea, 1869)

Superfamily Etherioidea

Family Mycetopodidae

Subfamily Mycetopodinae

Mycetopoda subsinuata (Sowerby, 1868)

Subfamily Anodontitinae

Anodontites montezuma (Lea, 1841) [E]
Anodontites tehuantepecensis (Crosse & Fischer, 1893)
Anodontites tortilis (Lea, 1857)
Anodontites trapesialis (Lamarck, 1819)

Cuban Region

CUBA PROVINCE

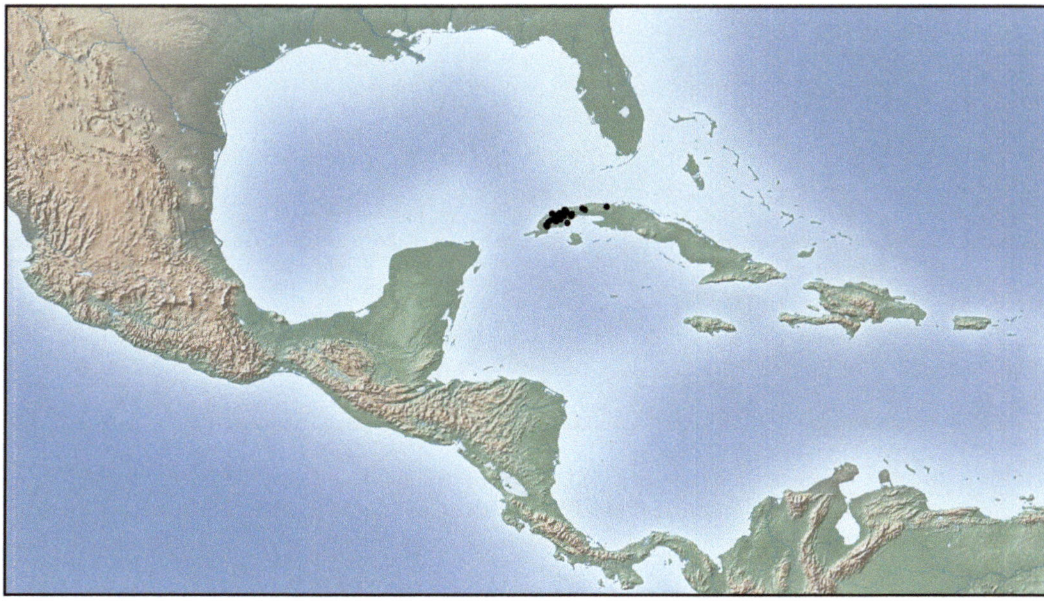

MAP 18 Collection sites in Cuba Province.

TABLE 15
Freshwater Mussels of Cuba Province.
Total: 2 Species (Both Native and Endemic)

Order Unionoida
Superfamily Unionoidea
Family Unionidae
Subfamily Ambleminae
Tribe Popenaiadini
Nephronaias gundlachi (Dunker, 1858)
Nephronaias scamnata (Morelet, 1849)

In the Caribbean, freshwater mussels have been found only on the islands of Cuba and Trinidad (Johnson 1981). The two Cuban species, *Nephronaias scamnata* (Morelet, 1849) and *Nephronaias gundlachi* (Dunker, 1858), are found primarily in streams in the provinces of Artemisa and Pinar del Río in the western-most part of the island (Map 18, Figures 38 and 39). Both are endemic to the island (Table 15). There are three lots of *N. gundlachi* from La Habana ($n=1$), Matanzas ($n=2$) provinces. No specimens were available for study post-1953, and their status and phylogenetic relationships are unknown.

DOI: 10.1201/9781003314448-20

FIGURE 38 Río Cuyaguateje (Caribbean Dr.), Guane, Pinar del Rio, Cuba. 22.1899, -84.0811. Habitat for *Nephronaias scamnata* (Morelet, 1849) and *Nephronaias gundlachi* (Dunker, 1858). (Photo by Adrián González-Guillén).

FIGURE 39 Río Taco Taco (Caribbean Dr.), Aspiro, Pinar del Rio, Cuba. 22.7059, -83.1753. Habitat for *Nephronaias scamnata* (Morelet, 1849) and *Nephronaias gundlachi* (Dunker, 1858). (Photo by Adrián González-Guillén).

Johnson (1981) summarized the information available for the freshwater mussels found on Cuba. He stated that "Two species of Recent Unionacea occur in Pinar del Rio, the westernmost province of Cuba; one is represented in the fossil record in the province of Habana. They belong to the genera found in North and Central America." Johnson placed *Nephronaias gundlachi* (Dunker, 1858) in the genus *Villosa* "because it is sexually dimorphic, and its shell is similar to that of *V. lienosa* (Conrad) of the Gulf Coastal and Apalachicolan regions of the United States." No one has tested that assumption with modern phylogenetic methods, and it is unknown where the Cuban species fit in a larger phylogenetic context. Johnson (1981:269) speculated that both species may have arrived there as a result of Continental drift.

Part 5

Species Accounts

Species Accounts

Species in this work are arranged by the higher classification following Graf and Cummings (2021:14). For each currently recognized species, the type species is listed first and then chronologically by synonymous taxa. The currently recognized name appears in bold at the top, followed by the Province(s) where it occurs, a general statement on its distribution and conservation status, information on anatomy and/or biology if known, reported adult size, a list of similar species that co-occur in the same drainage or locality with the species, and remarks on distribution and/or taxonomy. The species illustrated in the plates are not to scale, but the sizes are given as "(xx mm L, xx mm H)" in the plate legend if known.

For each nomen, the citation of the paper of the original description, verbatim type locality, type locality, and the institutions housing type specimens is given. A taxonomic history for each name considered a synonym under each currently recognized valid species is also given. To avoid any ambiguity, full citations are used for cited literature, and page numbers are included with references in the text. Citations with the author separated by a comma denote taxonomic descriptions and those without a comma are references found in the literature cited section. A description of the information contained in each species account is given below.

GENUS SPECIES (AUTHOR, DATE)

Province(s): The province or provinces in which the species occurs.

Distribution and Status: General comments on the distribution and the conservation status applied to a species by Williams et al. (1993), IUCN (2024), NatureServe (2024), or other published sources. If a species has no status designated, it is given as not assessed.

Anatomy: A short description of the shell and soft parts if known.

Biology: Remarks on the life history, reproduction, habitat, etc.

Adult Size: Reported size in the literature.

Similar Species: Similar species that co-occur in the same drainage or locality.

Remarks: Additional remarks on the taxonomy, types, or distribution of the taxon. Incorrect spellings or dates of publication are mentioned here if confusion might be caused by their omission.

Nominal Species: The name as originally published is given, followed by the complete bibliographic citation of the paper where it first appeared. The citation is followed by the page number where the nomen first appears, followed by a reference to the original figure or figures if published. If no figure was published in the original description but appeared in a later publication, it is referenced here.

Verbatim Type Locality: The type locality as published in the original description, in quotes with the original orthography.

Type Locality: The type locality as derived from the published locality and additional data from labels, catalogues, and/or subsequent publications, with an estimate of the location expressed as digital latitude/longitude given in brackets. Letters that precede the coordinates are indicative of the presumed accuracy of the locality: W = Water Body; D = Division (state, province, etc.); L = specific

DOI: 10.1201/9781003314448-22

locality. Those taxa described from only a country or large river (e.g., Río Papaloapán or México) were not assigned letters or coordinates.

Type Specimens: Freshwater mussel systematics has too often proceeded without explicit reference to specimens or types. In the light of this and the need for rigorous systematic research, the identification of type localities and type specimens is crucial, and we have attempted to provide reference to all that we could locate. The type specimens included in this work include primary types (holotypes, lectotype, and syntype) and secondary types (paratypes and paralectotypes). The museum housing the primary type is followed by the museum(s) housing secondary types. The number of specimens in each lot is indicated if known. Data and location of type material was compiled from specimens examined, the published literature, and online collection catalogues. References to papers citing types are also included here. Museum acronyms used follow Sabaj (2020), with some minor discrepancies.

Various terms have been used to denote primary types over the years: type, holotype, lectotype, types, cotype(s), or syntype(s). Since 2000, the terms used to denote primary types are restricted to three categories: holotype, lectotype, and syntype(s). We owe much to the late Richard I. Johnson (1925–2020) of the Museum of Comparative Zoology, Harvard who published over a dozen type catalogues over his illustrious career (Johnson 1952, 1953, 1956, 1969, 1971, 1972a, 1973, 1974a,b, 1975, 1979, 1980; Johnson and Baker 1973) and identified or selected types in numerous other publications. (e.g., Johnson 1970, 1999a). However, in many cases, Johnson identified a "holotype" when in fact he was making a lectotype designation (indirect lectotype selection by inference of holotype, ICZN Art. 74.6).

74.5. **Lectotype designations before 2000**. In a lectotype designation made before 2000, either the term "lectotype," or an exact translation or equivalent expression (e.g. "the type"), must have been used, or the author must have unambiguously selected a particular syntype to act as the unique name-bearing type of the taxon. When the original work reveals that the taxon had been based on more than one specimen, a subsequent use of the term "holotype" does not constitute a valid lectotype designation unless the author, when wrongly using that term, explicitly indicated that he or she was selecting from the type series that particular specimen to serve as the name-bearing type.

74.6. **Fixation of lectotype by inference of "holotype" or "the type" before 2000.** When it has been accepted that a nominal species-group taxon was based on a single specimen and the original description neither implies nor requires that there were syntypes, and if it is considered subsequently that the original description was based on more than one specimen, the first author to have published before 2000 the assumption that the species-group taxon was based upon a single type specimen is deemed to have designated that specimen as the lectotype (International Commission on Zoological Nomenclature, 2025).

74.6.1. The inference that the specimen is a "holotype" or "the type."
74.6.1.1. May be by reference to an illustration or description of a specimen [Art. 74.4].
74.6.1.2. Must be individual in accordance with Article 74.3.

Taxonomic History: A list of the primary synonyms is given below each currently recognized valid species and includes the same information as was given for each currently recognized valid species: description, type locality, types, etc. Subsequent name combinations are also given for each nomen. The list of taxonomic name combinations is not comprehensive (particularly for wide ranging species), and the taxonomic opinions given are those from major revisions (e.g., Simpson 1900; Frierson 1927; Haas 1969), monographs (e.g., Sowerby 1866–1870; von Martens 1890–1901; etc.), or papers citing mussels found in Central America.

Dates of Publication: In general, the text descriptions of new taxa were published in the same year as the corresponding illustration(s), but some illustrations were delayed by one or more years. Coan and Kabat (2024a:84–85) provided detailed collations of von Martens (1890–1901) land and freshwater Mollusca *In* Godman and Salvin's *Biologia Centrali-Americana* and Herrmannsen's (1846–1852) *Indicis Generum Malacozoorum Primordia*, and many other works. The von Martens collation was taken from the wrappers in a set in the Museum of Comparative Zoology Library. Pages 596–655 comprise a "Supplement" with corrections to the earlier-published sections, with additional corrections on page xxviii of the Preface, which also includes (pages vi–xii) short biographical sketches of the collectors in México and Central America. Dates for other works were obtained from the original publication or other compilations (Fox 1913; Winckworth 1936; Duncan 1937; Smith and England 1937; Johnson 1968; Coan and Harasewych 1993; Crosnier and Clark 1998; Kabat and Boss 1997; Welter-Schultes 1999; Kabat and Johnson 2008; Kabat and Coan 2016; Coan and Kabat 2017; Coan and Kabat 2024b)

Species Accounts for Central America

Order Unionoida
Superfamily Unionoidea Rafinesque, 1820
Family Unionidae Rafinesque, 1820
Subfamily Unioninae Rafinesque, 1820
Tribe Anodontini Rafinesque, 1820
Genus *Brachyanodon* Crosse & Fischer, 1894

Brachyanodon Crosse & Fischer, 1894. *In* P. Fischer & H. Crosse, 1894. Études sur les mollusques terrestres et fluviatiles du Mexique et du Guatemala. *In* M.H.–Milne Edwards (ed.). *Mission Scientifique au Mexique et dans l'Amérique centrale. Recherches Zoologiques pour servir á l'Histoire de la Faune de l'Amérique Centrale et du Mexique.* Septiéme Partie. 15th Livraison. Feuilles 62–72, pp. 489–576, plates 59–62, Imprimerie Nationale, Paris Volume 2. 731 pp. Page 517. Introduced as a subgenus of *Anodonta* Lamarck, 1799.

Diagnosis: "Coquille à sommet peu saillant; diamètre antéro-postérieur court; bord dorsal ascendant en arrière des sommets (*A. Chapalensis*, Crosse et Fischer; *A. coarctata*, Anton; *A. exilior*, Lea; *A. viridana*, Clessin; *A. Chalcoensis*, Crosse et Fischer)." [Shell with slightly projecting beak; short anteroposterior diameter; ascending dorsal edge behind the apices (*A. Chapalensis*, Crosse and Fischer; *A. coarctata*, Anton; *A. exilior*, Lea; *A. viridana*, Clessin; *A. Chalcoensis*, Crosse and Fischer)].

Type Species: *Anodonta chapalensis* Crosse & Fischer, 1892, by subsequent designation (Thiele 1934:832).

Taxonomic History:

Anodonta (s.s.) Lamarck, 1799 — von Martens (1900b, 1900c), Frierson (1927)
Anodonta Lamarck, 1799 — Simpson (1900; 1914), Thiele (1934), Modell (1964), Williams et al. (2008)
Anodonta (*Brachyanodon*) Crosse & Fischer, 1894 — Haas (1969)
Brachyanodon Crosse & Fischer, 1894 — Bouchet & Rocroi (2010)

Distribution and Remarks: The genus *Brachyanodon* Crosse & Fischer, 1894 is monotypic and its sole representative, *Brachyanodon impura* (Say, 1829) is endemic to Central México. *Brachyanodon impura* (Say, 1829) is native in a few endorheic basins and various Pacific drainages in Mesa Central and Balsas Province (primarily in the Ameca and Río Grande de Santiago basins). It has also been reported from the ríos Valles (FMNH 11631) and Tula (USNM 134837) in the Pánuco basin, but those records are regarded as suspect.

DOI: 10.1201/9781003314448-23

PLATE 7 ***Brachyanodon impura* (Say, 1829). Top Left**: *Anodonta exilior* Lea, 1871. Lectotype, USNM 86730 (46 mm L, 33 mm H). México. **Top Right**: *Anodonta chapalensis* Crosse & Fischer, 1893. Syntype, MNHN-IM-2000-1657 (42 mm L, 32 mm H). Lago de Chapala (Pacific–Río Grande de Santiago Dr.), Jalisco, México [W = 20.3049, -102.7781]. **Bottom Left**: *Anodonta chalcoensis* Crosse & Fischer, 1893. Lectotype, MNHN-IM-2000-1656 (51 mm L, 30 mm H). Lake Chalco (Endorheic Dr.), Ciudad de México, México [W = 19.2635, -98.9701]. **Bottom Right**: *Anodonta richardsoni* von Martens, 1900. Syntype, MCZ 6808 (59 mm L, 36 mm H). Río Ameca (Pacific Dr.), Jalisco, México [W = 20.5546, -104.1122].

Brachyanodon impura (Say, 1829)

Plate 7, Map 19

Province(s): Mesa Central and Balsas.

Distribution and Status: Widely distributed on the central high plateau of México and three Pacific drainages: the Ameca, Río Grande de Santiago, and San Pedro Mezquital. There are two lots from the Pánuco basin from the early 20th century (one from the Río Valles and one from the Río Tula), both from normally reliable collectors (Seth E. Meek, Edward W. Nelson, and Edward A. Goldman), but no records before or since. The Pánuco records are plotted here but remain questionable. There is also a lot (ZMB/MOLL.220177, $n=4$) from "Veracruz" without a specific locality that was figured in von Martens (1900f, pl. 40, figs. 8–11, as *Anodonta exilior*). Largely extirpated from lakes and streams around México City and environs (Lago de Xochimilco). Still extant in Lago de Chapala, in Jalisco, México. DD—Data Deficient (IUCN 2024); Not assessed (NatureServe 2024).

Anatomy: Shell moderately large, thin, and round to oblong or ovate and inflated. No lateral or pseudocardinal teeth. Beak almost flush with the hinge line and often eroded. Periostracum tan to dark brown. Nacre white, pink, or salmon.

Biology: Unknown.

Adult Size: Length to 15 cm. Say's type was small measuring only 3.5 cm.

Similar Species: The edentate *Anodontites trapesialis* (Lamarck, 1819) also occurs in Pacific drainages and in Laguna Valencia in the endorheic Laguna de Atotonico basin in Mesa Central and Balsas Province. *Brachyanodon impura* (Say, 1829) and *Anodontites trapesialis* co-occur in the Ameca and in Lago de Chapala in Jalisco, México. They can be separated by the presence of a brown triangular-shaped blotch at the posterior end of the hinge called a ligamental sinus, which is present in the genus *Anodontites* (Figure 2). *Anodontites jaliscoensis* Pilsbry, 1920 is also present in Mesa Central and Balsas Province, but it is known only from the type locality in the Río Armeria (Pacific Dr.), Tolimán, Jalisco, México and possibly in a muddy creek (also Pacific Dr.), near Singayta, Nayarit, México.

MAP 19 Distribution of *Brachyanodon impura* (Say, 1829).

Remarks: This species has never been included in a published phylogeny, but COI sequences available on Genbank suggest that the species is sister to a group of other Pacific Coast anodontines (e.g., *A. californiensis*, *A. nuttalliana*, and *A. oregonensis*).

Anodonta impura Say, 1829. Descriptions of some new terrestrial and fluviatile shells of North America. (cont.). The Disseminator of Useful Knowledge; containing hints to the youth of the U. States, from the School of Industry 2(23):355–356. 18 November 1829. Page 355. Not figured.

Verbatim Type Locality: "Mexico…ditch beside the road between the Capitol and Tacubya [sic]."

Type Locality: Ditch beside the road between México City and Tacubaya, Ciudad de México, México [L = 19.4296, -99.1614].

Type Specimens: Unknown. Presumed lost.

Remarks: Although the type of *Anodonta impura* is lost, there are specimens from Lago de Xochimilco (AMNH 139091, COMA 1050 and UMMZ 103247) and Lago de Texcoco (ANSP 85330), Ciudad de México, México, in the vicinity of the published type locality that could be selected as a neotype. *Anodonta impura* Servain, 1882 described from Germany is a primary homonym of *Anodonta impura* Say, 1829.

Taxonomic History:

Anodonta impura Say, 1829 — Ferussac (1835), Say (1840), Conrad (1853), Fischer & Crosse (1894a), Simpson (1914), Contreras (1930), Morrison (1968), Contreras-Arquieta (2000), Graf & Cummings (2007; 2021), Cummings & Graf (2010), Preece et al. (2021), Kiser et al. (2022), Czaja et al. (2023)
Anodonta (s.s.) *chalcoensis* Crosse & Fischer, 1893 — von Martens (1900d)
Anodonta henryana Lea, 1857 — Simpson (1900)
Anodonta cygnea impura Say, 1829 — Hannibal (1912a; 1912b) [in part]
Anodonta (s.s.) *impura* Say, 1829 — Frierson (1927)
Anodonta marginata impura Say, 1829 — Modell (1945)
Anodonta (*Brachyanodon*) *impura* Say, 1829 — Haas (1969)

Unio (*Anodonta*) *coarctata* Anton, 1839. Verzeichniss der conchylien welche sich in der Sammlung von Hermann Eduard Anton befinden. Eduard Anton, Halle 110 pp. Page 16. Figured in Küster (1838:34, pl. 8, fig. 2) and von Martens (1900f, pl. 40, figs. 1–7, ZMB/MOLL.305849, from Lago de Chapala).

Verbatim Type Locality: "Mexiko."

Type Locality: México.

Type Specimens: Unknown. Presumed lost.

Remarks: *Anodonta coarctata* Potiez & Michaud, 1844 from France is a secondary homonym of *Anodonta coarctata* Anton, 1839. Renamed *Anodonta potiezi* Bourguignat, 1881. A synonym of *Anodonta anatina* (Linnaeus, 1758).

Taxonomic History:

Anodonta coarctata (Anton, 1839) — Lea (1870), Simpson (1900; 1914), de Leon (1912), Haas (1929), Contreras-Arquieta (2000), Cruz & Quesada (2017)
Anodonta (*Brachyanodon*) *coarctata* (Anton, 1839) — Fischer & Crosse (1894a)
Anodonta (s.s.) *coarctata* (Anton, 1839) — von Martens (1900c; 1901c)
Anodonta cygnea impura Say, 1829 — Hannibal (1912a)
Anodonta (s.s.) *impura* Say, 1829 — Frierson (1927)
Anodonta (*Brachyanodon*) *impura* Say, 1829 — Haas (1969)

Anodon glabrus Sowerby, 1870. Monograph of the genus *Anodon*. *In* L. Reeve and G.B. Sowerby (eds.). *Conchologia Iconica* 17:[unpaginated]+Plates 1, 25–37. Plate 25, species 97.

Verbatim Type Locality: "Mexico."

Type Locality: México.

Type Specimens: Unknown. Presumed lost. Not found in the NHMUK.

Remarks: Attributed by Sowerby to "Valenciennes." Various authors have differed on their opinions as to what *Anodon glabrus* Sowerby, 1870 is. Some considered it as a member of the family Mycetopodidae (as *Glabaris* or *Anodontites*), while others placed it in the Unionidae (as *Anodonta* or *Brachyanodon*). Sowerby's figure is of the outside of the shell only. It does have beak sculpture, which is lacking in *Anodontites*, and we place it here following Haas (1969:362).

Taxonomic History:

Anodonta (*Brachyanodon*) *viridana* Clessin, 1876 — Fischer & Crosse (1894a)
Anodonta (s.s.) *exilior* Lea, 1871 — von Martens (1900d)
Glabaris (s.s.) *inæquivalvis* [sic] (Lea, 1868) — Simpson (1900)
Glabaris cylindraceus (Lea, 1838) — de Leon (1912)
Anodontites (s.s.) *inæquivalvis* [sic] (Lea, 1868) — Simpson (1914)
Glabaris glabrus Simpson, 1900 — Frierson (1927)
Anodonta (s.s.) *impura* Say, 1829 — Frierson (1927)
Anodonta (*Brachyanodon*) *impura* Say, 1829 — Haas (1969)

Anodonta exilior Lea, 1871. Descriptions of three new species of exotic Unionidae. *Proceedings of the Academy of Natural Sciences of Philadelphia* 23(1871):188. Figured in Lea (1874a JANSP Vol. 8:24; 1874b Obs.13:28, pl. 7, fig. 21) and von Martens (1900f, pl. 40, figs. 8–11, ZMB/MOLL.220177).

Verbatim Type Locality: "Mexico."

Type Locality: México.

Type Specimens: Lectotype, USNM 86730 designated by Johnson and Baker (1973:154; Johnson 1974a:54, both as the holotype). Paralectotypes, ANSP 125882 (*n*=2) (Johnson and Baker 1973:154, as paratypes).

Remarks: The four specimens figured in von Martens (1900f, pl. 40, figs. 8–11, ZMB/MOLL.220177), supposedly from Veracruz, are the only records from that state and the locality data may be in error.

Taxonomic History:

Anodonta (*Brachyanodon*) *exilior* Lea, 1871 — Fischer & Crosse (1894a)
Anodonta (s.s.) *exilior* Lea, 1871 — von Martens (1900d; 1901c)
Anodonta exilior Lea, 1871 — Simpson (1900; 1914), de Leon (1912), Contreras-Arquieta (2000)
Anodonta cygnea impura Say, 1829 — Hannibal (1912a)
Anodonta (s.s.) *impura* Say, 1829 — Frierson (1927)
Anodonta impura Say, 1829 — Contreras (1930)
Anodonta (*Brachyanodon*) *impura* Say, 1829 — Haas (1969)
Anodonta exilior Lea, 1871 — Cruz & Quesada (2017)

Anodonta viridana Clessin, 1876. Die Gattung *Anodonta* nebst den übrigen najaden mit unvollkommenem Schloss. *In* Abbildungen nach der Natur mit Beschreibungen. Systematisches Conchylien-Cabinet von Martini und Chemnitz. Küster edition. 9(Part 1):287 pp., pls. 20–87, A, 11. Page 226, pl. 75, fig. 5.

Verbatim Type Locality: "Mexiko."

Type Locality: México.

Type Specimens: Clessin's types were in the Stuttgart Museum, but the museum was totally destroyed in World War II (Tomlin 1947:288). As with *Anodon glabrus* Sowerby, 1870, the authors have differed on the placement of *Anodonta viridana* Clessin, 1876 in the family Mycetopodidae (as *Glabaris* or *Anodontites*) or Unionidae (as *Anodonta* or *Brachyanodon*). Clessin's figure is of the outside of the shell only and lacks any of the diagnostic features to place it in one family or the other, and there is no known type to examine to make a determination.

Taxonomic History:

Anodonta (Brachyanodon) viridana Clessin, 1876 — Fischer & Crosse (1894a)
Anodonta (s.s.) *exilior* Lea, 1871 — von Martens (1900d)
Anodontites (s.s.) *inæquivalvis* [sic] (Lea, 1868) — Simpson (1900)
Glabaris (s.s.) *inæquivalvis* [sic] (Lea, 1868) — Simpson (1914)
Anodonta (s.s.) *impura* Say, 1829 — Frierson (1927)
Anodonta (Brachyanodon) impura Say, 1829 — Haas (1969)

Anodonta chapalensis Crosse & Fischer, 1893. Diagnoses Molluscorum Reipublicæ Mexicanæ et Guatemalæ incolarum. *Journal de Conchyliologie* 40(3):294–296. Page 295. Figured in Fischer and Crosse (1894a:518, pl. 64, figs. 8–11) and von Martens (1900f, pl. 40, figs. 1–7, as *Anodonta* (s.s.) *coarctata* Anton, 1839).

Verbatim Type Locality: "in lacu Chapala, in provincia Jalisco dicta, Reipublicæ Mexicanæ."

Type Locality: Lago de Chapala (Pacific–Río Grande de Santiago Dr.), Jalisco, México [W=20.3049, -102.7781].

Type Specimens: Syntypes, MNHN-IM-2000-1657 (*n*=5).

Remarks: This is the type species of *Brachyanodon* Crosse & Fischer, 1894.

Taxonomic History:

Anodonta (Brachyanodon) chapalensis Crosse & Fischer, 1893 — Fischer & Crosse (1894a)
Anodonta (s.s.) *coarctata* Anton, 1839 — von Martens (1900c; 1901c)
Anodonta coarctata Anton, 1839 — Simpson (1900; 1914), de Leon (1912)
Anodonta cygnea impura Say, 1829 — Hannibal (1912a)
Anodonta (s.s.) *impura* Say, 1829 — Frierson (1927)
Anodonta chapalensis Crosse & Fischer, 1893 — Haas (1929), Thiele (1934) Contreras-Arquieta (2000)
Anodonta (Brachyanodon) impura Say, 1829 — Haas (1969)

Anodonta chalcoensis Crosse & Fischer, 1893. Diagnoses molluscorum novorum, Reipublicæ Mexicanæ incolarum. *Journal de Conchyliologie* 41(2):110–111. Page 110. Figured in Fischer and Crosse (1894a:522, pl. 64, figs. 7–7a).

Verbatim Type Locality: "in lacu Chalco, propè civitatem Mexico."

Type Locality: Lago de Chalco (Endorheic Dr.), Ciudad de México, México [W=19.2635, -98.9701].

Type Specimens: Lectotype, MNHN-IM-2000-1656, here designated (the figured specimen).

Taxonomic History:

Anodonta (Brachyanodon) chalcoensis Crosse & Fischer, 1893 — Fischer & Crosse (1894a)
Anodonta (s.s.) *chalcoensis* Crosse & Fischer, 1893 — von Martens (1900d; 1901c)
Anodonta exilior Lea, 1871 — Simpson (1900; 1914), de Leon (1912)

Anodonta cygnea impura Say, 1829 — Hannibal (1912a)
Anodonta (s.s.) *impura* Say, 1829 — Frierson (1927)
Anodonta impura Say, 1829 — Contreras (1930)
Anodonta (*Brachyanodon*) *impura* Say, 1829 — Haas (1969)
Anodonta chalcoensis Crosse & Fischer, 1893 — Cruz & Quesada (2017)

Anodonta (*Anodonta*) *richardsoni* von Martens, 1900. Land and freshwater Mollusca. *In* F.D. Godman and O. Salvin (eds.). *Biologia Centrali-Americana* Part 157, Fascile 67:529–536, plates 36–37. Page 529 and Part 159, Fascile 72–75:569–600, plates 40–41. Plate. 41, figs. 1–1a, 2–2a.

Verbatim Type Locality: "N.W. Mexico: Rio Ameca, State of Jalisco."

Type Locality: Río Ameca (Pacific Dr.), Jalisco, México [W=20.5546, -104.1122].

Type Specimens: Syntypes, MCZ 6808 (*n*=3) (Johnson 1956:132, as cotypes); ZMB/MOLL.108856 (*n*=6 and$^{13}/_{2,}$ 3 figured).

Taxonomic History:

Anodonta richardsoni von Martens, 1900 — Simpson (1914), Haas (1929), Cruz & Quesada (2017)
Anodonta (s.s.) *richardsoni* von Martens, 1900 — Frierson (1927)
Anodonta (*Brachyanodon*) *impura* Say, 1829 — Haas (1969)

Order Unionoida
Superfamily Unionoidea Rafinesque, 1820
Family Unionidae Rafinesque, 1820
Subfamily Unioninae Rafinesque, 1820
Tribe Anodontini Rafinesque, 1820
Genus *Pyganodon* Crosse & Fischer, 1894

Pyganodon Crosse & Fischer, 1894. *In* P. Fischer and H. Crosse. 1894. Études sur les mollusques terrestres et fluviatiles du Mexique et du Guatemala. *In* M.H.–Milne Edwards (ed.). *Mission Scientifique au Mexique et dans l'Amérique centrale. Recherches Zoologiques pour servir á l'Histoire de la Faune de l'Amérique Centrale et du Mexique.* 15th Livraison. Feuilles 62–72, pp. 489–576, plates 59–62, Imprimerie Nationale, Paris Volume 2. 731 pp. Page 518. Introduced as subgenus of *Anodonta* Lamarck, 1799.

Diagnosis: "Sommets très-saillants; coquille de grande taille (*A. globosa*, Lea; *A. Tabascensis*, Morelet; *A. Nopalatensis*, Sowerby; *A. Grijalvæ*, Morelet; *A. glauca*, Valenciennes)." [Very prominent beaks; large shell (*A. globosa*, Lea; *A. Tabascensis*, Morelet; *A. Nopalatensis*, Sowerby; *A. Grijalvæ*, Morelet; *A. glauca*, Valenciennes)].

Type Species: *Anodonta globosa* Lea, 1841, by subsequent designation (Frierson 1927:9).

Taxonomic History:

Anodonta (*Patularia*) Swainson, 1840 — von Martens (1900c, 1900d)
Anodonta (*Pyganodon*) Crosse & Fischer, 1894 — Frierson (1927), Haas (1969), Johnson (1970; 1999a), Heard (1975)
Anodonta Lamarck, 1799 — Thiele (1934)
Pyganodon Crosse & Fischer, 1894 — Modell (1942; 1945; 1964), Graf & Cummings (2007; 2021), Lopes-Lima et al. (2019), Pfeiffer et al. (2019b), Tiemann et al. (2020), Bieler (2021), Nagel (2025)

Distribution & Remarks: The genus *Pyganodon*, Crosse & Fischer, 1894, is widely distributed from Northeastern North America south to Central México (Map 20). However, this warrants further investigation to determine if all members currently assigned to the genus are monophyletic. There is a wide disjunction between *Pyganodon globosa* (Lea, 1841) of the Papaloapán and Usumacinta and *Pyganodon grandis* (Say, 1829) of North America. If they are not monophyletic, a new genus will be needed for the five species from eastern North America: *P. grandis* (Say, 1829), *P. cataracta* (Say, 1817), *P. fragilis* (Lamarck, 1819), *P. gibbosa* (Say, 1824), and *P. lacustris* (Lea, 1857).

We recognize *Pyganodon globosa* (Lea, 1841) as a full species, distinct from the widely disjunct *P. grandis* (Say, 1829). *Pyganodon globosa* (Lea, 1841) is the only native species of the subfamily Unioninae, tribe Anodontini, in any Gulf of Mexico drainage south of the Río Grande, and it is the southernmost member of the tribe in the Western Hemisphere occurring as far south as Tabasco México. *Pyganodon globosa* is found in both the Pánuco and Papaloapán provinces, and if it is revealed that each basin has its own species, *Pyganodon grijalvae* (Morelet, 1884) would be the name applied to the Usumacinta population.

Distribution & Remarks: The genus *Pyganodon*, Crosse & Fischer, 1894, is widely distributed from Northeastern North America south to Central México. Representatives of this genus from south of the Rio Grande have never been included in a molecular phylogenetic analysis, and the monophyly of this genus and the extent of its southern geographic distribution are worthy of further investigation.

We recognize *Pyganodon globosa* (Lea, 1841) as a full species, distinct from the widely disjunct *P. grandis* (Say, 1829). *Pyganodon globosa* (Lea, 1841) is the only native species of the subfamily Unioninae, tribe Anodontini in any Gulf of Mexico drainage south of the Río Grande, and it is the southernmost member of the tribe in the Western Hemisphere occurring as far south as Tabasco, México.

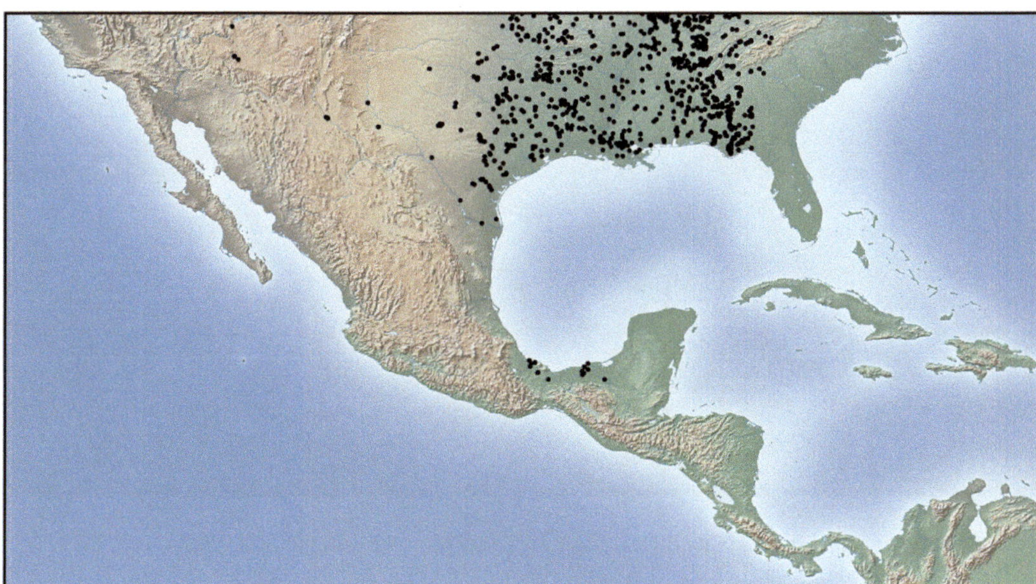

MAP 20 Distribution of *Pyganodon* Crosse & Fischer, 1894.

PLATE 8 ***Pyganodon globosa*** **(Lea, 1841). Top Left**: *Anodonta globosa* Lea, 1841. Lectotype, USNM 86701 (100 mm L, 73 mm H). Concha Lake (Papaloapán Dr.), near Tlacotalpan, Veracruz, México [L = 18.6089, -95.6590]. **Top Right**: *Anodonta grijalvae* Morelet, 1884. Lectotype, NHMUK 1893.2.4.2007 (150 mm L, 96 mm H). Río dos Ídolos (Grijalva Dr.), Tabasco, México [W = 18.2772, -92.6481]. **Bottom Left**: *Anodonta tabascensis* Morelet, 1884. Lectotype, NHMUK 1893.2.4.2006 (175 mm L, 115 mm H). Tabasco (Grijalva Dr.?), México [D = 18.4205, -92.8478]. **Bottom Right**: *Anodon nopalatensis* Sowerby, 1867. Holotype, NHMUK 1965189 (149 mm L, 109 mm H). Río Nopalapan (Papaloapán–San Juan Dr.), Veracruz, México [W = 18.1136, -95.3264].

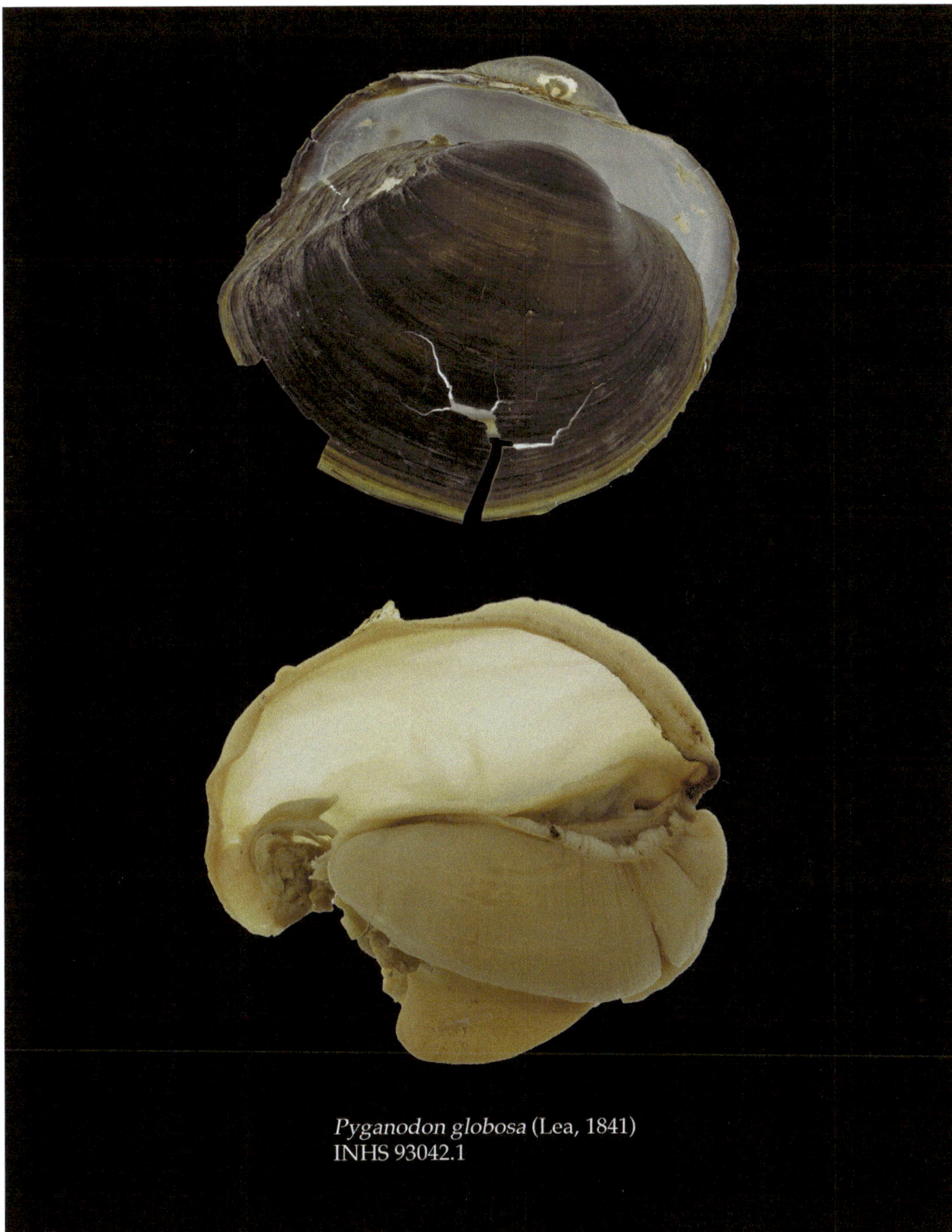

PLATE 9 ***Pyganodon globosa*** **(Lea, 1841).** INHS 93042.1 (100 mm L, 80 mm H). trib. Rio San Juan (Rio Papaloapán Dr.), Nopalapan, Veracruz, México [18.1138, -95.3265]. Gravid female with eggs collected on 4 December 2022.

Pyganodon globosa (Lea, 1841)

Plates 8–9, Map 21

Province(s): Papaloapán; Usumacinta.

Distribution and Status: Found in the Coatzacoalcos, Papaloapán, and Usumacinta basins in central and southern México. Not assessed (IUCN 2024).

Anatomy: Shell large, thin, smooth, round, and globose. Beaks prominent and elevated. Beak cavity deep. No lateral or pseudocardinal teeth. Periostracum green to black and faintly rayed when young. Nacre bluish white and iridescent.

Biology: Found in mud and sand in standing or slow-flowing water. A female gravid with eggs was collected from a tributary to the Rio San Juan (Papaloapán Dr.), Nopalapan, Veracruz, México on 4 December 2022 (INHS 93042, Plate 9). No glochidia were observed. Morrison (1968:50) reported collecting *P. globosa* from a mud-bottom pasture pond near Villahermosa, Tabasco, México and noted that "its glochidia are now on record" but he did not provide a figure or state when they were collected.

Adult size: Length to 12 cm.

Similar Species: *Pyganodon globosa* (Lea, 1841) is found with *Anodontites cylindracea* (Lea, 1838) in the Papaloapán and Usumacinta basins and with *Anodontites lurulenta* (Morelet, 1849) in the Usumacinta. Both species of *Anodontites* have a triangular ligamental sinus that is lacking in *P. globosa* (Figure 2).

MAP 21 Distribution of *Pyganodon globosa* (Lea, 1841).

Anodonta globosa Lea, 1841. On fresh water and land shells (cont.). *Proceedings of the American Philosophical Society* 2(17):30–34. Page 31. Figured in Lea (1842a TAPS Vol. 8:241; 1842b Obs. 3:79, pl. 24, fig. 56), Fischer and Crosse (1894a:529, pl. 68, figs. 1–1a) and Clessin (1875, pl. 65, figs. 3–4).

Verbatim Type Locality: "Concha Lake, near Tlocatalpam [sic], Mex."

Type Locality: Concha Lake (Papaloapán Dr.), near Tlacotalpan, Veracruz, México [L = 18.6089, -95.6590].

Type Specimens: Lectotype, USNM 86701 designated by Johnson and Baker (1973:156; Johnson 1974:64, both as the holotype). Paralectotype, ANSP 41262 (*n* = 1) (Johnson and Baker 1973:156, as a paratype).

Taxonomic History:

Margarita (Anodonta) globosa Say — Lea (1836) *nomen nudem*
Margaron (Anodonta) globosa (Lea, 1841) — Lea (1852; 1870)
Anodonta globosa Lea, 1841 — Hanley (1856), Adams & Adams (1857), Clessin (1875), Simpson (1900; 1914), de Leon (1912), H.B. Baker (1922), Thiele (1934), Contreras-Arquieta (2000), Cruz & Quesada (2017)
Anodon globosa (Lea, 1841) — Sowerby (1870)
Anodonta (Pyganodon) globosa Lea, 1841 — Fischer & Crosse (1894a), Morrison (1968), Haas (1969)
Anodonta (Patularia) globosa Lea, 1841 — von Martens (1900d, 1901c)
Anodonta globosa globosa Lea, 1841 — H.B. Baker (1922)
Anodonta (Pyganodon) grandis-globosa Lea, 1841 — Frierson (1927)
Anodonta grandis globosa Lea, 1841 — Haas (1929)
Anodonta (Pyganodon) grandis globosa Lea, 1841 — Haas (1969)
Pyganodon grandis (Say, 1829) — Graf & Cummings (2007; 2021) [part], Czaja et al. (2023)

Anodonta grijalvæ Morelet, 1884. Deux Naïades inédites du Tabasco. *Journal de Conchyliologie* 32(2):122–125. Page 123. Figured in Fischer and Crosse (1894a:532, pl. 59, fig. 1).

Verbatim Type Locality: "Rio dos Idolos, un des bras du fleuve Grijalva, dans le Tabasco."

Type Locality: Río dos Ídolos (Gulf of México–Grijalva Dr.), Tabasco, México [W = 18.2772, -92.6481].

Type Specimens: Lectotype, NHMUK 1893.2.4.2007 (*n* = ½) designated by Johnson (1971:85, incorrectly as NHMUK 1893.2.4.2009, which is the holotype of *Unio calamitarum* var. γ, *arcuans* Fischer & Crosse, 1894; Breure et al. 2018:299, figs. 469–470).

Remarks: Frierson (1927:16, citing E. Smith) pointed out that *Anodonta grijalvae* Morelet, 1884 was a true "*Anodonta*" and that it lacked the triangular ligamental sinus on the hinge margin, which would place it in *Anodontites* (Mycetopodidae).

Taxonomic History:

Anodonta grijoloae [sic] Morelet, 1884 — Paetel (1890)
Anodonta (Pyganodon) grijalvæ [sic] Morelet, 1884 — Fischer & Crosse (1894a)
Anodonta (Patularia) grijalvae Morelet, 1884 — von Martens (1900d)
Glabaris (s.s.) *grijalvae* (Morelet, 1884) — Simpson (1900)
Glabaris grijalvæ [sic] (Morelet, 1884) — de Leon (1912)
Anodontites (s.s.) *grijalvae* (Morelet, 1884) — Simpson (1914)
Anodonta grandis grijalvae Morelet, 1884 — Frierson (1922)
Anodonta (Pyganodon) grandis-grijalvae Morelet, 1884 — Frierson (1927)
Anodonta grandis grijvalvae [sic] Morelet, 1884 — Haas (1929)
Anodonta (Pyganodon) grandis tabascensis Morelet, 1884 — Haas (1969)
Anodonta grijalvae Morelet, 1884 — Cruz & Quesada (2017)

Anodonta tabascensis Morelet, 1884. Deux Naïades inédites du Tabasco. *Journal de Conchyliologie* 32(2):122–125. Page 124. Figured in Fischer and Crosse (1894a:530, pl. 62, fig. 1).

Verbatim Type Locality: "Marias du Tabasco."

Type Locality: Tabasco (Gulf of México–Grijalva Dr.?), México [D=18.4205, -92.8478].

Type Specimens: Lectotype, NHMUK 1893.2.4.2006 designated by Johnson (1971:95, incorrectly as NHMUK 93.2.4.2008 which is the holotype of *Unio calamitarum* var. *β, prolongata* Fischer & Crosse, 1894, and as the holotype; Breure et al. 2018:447, figs. 1119–1120).

Taxonomic History:

Anodonta (Pyganodon) tabascensis Morelet, 1884 — Fischer & Crosse (1894a)
Anodonta (Patularia) tabascensis Morelet, 1884 — von Martens (1900d)
Anodonta tabascoensis [sic] Morelet, 1884 — Simpson (1900), de Leon (1912), Cruz & Quesada (2017)
Anodonta tabascensis Morelet, 1884 — Simpson (1914), H.B. Baker (1922)
Anodonta (Pyganodon) grandis-tabascensis Morelet, 1884 — Frierson (1927)
Anodonta (Pyganodon) grandis tabascensis Morelet, 1884 — Haas (1969)

Anodon nopalatensis Sowerby, 1867. Monograph of the genus *Anodon*. *In* L. Reeve and G.B. Sowerby (eds.). *Conchologia Iconica* 17:[unpaginated]+Plates 2–19. Plate 16, species 58. Also figured in Clessin (1876, pl. 73, fig. 3).

Verbatim Type Locality: "River Nopalata [sic], Mexico."

Type Locality: Río Nopalapan (Gulf of México–Papaloapán–San Juan Dr.), Veracruz, México [W=18.1136, -95.3264].

Type specimen: Holotype by monotypy, NHMUK 1965189 (n=½) (Johnson 1971:89).

Remarks: Regarding the type locality, von Martens (1900c:533) noted that "Dr. Seler, of Berlin, who is well versed in Mexican geography and antiquities, tells me that Nopal-apa or Nopal-apan signifies Indian-fig water in the Mexican language, and that there are three localities of that name in Mexico—one on the central plateau, one in the State of Vera Cruz, and one in Tabasco. The second of these, situated in the district of Arayucan, in the southern part of Vera Cruz, on the river San Juan, is probably the correct one." Simpson (1900:646) stated that "Lea's *globosa* is a young shell. Sowerby's *nopalatensis* is the adult. There is a magnificent specimen in the collection of the Philadelphia Academy of Natural Sciences, 7½ inches long and 5½ inches high." von Martens (1901c:654) seems to agree stating that "*A. globosa* is the young and *A. nopalatensis* the adult state of the same species; but the name *A. globosa* must be retained, on account of priority."

Taxonomic History:

Anodonta nopalatensis (Sowerby, 1867) — Clessin (1876), Cruz & Quesada (2017)
Margaron (Anodonta) nopalatensis (Sowerby, 1867) — Lea (1870)
Anodonta (Pyganodon) nopalatensis (Sowerby, 1867) — Fischer & Crosse (1894a)
Anodonta (Patularia) nopalatensis (Sowerby, 1867) — von Martens (1900d, 1901c)
Anodonta (Anodonta) globosa Lea, 1841 — von Martens (1901c)
Anodonta globosa Lea, 1841 — Simpson (1900; 1914), de Leon (1912)
Anodonta globosa nopalatensis (Sowerby, 1867) — H.B. Baker (1922)
Anodonta (Pyganodon) grandis-globosa Lea, 1841 — Frierson (1927)
Anodonta (Pyganodon) grandis globosa Lea, 1841 — Haas (1969)

Order Unionoida
Superfamily Unionoidea Rafinesque, 1820
Family Unionidae Rafinesque, 1820
Subfamily Unioninae Rafinesque, 1820
Tribe Anodontini Rafinesque, 1820
Genus *Sinanodonta* Modell, 1945

Sinanodonta Modell, 1945. Die Anodontinae, Ortm. emend. (Najad., Mollusca). Eine Studie über die Zusammenhänge von Klimazonen und Entwicklungsgeschichte (Klimazonentheorie). *Jenaische Zeitschrift für Medizin und Naturwissenschaft* 78(1):58–100. Page 96.

Diagnosis: "Der Name wurde von bereits in verschiedenen Arbeiten verwendet, deren Drucklegung sich aber durch die Kriegsereignisse verzögert hat, ich wiederhole daher hier die Diagnose: Gattungstypus: *Sinanodonta woodiana* Lea 1834. Schloßgegon Najaden mit bauchiger Schale von Kugeliger bis flache, etwas schägliegende, gegen des Ende starker werdende Doppelbogon, die hinten kurz abgestutzt erscheinen. Verbreitung: Ost-asian, Mesopotamien, Miozän von Kamtschatka. Pliozän von Mittel-und Oeseuropa."

"The name has already been used in various works, but their publication was delayed by the war events, so I repeat the diagnosis here: Genus type: *Sinanodonta woodiana* Lea 1834. Toothless Naiads with a bulbous shell from spherical to flat, slightly sloping, becoming stronger towards the end and appearing short at the back. Distribution: East Asia, Mesopotamia, Miocene of Kamchatka. Pliocene of Central and Eastern Europe."

Type Species: *Symphynota woodiana* Lea, 1834, by original designation.

Taxonomic History:

Margarita (*Anodonta*) Lamarck, 1799 — Lea (1836)
Margaron (*Anodonta*) Lamarck, 1799 — Lea (1852; 1870)
Anodonta (*Sinanodonta*) Modell, 1945 — Adams & Adams (1857)
Anodonta Lamarck, 1799 — Simpson (1900; 1914), Thiele (1934) [in part]
Anodonta (s.s.) Lamarck, 1799 — Haas (1969)
Sinanodonta Modell, 1945 — Modell (1964), Graf & Cummings (2007; 2021), Pfeiffer & Graf (2013), Lopes-Lima et al. (2017; 2020), Williams et al. (2017), Pfeiffer et al. (2019b), Nagel (2025)

PLATE 10 ***Sinanodonta woodiana*** **(Lea, 1834). Top:** *Symphynota woodiana* Lea, 1834. Holotype, USNM 86380 (89 mm L, 56 mm H). Canton, Guangdong Province, China [L=23.1166, 113.2500]. **Bottom:** INHS 2727. Ponds (Pacific–Tempisque Dr.), 10 km S Cañas, Estacion Exp. Enrique Jimenez Nunez, Guanacaste Province, Costa Rica [L=10.3333, -85.1333].

Sinanodonta woodiana (Lea, 1834)

Plate 10, Map 22

Province(s): Gulf of Honduras & Northern Caribbean; Oaxaca, México to Colombia.

Distribution and Status: Widely introduced around the world: Europe, North America, and Asia. Introduced in Caribbean and Pacific drainages in Central America in Costa Rica, El Salvador, Guatemala, and Panamá. LC—Least Concern (IUCN 2024); GNR—No Status Rank (NatureServe 2024).

Anatomy: Shell large, thin, oval, and inflated with swollen beaks. Posterior slope somewhat carinate. Ligament long and somewhat thick. Beaks inflated and the beak cavity shallow. Periostracum dull, dark brown with indistinct rays. Nacre pearly white and iridescent.

Biology: Villalobos and Cruz (1984a; 1984b) studied age and growth in Guanacaste Province, Costa Rica. Cruz and Quesada (2017) later described and figured the glochidia of a species they called *Anodonta luteola* from the same population. The mussels at the facility were later determined to be *Sinanodonta woodiana* (Lea, 1834) and identified in a cox1 DNA barcoding survey of mussels from the aquaculture facility in Guanacaste, Costa Rica (Bauer et al. 2021).

Adult Size: Length to 15 cm.

Similar Species: *Sinanodonta woodiana* (Lea, 1834) co-occurs with other edentate species in the Provinces where it has been found, but not in the same localities.

MAP 22 Distribution of *Sinanodonta woodiana* (Lea, 1834) in Central America.

Symphynota woodiana Lea, 1834. Observations on the naiades; and descriptions of new species of that and other families. *Transactions of the American Philosophical Society* 5(N.S.):23–119 + 19 plates. Page 42, Plate. 5, fig. 13.

Verbatim Type Locality: "Canton."

Type Locality: Canton, Guangdong Province, China [L=23.1166, 113.2500].

Type Specimens: Holotype by monotypy, USNM 86380.

Remarks: Johnson and Baker (1973:175) and Johnson (1974a:158) stated that the figured type was lost.

Taxonomic History:

Margarita (Anodonta) woodiana (Lea, 1834) — Lea (1836)
Anodonta woodiana (Lea, 1834) — Simpson (1900; 1914)
Margaron (Anodonta) woodiana (Lea, 1834) — Lea (1870)
Anodonta (Symphynota) woodiana (Lea, 1834) — Adams & Adams (1857)
Anodon woodianus (Lea, 1834) — Sowerby (1870)
Sinanodonta woodiana woodiana (Lea, 1834) — Modell (1945)
Anodonta (s.s.) *woodiana woodiana* (Lea, 1834) — Haas (1969)
Sinanodonta woodiana (Lea, 1834) — Graf & Cummings (2007; 2021), Pfeiffer & Graf (2013), Williams et al. (2017), Lopes-Lima et al. (2017), Pfeiffer et al. (2019a), Bauer et al. (2021)
Glabaris luteolus (Lea, 1858) — Villalobos & Cruz (1984a–b)
Anodonta luteola Lea, 1858 — Cruz & Quesada (2017)
Sinanodonta cf. *woodiana* (Lea, 1834) — Lopes-Lima et al. (2020)

Order Unionoida
Superfamily Unionoidea Rafinesque, 1820
Family Unionidae Rafinesque, 1820
Subfamily Unioninae Rafinesque, 1820
Tribe Anodontini Rafinesque, 1820
Genus *Utterbackia* F.C. Baker, 1927

Utterbackia F.C. Baker, 1927. On the division of the Sphaeriidae into two subfamilies: and the descriptions of a new genus of Unionidae, with descriptions of new varieties. *American Midland Naturalist* 10(7):220–223. Page 222. As *Utterbachia.*

Diagnosis: "The shape of the umbones is quite different, No *Anodonta* having such flattened beaks. The animal is hermaphroditic and there is no parasitic stage in the development of the glochidium, the process taking place in the marsupium and the young mussel discharged when ready to begin independent life. These characters differentiate *imbecillis* from all other *Anodonta* at present known" (Baker 1927:222).

Type Species: *Anodonta imbecillis* Say, 1829, by original designation.

Taxonomic History:

Anodonta Lamarck, 1799 — Thiele (1934)
Anodonta (s.s.) Lamarck, 1799 — Haas (1969)
Anodonta (*Utterbackia*) F.C. Baker, 1927 — Johnson (1970; 1999a), Heard (1975)
Utterbackia F.C. Baker, 1927 — Baker (1928), Modell (1964), Cummings & Mayer (1992), Williams et al. (1993), Graf & Cummings (2007; 2021), Watters et al. (2009; 2017), Cummings & Graf (2010), Inoue et al. (2014; 2018), Lopes-Lima et al. (2017), Pfeiffer et al. (2019b), Tiemann et al. (2020), Bieler (2021), Nagel (2025)

? Nayadina de Gregorio, 1914. Su taluni molluschi di acqua dolce di America. 31–72+plates 3–12. Page 64.

Type Species: *Anodonta* (*Nayadina*) *venusta* de Gregorio, 1914, by monotypy.

Remarks: If *Anodonta venusta* is a synonym of *A. suborbiculata* Say, 1831, then Gregorio's generic nomen *Nayadina* will have priority over *Utterbackia* (assuming that *imbecillis* and *suborbiculata* belong to the same genus-group) or *Utterbackiana* (assuming that *imbecillis* and *suborbiculata* belong to different genus-groups). However, this would have no effect on generic nomenclature because *Nayadina* is preoccupied. Homonym of the fossil bivalve *Nayadina* Munier-Chalmas, 1864, Family Mallaeidae.

Distribution & Remarks: The genus *Utterbackia*, with three currently recognized species, is widespread and common in eastern North America and ranges from Canada south and east to the Río Grande (Graf and Cummings 2021:15). *Utterbackia imbecillis* (Say, 1829) has been introduced into both the Pánuco and Papaloapán drainages in Central America.

PLATE 11 ***Utterbackia imbecillis* (Say, 1829). Top**: *Anodon horda* Gould, 1855. Lectotype, USNM 678301 (89 mm L, 38 mm H). Comanche Creek (Colorado (TX)–Llano Dr.), Mason County, Texas, USA [W=30.7501, -99.2285]. **Bottom**: *Anodonta henryana* Lea, 1857. Lectotype, USNM 86692 (71 mm L, 38 mm H). Río Grande del Norte, Matamoros, Tamaulipas, México [L=25.8829, -97.4769].

Utterbackia imbecillis (Say, 1829)

Plates 11, Map 23

Province(s): Río Grande del Norte. Introduced in Colorado, Pánuco, Papaloapán provinces.

Distribution and Status: *Utterbackia imbecillis* is widespread in eastern North America. It is native to the Río Grande del Norte basin in Texas, and in Chihuahua, Coahuila, Nuevo León, and Tamaulipas, México. *Utterbackia imbecillis* has been present in portions of the Colorado River since at least the 1980s and is known from two locations: in the lower Bill Williams River National Wildlife Refuge, near Lake Havasu (USNM 892027) and Mittry Lake (Terry Myers, pers. comm. 2023), and were likely introduced during the stocking of non-native fishes (Myers 2009). It was also introduced further south in the Pánuco and Papaloapán basins in San Luis Potosí, Tamaulipas, and Veracruz, México. CS—Currently Stable (Williams et al. 1993:14); LC—Least Concern (IUCN 2024); G5—Secure (NatureServe 2024).

Anatomy: Shell smooth, shiny, extremely thin, transparent in young individuals, elongate, oblong, compressed in juveniles, becoming inflated in adults. Anterior end rounded; posterior end somewhat pointed. Ventral margin straight to moderately rounded. Beaks placed well forward, flattened, and not rising above the hinge line. Beak cavity very shallow. Beak sculpture double-looped, with 5–6 fine, wavy ridges. Periostracum light yellow in young and on the umbo of adults, the rest of the shell is brownish green with fine green rays. Nacre bluish white and iridescent.

Biology: Ortmann (1912:291–292, as *A. henryana*) examined seven specimens from a tributary of the Río Grande, Mercedes, Hidalgo Co., Texas, collected on 15 May 1907. He noted that they were all gravid with glochidia and were 0.29 mm in length and 0.30 mm in height. Gravid females have also been collected from the Río Valles (Pánuco Dr.), Álvaro Obregón, at confluence of the ríos Valles and Tampaón, San Luis Potosí, México on 9 December 2018 (INHS 90089, Plate 11), the Río Carrizal (Pánuco Dr.), Nuevo Progreso, Tamaulipas, México on 27 January 1982 (USNM 1282737), and the Río Salado de Nadadores (Río Grande del Norte Dr.), 0.5 mi S Progresso, Coahuila, México on 12 April 1981 (USNM 1282735). Watters and O'Dee (1998) conducted a study to determine whether inexpensive and easily maintained exotic fishes, amphibians, and crustaceans could serve as hosts for native North American mussels. *Utterbackia imbecillis* successfully metamorphosed on 30 species of exotic fishes or non-piscine hosts in various families including Cichlidae, Characidae, Cyprinidae, and Fundulidae. Nine of these hosts produced metamorphosed juveniles at a level comparable to or above that of Largemouth Bass, *Micropterus salmoides* (Lacepède, 1802). Three exotic host fishes previously reported in the literature, guppy (*Poecilia reticulata* Peters, 1859), goldfish (*Carassius auratus* Linnaeus, 1759), and green swordtail (*Xiphophorus helleri* Heckel, 1848), also served as hosts. However, none resulted in a percentage of metamorphosis comparable to largemouth bass or belontiids.

Adult Size: Length to 10 cm.

Similar Species: *Pyganodon grandis* (Say, 1829) and *Anodonta californiensis* (Lea, 1852).

Remarks: This species has an extensive synonymy. The names below only include the type species and those species described from the southwestern USA and Central America. *Utterbackia imbecillis* has been found to be sister to both *Utterbackia peninsularis* Bogan & Hoeh, 1995 (Hoeh et al. 1995; Inoue et al. 2014; Lopes-Lima et al. 2017; Smith et al. 2018) and *Utterbackia peggyae* (Johnson, 1965) (Hoeh 1990; Zanatta et al. 2007).

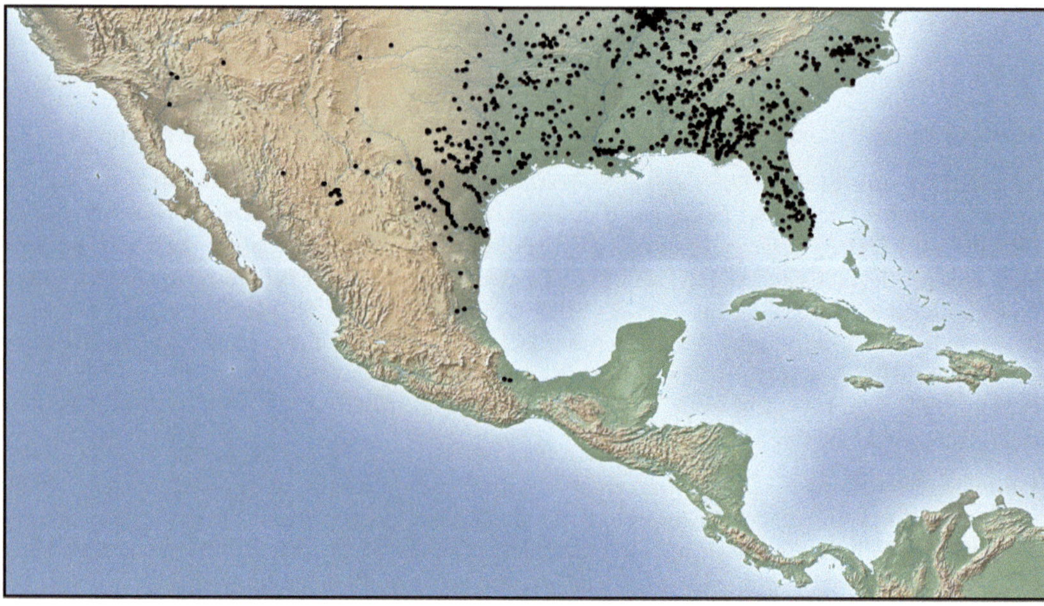

MAP 23 Distribution of *Utterbackia imbecillis* (Say, 1829).

Anodonta imbecillis Say, 1829. Descriptions of some new terrestrial and fluviatile shells of North America. (cont.). The Disseminator of Useful Knowledge; containing hints to the youth of the U. States, from the School of Industry 2(23):355–356. 18 November 1829. Page 355. Not figured. Figured in Sowerby (1870, pl. 27, sp. 102).

Verbatim Type Locality: "Wabash River."

Type Locality: Wabash River (Gulf of México–Mississippi–Ohio Dr.), Indiana, USA [W=38.1345, -87.9407].

Type Specimens: Unknown. Presumed lost.

Remarks: Probably described from the Wabash River basin near New Harmony, Posey County, Indiana. The Neotype designation of SMF 4301 by Haas (1930:326) is invalid due to the selection of a specimen from "North America" and not from the Wabash River. Evidence that the neotype came as nearly as practicable from the original type locality is a qualifying condition under Article 75.3.6 of the International Code of Zoological Nomenclature (ICZN).

Taxonomic History:

Anodonta imbecillis Say, 1829 — Ferussac (1835), Say (1840), Simpson (1914), de Leon (1912), Howells et al. (1996), Contreras-Arquieta (2000)

Margaron (Anodonta) imbecillis (Say, 1829) — Lea (1852; 1870)

Anodonta imbecilis [sic] Say, 1829 — Adams & Adams (1857), Neck & Metcalf (1988)

Anodon imbecillis (Say, 1829) — Sowerby (1870)

Utterbackia imbecillis (Say, 1829) — F.C. Baker (1927), Graf & Cummings (2007; 2021), Cummings & Graf (2010), Williams et al. (2017), Kiser et al. (2022), Czaja et al. (2023)

Anodonta (*Lastena*) *imbecillis* Say, 1829 — Haas (1930)
Anodonta (*Utterbackia*) *imbecilis* [sic] Say, 1829 — Johnson (1970), Heard (1975)
Anodonta (*Utterbackia*) *imbecillis* Say, 1829 — Johnson (1999a)

Anodon horda Gould, 1855. New species of land and fresh-water shells from western (N.) America. (Cont.). *Proceedings of the Boston Society of Natural History* 5(15):228–229. Page 229. Figured in Sowerby (1867, pl. 18, sp. 66) and Johnson (1975:30, pl. 2, fig. 1).

Verbatim Type Locality: "Comanche Creek, Texas."

Type Locality: Comanche Creek (Colorado (TX)–Llano Dr.), Mason County, Texas [W=30.7501, -99.2285].

Type Specimens: Lectotype, NHMUK 196465 designated by Johnson (1964:88, as uncatalogued; Johnson 1971:85, as a paratype). Paralectotype, USNM 678301 (Johnson 1970:363; 1971:85; 1975:30, 54, pl. 2, fig. 1, all as the holotype).

Remarks: There is some confusion regarding the lectotype of *Anodon horda* Gould, 1855. Johnson (1964:83) designated the specimen figured in Sowerby (1867, pl. 18, sp. 66) (NHMUK 196465, uncatalogued at the time) as the lectotype. Sowerby did not give any locality information, but the catalogue entry reads "Comanche Creek, Texas." It is unknown if Johnson simply presumed that was where it was from or if in fact it was collected there. Johnson subsequently (1970:363; 1971:85; 1975:30) stated that the "measured holotype" was USNM 678301 and that he "selected a lectotype for this species before the type was located; it is now invalid." According to Article 74.1.1 of the Code, "The valid designation of a lectotype fixes the status of the specimen as the sole name-bearing type of that nominal taxon; no later designation of a lectotype has any validity," and the designation of NHMUK 196465 stands.

An annotated label with the paralectotype at the USNM reads [trib. of the Pecos, Pecos Co.]. Presumably, those data were added by Johnson when he compiled his catalogue of Gould's names. Subsequent authors have placed the type locality as being near the town of Mason, in Mason County, Texas (Johnson 1970:363; Johnson 1975:30; Taylor 1967:153).

Taxonomic History:
Anodon hordeum [sic] Gould, 1855 — Sowerby (1867)
Margaron (*Anodonta*) *imbecillis* (Say, 1829) — Lea (1870)
Anodonta imbecillis Say, 1829 — Simpson (1900; 1914), de Leon (1912), Howells et al. (1996)
Anodonta (*Lastena*) *ohiensis-horda* Gould, 1855 — Frierson (1927)
Anodonta ohiensis horda Gould, 1855 — Strecker (1931)
Anodonta (*Lastena*) *ohiensis* Rafinesque, 1820 — Haas (1969)
Anodonta (*Utterbackia*) *imbecilis* [sic] Say, 1829 — Johnson (1970)
Utterbackia imbecillis (Say, 1829) — Williams et al. (2008)
Anodonta horda Gould, 1855 — Contreras-Arquieta (2000)

Anodonta henryana Lea, 1857. Description of six new species of fresh water and land shells of Texas and Tamaulipas, from the collection of the Smithsonian Institution. *Proceedings of the Academy of Natural Sciences of Philadelphia* 9:101–102. Page 102, as *Anodonta henryiana*. Figured in Lea (1860b JANSP Vol. 4, Part 4(13):373; 1860d Obs. 8:55, pl. 66, fig. 198).

Verbatim Type Locality: "Matamoras, Tamaulipas, Mexico."

Type Locality: Río Grande del Norte (Gulf of Mexico Dr.), Matamoros, Tamaulipas, México [L=25.8829, -97.4769].

Type Specimens: Lectotype, USNM 86692 designated by Johnson (1970:363; 1974a:69, both as the holotype; Johnson 1999a:14, 52, pl. 1, fig. 1). Also, ANSP A14548 ($n=2$, soft parts in alcohol).

Remarks: According to Bieler (2021:12), Lea's 1860 correction of the original spelling of *henryiana* to *henryana* does not fulfill the requirements of a demonstrably intentional emendation under Art. 33.2 of the International Code of Zoological Nomenclature (Lea referenced the original publication, but did not cite the original spelling). Although almost all subsequent authors followed the secondary spelling (see synonymy below), in his opinion, it did not meet the level of prevailing usage as he interpreted the Code (1999). Bieler considered *Anodonta henryana* an incorrect subsequent spelling (note that an unjustified emendation would have resulted in a junior homonym under Art. 58.15).

Taxonomic History:

Anodonta henryana Lea, 1857 — Lea (1860b), Clessin (1874b), Paetel (1890), Simpson (1896; 1900; 1914), Ortmann (1912), Morrison (1968), Heard (1975), Contreras-Arquieta (2000), Cruz & Quesada (2017)
Margaron (Anodonta) henryana (Lea, 1857) — Lea (1870)
Anodon henryana (Lea, 1857) — Sowerby (1870)
Anodonta (Mesanodon) henryana Lea, 1857 — Fischer & Crosse (1894a)
Anodonta (s.s.) *henryana* Lea, 1857 — von Martens (1900d)
Anodonta heuryana [sic] Lea, 1857 — de Leon (1912)
Utterbackia henryana (Lea, 1857) — F.C. Baker (1927)
Anodonta (Lastena) ohiensis Rafinesque, 1820 — Haas (1969)
Anodonta (Utterbackia) imbecilis [sic] Say, 1829 — Johnson (1970)

Order Unionoida
Superfamily Unionoidea Rafinesque, 1820
Family Unionidae Rafinesque, 1820
Subfamily Ambleminae Rafinesque, 1820
Tribe Lampsilini von Ihering, 1901
Genus *Actinonaias* Crosse & Fischer, 1894

Actinonaias Crosse & Fischer, 1894. *In* P. Fischer and H. Crosse. 1894. Études sur les mollusques terrestres et fluviatiles du Mexique et du Guatemala. *In* M.H.–Milne Edwards (ed.). *Mission Scientifique au Mexique et dans l'Amérique centrale. Recherches Zoologiques pour servir á l'Histoire de la Faune de l'Amérique Centrale et du Mexique.* 15th Livraison. Feuilles 62–72, pp. 489–576, plates 59–62, Imprimerie Nationale, Paris Volume 2. 731 pp. Page 556. Introduced as a subgenus of *Unio* Philipsson, 1788.

Diagnosis: "Coquille subovale, très-épaisse, renflée; surface ornée de rayons étroits, nombreux; dents latérales antérieures très-fortes (*U. Sapotalensis*, Lea)." [Shell subovate, very thick, swollen; surface adorned with numerous narrow rays; very strong anterior lateral teeth (*U. Sapotalensis*, Lea)].

Type Species: *Unio sapotalensis* Lea, 1841, by original designation.

Taxonomic History:

Unio (Lampsilis) Rafinesque, 1820 — von Martens (1900a) [in part]
Nephronaias Crosse & Fischer, 1894 — Simpson (1900; 1914), von Ihering (1901) [in part]
Lampsilis Rafinesque, 1820 — Pilsbry (1910), Simpson (1914) [in part]
Actinonaias Crosse & Fischer, 1894 — Frierson (1917), Thiele (1934), Morrison (1968), Haas (1969), Graf & Cummings (2007; 2021), Pfeiffer et al. (2019b)
Lampsilis (Actinonaias) Crosse & Fischer, 1894 — Frierson (1927), Haas (1929)
Lampsilis (Friersonia) Ortmann (1912) — Haas (1929)

Graphonaias Crosse & Fischer, 1894. *In* P. Fischer and H. Crosse. 1894. Études sur les mollusques terrestres et fluviatiles du Mexique et du Guatemala. *In* M.H.–Milne Edwards (ed.). *Mission Scientifique au Mexique et dans l'Amérique centrale. Recherches Zoologiques pour servir á l'Histoire de la Faune de l'Amérique Centrale et du Mexique*. 15th Livraison. 62–72, pp. 489–576, plates 59–62, Imprimerie Nationale, Paris Volume 2. 731 pp. Page 557. Introduced as a subgenus of *Unio* Philipsson, 1788.

Diagnosis: "Coquille oblongue, allongée, peu convexe, inéquilatérale, lisse, souvent rayonnée; côté postérieur arrondi ou subtronqué (*U. Medellinus*, Lea)." [Shell oblong, elongated, slightly convex, unequal, smooth, often radiated; posterior side rounded or subtruncated (*U. Medellinus*, Lea)].

Type Species: *Unio medellinus* Lea, 1838, by original designation.

Taxonomic History:

Actinonaias Crosse & Fischer, 1894 — Thiele (1934), H.B. Baker (1922)
Nephronaias (s.s.) Crosse & Fischer, 1894 — Haas (1969)

Distribution & Remarks: The genus *Actinonaias* is widely distributed along the Gulf Coast of México (Map 24). Various nomenclatorial changes were made in the genus *Actinonaias* since the 2021 checklist of Graf and Cummings. They recognized seven species in the genus in Central America. We sunk *Actinonaias computata* (Crosse & Fischer, 1894) into *Actinonaias sapotalensis* (Lea, 1841) and *Actinonaias undivaga* (Pilsbry, 1910) into *Actinonaias signata* (Pilsbry, 1910). We also placed *Actinonaias moctezumensis* (Pilsbry, 1910) under *Disconaias fimbriata* (Frierson, 1927). The four remaining species recognized here as valid are all endemic to Gulf Coast of México drainages from the Río Sabinas (Pánuco–Tamesí basin) in Tamaulipas south to the Coatzacoalcos basin in Veracruz, México. No phylogenetic analysis has ever tested the monophyly (or non-monophyly) of the species formerly classified as *Actinonaias*. However, the biogeography of the two groups of species (*Actinonaias sapotalensis*, etc. of Central America vs. "*Actinonaias*" *ligamentina* (Lamarck, 1819) and "*A.*" *pectorosa* (Conrad, 1834) of North America) represents an unprecedented disjunction between the Mesoamerican and Nearctic faunal provinces. Separating the latter two species into *Ortmanniana* and leaving *Actinonaias* to *A. sapotalensis* and other Central American species makes biogeographical sense (Graf and Cummings 2021:18).

MAP 24 Distribution of the genus *Actinonaias* Crosse & Fischer, 1894.

PLATE 12 ***Actinonaias sapotalensis* (Lea, 1841). Top Left:** *Unio sapotalensis* Lea, 1841. Holotype, USNM 85072 (56 mm L, 36 mm H). Río Papaloapán, Tlacotalpan, Veracruz, México [L = 18.6089, -95.6590]. **Middle Right**: *Unio computatus* Crosse & Fischer, 1894. Lectotype, MNHN-IM-2000-1662 (77 mm L, 52 mm H). Río Coatzacoalcos (Gulf of Mexico Dr.), Coatzacoalcos, Veracruz, México [L = 18.1500, -94.4166]. **Bottom Left:** *Lampsilis ruthveni* H.B. Baker, 1922. Holotype, UMMZ 34891 (59 mm L, 38 mm H). Río San Juan (Papaloapán Dr.), Cuatotolapam, Veracruz, México [L = 18.1241, -95.2998].

Actinonaias sapotalensis (Lea, 1841)

Plate 12, Map 25

Province(s): Papaloapán.

Distribution and Status: Known from the Coatzacoalcos and Papaloapán basins in Veracruz, México. Not assessed (IUCN 2024). *Actinonaias sapotalensis* was found at six sites in the Papaloapán basin in 1982, but no live specimens were collected in a limited survey of the drainage in 2022.

Anatomy: Shell smooth, solid, elliptical, and somewhat inflated. Beaks low and scarcely prominent. Beak cavity not deep. Left valve with two stumpy pseudocardinals and two strong laterals. Right valve with one pseudocardinal and one lateral. Periostracum yellow or chestnut-colored with radiating green rays. Nacre white or pink and iridescent.

Ortmann noted that the soft parts were "In every respect like the two preceding species [*Ortmanniana ligamentina* and *O. pectorosa*]. Mantle-connection between anal and supra-anal shorter than anal, the latter with crenulations. the branchial with papillae. Inner mantle edge of the female in front of branchial very slightly lamellate, with fine crenulations, and a black streak accompanying it. Palpi with posterior margins connected at base. Inner lamina of inner gills entirely connected with abdominal sac. Marsupium formed by numerous (twenty and more) water-tubes, occupying a section of the posterior half of the outer gill. In the sterile females at hand, the margin of the marsupium is rounded, projecting slightly beyond the edge of the gill, beaded, and marked with brownish black pigment. Placentae and glochidia unknown. Color whitish, with the margin of the mantle brown and blackish in the region of the anal and branchial, a blackish streak in front of the branchial, and black pigment on the edge of the marsupium."

MAP 25 Distribution of *Actinonaias sapotalensis* (Lea, 1841).

Biology: Ortmann (1912:326–327, text fig. 21, as *Nephronajas sapotalensis*) reported three males and two sterile females, collected on 23 July 1910 from the Río Hueyapan, Hacienda de Cuatotolapam, Alayucan Municipality, Veracruz, México.

Adult Size: Length to 6 cm.

Similar Species: *Actinonaias medellina* (Lea, 1838).

Unio sapotalensis Lea, 1841. On fresh water and land shells (cont.). *Proceedings of the American Philosophical Society* 2(17):30–34. Page 30. Figured in Lea (1842a TAPS Vol. 8:233; 1842b Obs. 3:71, pl. 21, fig. 47) and Sowerby (1868, pl. 91, sp. 495).

Verbatim Type Locality: "Sapotal River, near Tlocatalpam [sic], Mex."

Type Locality: Río Papaloapán (Gulf of Mexico Dr.), Tlacotalpan, Veracruz, México [L = 18.6089, -95.6590].

Type specimen: Holotype by monotypy, USNM 85072 (Johnson and Baker 1973:169; Johnson 1974a:129, incorrectly as USNM 84072).

Remarks: Ortmann (1912:326) writing about specimens he received from A.G. Ruthven from the Río Hueyapan (CM 61.5000) stated that "B. Walker writes to me about these specimens: they 'agree exactly with Crosse and Fischer's figure of their *computatus*, which according to von Martens is probably only a variety of *sapotalensis*, differing mainly in having the pseudocardinals slender, while in *sapotalensis* they are heavy.' The type locality of *sapotalensis* is Sapotal River, near Tlocatalpam [sic], Mexico. This is not far from the locality of my specimens, and in the same general drainage system (Papaloapán and San Juan Rivers)." Tlacotalpan is on the Río Papaloapán.

Taxonomic History:

Margaron (*Unio*) *sapotalensis* (Lea, 1841) — Lea (1852; 1870)
Unio sapotalensis Lea, 1841 — Conrad (1853), Sowerby (1868)
Unio (*Hyridella*) *sapotalensis* Lea, 1841 — Adams & Adams (1857)
Unio (*Actinonaias*) *sapotalensis* Lea, 1841 — Fischer & Crosse (1894b)
Unio (*Lampsilis*) *sapotalensis* Lea, 1841 — von Martens (1900b)
Nephronaias sapotalensis (Lea, 1841) — Simpson (1900; 1914), de Leon (1912)
Nephronajas [sic] *sapotalensis* (Lea, 1841) — Ortmann (1912)
Actinonaias sapotalensis (Lea, 1841) — Frierson (1917), Thiele (1934), Haas (1969), Graf & Cummings (2007; 2021)
Lampsilis (*Actinonaias*) *sapotalensis* (Lea, 1841) — Frierson (1927)
Actinonaias sapotalensis sapotalensis (Lea, 1841) — H.B. Baker (1922),

Unio computatus Crosse & Fischer, 1894. Diagnoses molluscorum novorum, reipublicæ Mexicanæ incolarum. *Journal de Conchyliologie* 41(4):293–297. Page 295. Figured in Fischer and Crosse (1894b:601, pl. 67, figs. 2–2a).

Verbatim Type Locality: "Goatzalcoalcos, in provincia Vera Cruz dicta, republicæ Mexicanæ."

Type Locality: Río Coatzacoalcos (Gulf of Mexico Dr.), Coatzacoalcos, Veracruz, México [L = 18.1500, -94.4166].

Type Specimens: Lectotype, MNHN-IM-2000-1662, here designated (the figured specimen).

Remarks: von Martens (1900b:513) noted that "Perhaps only a variety of the preceding [*sapotalensis*], from which it differs in the larger size, the more swollen shell, and the more feeble cardinal teeth." The date of publication is often cited as 1893 but is 1894 according to Winckworth (1936:156).

Taxonomic History:

Unio (Actinonaias) computatus Crosse & Fischer, 1893 — Fischer & Crosse (1894b)
Unio (Lampsilis) computatus Crosse & Fischer, 1893 — von Martens (1900b)
Nephronaias computata (Crosse & Fischer, 1893) — Simpson (1914)
Actinonaias sapotalensis computata (Crosse & Fischer, 1893) — H.B. Baker (1922)
Lampsilis (Actinonaias) computata (Crosse & Fischer, 1893) — Frierson (1927)
Actinonaias computata (Crosse & Fischer, 1893) — Haas (1969), Graf & Cummings (2021), Czaja et al. (2023)
Actinonaias computatus (Crosse & Fischer, 1893) — Graf & Cummings (2007)

Lampsilis ruthveni H.B. Baker, 1922. The Mollusca collected by the University of Michigan-Walker Expedition in Southern Vera Cruz, México. I. Occasional Papers of the Museum of Zoology, University of Michigan No. 106. 95 pp. Page 29, pl. 11, fig. 53 (hinge); pl. 12, fig. 53 (Exterior view of left valve and hinge armature of right; dorsal view of entire shell); pl. 13, figs. 51–54, fig. 51, as the "type (female?) shell."

Verbatim Type Locality: "Rio San Juan."

Type Locality: Río San Juan (Gulf of México–Papaloapán Dr.), Cuatotolapam, Veracruz, México [L=18.1241, -95.2998].

Type Specimens: Holotype, UMMZ 34891 by original designation (as the "type" without a catalogue number; Johnson and Baker 1973:169; Johnson 1979:34). Paratypes, ANSP 133696 (*n*=2) (Johnson and Baker 1973:169); MCZ 319588 (*n*=1½), ex. UMMZ 34892); UMMZ 34892 (*n*=1½).

Remarks: Baker (1922:29) remarked that he had "seven specimens, including odd valves (2)" and figured a "type" but did not give catalogue numbers.

Taxonomic History:

Lampsilis (Actinonaias) ruthveni H.B. Baker, 1922 — Frierson (1927)
Actinonaias computata (Crosse & Fischer, 1893) — Haas (1969)

PLATE 13 ***Actinonaias coyensis*** **(Pilsbry, 1910). Top Left**: *Lampsilis coyensis* Pilsbry, 1910. Lectotype, ANSP 99522a (36 mm L, 21 mm H). **Top Right & Bottom**: Paralectotypes, INHS 21726 & INHS 15231. All, Río Coy (Pánuco Dr.), San Luis Potosí, México [W=21.7497, -98.9573]. Specimen sizes range from 15 to 30 mm.

Actinonaias coyensis (Pilsbry, 1910)

Plate 13, Map 26

Province(s): Pánuco.

Distribution and Status: Endemic to the Pánuco basin in San Luis Potosí and Tamaulipas, México. *Actinonaias coyensis* was common in a survey of the Pánuco basin, San Luis Potosí, México, in 1982 and 2018. Not assessed (IUCN 2024).

Anatomy: Shell moderately thick and oblong. Beaks low and located at the anterior fourth of the shell with minute, recurved tips sculptured with a few coarse wrinkles. Anterior end rounded. Shell surface smooth but not glossy. Periostracum greenish yellow, indistinctly marked with green rays. Near the beaks the rays are very distinct on a light background. Nacre white and somewhat thick anteriorly, thin, blue and iridescent in the posterior half.

Biology: A female gravid with eggs only was collected from the Rio Valles, Estacion Micos, 21 km NW of Valles, San Luis Potosí, México on 8 December 2018 (INHS 90393).

Adult Size: Length to 3.5 cm.

Similar Species: *Actinonaias medellina* (Lea, 1838) and *Actinonaias signata* (Pilsbry, 1910).

MAP 26 Distribution of *Actinonaias coyensis* (Pilsbry, 1910).

Lampsilis coyensis Pilsbry, 1910. Unionidae of the Panuco River system, Mexico. *Proceedings of the Academy of Natural Sciences of Philadelphia* 61(3):532–539 + 3 plates. Page 538, plate 27, figs. 3–4.

Verbatim Type Locality: "Coy River, State of San Luis Potosi, Mexico."

Type Locality: Río Coy (Gulf of México–Pánuco–Tampaón Dr.), San Luis Potosí, México [W = 21.7497, -98.9573].

Type Specimens: Lectotype, ANSP 99522a designated by Johnson and Baker (1973:152, as the holotype); Paralectotypes, ANSP 99522 (*n*=4), FMNH 68098 (*n*=1); INHS 15231 (*n*=5); INHS 21726 (*n*=1).

Remarks: Pilsbry (1910:538) gave only a single set of measurements and figured a single specimen, but he did not explicitly indicate a "type." He mentioned other specimens from the same collection.

Taxonomic History:

Lampsilis coyensis Pilsbry, 1910 — Simpson (1914)
Lampsilis (*Actinonaias*) *coyensis* Pilsbry, 1910 — Frierson (1927)
Lampsilis (*Friersonia*) *coyensis* Pilsbry, 1910 — Haas (1929)
Actinonaias coyensis (Pilsbry, 1910) — Haas (1969), Graf & Cummings (2007; 2021), Kiser et al. (2022), Czaja et al. (2023)

Lampsilis (*Friersonia*) *vallesensis* Haas, 1929. Bemerkungen über mittelamerikanische Najaden. *Senckenbergiana* 11(5/6):310–344. Page 317.

Remarks: *Nomen nudem.* "Die von mir untersuchten Stücke waren als var. *vallesensis* Pilsbry der *coyensis* bestimmt; meines Wissens ist aber ein solcher Varietätsnamen niemals veröffentlicht worden." Haas (1929:317) noted that the specimens he studied were designated as var. *vallesensis* Pilsbry of *coyensis*, but to his knowledge, no such variety name had ever been published.

PLATE 14 ***Actinonaias medellinus* (Lea, 1838). Top**: *Unio medellinus* Lea, 1838. Lectotype, ANSP 9679 (58 mm L, 33 mm H). **Bottom**: *Unio veracruzensis* Lea, 1868. Holotype, USNM 25432 (30 mm L, 18 mm H). Both, Río Medellin (Jamapa Dr.), Veracruz, Veracruz, México [L = 19.0574, -96.1550].

PLATE 15 ***Actinonaias medellina*** **(Lea, 1838). Top:** INHS 90159.1 (90 mm L, 52 mm H) and **Bottom:** INHS 90159.2 (64 mm L, 37 mm H). Rio Tampaón (Pánuco Dr.), 4.2 km ESE of Tamuin, San Luis Potosí México [L=21.9923, -98.7418]. 15 December 2018.

PLATE 16 ***Actinonaias medellina*** **(Lea, 1838). Top:** INHS 90159.4 (30 mm L, 17 mm H). **Bottom:** INHS 90159.5 (41 mm L, 25 mm H). Both, Rio Tampaón (Pánuco Dr.), 4.2 km ESE of Tamuin, San Luis Potosí México [L = 21.9923, -98.7418]. 15 December 2018.

PLATE 17 ***Actinonaias medellina*** **(Lea, 1838). Top:** INHS 90090.2 (66 mm L, 34 mm H). Rio Valles (Pánuco Dr.), Álvaro Obregón, at confluence of the rios Valles and Tampaón [L=21.8461, -98.9438]. 9 December 2018. **Bottom:** INHS 90203.11 (43 mm L, 25 mm H). Canal of Rio Tampaón (Pánuco Dr.), 23 km E of Tamuin [L=21.9692, -98.5596]. 15 December 2018. Both, San Luis Potosí, México.

Actinonaias medellina (Lea, 1838)

Plates 14–17, Map 27

Province(s): Pánuco; Papaloapán.

Distribution and Status: Widespread in central Gulf of Mexico drainages in México from the Pánuco basin south to the Papaloapán drainage in Veracruz. Not assessed (IUCN 2024). *Actinonaias medellina* was common in surveys of the Pánuco and Papaloapán basins in San Luis Potosí and Veracruz, México in 1982 and at a few sites 2017–2018 and 2022.

Anatomy: Shell oval to slightly elongate and moderately thick. Occasionally compressed in juveniles and moderately inflated in adults. The ventral margin is typically straight in adults and slightly convex in juveniles. Left valve with two pseudocardinals and one lateral. Right valve with one pseudocardinal and one lateral. Periostracum yellowish green with green rays particularly in young individuals. Nacre white and iridescent, with a yellowish or rosy hue in the center.

Biology: Gravid females with fully developed glochidia were collected from México in the Río Panucó drainage at five localities: the Río Tampaón, Álvaro Obregón, at confluence of the ríos Valles and Tampaón on 30 January 1982 (USNM 1282277); the Rio Valles, Álvaro Obregón, at confluence of the ríos Valles and Tampaón on 9 December 2018 (INHS 90090, Plate 17); the Río Valles, Estacion Micos, 21 km NW of Valles on 8 December 2018 (INHS 90398) (all in San Luis Potosí México); and the Río Tempoal, at Mex. Rt. 105 between Tempoal and Platón Sánchez, Veracruz, on 8 February 1982 (USNM 128288). Gravid females were also found in the Río Pantepec (a direct tributary to the Gulf of México), at Alamo, Veracruz, on 23 February 1982 (USNM 1282434), and from a tributary to the Río Actopan in the Río Papaloapán basin, 3 km above the Mex. Rt. 180 bridge, above La Gloria, Veracruz, México on 10 February 1982 (USNM 1282290).

Adult Size: Length to 7.5 cm.

Similar Species: Compare to *Actinonaias coyensis* (Pilsbry, 1910) and *Actinonaias signata* (Pilsbry, 1910) in the Pánuco and to *Actinonaias sapotalensis* in the Papaloapán.

MAP 27 Distribution of *Actinonaias medellina* (Lea, 1838).

Unio medellinus Lea, 1838. Description of new freshwater and land shells. *Transactions of the American Philosophical Society* 6(N.S.):1–154+plates 1–24. Page 39, plate 12, fig. 34. Also figured in Sowerby (1866, pl. 33, sp. 171) and von Martens (1900e, pl. 39, figs. 7, 7a–b, ZMB/MOLL.305847).

Verbatim Type Locality: "River Medellin, near Vera Cruz."

Type Locality: Río Medellin (Gulf of México–Jamapa Dr.), Veracruz, Veracruz, México [L=19.0574, -96.1550].

Type Specimens: Lectotype, ANSP 9679 designated by Johnson and Baker (1973:161, as the holotype). Paralectotypes, MCZ 178881 (*n*=2) (Johnson 1956:124, as paratypes).

Taxonomic History:

Margarita (Unio) medellinus — Lea (1836) *nomen nudem*
Unio medellinus Lea, 1838 — Hanley (1843; 1856), Küster (1861), Sowerby (1866)
Margaron (Unio) medellinus (Lea, 1838) — Lea (1852; 1870)
Unio (Hyridella) medellinus Lea, 1838 — Adams & Adams (1857)
Unio (Graphonaias) medellinus Lea, 1838 — Fischer & Crosse (1894a)
Unio (Elliptio) medellinus Lea, 1838 — von Martens (1900b)
Nephronaias medellinus (Lea, 1838) — Simpson (1900), von Ihering (1901), de Leon (1912)
Nephronaias medellina (Lea, 1838) — Simpson (1914)
Actinonaias medellinus (Lea, 1838) — Frierson (1917)
Lampsilis (Actinonaias) meddellina [sic] (Lea, 1838) — Frierson (1927)
Lampsilis (Actinonaias) medellina (Lea, 1838) — Haas (1929)
Actinonaias medellina (Lea, 1838) — Thiele (1934), Haas (1969), Graf & Cummings (2007; 2021), Kiser et al. (2022), Czaja et al. (2023)

Unio veracruzensis Lea, 1868. Descriptions of four new species of exotic Unionidae. *Proceedings of the Academy of Natural Sciences of Philadelphia* 20:150. Page 150. Figured in Lea (1868b JANSP Vol. 6:320; 1869a Obs.12:80, pl. 52, fig. 133).

Verbatim Type Locality: "Vera Cruz, Mexico."

Type Locality: Río Medellin (Gulf of México–Jamapa Dr.), Veracruz, Veracruz, México [L=19.0574, -96.1550].

Type specimen: Holotype by monotypy, USNM 25432 (Johnson 1974a:153).

Remarks: Although everyone after Simpson (1900) has considered this taxon as a synonym of *Unio umbrosus* Lea, 1856, we agree with von Martens (1900b:518) who wrote that "*U. veracruzensis* Lea, is perhaps the young state of *U. medellinus*; it often happens in this genus, and especially in the present species, that the ventral margin is more convex in young specimens than in old ones." *Unio veracruzensis* has obvious green rays on the shell that are lacking in *Mesonaias umbrosa*.

Taxonomic History:

Margaron (Unio) veracruzensis (Lea, 1868) — Lea (1870)
Unio (Graphonaias) veracruzensis Lea, 1868 — Fischer & Crosse (1894a)
Unio (Elliptio) medellinus Lea, 1838 — von Martens (1900b)
Lampsilis (Proptera) umbrosus Lea, 1856 — Simpson (1900)
Lampsilis umbrosus Lea, 1856 — de Leon (1912)
Lampsilis (Proptera) umbrosa Lea, 1856 — Simpson (1914)
Lampsilis (Cyrtonaias) umbrosa Lea, 1856 — Frierson (1927)
Lampsilis (Cyrtonaias) umbrosa umbrosa Lea, 1856 — Haas (1969)

PLATE 18 ***Actinonaias signata*** **(Pilsbry, 1910). Top:** *Lampsilis signata* Pilsbry, 1910. Paralectotype, ANSP 99525. Río Valles, Río Valles Falls, 4.8 km S Valles [L=21.9372, -98.9893]. **Bottom:** *Lampsilis undivaga* Pilsbry, 1910. Lectotype, ANSP 99520a (54 mm L, 30 mm H). Río Valles, 3.2 km above Micos [L=22.1163, -99.1663]. Both, Pánuco basin, San Luis Potosí, México.

Actinonaias signata (Pilsbry, 1910)

Plate 18, Map 28

Province(s): Pánuco.

Distribution and Status: Found in the Pánuco and adjacent basins in San Luis Potosí, Tamaulipas, and Veracruz, México. *Actinonaias signata* was common in Arroyo los Gatos (Pánuco Dr.), Nuevo Morelos, Tamaulipas, México, in 1978 and in the Rió Valles (Pánuco Dr.), Estacion Micos, 21 km NW of Valles, San Luis Potosí, México in 2018. Not assessed (IUCN 2024).

Anatomy: Shell oblong and twice as long as wide. Valves rather thick in the anterior half, becoming abruptly thinner posteriorly. Anterior end rounded; posterior end truncated. Beaks low and situated near the anterior fourth of the shell with acute fine radial wrinkles on the posterior-dorsal slope. Beaks sculptured with two radial series of short concentric wrinkles separated by a slight depression. Pseudocardinal teeth stout, the anterior one in the left valve wedge-shaped, the posterior one lying directly under the beak. Lateral teeth double. Right valve with a stout, rugose pseudocardinal and a very small, compressed laminar tooth above it. Lateral tooth single. Periostracum dark brown to black, with numerous faint green rays, fading out towards the ventral edge. Beaks with distinct green rays in some individuals. Nacre purple or bluish white, and iridescent posteriorly.

Biology: Gravid females were collected from the Río Valles, Estacion Micos, San Luis Potosí, México on 8 December 2018 (INHS 90394).

Adult Size: Length to 7.5 cm.

Similar Species: *Actinonaias coyensis* (Pilsbry, 1910) and *Actinonaias medellina* (Lea, 1838).

MAP 28 Distribution of *Actinonaias signata* (Pilsbry, 1910).

Lampsilis signata Pilsbry, 1910. Unionidae of the Panuco River system, Mexico. *Proceedings of the Academy of Natural Sciences of Philadelphia* 61(3):532–539 + 3 plates. Page 536, pl. 25, figs. 5–6.

Verbatim Type Locality: "Valles River, 'near a small falls about three miles below Valles, in the muddy banks of a pool caused by low water. Feb. 4, 1909' (A.A.H.) A few smaller specimens were taken in the Moctezuma River, Casas Viejas River and Tampamolon Creek."

Type Locality: Río Valles (Gulf of México–Pánuco–Tampaón Dr.), Río Valles Falls, 4.8 km S Valles, San Luis Potosí, México [L=21.9372, -98.9893].

Type Specimens: Lectotype, ANSP 99529a (from a lot of 6 valves) designated and type locality fixed by Johnson and Baker (1973:170, as the holotype). Paralectotypes, ANSP 99525 (*n*=27); ANSP 99528 (*n*=2); ANSP 99529 (*n*=2); CHAS 25171 (*n*=22); FMNH 21849 (*n*=4); FMNH 68332 (*n*=4); FMNH 68416 (*n*=1); INHS 82201 (*n*=6); MCZ 192367 (*n*=10) (Johnson 1956:135, as cotypes), all from the type locality. ANSP 99530 (*n*=1); INHS 21747 (*n*=2); INHS 21748 (*n*=4), from the Río Casas Viejas, San Luis Potosí. ANSP 99527 (*n*=6); INHS 21751 (*n*=1); INHS 21752 (*n*=3), from Tampamolón Creek, San Luis Potosí. INHS 21750 (*n*=1), from the Río Moctezuma, S of Tampamolón.

Remarks: Pilsbry indicated that he had multiple specimens, although he did not designate any of them as the "type" as he had done for other species he described in his paper. He figured an adult specimen (pl. 25, figs. 5–6; Plate 18 above) and "A young shell with the corrugation unusually well developed" (Pilsbry 1910:536, pl. 26, figs. 5–6).

Taxonomic History:

Lampsilis signata Pilsbry, 1910 — Simpson (1914)
Lampsilis (Actinonaias) signata Pilsbry, 1910 — Frierson (1927)
Lampsilis (Friersonia) signata Pilsbry, 1910 — Haas (1929)
Actinonaias signata (Pilsbry, 1910) — Haas (1969), Graf & Cummings (2007; 2021), Kiser et al. (2022), Czaja et al. (2023)

Lampsilis undivaga Pilsbry, 1910. Unionidae of the Panuco River system, Mexico. *Proceedings of the Academy of Natural Sciences of Philadelphia* 61(3):532–539+3 plates. Page 536, pl. 26, figs. 1–4. Lectotype, figs. 3–4.

Verbatim Type Locality: "Valles River two miles above Mecos, in a muddy bank just above rapids."

Type Locality: Río Valles (Gulf of México–Pánuco–Tampaón Dr.), 3.2 km above Micos, San Luis Potosí, México [L=22.1163, -99.1663].

Type Specimens: Lectotype, ANSP 99520a designated by Johnson and Baker (1973:173). Paralectotypes, ANSP 99520 (*n*=27), CHAS 25173 (*n*=18); FMNH 21848 (*n*=2); FMNH 68038 (*n*=4); INHS 82205 (*n*=12); MCZ 30633 (*n*=3) (Johnson 1956:140, as cotypes).

Taxonomic History:

Lampsilis undivaga Pilsbry, 1910 — Simpson (1914)
Lampsilis (Actinonaias) undivaga Pilsbry, 1910 — Frierson (1927)
Lampsilis (Friersonia) undivaga Pilsbry, 1910 — Haas (1929)
Actinonaias undivaga (Pilsbry, 1910) — Haas (1969), Graf & Cummings (2007; 2021), Czaja et al. (2023)

Order Unionoida
Superfamily Unionoidea Rafinesque, 1820
Family Unionidae Rafinesque, 1820
Subfamily Ambleminae Rafinesque, 1820
Tribe Lampsilini von Ihering, 1901
Genus *Arotonaias* von Martens, 1900

Arotonaias von Martens, 1900. Land and freshwater Mollusca. *In* F.D. Godman and O. Salvin (eds.). *Biologia Centrali-Americana* Part 155, Fascile 63–65:497–520. Page 480, 497. Introduced as a subgenus of *Unio* Philipsson, 1788.

Diagnosis: "Shell triangular, rather compressed, and of comparatively small size, with strong concentric, often somewhat irregular, sculpture and strong teeth. In some species the summits are unequal."

Type Species: *Unio cyrenoides* Philippi, 1847, by subsequent designation (Simpson 1914:311, as *Artonaias*).

Remarks: Johnson (1999b:66) credited Frierson (1927:11) with the type designation of *Arotonaias* von Martens, 1900.

Taxonomic History:

Plagiola (*Ptychoderma*) Simpson, 1900 — Simpson (1900)
Plagiola (*Artonaias*) [sic] von Martens, 1900 — Simpson (1914)
Plagiola Rafinesque, 1819 — Goodrich & van der Schalie (1937)
Arotonaias von Martens, 1900 — Frierson (1927), Modell (1942), Graf & Cummings (2007; 2021), Pfeiffer et al. (2019b), Bieler (2021)
Pachynaias Crosse & Fischer, 1894 — Haas (1929; 1969), Thiele (1934)
Ellipsaria (*Pachynaias*) Crosse & Fischer, 1894 — Modell (1964)

Ptychoderma Simpson, 1900. Synopsis of the naiades, or pearly fresh-water mussels. *Proceedings of the United States National Museum* 22(1205):501–1044. Page 606. Introduced as a subgenus of *Plagiola* Rafinesque, 1819.

Diagnosis: "Shell triangularly ovate or rounded, solid, bluntly angled before, more sharply angular behind, with a tolerably well-marked posterior ridge; surface strongly and irregularly concentrically sulcate; beaks small but rather prominent, sculptured with fine, irregular, broken ridges, which are somewhat doubly looped, the front loop being larger and more rounded; epidermis olive to tawny, wrinkled, sometimes having faint rays; pseudocardinals compressed, ragged; hinge plate narrow; laterals short, slightly curved, obliquely striated; muscle scars rather shallow, the posterior round; female shell produced in post-basal region.

Animal with small branchiae, inner united the whole length to the abdominal sac, wider than the outer in front, narrower behind; marsupium occupying the posterior part of the outer gills in from eleven to twenty distinct ovisacs which are rounded and dark below, the whole having a decided, parallel sulcus inside and out near its base; mantle thin, papery, with a wide, thickened, double edge, which is sometimes crenate; branchial opening large fringed; anal opening smooth or with only the faintest crenulations."

Type Species: *Unio cyrenoides* Philippi, 1847, by original designation.

Taxonomic History:

Plagiola (*Artonaias*) [sic] von Martens, 1900 — Simpson (1914)
Pachynaias Crosse & Fischer, 1894 — Thiele (1934)
Ellipsaria (*Pachynaias*) Crosse & Fischer, 1894 — Modell (1964)

Reticulatus Frierson, 1927. A classification and annotated check list of the North American naiades. Baylor University Press, Waco, Texas. 111 pp. Page 38. Introduced as a subgenus of *Elliptio* Rafinesque, 1819.

Diagnosis: As with *Popenaias*, Frierson provided no diagnosis for his genus *Reticulatus*. The diagnosis of *Nephronaias reticulatus* Simpson 1900—"Shell evenly elliptical, subinflated, rather solid, slightly biangulate behind; surface decidedly sulcate and sculptured throughout with radiating ridges or slight furrows, which cut the sulcations into loops, and give the shell a reticulated appearance; epidermis tawny; pseudo-cardinals stumpy; laterals rather heavy and granular; muscle scars distinct; nacre lurid, thicker in front," contains no characters that make it distinct and we place it in the synonymy of *Nephronaias*.

Type Species: *Nephronaias reticulatus* Simpson, 1900, by original designation.

Taxonomic History:

Arotonaias von Martens 1900 — von Martens (1900e)
Reticulata [sic] Frierson, 1927 — Frierson (1927), Johnson (1972)
Elliptio Rafinesque, 1819 — Thiele (1934)
Reticulatus Frierson, 1927 — Modell (1942), Graf & Cummings (2007; 2021), Lopes-Lima et al. (2017), Pfeiffer et al. (2019b)
Nephronaias (*Reticulatus*) Frierson, 1927 — Haas (1969)

Distribution & Remarks: Graf and Cummings (2021:17) recognized three species in the genus *Arotonaias* (Ambleminae: Lampsilini). We found that *Unio imbricatus* Mörch, 1861 was pre-occupied and is a now a synonym of *Arotonaias encarpa* (Lea, 1868). *Arotonaias* now contains four species, three of which are endemic to the Río San Juan basin (Caribbean Dr.) in Nicaragua. *Arotonaias reticulata* (Simpson, 1900) is found in the Coco basin in Nicaragua and the Patuca drainage of Honduras (Map 29). There are no published studies to date regarding the validity, relationships, and monophyly of the members of *Arotonaias*.

MAP 29 Distribution of the genus *Arotonaias* von Martens, 1900.

PLATE 19 ***Arotonaias cyrenoides*** **(Philippi, 1847). Top Left:** *Unio newcombianus* Lea, 1856. Lectotype, USNM 84419 (34 mm L, 28 mm H). ***Arotonaias gabbiana*** **(Lea, 1868). Top Right:** *Unio gabbianus* Lea, 1868. Lectotype, USNM 84387 (39 mm L, 27 mm H). **Bottom Left:** *Unio nicaraguensis* Lea, 1868. Lectotype, USNM 84390 (34 mm L, 25 mm H). ***Arotonaias encarpa*** **(Lea, 1868). Bottom Right:** *Unio encarpus* Lea, 1868. Lectotype, USNM 84386 (36 mm L, 25 mm H). All from Lake Nicaragua (Caribbean–San Juan Dr.), Río San Juan Department, Nicaragua [W = 11.1350, -84.8367].

Arotonaias cyrenoides (Philippi, 1847)

Plate 19, Map 30

Province(s): San Juan.

Distribution and Status: Endemic to the Río San Juan Basin in Nicaragua with all records coming from Lake Nicaragua and adjacent Lake Managua except for one record (UF 93197) consisting of two half valves from Río Escondido, San Juan del Sur on the Pacific coast of Nicaragua. Not assessed (IUCN 2024).

Anatomy: Shell solid, triangular, and rather convex. Surface of the shell sculptured with slightly corrugated, concentric ridges. Beak cavity shallow. Ventral margin much arcuated. Left valve with two ragged pseudocardinals and two curved laterals. Right valve with two pseudocardinals and one lateral. Periostracum uniform tan or brown with impressed radial lines, especially behind. Nacre pale red in the type, while specimens from Lake Managua are white.

Biology: Unknown.

Adult Size: Length to 5 cm.

Similar Species: *Arotonaias encarpa* (Lea, 1868) and *Arotonaias gabbiana* (Lea, 1868).

MAP 30 Distribution of *Arotonaias cyrenoides* (Philippi, 1847).

Unio cyrenoides Philippi, 1847. Testaceorum novorum centuria. *Zeitschrift für Malakozoologie* 4(6):84–96. Page 93. Figured in Philippi (1848:79, pl. 5, fig. 1), Küster (1862:285, pl. 96, fig. 1).

Verbatim Type Locality: "Lacus Nicaraguae."

Type Locality: Lake Nicaragua (Caribbean–San Juan Dr.), Río San Juan Department, Nicaragua [11.1350, -84.8367].

Type Specimens: Holotype by monotypy, MNHNS 50199 (Coan and Kabat 2017:70).

Remarks: Haas (1929:319) referred to specimens from Lake Nicaragua (SMF 12785, $n=2$) as *Pachynaias rugososulcata*, which is a synonym of *Pachynaias spheniopsis* (Morelet, 1849) and found only in the Usumacinta basin in Guatemala. Another label with the specimens reads *Plagiola nicaraguensis*, which is a synonym of *Arotonaias gabbiana* (Lea, 1868), but they are *Arotonaias cyrenoides* (Philippi, 1847).

Taxonomic History:

Unio cyrenoides Philippi, 1847 — Philippi (1848), Conrad (1853), Küster (1862), Tate (1870)
Margaron (*Unio*) *cyrenoides* (Philippi, 1847) — Lea (1852; 1870)
Unio (*Lampsilis*) *cyrenoides* Philippi, 1847 — Adams & Adams (1857)
Unio (*Arotonaias*) *cyrenoides* Philippi, 1847 — von Martens (1900b)
Plagiola (*Ptychoderma*) *cyrenoides* (Philippi, 1847) — Simpson (1900)
Plagiola (*Arotonaias*) *cyrenoides* (Philippi, 1847) — Simpson (1914)
Arotonaias cyrenoides (Philippi, 1847) — Frierson (1927), Johnson (1999b, *Unio gabbianus* von Martens, 1900 (non Lea), Graf & Cummings (2007; 2021), Coan & Kabat (2017), Bieler (2021)
Pachynaias cyrenoides (Philippi, 1847) — Haas (1929)

Unio newcombianus Lea, 1856. Description of four new species of exotic Uniones. *Proceedings of the Academy of Natural Sciences of Philadelphia* 8(3):103. Figured in Lea (1858a JANSP Vol. 3:312; 1858b Obs. 6:32, pl. 30, fig. 27) and von Martens (1900a, pl. 29, figs. 1, 1a–b, ZMB/MOLL.305825).

Verbatim Type Locality: "Lake Nicaragua."

Type Locality: Lake Nicaragua (Caribbean–San Juan Dr.), Río San Juan Department, Nicaragua [W=11.1350, -84.8367].

Type Specimens: Lectotype, USNM 84419, here designated.

Remarks: Lea (1856) did not initially mention how many specimens he had before him. He later (1858:312) stated that there were specimens in "My cabinet and cabinets of Dr. Newcomb, Albany, and C. M. Wheatley, New York" and that he had "three specimens before me." Johnson and Baker (1973:163) and Johnson (1974a:98) stated that the figured type was not in the USNM and considered lost. They also identified "syntypes ANSP 126425 mentioned as in the Wheatley Collection" without giving the number of specimens in the lot. They also noted that there were two syntypes in the Newcomb Collection, Cornell University, Ithaca, New York. These are now PRI 10023 ($n=2$) (formerly Cornell University 15470). ANSP 126425 contains four specimens. Clearly not all of the specimens mentioned by Johnson and Baker (1973:163) and Johnson (1974a:98) are part of the original group described by Lea and given that it is impossible to tell which ones he had, all of the "syntypes" mentioned above are best considered as topotypes.

Lea (1868a:295) referred to this species of "*Newcombii* (nobis)" which Bieler (2021:15) interpreted as an incorrect subsequent spelling of *Unio newcombianus*. In the following year, Lea (1869b:22, errata et corrigenda) corrected the spelling of *newcombii* to *newcombianus*, which confirmed the usage of the name *Unio newcombii* as erroneous.

Taxonomic History:

Unio newcombianus Lea, 1856 — Lea (1858a–b), Sowerby (1868), Tate (1870), Wright (1888)
Unio Newcombii [sic] Lea, 1856 — Lea (1868a; 1869a–b)
Margaron (*Unio*) *newcombianus* (Lea, 1856) — Lea (1870)
Unio (*Arotonaias*) *newcombianus* Lea, 1856 — von Martens (1900b)
Plagiola (*Ptychoderma*) *cyrenoides* (Philippi, 1847) — Simpson (1900)

Plagiola (*Arotonaias*) *cyrenoides* (Philippi, 1847) — Simpson (1914)
Arotonaias cyrenoides (Philippi, 1847) — Frierson (1927), Johnson (1999b), Bieler (2021)

Unio newcombii Lea, 1868. New Unionidae, Melanidae, etc., chiefly of the United States. *Journal of the Academy of Natural Sciences of Philadelphia Second Series* Vol. 6, Part 3(6):249–302+pls 29–45. Page 295, errata.
Remarks: Error for *Unio newcombianus* Lea, 1856.

Unio sagrinatus Sowerby, 1868. Monograph of the genus *Unio In* L. Reeve and G.B. Sowerby (eds.). *Conchologia Iconica* 16: [unpaginated]+Plates 1, 61–96. Plate 67, species 345.

Verbatim Type Locality: "Hab. —"

Type Locality: Nicaragua.

Type Specimens: Unknown. Presumed lost. Not found in the NHMUK.

Remarks: In the errata in Sowerby, 1868 "For *U. sagrinatus* — read *U. Newcombianus*, Lea... The habitat is Nicaragua."

Taxonomic History:
Unio newcombianus Lea, 1856 — Sowerby (1868)
Margaron (*Unio*) *newcombianus* (Lea, 1856) — Lea (1870)
Plagiola (*Ptychoderma*) *cyrenoides* (Philippi, 1847) — Simpson (1900)
Unio (*Arotonaias*) *newcombianus* Lea, 1856— von Martens (1900b)
Plagiola (*Arotonaias*) *cyrenoides* (Philippi, 1847) — Simpson (1914)
Arotonaias cyrenoides (Philippi, 1847) — Frierson (1927), Johnson (1999b)

Arotonaias encarpa (Lea, 1868)

Plate 19, Map 31

Province(s): San Juan.

Distribution and Status: Endemic to Lake Nicaragua (Río San Juan Dr.) Nicaragua. Not assessed (IUCN 2024).

Anatomy: Shell solid, rather inflated, ovate with a well-developed posterior ridge. Beaks moderately full and beak cavity not deep. Left valve with two pseudocardinals and two laterals. Right valve with one large pseudocardinal and a vestigial second one above it, and one lateral. Surface covered with fine concentric ridges that are minutely waved, especially in the anterior end. Periostracum olive or brown. Nacre dull and brown or olive colored.

Biology: Unknown.

Adult Size: Length to 4 cm.

Similar Species: *Arotonaias cyrenoides* (Philippi, 1847) and *Arotonaias gabbiana* (Lea, 1868).

MAP 31 Distribution of *Arotonaias encarpa* (Lea, 1868).

Unio (*Plagiola*?) *imbricatus* Mörch, 1861. Beiträge zur Molluskenfauna Central-Amerika's. (cont.). Malakozoologische Blätter 8:27–48. Page 205. Figured in von Martens (1901a, pl. 43, figs. 6, 6a–b, Mörch's type).

Verbatim Type Locality: "San Juan de Nicaragua."

Type Locality: Río San Juan, Nicaragua (Caribbean Dr.), Río San Juan Department, Nicaragua [W=11.0702, -84.6098].

Type Specimens: Unknown. Presumed lost.

Remarks: Name preoccupied by *Unio imbricatus* Hislop, S. 1860. Description of fossil shells, from the above-described deposits. Quarterly Journal of the Geological Society of London 16:166–181. Page 175, pl. 7, figs. 27a–c.

Taxonomic History:

Unio (*Plagiola*?) *imbricatus* Mörch, 1861 — Lea (1870)
Unio imbricatus Mörch, 1861 — Wright (1888)
Unio (*Arotonaias*) *imbricatus* Mörch, 1861 — von Martens (1900b)
Plagiola imbricatus (Mörch, 1861) — Simpson (1900)
Plagiola (*Arotonaias*) *imbricata* (Mörch, 1861) — Simpson (1914)
Arotonaias imbricata (Mörch, 1861) — Frierson (1927), Johnson (1999b), Graf & Cummings (2007; 2021)
Pachynaias imbricata (Mörch, 1861) — Haas (1929) [in part]

Unio encarpus Lea, 1868. Description of nine species of Unionidae from Lake Nicaragua, Central America. *Proceedings of the Academy of Natural Sciences of Philadelphia* 20:94–96. Page 95. Figured in Lea (1868a JANSP Vol. 6:294; 1869a Obs. 12:54, pl. 42, fig. 105).

Verbatim Type Locality: "Lake Nicaragua."

Type Locality: Lake Nicaragua (Caribbean–San Juan Dr.), Río San Juan Department, Nicaragua [W = 11.1350, -84.8367].

Type Specimens: Lectotype, USNM 84386 designated by Johnson and Baker (1973:154; Johnson 1974:98, both as the holotype). Paralectotype, ANSP 56544 (Johnson and Baker 1973:154, as a paratype).

Taxonomic History:

Margaron (*Unio*) *encarpus* (Lea, 1868) — Lea (1870)
Unio encarpus Lea, 1868 — Wright (1888)
Unio (*Arotonaias*) *imbricatus* Mörch, 1861 — von Martens (1900b)
Plagiola (*Ptychoderma*) *encarpus* (Lea, 1868) — Simpson (1900)
Plagiola (*Arotonaias*) *imbricata* (Mörch, 1861) — Simpson (1914)
Arotonaias imbricata (Mörch, 1861) — Frierson (1927), Johnson (1999b)

Arotonaias gabbiana (Lea, 1868)

Plate 19, Map 32

Province(s): San Juan.

Distribution and Status: Endemic to the Río San Juan Basin in Nicaragua with all records coming from Lake Nicaragua and adjacent Lake Managua. Not assessed (IUCN 2024). Goodrich & van der Schalie (1937:46) reported *Plagiola nicaraguensis* [= *Arotonaias gabbiana* (Lea, 1868)] from Guatemala and stated that "Its occurrence there extends its range considerably north, since hitherto it has only been reported from Nicaragua" This locality is 850 km from Lake Nicaragua. We have examined the specimen in question (UMMZ 65540) and it does resemble other members of the genus *Arotonaias*. This certainly warrants further investigation.

Anatomy: Shell solid, triangular and somewhat inflated, with a well-developed, slightly rounded posterior ridge, obliquely truncated anteriorly. Beaks prominent with sculpture consisting of a few fine corrugations. Left valve with two pseudocardinals and two laterals, right valve with two pseudocardinals, the upper compressed and small, and one lateral. Periostracum yellowish green or dark olive with faint rays. Nacre silver-white and iridescent.

Biology: Unknown.

Adult Size: Length to 4.5 cm.

Similar Species: *Arotonaias cyrenoides* (Philippi, 1847) and *Arotonaias encarpa* (Lea, 1868).

MAP 32 Distribution of *Arotonaias gabbiana* (Lea, 1868).

Unio gabbianus Lea, 1868. Description of nine species of Unionidae from Lake Nicaragua, Central America. *Proceedings of the Academy of Natural Sciences of Philadelphia* 20:94–96. Page 95. Figured in Lea (1868a JANSP Vol. 6:295; 1869a Obs. 12: 55, pl. 43, fig. 106) and von Martens (1900e, pl. 39, figs, 5, 5a–b, ZMB/MOLL.305846).

Verbatim Type Locality: "Lake Nicaragua, Cent. Amer."

Type Locality: Lake Nicaragua (Caribbean–San Juan Dr.), Río San Juan Department, Nicaragua [W=11.1350, -84.8367].

Type Specimens: Lectotype, USNM 84387 designated by Johnson and Baker (1973:163; Johnson 1974:98, both as the holotype). Paralectotype, ANSP 56542 (Johnson and Baker 1973:163, as a paratype).

Taxonomic History:

Margaron (*Unio*) *gabbianus* (Lea, 1868) — Lea (1870)
Unio gabbiana Lea, 1868 — Tate (1870)
Unio gabbianus Lea, 1868 — Wright (1888)
Unio (*Arotonaias*) *gabbianus* Lea, 1868 — von Martens (1900b)
Plagiola (*Ptychoderma*) *nicaraguensis* (Lea, 1868) — Simpson (1900)
Plagiola (*Arotonaias*) *nicaraguensis* (Lea, 1868) — Simpson (1914)
Arotonaias gabbiana (Lea, 1868) — Frierson (1927), Johnson (1999b)
Pachynaias rugososulcata (Lea, 1866) — Haas (1929) [in part]
Arotonaias gabbianus (Lea, 1868) — Graf & Cummings (2021)

Unio nicaraguensis Lea, 1868. Description of nine species of Unionidae from Lake Nicaragua, Central America. *Proceedings of the Academy of Natural Sciences of Philadelphia* 20:94–96. Page 95. Figured in Lea (1868a JANSP Vol. 6:296; 1869a Obs. 12:56, pl. 43, fig. 107).

Verbatim Type Locality: "Lake Nicaragua."

Type Locality: Lake Nicaragua (Caribbean–San Juan Dr.), Río San Juan Department, Nicaragua [W=11.1350, -84.8367].

Type Specimens: Lectotype, USNM 84390 designated by Johnson and Baker (1973:163; Johnson 1974:98, both as the holotype). Paralectotypes, ANSP lost (Johnson and Baker 1973:163, as a paratypes).

Taxonomic History:

Margaron (*Unio*) *nicaraguensis* (Lea, 1868) — Lea (1870)
Plagiola (*Ptychoderma*) *nicaraguensis* (Lea, 1868) — Simpson (1900)
Plagiola (*Arotonaias*) *nicaraguensis* (Lea, 1868) — Simpson (1914)
Plagiola nicaraguensis (Lea, 1868) — Goodrich & van der Schalie (1937)
Arotonaias gabbiana (Lea, 1868) — Johnson (1999b)
Arotonaias nicaraguensis (Lea, 1868) — Graf & Cummings (2007)
Arotonaias gabbianus (Lea, 1868) — Graf & Cummings (2021)

PLATE 20 ***Arotonaias reticulata*** **(Simpson, 1900). Top:** Holotype, USNM 152964 (50 mm L, 33 mm H). Río Patuca (Caribbean Dr.), Honduras [W = 15.7833, -84.2500]. **Bottom:** *Nephronaias elvae* Walker, 1924. Holotype, USNM 515798 (62 mm L, 44 mm H). Río Coco (Caribbean Dr.), Región Autónoma del Atlántico Norte, Nicaragua [W = 15.0000, -83.1666].

Arotonaias reticulata (Simpson, 1900)

Plate 20, Map 33

Province(s): Gulf of Honduras & Northern Caribbean.

Distribution and Status: Known only from the Patuca drainage in Honduras and the Coco basin in Nicaragua Not assessed (IUCN 2024).

Anatomy: Shell elliptical, slightly inflated, rather solid, and slightly pointed posteriorly. Beaks slightly elevated. Left valve with two small pseudocardinals and two curved laterals. Right valve with one solid pseudocardinal, with a small compressed tooth above it, and one lateral. Surface of the shell rayless and sculptured throughout with fine radiating ridges or slight furrows. Periostracum light brown or chestnut. Nacre white.

Biology: Unknown.

Adult Size: Length to 5 cm.

Similar Species: The quadrate shaped *Nephronaias dysonii* (Lea, 1859) is also found in the Patuca drainage and the elliptical *Nephronaias flucki* Bartsch, 1906 co-occurs in the Río Coco.

MAP 33 Distribution of *Arotonaias reticulata* (Simpson, 1900).

Nephronaias reticulatus Simpson, 1900. New and unfigured Unionidae. *Proceedings of the Academy of Natural Sciences of Philadelphia* 52:74–86+5 plates. Page 77, pl. 2, fig. 3.

Verbatim Type Locality: "Patook river, Honduras."

Type Locality: Río Patuca (Caribbean Dr.), Honduras [W=15.7833, -84.2500].

Type Specimens: Holotype by monotypy, USNM 152964 (Johnson 1975:18).

Remarks: As mentioned above Frierson provided no diagnosis for his genus *Reticulatus*. Simpson's diagnosis of *Nephronaias reticulatus* contains no characters that makes the genus *Reticulatus* distinct, and we place it in *Arotonaias* following von Martens (1900e:649).

Taxonomic History:

Unio (*Arotonaias*) *reticulatus* Simpson, 1900 — von Martens (1901c)
Nephronaias reticulatus Simpson, 1900 — Simpson (1900b)
Arotonaias reticulatus — von Martens (1901c)
Nephronaias reticulata Simpson, 1900 — Simpson (1914)
Elliptio (*Reticulatus*) *reticulatus* (Simpson, 1900) — Frierson (1927)
Elliptio reticulata (Simpson, 1900) — Thiele (1934)
Nephronaias (*Reticulatus*) *reticulata* Simpson, 1900 — Haas (1969)
Reticulatus reticulatus (Simpson, 1900) — Graf & Cummings (2007; 2021)

Nephronaias elvae Walker, 1924. Two new naiades from Nicaragua. *The Nautilus* 38(2):51–54 + plate 1. Page 52, pl. 1, figs. 1–2.

Verbatim Type Locality: "Wanks River or Rio Coco, Nicaragua."

Type Locality: Río Coco (Caribbean Dr.), Región Autónoma del Atlántico Norte, Nicaragua [W = 15.0000, -83.1666].

Type Specimens: Holotype by monotypy, USNM 515798. Presumed lost by Johnson (1979:32).

Remarks: Walker (1924:52) wrote about the similarity of *Nephronaias elvae* to *N. reticulatus* and discussed his conversations with Frierson and Marshall: "Mr. L.S. Frierson, to whom a photograph was submitted, suggested that it was very close to *Nephronaias reticulata* Simp., if not identical with it." Marshall had a similar first impression but stated that "but when I came to examine the sculpture, I found that, although Simpson's shell is only about two-thirds as large as yours, the sculpture on his is two or three times as coarse." Marshall spoke with Bartsch and quoted in Walker said that "between us, we came to the conclusion that yours is probably different from Simpson's. Your shell is much more round than Simpson's" and that "It would seem, therefore, that the present form is well entitled to a name, even though a larger series might afford connecting links." Only a handful of specimens were available at the time Walker described *N. elvae*. We examined a series collected by Fred Thompson (UF 229376, $n = 11$) in which a wide range of shell shapes were present from round to oval, and we provisionally place *Nephronaias elvae* Walker, 1924 under *Arotonaias reticulata* (Simpson, 1900). Haas (1929:318) included specimens he received from Karl Sapper SMF 12781 ($n = 3$, from the Río Coco) and SMF 12780 ($n = 1^3/_2$ from, Rio Del Meico, both in Nicaragua) as *Pachynaias imbricata* (Mörch, 1861). There are two localities on the original label that accompanies SMF 12780: Río Del Meico and Río Coco. We could not find Río Del Meico on any map or in any gazetteer, and the locality is unknown; they could very well both be from the Río Coco.

Taxonomic History:

Nephronaias (*Reticulatus*) *elvae* Walker, 1924 — Haas (1969)
Pachynaias imbricata (Mörch, 1861) — Haas (1929) [in part]
Reticulatus elvae (Walker, 1924) — Graf & Cummings (2007; 2021)

Order Unionoida
Superfamily Unionoidea Rafinesque, 1820
Family Unionidae Rafinesque, 1820
Subfamily Ambleminae Rafinesque, 1820
Tribe Lampsilini von Ihering, 1901
Genus *Delphinonaias* Crosse & Fischer, 1894

Delphinonaias Crosse & Fischer, 1894. *In* P. Fischer and H. Crosse. 1894. Études sur les mollusques terrestres et fluviatiles du Mexique et du Guatemala. *In* M.H.–Milne Edwards (ed.). *Mission Scientifique au Mexique et dans l'Amérique centrale. Recherches Zoologiques pour servir á l'Histoire de la Faune de l'Amérique Centrale et du Mexique*. 15th Livraison. 62–72, pp. 489–576, plates 59–62, Imprimerie Nationale, Paris Volume 2. 731 pp. Page 555. Introduced as subgenus of *Unio* Philipsson, 1788.

Diagnosis: "Forme très-aplatie; bord dorsal elevé en une grande aile, en arrière du sommet des valves. Dents latérales antérieures comprimées (*U. delphinulus*, Morelet)." [Very flattened shape; dorsal edge raised in a large wing behind the top of the valves. Compressed anterior lateral teeth (*U. delphinulus*, Morelet)].

Type Species: *Unio delphinulus* Morelet, 1849, by original designation.

Taxonomic History:

Unio (*Metaptera*) Rafinesque, 1820 — von Martens (1900b)
Lampsilis (*Proptera*) Rafinesque, 1819 — Simpson (1900; 1914)
Actinonaias Crosse & Fischer, 1894 — H.B. Baker (1922)
Lampsilis (*Delphinonais*) [sic] Crosse & Fischer, 1894 — Frierson (1927)
Lampsilis (*Delphinonaias*) Crosse & Fischer, 1894 — Haas (1929; 1969)
Lampsilis (sec. *Delphinonaias*) Crosse & Fischer, 1894 — Thiele (1934)
Leptodea (*Delphinonaias*) Crosse & Fischer, 1894 — Modell (1964)
Delphinonaias Crosse & Fischer, 1894 — Modell (1942), Graf & Cummings (2007; 2021), Lopes-Lima et al. (2017), Pfeiffer et al. (2019a–b)

Phyllonaias Crosse & Fischer, 1894. *In* P. Fischer and H. Crosse. 1894. Études sur les mollusques terrestres et fluviatiles du Mexique et du Guatemala. *In* M.H.–Milne Edwards (ed.). *Mission Scientifique au Mexique et dans l'Amérique centrale. Recherches Zoologiques pour servir á l'Histoire de la Faune de l'Amérique Centrale et du Mexique* 15th Livraison. 62–72, pp. 489–576, plates 59–62, Imprimerie Nationale, Paris Volume 2. 731 pp. Page 555. Introduced as subgenus of *Unio* Philipsson, 1788.

Diagnosis: "Coquille aplatie; bord dorsal élevé en arrière et formant un aileron rudimentaire. Dents latérales antérieures comprimées (*U. paludosus*, Morelet)." [Flattened shell; dorsal edge raised behind and forming a rudimentary wing. Anterior lateral teeth usually compressed (*U. paludosus*, Morelet)].

Type Species: *Unio paludosus* Morelet, 1849, by original designation.

Taxonomic History:

Actinonaias Crosse & Fischer, 1894 — H.B. Baker (1922)
Lampsilis (*Delphinonais*) [sic] Crosse & Fischer, 1894 — Frierson (1927)
Lampsilis (sec. *Phyllonaias*) Crosse & Fischer, 1894 — Thiele (1934)
Leptodea Rafinesque, 1820 — Goodrich & van der Schalie (1937)
Lampsilis (*Phyllonaias*) Crosse & Fischer, 1894 — Haas (1969)

Distribution: The genus *Delphinonaias* contains two species, both of which are found in drainages south of the Yucatán Peninsula in the Usumacinta basin in México and Guatemala and in a few streams that drain into the Caribbean Sea in Belize and Honduras (Map 33).

Remarks: There are no published studies to date regarding the validity, relationships, and monophyly of the members of *Delphinonaias*.

MAP 34 Distribution of the genus *Delphinonaias* Crosse & Fischer, 1894.

PLATE 21 ***Delphinonaias largillierti*** **(Philippi, 1847). Top Left:** *Unio largillierti* Philippi, 1847. Neotype, MCZ 155569. **Middle Right:** *Unio paludosus* Morelet, 1849. Lectotype, NHMUK 1893.2.4.2024 (57 mm L, 31 mm H). Both, Río Usumacinta (Grijalva Dr.), Boca de San Gerónimo, Tabasco, México [L = 17.9016, -91.9131]. **Middle Left:** *Unio delphinulus* Morelet, 1849, Lectotype, NHMUK 1893.2.4.2.1588 (60 mm L, 26 mm H). Lago Petén Itzá (Usumacinta Dr.), Flores, El Petén Department, Guatemala [L = 16.9461, -89.9154]. **Bottom Right:** *Unio planivalvis* Morelet, 1851. Lectotype, NHMUK 1893.2.4.2.2025 (66 mm L, 37 mm H). Also, from Boca de San Gerónimo.

FIGURE 40 *Delphinonaias largillierti* (Philippi, 1847). Laguna de Catazajá (Usumacinta Dr.), Catazajá, Chiapas, Mexico [L=17.7297, -92.0183]. 2 December 2019.

Delphinonaias largillierti (Philippi, 1847)

Plate 21, Figure 40, Map 35

Province(s): Usumacinta; Gulf of Honduras & Northern Caribbean.

Distribution and Status: Found in the Grijalva–Usumacinta drainage in Chiapas and Tabasco, México and El Petén Department, Guatemala east to Belize in a few rivers draining to the Caribbean. Johnson (1951a:79) gave the range as "Southern Mexico, Northern Guatemala and Honduras." There are two lots [AMS 126902 ($n=3$) and USNM 83936 ($n=2$)] from "Honduras" without specific locality data, and these are regarded as spurious until new data can verify its presence there. Live individuals were found in Laguna de Catazajá (Usumacinta Dr.), Catazajá, and the Río Chacamax at La Libertad in Chiapas, México in 2019. Not assessed (IUCN 2024).

Anatomy: Shell triangular, elongate, compressed and thin with a prominent dorsal wing. Posterior ridge well developed. Anterior end rounded and obliquely truncated posteriorly. Beaks low, not inflated, with delicate, double-looped sculpture. Beak cavity shallow. The left valve with two pseudocardinals and two lamellar laterals. Right valve with two pseudocardinals and one lateral. Shell surface covered with concentric growth lines, slightly radiately sculptured on the umbo. Periostracum yellow or greenish olive. Nacre white or light pink and iridescent.

Biology: Found in mud and sand among vegetation in slow-flowing water.

Adult Size: Length to 8 cm.

Similar Species: *Delphinonaias scutulata* (Morelet, 1849) is syntopic with *D. largillierti* (Philippi, 1847) in the Usumacinta in Tabasco, México. Both are relatively thin-shelled and possess a conspicuous dorsal wing, but *Delphinonaias largillierti* lacks pustules on the surface of the shell that are present in *D. scutulata*.

MAP 35 Distribution of *Delphinonaias largillierti* (Philippi, 1847) (black) and *Delphinonaias scutulata* (Morelet, 1849) (red).

Unio largillierti Philippi, 1847. Testaceorum novorum centuria. *Zeitschrift für Malakozoologie* 4(6):84–96. Page 94. Not figured.

Verbatim Type Locality: "Yucatan."

Type Locality: Río Usumacinta (Gulf of México–Grijalva Dr.), Boca de San Gerónimo, Tabasco, México [L = 17.9016, -91.9131].

Type Specimens: Neotype, MCZ 155569 designated and type locality fixed by Johnson (1951a:79, as the lectotype). Paraneotype, MCZ 184531 (*n* = 1) (Johnson 1951a:79; Johnson 1956:121, both as a paratype).

Remarks: Johnson restricted the type locality to "San Geronimo, Campeche, Mexico." Boca de San Gerónimo is actually in Tabasco, México, near the border with Campeche at the southern end of the Yucatán Peninsula.

Taxonomic History:

Barbala largillierti (Philippi, 1847) — Adams & Adams (1857)

Unio (*Metaptera*) *largillierti* Philippi, 1847 — von Martens (1900, *nomen dubium*)

Lampsilis (*Proptera*) *largillierti* (Philippi, 1847) — Simpson (1900; 1914)
Lampsilis largillierti (Philippi, 1847) — de Leon (1912)
Leptodea largillierti (Philippi, 1847) — Johnson (1951a)
Potamilus largillierti (Philippi, 1847) — Graf & Cummings (2007), Coan & Kabat (2017)
Delphinonaias largillierti (Philippi, 1847) — Graf & Cummings (2021), Ramírez et al. (2022), Czaja et al. (2023)

Unio paludosus Morelet, 1849. Testacea novissima insulae Cubanae, et Amer. Centralis. Librairie de L'Académie Nationale de Medecine, Paris Part 1:1–31. Page 30. Figured in Fischer and Crosse (1894a:559, pl. 59, fig. 3).

Verbatim Type Locality: "prope S. Geronimo Yucataneorum."

Type Locality: Río Usumacinta (Gulf of México–Grijalva Dr.), Boca de San Gerónimo, Tabasco, México [L = 17.9016, -91.9131].

Type Specimens: Lectotype, NHMUK 1893.2.4.2024 designated by Johnson (1971:90, as the holotype; Breure et al. 2018:381, figs. 823–824).

Remarks: Johnson (1951a:77) stated that the type locality was "swamps near San Geronimo, Yucatan [now Campeche, México]." There are no verified specimens from the Yucatán near Campeche, and the type locality is almost certainly Boca de San Gerónimo, Tabasco, México.

Taxonomic History:

Unio (*Phyllonaias*) *paludosus* Morelet, 1849 — Fischer & Crosse (1894b)
Unio (*Metaptera*) *paludosus* Morelet, 1849 — von Martens (1900b)
Lampsilis (*Proptera*) *paludosus* (Morelet, 1849) — Simpson (1900)
Lampsilis paludosus (Morelet, 1849) — de Leon (1912)
Lampsilis (*Proptera*) *paludosa* (Morelet, 1849) — Simpson (1914)
Actinonaias paludosus (Morelet, 1849) — H.B. Baker (1922)
Lampsilis (*Delphinonaias*) *paludosa* (Morelet, 1849) — Frierson (1927)
Lampsilis (sec. *Phyllonaias*) *paludosus* (Morelet, 1849) — Thiele (1934)
Leptodea paludosa (Morelet, 1849) — Goodrich & van der Schalie (1937)
Leptodea largillierti (Philippi, 1847) — Johnson (1951a)
Lampsilis (*Phyllonaias*) *paludosa* (Morelet, 1849) — Haas (1969)
Delphinonaias paludosa (Morelet, 1849) — Graf & Cummings (2007)
Disconaias largillierti (Philippi, 1847) — Graf & Cummings (2021)

Unio delphinulus Morelet, 1849. Testacea novissima insulae Cubanae, et Amer. Centralis. Librairie de L'Académie Nationale de Medecine, Paris Part 1:1–31. Page 31. Figured in Fischer and Crosse (1894a:557, pl. 63, figs. 2, 2a–b).

Verbatim Type Locality: "flum. Usumasinta et lacum Petenensem."

Type Locality: Lago Petén Itzá (Gulf of México–Grijalva–Usumacinta Dr.), Flores, El Petén Department, Guatemala [L = 16.9461, -89.9154].

Type Specimens: Lectotype, NHMUK 1893.2.4.1588 (Johnson 1971:82, as the probable holotype; Breure et al. 2018:267, figs. 330–331, as syntypes), here designated and type locality fixed. Paralectotypes, MNHN-IM-2000-1672 (*n* = 5) from "marecages de l'Usumacinta" (Johnson 1951a:79, as cotypes; Breure et al. 2018:267).

Remarks: Breure et al. (2018:267) listed ANSP 56512 (*n* = 3) from "Marias de S. Geronimo, Yucatan" as "possible syntypes" "and that they were '*ex* Morelet,' presumably acquired from Fulton, but they are not from the published type locality."

Taxonomic History:

Unio delphinulus Morelet, 1849 — Hanley (1856), Sowerby (1866), H.B. Baker (1922)
Margaron (Unio) delphinulus (Morelet, 1849) — Lea (1870)
Unio (Delphinonaias) delphinulus Morelet, 1849 — Fischer & Crosse (1894b)
Unio (Metaptera) delphinulus Morelet, 1849 — von Martens (1900b)
Lampsilis (Proptera) delphinulus (Morelet, 1849) — Simpson (1900; 1914)
Actinonaias delphinulus (Morelet, 1849) — H.B. Baker (1922)
Lampsilis (Delphinonaias) delphinulus (Morelet, 1849) — Frierson (1927), Haas (1929; 1969)
Lampsilis (sec. *Delphinonaias*) *delphinulus* (Morelet, 1849) — Thiele (1934)
Leptodea paludosa (Morelet, 1849) — Goodrich & van der Schalie (1937)
Leptodea largillierti (Philippi, 1847) — Johnson (1951a)
Delphinonaias delphinulus (Morelet, 1849) — Graf & Cummings (2007)
Disconaias largillierti (Philippi, 1847) — Graf & Cummings (2021), Ramírez et al. (2022)

Unio planivalvis Morelet, 1851. Testacea novissima insulae Cubanae, et Amer. Centralis. Librairie de L'Académie Nationale de Medecine, Paris Part 2:1–30. Page 24. Figured in Fischer and Crosse (1894a:560, pl. 59, fig. 2).

Verbatim Type Locality: "in paludibus fluminis Usumacinta vicinis."

Type Locality: Río Usumacinta (Gulf of México–Grijalva Dr.), Boca de San Gerónimo, Tabasco, México [L = 17.9016, -91.9131].

Type Specimens: Lectotype, NHMUK 1893.2.4.2025 designated by Johnson (1971:91, as the holotype; Breure et al. 2018:390, figs. 863–864).

Remarks: The label with the lectotype reads "Marais de S. Geronimo."

Taxonomic History:

Unio (Phyllonaias) planivalvis Morelet, 1851 — Fischer & Crosse (1894b)
Unio (Metaptera) paludosus var. *planivalvis* Morelet, 1851 — von Martens (1900b)
Lampsilis (Proptera) planivalvis (Morelet, 1851) — Simpson (1900; 1914)
Lampsilis planivalvis (Morelet, 1851) — de Leon (1912)
Lampsilis (Delphinonaias) planivalvis (Morelet, 1851) — Haas (1929)
Leptodea paludosa (Morelet, 1849) — Goodrich & van der Schalie (1937)
Leptodea largillierti (Philippi, 1847) — Johnson (1951a)
Lampsilis (Phyllonaias) planivalvis (Morelet, 1851) — Haas (1969)
Delphinonaias planivalvis (Morelet, 1851) — Graf & Cummings (2007), Ramírez et al. (2022), Czaja et al. (2023)
Disconaias largillierti (Philippi, 1847) — Graf & Cummings (2021)

PLATE 22 ***Delphinonaias scutulata* (Morelet, 1849). Top Left:** *Unio scutulatus* Morelet, 1849. Lectotype, NHMUK 1893.2.4.2014 (46 mm L, 29 mm H). **Middle Right**: *Unio scutulatus* var. *β*, *obliterata* Fischer & Crosse, 1894. Lectotype, NHMUK 1893.2.4.2017 (45 mm L, 27 mm H). **Bottom Left**: *Unio scutulatus* var. *γ*, *secabilis* Fischer & Crosse, 1894. Lectotype, NHMUK 1893.2.4.2016 (50 mm L, 30 mm H). All, Río Usumacinta (Grijalva Dr.), Boca de San Gerónimo, Tabasco, México [L = 17.9016, -91.9131].

Delphinonaias scutulata (Morelet, 1849)

Plate 22, Map 35

Province(s): Usumacinta.

Distribution and Status: Known only from the type locality—the Río Usumacinta (Gulf of México–Grijalva Dr.), Boca de San Gerónimo, Tabasco, México. A visit to the type locality in 2019 found the site heavily polluted, and it is possibly extirpated from that location but may occur elsewhere the river. Not assessed (IUCN 2024).

Anatomy: Shell elliptical or rhomboid, compressed and thin with a well-developed posterior wing. The beaks are low and not inflated. Left valve with two pseudocardinals and two laterals. Right valve with two compressed pseudocardinals and one lateral. Shell surface with radiating pustules on the posterior half. Periostracum green or light olive green with a few faint rays. Nacre bluish white and iridescent.

Biology: Unknown.

Adult Size: Length to 5 cm.

Similar Species: *Delphinonaias largillierti* (Philippi, 1847) is similar in shape but lacks surface sculpturing.

Unio scutulatus Morelet, 1849. Testacea novissima insulæ Cubanæ, et Americæ Centralis. Librairie de L'Académie Nationale de Medecine, Paris Part 1:1–31. Page 30. Figured in Fischer and Crosse (1894a:561, pl. 59, fig. 4; pl. 67, fig. 6).

Verbatim Type Locality: "H cum præecedente." [prope S. Geronimo Yucataneorum].

Type Locality: Río Usumacinta (Gulf of México–Grijalva Dr.), Boca de San Gerónimo, Tabasco, México [L = 17.9016, -91.9131].

Type Specimens: Lectotype, NHMUK 1893.2.4.2014 designated by Johnson (1971:93, as the holotype; Breure et al. 2018:421, figs. 1006 (shell) and 1007 (label). Paralectotypes, NHMUK 1893.2.4.2015 ($n = 1$) and NHMUK 1893.2.4.2018 ($n = 1$) (Johnson 1971:93, as a paratype).

Remarks: Johnson stated that "NHMUK 1893.2.4.2015–16 were also figured, pl. 59, figs. 5 and 6, and named varieties obliterata and secabilis respectively." NHMUK 1893.2.4.2015 does not match the figure of *Unio scutulatus* var. *β*, *obliterata* Crosse & Fischer, 1894 and is a paralectotype of *Unio scutulatus* Morelet, 1849. Johnson (1971:93) listed 1893.2.4.2017–2018 as paratypes. NHMUK 1893.2.4.2017 is the lectotype of *Unio scutulatus* var. *β*, *obliterata* Fischer & Crosse, 1894 (see below). Breure et al. (2018:421) stated that "paralectotypes NHMUK 1893.2.4.2015–2018 (4)." NHMUK 1893.2.4.2016 is the lectotype of *Unio scutulatus* var. γ, *secabilis* Fischer & Crosse. 1894 (see below).

Taxonomic History:

Unio (*Phyllonaias*) *scutulatus* Morelet, 1849 — Fischer & Crosse (1894b)
Unio (*Metaptera*) *scutulatus* Morelet, 1849 — von Martens (1900b)
Lampsilis (*Proptera*) *scutulatus* (Morelet, 1849) — Simpson (1900)
Lampsilis scutulatus (Morelet, 1849) — de Leon (1912)
Lampsilis (*Proptera*) *scutulata* (Morelet, 1849) — Simpson (1914)
Lampsilis (*Delphinonaias*) *scutulata* (Morelet, 1849) — Frierson (1927)
Lampsilis (*Phyllonaias*) *scutulata* (Morelet, 1849) — Haas (1969)
Delphinonaias scutulata (Morelet, 1849) — Graf & Cummings (2007; 2021), Czaja et al. (2023)

Unio scutulatus var. *β*, *obliterata* Fischer & Crosse. 1894. Études sur les mollusques terrestres et fluviatiles du Mexique et du Guatemala. 15th Livraison. 62–72, pp. 489–576, plates 59–62 in Recherches Zoologiques pour servir À l'Histoire de la Faune de l'Amérique Centrale et du Mexique, Imprimerie Nationale, Paris Volume 2. 731 pp. Page 561, pl. 59, fig. 5.

Verbatim Type Locality: "Mexique, dans les marais, au voisinage de San Geronimo, État de Yucatan."

Type Locality: Río Usumacinta (Gulf of México–Grijalva Dr.), Boca de San Gerónimo, Tabasco, México [L = 17.9016, -91.9131].

Type Specimens: Lectotype, NHMUK 1893.2.4.2017, here designated.

Taxonomic History:

Unio (*Metaptera*) *scutulatus* Morelet, 1849 — von Martens (1900b)

Unio scutulatus var. *γ*, *secabilis* Fischer & Crosse. 1894. Études sur les mollusques terrestres et fluviatiles du Mexique et du Guatemala. 15th Livraison. Feuilles 62–72, pp. 489–576, plates 59–62 in Recherches Zoologiques pour servir À l'Histoire de la Faune de l'Amérique Centrale et du Mexique, Imprimerie Nationale, Paris Volume 2. 731 pp. Page 561, pl. 59, fig. 6.

Verbatim Type Locality: "Mexique, dans les marais, au voisinage de San Geronimo, État de Yucatan."

Type Locality: Río Usumacinta (Gulf of México–Grijalva Dr.), Boca de San Gerónimo, Tabasco, México [L = 17.9016, -91.9131].

Type Specimens: Lectotype, NHMUK 1893.2.4.2016, here designated.

Taxonomic History:

Unio (*Metaptera*) *scutulatus* Morelet, 1849 — von Martens (1900b)

Order Unionoida
Superfamily Unionoidea Rafinesque, 1820
Family Unionidae Rafinesque, 1820
Subfamily Ambleminae Rafinesque, 1820
Tribe Lampsilini von Ihering, 1901
Genus *Disconaias* Crosse & Fischer, 1894

Disconaias Crosse & Fischer, 1894. *In* P. Fischer and H. Crosse. 1894. Études sur les mollusques terrestres et fluviatiles du Mexique et du Guatemala. *In* M.H.–Milne Edwards (ed.). *Mission Scientifique au Mexique et dans l'Amérique centrale. Recherches Zoologiques pour servir á l'Histoire de la Faune de l'Amérique Centrale et du Mexique.* 15th Livraison. Feuilles 62–72, pp. 489–576, plates 59–62, Imprimerie Nationale, Paris Volume 2. 731 pp. Page 556. Introduced as subgenus of *Unio* Philipsson, 1788.

Diagnosis: "Coquille grande, ovale ou subtrigone, très-aplatie, striée concentriquement; dents latérales antérieures fortes; dent cardinale très-large et sillonnée (*U. discus*)." [Shell large, oval or subtrigonal, very flattened, concentrically ribbed; strong anterior lateral teeth; very large and furrowed cardinal tooth (*U. discus*)].

Type Species: *Unio discus* Lea, 1838, by original designation.

Taxonomic History:

Unio (*Lampsilis*) — von Martens (1900b)
Unio (sec. *Elliptio*) Rafinesque, 1819 — Simpson (1900; 1914) [in part]
Lampsilis Rafinesque, 1820 — Pilsbry (1910)
Lampsilis (*Disconaias*) Crosse & Fischer, 1894 — Frierson (1927), Haas (1929; 1940; 1969)
Lampsilis (sec. *Disconaias*) Crosse & Fischer, 1894 — Thiele (1934)
Disconaias Crosse & Fischer, 1894 — Morrison (1968), Johnson (1998), Contreras-Arquieta (2000), Pfeiffer et al. (2919b), Tiemann et al. (2020), Graf & Cummings (2021), Nagel (2025)

Lapidosus Simpson, 1900. Synopsis of the naiades, or pearly fresh-water mussels. *Proceedings of the United States National Museum* 22(1205):501–1044. Page 700.

Diagnosis: "Shell inequilateral, subtriangular ovate, large, solid, slightly inflated, with a double posterior ridge, triangulate behind; beaks high, sculpture not seen; disks strongly, irregularly sulcate; epidermis olive; three thick pseudocardinals in the right valve and two in the left; laterals strong; anterior muscle scars deep, tripartite; posterior scars deep, distinct; nacre rosy white, iridescent behind. Animal unknown."

Type Species: *Unio lapidosus* Kobelt, 1838, by original designation.

Distribution & Remarks: Five species were recognized in the recent checklist by Graf and Cummings (2021:17): *Disconaias disca*, *D. conchos*, *D. fimbriata*, *D. novileonis*, and *D. purpuriata*. We sunk *Actinonaias moctezumensis* (Pilsbry, 1910) and *Disconaias novileonis* (Pilsbry, 1910) into *Disconaias fimbriata* (Frierson, 1907) and now recognize two or possibly three species in the genus. *Disconaias disca* is endemic to Panucó Province in San Luis Potosí, Tamaulipas, and Veracruz, Mexicó. *Disconaias fimbriata* is found primarily in Pánuco Province with the exception of a population in the Actopan basin in Papaloapán Province. *Disconaias conchos* known only from the ríos Conchos and Salado de Nadadores in the Río Grande del Norte basin of México is of uncertain placement. A survey conducted in 2018 did not find it alive and it is possibly extinct (Tiemann et al. 2020:180).

MAP 36 Distribution of the genus *Disconaias* Crosse & Fischer, 1894.

PLATE 23 ***Disconaias disca* (Lea, 1838). Top Left:** *Unio discus* Lea, 1838. Holotype, USNM 85341 (132 mm L, 84 mm H). Río Moctezuma (Pánuco Dr.), México [W=21.9672, -98.5618]. **Top Right**: *Unio panacoensis* von dem Busch, 1843. Lectotype, MCZ 178882 (144 mm L, 82 mm H). Río Pánuco, Tampico, Tamaulipas, México [L=22.2166, -97.8500]. **Middle Left:** *Unio mexicanus* Sowerby, 1867. Holotype, NHMUK 1879.2.26.249 (114 mm L, 71 mm H). México**. Middle Right:** *Unio lapidosus* Kobelt, 1893. Holotype, ZMB/MOLL.17236 (126 mm L, 85 mm H). Locality unknown. **Bottom Left:** *Unio* (*Lampsilis*) *discus* var. *connectens* von Martens, 1900. Lectotype, ZMB/MOLL.305852 (130 mm L, 85 mm H). Río Pánuco, Tampico, Tamaulipas, México [L=22.2078, -97.8482].

PLATE 24 ***Disconaias disca* (Lea, 1838).** INHS 90182.3 (137 mm L, 87 mm H). Rio Tampaón (Pánuco Dr.), Álvaro Obregón, at the confluence of the ríos Valles and Tampaón, San Luis Potosí México [L=21.8448, -98.9437]. 9 December 2018.

PLATE 25 ***Disconaias disca* (Lea, 1838).** INHS 90160.1 (91 mm L, 62 mm H). Rio Tampaón (Pánuco Dr.), 4.2 km ESE of Tamuin, San Luis Potosí México [L=21.9923, -98.7418]. 15 December 2018.

PLATE 26 ***Disconaias disca*** **(Lea, 1838).** INHS 90160.2 (88 mm L, 56 mm H). Rio Tampaón (Pánuco Dr.), 4.2 km ESE of Tamuin, San Luis Potosí México [L = 21.9923, -98.7418]. 15 December 2018.

Disconaias disca (Lea, 1838)

Plates 23–26, Map 37

Province(s): Pánuco.

Distribution and Status: Endemic to the Pánuco basin in Tamaulipas, San Luis Potosí, and Veracruz, México. Not assessed (IUCN 2024). *Disconaias disca* was found alive at two sites in the Río Guayalejo in Tamaulipas, México, one in the in Río Tempoal in Veracruz (both in 1982) and two in the Río Tampaón in San Luis Potosí, México in 2018. All Pánuco Dr.

Anatomy: Shell ovate to triangular, heavy, solid, and somewhat compressed. Anterior end broadly rounded, posterior end more broadly rounded and subtruncate. Dorsal margin broadly curved. Ventral margin rounded anteriorly and straight posteriorly. Beaks slightly elevated above the hinge line, located in the anterior quarter of the shell, and without observed sculpture. Beak cavity very shallow. Left valve with a strong, triangular pseudocardinal and one lateral. Right valve with two strong pseudocardinals and two short laterals. Periostracum yellowish green becoming brown in older individuals. Nacre white, salmon, or deep purple and iridescent.

Biology: Gravid and partially gravid females were collected from the Rio Tampaón, at Álvaro Obregón, at confluence of the ríos Valles and Tampaón, on 9 December 2018 (INHS 90182, Plate 24), 4.2 km ESE of Tamuin 15 December 2018 (INHS 90160, Plates 25–26), both in San Luis Potosí, México. Also in the Río Tempoal, between Tempoal and Platón Sánchez, Veracruz, México on 8 February 1982 (USNM 1282362–64).

Adult Size: Length to 15 cm.

Similar Species: *Disconaias fimbriata* (Frierson, 1907) is syntopic with *Disconaias disca* in Pánuco Province, but it is typically smaller, much thinner, and more elongate than *D. disca*.

MAP 37 Distribution of *Disconaias disca* (Lea, 1838).

Unio discus Lea, 1838. Description of new freshwater and land shells. *Transactions of the American Philosophical Society* 6(N.S.):1–154+plates 1–24. Page 74, pl. 18, fig. 57. Also figured in Lea (1838c Obs. 3:74, pl. 18, fig. 57).

Verbatim Type Locality: "India."

Type Locality: Río Moctezuma (Pánuco Dr.), San Luis Potosí, México [W=21.9672, -98.5618].

Type specimen: Holotype by monotypy, USNM 85341 (Johnson 1974a:47; Johnson 1998:434, pl. 25, fig. 1).

Remarks: The published type locality is "India." A label with the holotype (presumable in the hand of C.T. Simpson) reads "found in sand ballast, Phila. Pa." "Bought of [sic] B. Tanner, Loc. E. Indies? is penciled in the valve, but it is a Mexican species." Lea (1870:50) later stated that it was from the "Moctizuma [sic] River, Central America." We concur that it is from México and restrict the type locality to the Río Moctezuma (Pánuco Dr.), San Luis Potosí, México.

Taxonomic History:

Margarita (*Unio*) *discus* — Lea (1836) *nomen nudem*
Unio discus Lea, 1838 — Hanley (1843; 1856), Conrad (1853), Sowerby (1868), Wright (1888), de Leon (1912)
Margaron (*Unio*) *discus* (Lea, 1838) — Lea (1852; 1870)
Unio (*Hyridella*) *discus* Lea, 1838 — Adams & Adams (1857)
Unio (*Disconaias*) *discus* Lea, 1838 — Fischer & Crosse (1894b)
Unio (*Lampsilis*) *discus* Lea, 1838 — von Martens (1900b)
Unio (sec. *Elliptio*) *discus* Lea, 1838 — Simpson (1900; 1914)
Lampsilis discus (Lea, 1838) — Pilsbry (1910)
Actinonaias disca disca (Lea, 1838) — H.B. Baker (1922)
Lampsilis (*Disconaias*) *discus* (Lea, 1838) — Frierson (1927), Haas (1929; 1940; 1969)
Lampsilis (sec. *Disconaias*) *discus* (Lea, 1838) — Thiele (1934)
Disconaias disca (Lea, 1838) — Johnson (1998), Contreras-Arquieta (2000), Graf & Cummings (2007; 2021), Kiser et al. (2022), Nagel (2025)
Disconaias discus (Lea, 1838) — Czaja et al. (2023)

Unio panacoensis von dem Busch in Philippi, 1843. *Unio* Tab. II. Abbildungen und Beschriebungen neuer oder wenig gekannter Conchylien. Cassel. Theodor Fischer 1:75, pl. 2. Also figured in von Martens (1900b, pl. 33, figs. 2, 2a–b, ZMB/MOLL.305835).

Verbatim Type Locality: "Flumen Panaco prope Tampico."

Type Locality: Río Pánuco (Gulf of Mexico Dr.), Tampico, Tamaulipas, México [L=22.2166, -97.8500].

Type Specimens: Lectotype, MCZ 178882 designated by Johnson (1998:434, 450, pl. 25, fig. 2; Johnson 1956:128, as one of three paratypes). Paralectotypes, NHMUK 1965183 ($n=1$) (Johnson 1971:90, as a paratype); MCZ 316163 ($n=2$, likely from MCZ 178882); SMF 12767 ($n=1$) (Zilch 1967:120, as a possible syntype; Nagel 2925:165, as a possible paralectotype); USNM 85340 ($n=1$) (Johnson 1975:33, as a paratype).

Taxonomic History:

Unio panacoensis von dem Busch in Philippi, 1843 — Küster (1861), H.B. Baker (1922)
Margaron (*Unio*) *discus* (Lea, 1838) — Lea (1852; 1870)
Unio (*Disconaias*) *discus* Lea, 1838 — Fischer & Crosse (1894a)
Unio (*Lampsilis*) *discus* var. *panucoensis* [sic] von dem Busch in Philippi, 1843 — von Martens (1900b; 1901c)

Unio (sec. *Elliptio*) *discus* Lea, 1838 — Simpson (1900; 1914)
Lampsilis discus (Lea, 1838) — Pilsbry (1910)
Unio discus Lea, 1838 — de Leon (1912, as *Unio panamensis*)
Lampsilis (*Disconaias*) *panucoensis* [sic] (von dem Busch in Philippi, 1843) — Frierson (1927)
Lampsilis (*Disconaias*) *discus* (Lea, 1838) — Zilch (1967)
Lampsilis (*Disconaias*) *panacoensis* (von dem Busch in Philippi, 1843) — Haas (1969)
Disconaias disca (Lea, 1838) — Johnson (1998), Graf & Cummings (2021), Nagel (2025)
Disconaias panacoensis (von dem Busch in Philippi, 1843) — Graf & Cummings (2007), Czaja et al. (2023)

Unio mexicanus Sowerby, 1867. Monograph of the genus *Unio*. *In* L. Reeve and G.B. Sowerby (eds.). *Conchologia Iconica* 16:[unpaginated]+plates 55–60. Plate 55, species 281.

Verbatim Type Locality: "Mexico."

Type Locality: México.

Type specimen: Holotype by monotypy, NHMUK 1879.2.26.249 (Johnson 1971:88; Johnson 1998:434, incorrectly as NHMUK 79.2.26.281).

Remarks: Primary homonym of *Unio mexicanus* Philippi, 1847.

Taxonomic History:

Margaron (*Unio*) *discus* (Lea, 1838) — Lea (1870)
Unio (*Disconaias*) *discus* Lea, 1838 — Fischer & Crosse (1894a)
Unio (*Lampsilis*) *discus* var. *connectens* von Martens, 1900 — von Martens (1900b)
Unio (sec. *Elliptio*) *discus* Lea, 1838 — Simpson (1900; 1914)
Unio discus Lea, 1838 — de Leon (1912)
Lampsilis (*Disconaias*) *panucoensis* [sic] (von dem Busch in Philippi, 1843) — Frierson (1927)
Lampsilis (*Disconaias*) *panacoensis* (von dem Busch in Philippi, 1843) — Haas (1969)
Disconaias disca (Lea, 1838) — Johnson (1998)

Unio lapidosus Kobelt, 1893. Diagnosen neuer palaearctischer Arten. *Nachrichtsblatt der Deutschen Malakozoologischen Gesellschaft* 25:150–153. Page 151. Figured in Kobelt (1893:90, pl. 176, fig. 1120).

Verbatim Type Locality: "in fluvio Euphrates."

Type Locality: Unknown.

Type specimen: Holotype by monotypy, ZMB/MOLL.17236 (Johnson 1998:434, as the figured type, ZMB without a catalogue number).

Remarks: The original published type locality is the Euphrates River. After examining the holotype, we concur with von Martens (1900b:511) and Haas (1940:141) that it is *Disconaias disca* and that the locality data are erroneous. Label with the specimen reads "Nordmexiko!" Most likely from the Río Pánuco (Gulf of Mexico Dr.) basin.

Taxonomic History:

Unio (*Lampsilis*) *discus* var. *panucoensis* [sic] von dem Busch in Philippi, 1843 — von Martens (1900b; 1901c)
Unio (sec. *Lapidosus*) *lapidosus* Kobelt, 1893 — Simpson (1900)
Lampsilis discus (Lea, 1838) — Pilsbry (1910)
Unio (sec. *Elliptio*) *discus* Lea, 1838 — Simpson (1914)

Lampsilis (*Disconaias*) *panucoensis* [sic] (von dem Busch in Philippi, 1843) — Frierson (1927)
Lampsilis (*Disconaias*) *discus* (Lea, 1838) — Haas (1940)
Lampsilis (*Disconaias*) *panacoensis* (von dem Busch in Philippi, 1843) — Haas (1969)
Disconaias disca (Lea, 1838) — Johnson (1998)

Unio (*Lampsilis*) *discus* var. *connectens* von Martens, 1900. Land and freshwater Mollusca. *In* F.D. Godman and O. Salvin (eds.). *Biologia Centrali-Americana* Part 155, Fascile 63–65:497–520, pls. 33–35. Page 510. Not figured.

Verbatim Type Locality: "Rio Panuco, near Tampico."

Type Locality: Río Pánuco (Gulf of Mexico Dr.), Tampico, Tamaulipas, México [L=22.2078, -97.8482].

Type Specimens: Lectotype, ZMB/MOLL.305852, here designated.

Remarks: In von Martens description, he compared "the specimen before me" from the Albers Collection to "*Unio panucoensis* [sic] (v.d. Busch), Küster, in Martini und Chemnitz Syst. Conch.-Cabinet ed. 2, *Unio*, p. 242, t. 81, fig. 1. *Unio mexicanus* Sowerby, 1867, pl. 55. fig. 281." He noted that it was "More elongate than Philippi's type of *U. panucoensis* [sic] the hinder part of the shell longer and more perpendicularly truncated, the anterior part as in *U. discus*. Sculpture similar. Inside pale orange or (as in the specimen before me) salmon-coloured. Probably only an individual variety of *U. panucoensis* [sic], although the general form seems to be rather different at first sight." "Küster says that the inside is either violet, flesh-coloured, or white; but as he does not distinguish this form from typical *panucoensis* it is not certain whether all these colours are to be found in it" (von Martens 1900b:510).

Taxonomic History:

Lampsilis discus (Lea, 1838) — Pilsbry (1910)
Unio (sec. *Elliptio*) *discus* Lea, 1838 — Simpson (1914)
Unio discus var. *connectans* [sic] von Martens, 1900 — Simpson (1914)
Lampsilis (*Disconaias*) *panucoensis* [sic] (von dem Busch in Philippi, 1843) — Frierson (1927)
Lampsilis (*Disconaias*) *panacoensis* (von dem Busch in Philippi, 1843) — Haas (1969)
Disconaias disca (Lea, 1838) — Johnson (1998)

PLATE 27 ***Disconaias fimbriata* (Frierson, 1927). Top Left:** *Lampsilis fimbriata* Frierson, 1927. Holotype, UMMZ 91161 (80 mm L, 47 mm H). Río Valles [W=21.9829, -99.0184]. **Middle Right**: *Lampsilis* (*Proptera*) *salinasensis* Simpson, 1908. Lectotype, USNM 163156 (70 mm L, 40 mm H). Río Valles, Valles [L=21.9829, -99.0184]. **Middle Left:** *Lampsilis moctezumensis* Pilsbry, 1910. Lectotype, ANSP 99517 (68 mm L, 38 mm H). Río Moctezuma, S of Tampamolón [L=21.5069, -98.7865]. **Bottom Right:** *Lampsilis* (*Disconaias*) *pilsbryana* Frierson, 1927. Holotype, UMMZ 91175 (66 mm L, 38 mm H). Río Valles, Micos [L=22.1004, -99.1529]. All, Pánuco basin, San Luis Potosí, México.

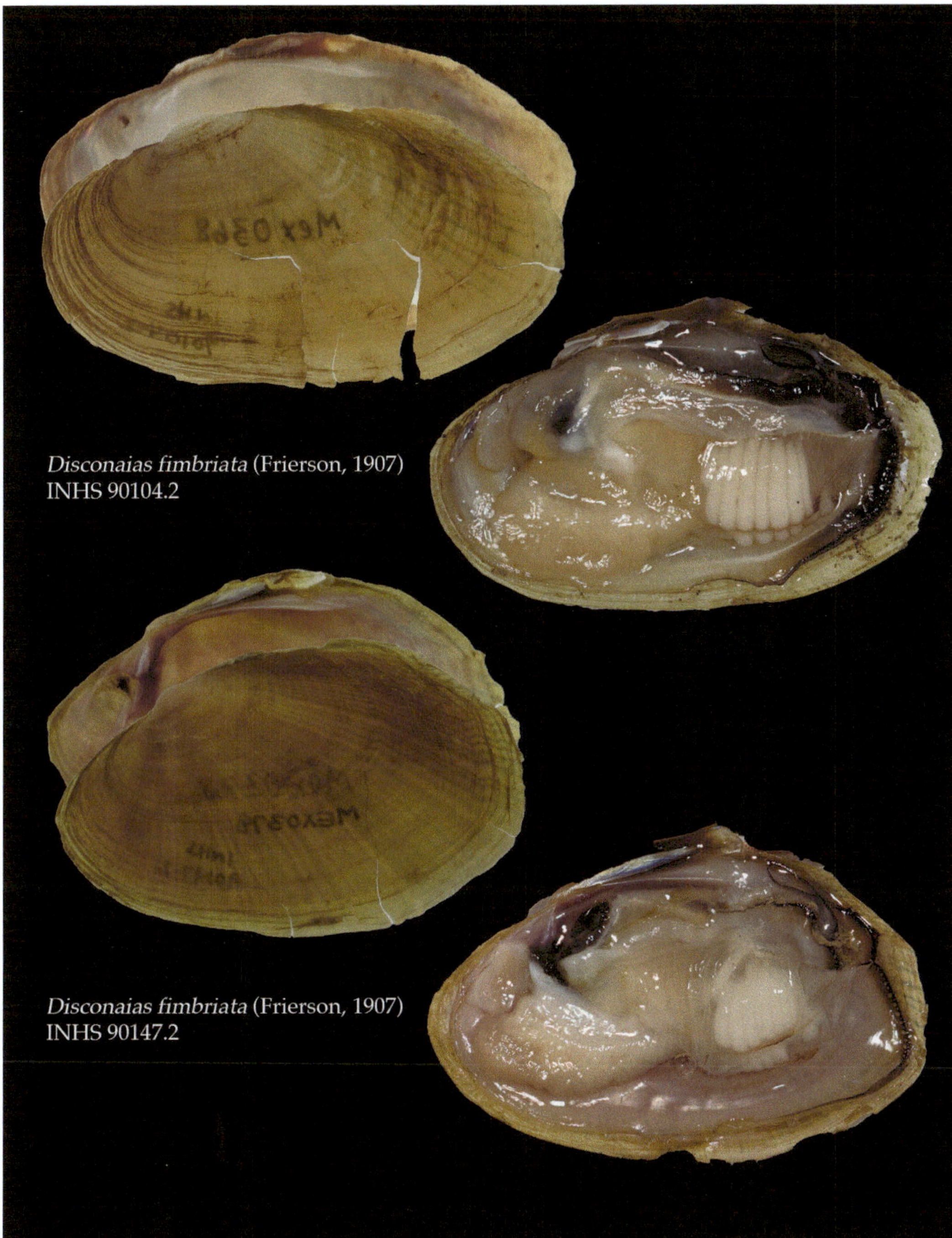

PLATE 28 ***Disconaias fimbriata*** **(Frierson, 1927). Top:** INHS 90104.2 (56 mm L, 32 mm H). Rio Amajac (Moctezuma Dr.), Tlacuilola, 7.4 km WSW Tamazunchale [21.2456, -98.8588]. 13 December 2018. **Bottom:** INHS 90147.2 (61 mm L, 38 mm H). Río Tampaón, Puente de Dios, 22.1 km SSW of Ciudad Valles [21.8069, -99.0745]. 13 December 2018. Both, Pánuco basin, San Luis Potosí, México.

PLATE 29 ***Disconaias fimbriata* (Frierson, 1927).** INHS 90146 (81 mm L, 53 mm H). Rio Tampaón (Pánuco Dr.), Puente de Dios, 22.1 km SSW of Ciudad Valles, San Luis Potosí, Mexico [L=21.8069, -99.0745]. 14 December 2018.

Disconaias fimbriata (Frierson, 1907)

Plates 27–29, Map 38

Province(s): Pánuco and Papaloapán.

Distribution and Status: Found in the Río Pánuco Basin in Tamaulipas and San Luis Potosí, and the Actopan drainage in Veracruz, México. *Disconaias fimbriata* was alive and common at nine sites in the Pánuco basin in 1982 and eight in 2018. Czaja et al. (2023:87) listed *D. fimbriata* as EN—Endangered by the IUCN, presumably under *Disconaias salinasensis* (Simpson, 1908).

Anatomy: Shell elliptical, moderately thin, and compressed. Anterior margin obtusely truncated. Ventral margin nearly straight, occasionally slightly curved, and convex. Beaks low and without sculpture. Beak cavity shallow. Hinge ligament stout and rather long. Lateral teeth stout, double in the left valve, and single in the right. Beak cavity shallow. Periostracum yellow or light brown in color and occasionally rayed. Nacre white, pink, or light purple.

In the gravid females of *Disconaias fimbriata*, the marsupium occupied the posterior half of the external gill except for the extreme anterior end (Plate 28). Seven to eight of the water tubes were charged with eggs and glochidia.

Biology: Ortmann (1912:332) reported that a gravid female (with eggs) was collected from the Río Valles, Valles, San Luis Potosí, México, in December 1906 or January 1907. He said that "the presence of eggs (beginning of breeding season) in 'winter' should be noted" and that no glochidia were present, but "the eggs are remarkably small (about 0.10 mm.), and this would indicate similarly small glochidia." He further stated that the structure of the soft parts was in every respect like that of the North American *Potamilus fragilis* (Rafinesque, 1820), but that other genera have a similar structure (Ortmann 1912:332).

Gravid and partially gravid females were collected from the Rio Amajac, 7.4 km WSW of Tamazunchale on 13 December 2018 (INHS 90104, Plate 28), the Rio Huichihuayán, 9 km WSW of Axtla de Terrazas on 10 December 2018 (INHS 90190), the Rio Tampaón, Puente de Dios, 22.1 km SSW of Ciudad Valles on 14 December 2018 (INHS 90146–47, Plates 28–29), and the Rio Valles, Estacion Micos, 21 km NW of Valles on 8 December 2018 (INHS 90395) (all in the Pánuco basin in San Luis Potosí, México). Also found gravid in Arroyo los Gatos, Nuevo Morelos, Tamaulipas, México collected on 13 November 1978 (USNM 1282377).

Adult Size: Length to 10 cm.

Similar Species: *Disconaias disca* (Lea, 1838) is more triangular in shape, stouter, and very thick.

Remarks: Regarding its supposed occurrence in the Río Grande basin, Johnson (1999a:8) stated that "*D. fimbriata* and *salinasensis* (*non,* Simpson) are misidentifications of *Potamilus metnecktayi* Johnson." See additional comments under *Lampsilis* (*Proptera*) *salinasensis* Simpson in Dall, 1908 below.

MAP 38 Distribution of *Disconaias fimbriata* (Frierson, 1907).

Lampsilis fimbriata Frierson, 1907. A new Mexican mussel, *Lampsilis fimbriata*. *The Nautilus* 21(8):86–87+1 plate. Page 86, pl. 12, two upper figures are the holotype. Also figured in Johnson (1972a:162, pl. 27, fig. 2).

Verbatim Type Locality: "Valles River."

Type Locality: Río Valles (Gulf of México–Pánuco–Tampaón Dr.), San Luis Potosí, México [W=21.9829, -99.0184].

Type Specimens: Holotype, UMMZ 91161, by original designation (Johnson 1971:84, incorrectly as UMMZ 191161; Johnson 1972a:162, pl. 27, fig. 2; Johnson and Baker 1973:155; Johnson 1975:28; 1979:32; 1998:434, 452, pl. 26, fig. 1, Parodiz and Tripp 1988:119; all incorrectly as UMMZ 87579). Paratypes, CHAS 22133 (*n*=2); INHS 82195 (*n*=6); INHS 82206 (*n*=3); NHMUK 1910.9.30.170–171 (*n*=2) (Johnson 1971:84); CM 61.4496 (*n*=1) (Brooks and Brooks 1931:175; Parodiz 1967:28; Parodiz and Tripp 1988:119); MCZ 167699 (*n*=3) (Johnson 1956:114; Johnson 1972a:144; Johnson 1998:434); ANSP 99547 (*n*=7), ANSP 99548 (*n*=5), ANSP 99549 (*n*=2).

Remarks: Frierson clearly designated one of the figured specimens as the "type specimen." He was, however, unclear about how many paratypes there were and ambiguous about the deposition of the figured specimen. Numerous citations listed above gave UMMZ 87579 as the "holotype," which is the Bryant Walker collection number. UMMZ 87579 is *Gyraulus* from Sri Lanka. Regarding paratypes, Johnson (1972a:144; 1975:28) identified USNM 207440 (*n*=1) as a paratype of *Lampsilis fimbriata*. Later in his discussion of the Mexican genus *Disconaias*, Johnson (1998:434–435) identified "2 paratypes USNM 207440" and figured one of the two specimens in the lot (plate 26, fig. 2) as a paratype of *Lampsilis fimbriata* Frierson. Later in the same paper, he called one of the specimens in this lot (USNM 207440) a paralectotype of *Lampsilis* (*Proptera*) *salinasensis* Simpson, 1908. The USNM online catalogue identifies it as the "holotype" of *Lampsilis fimbriata* Frierson, 1907. Further complicating matters is that Johnson considered *fimbriata* to be a synonym of *Unio discus* Lea, 1838. The specimen in question, USNM 207440, is *Disconaias disca*, which is distinct from *D. fimbriata*.

Taxonomic History:

Lampsilis fimbriata Frierson, 1907 — Pilsbry (1910)
Paraptera fimbriata (Frierson, 1907) — Ortmann (1912)
Lampsilis (Proptera) fimbriata Frierson, 1907 — Simpson (1914)
Actinonaias disca fimbriata (Frierson, 1907) — H.B. Baker (1922)
Lampsilis (Disconaias) fimbriata Frierson, 1907 — Frierson (1927), Haas (1929; 1969)
Potamilus fimbriatus (Frierson, 1907) — Neck & Metcalf (1988)
Potamilus salinasensis (Simpson, 1908) — Howells et al. (1996)
Disconaias disca (Lea, 1838) — Johnson (1998)
Disconaias fimbriata (Frierson, 1907) — Graf & Cummings (2007; 2021), Cummings & Graf (2010), Williams et al. (2017), Kiser et al. (2022), Czaja et al. (2023)

Lampsilis (Proptera) salinasensis Simpson in Dall 1908. Descriptions and figures of some land and fresh-water shells from Mexico, believed to be new. *Proceedings of the United States National Museum* 35(1642):177–182+plates 29–30. Page 181, pl. 30, fig. 3.

Verbatim Type Locality: "Salinas River, Coahuila, México, Nelson and Goldman; also from Valles River, Valles, México."

Type Locality: Río Valles (Gulf of México–Pánuco–Tampaón Dr.), Valles, San Luis Potosí, México [L=21.9829, -99.0184].

Type Specimens: Lectotype, USNM 163156 designated by Johnson (1998:435). Paralectotypes, USNM 196262 (n=1); USNM 251834 (n=1) (Johnson 1998:435).

Remarks: Simpson (in Dall 1908:181) stated that "The types are Cat. No. 163156, U.S.N.M." and that there were three specimens in the lot. There is only one specimen in that lot today and labeled as coming from the "Salinas R., Coahuila, Méx." Johnson (1975:19) also identified USNM 163156 (n=1) as the "figured holotype" and stated that it was from the "Salinas River, Coahuila." However, the locality data on the label is suspect. Johnson (1998:435, 452, pl. 26, fig. 3) later stated that he "inadvertently" selected USNM 163156 as the holotype and formally designated it as the lectotype but changed the locality to the Río Valles. He presented data (quoted below) that the locality on the label was wrong and that USNM 163156 was collected in the Río Valles, at Valles, San Luis Potosí, México. "As previously mentioned, Dall specifically stated that the material was collected in 1898. Taylor (1966:165) pointed out that Nelson and Goldman were not in Coahuila at all during that year. Peradventure with this in mind, Metcalf (1982:49) accepted what is regarded here as the 'corrected' label as authentic. Taylor was correct: Nelson and Goldman were not at Coahuila during 1898; they were actually at Valles during April 27–29 and May 27–29 of that year, and in the State of San Luis Potosi even longer. The original locality does not exist, and the corrected one appears to be a spurious interpolation, confirmed by the absence of any authentic published records of *Lampsilis (Proptera) salinasensis* from the Rio Grande system since the mussel was described almost a century ago. It is almost certain that it does not occur there. It is most probable that all of Simpson's materials came from the Panuco River System. The type locality has therefore been restricted to the second locality, Valles River, Valles, State of San Luis Potosi, Mexico" (Johnson 1998:436).

Ortmann (1912:332) had previously expressed the same opinion. "Further there is no doubt that *Lampsilis salinasensis* Dall (1909 [sic], p. 181, pl. 30, fig. 3) is the same species. This has been placed by Dall in Simpson's subgenus *Proptera* on account of shell-characters, and there is indeed much similarity of the shells, so that we may regard *P. fimbriata* as a *gracilis* less typically developed, with a stronger shell and better developed hinge-teeth. This species is certainly not a typical *Lampsilis* as shown by the absence of special structures on the edge of the mantle."

Regarding the deposition of the other two specimens in the original lot (USNM 163156), there are two specimens at Florida (UF 270338, ex. Simpson) labeled as "Salinas R. Mex.," the smaller of which may be from the original lot. There is also a likely paralectotype, USNM 251834 ($n=1$), collected by Nelson and Goldman and labeled as coming from "Salinas R., Salinas, Coahuila, México."

Remarks: Howells et al. (1996:103) placed this species in *Potamilus* and provided a map showing various records in the Río Grande del Norte. These are misidentified *Potamilus metnecktayi.*

Taxonomic History:

Lampsilis (Proptera) salinasensis Simpson, 1908 — Dall (1908)
Lampsilis fimbriata Frierson, 1907 — Pilsbry (1910)
Lampsilis (Proptera) fimbriata Frierson, 1907 — Simpson (1914)
Lampsilis (Disconaias) fimbriata Frierson, 1907 — Frierson (1927), Haas (1929; 1969), Taylor (1997)
Disconaias disca (Lea, 1838) — Johnson (1998)
Potamilus salinasensis (Simpson, 1908) — Howells et al. (1996), Haag (2012)
Disconaias salinasensis (Simpson, 1908) — Turgeon et al. (1988; 1998), Williams et al. (1993)
Potamilus salinasensis (Fimbriata, 1907) [sic] — Contreras-Arquieta (2000)
Disconaias fimbriata (Frierson, 1907) — Williams et al. (2017)

Lampsilis moctezumensis Pilsbry, 1910. Unionidae of the Panuco River system, México. *Proceedings of the Academy of Natural Sciences of Philadelphia* 61(3):532–539 + 3 plates. Page 534, pl. 25, figs. 1–2 (lectotype), figs. 3–4 (paralectotype).

Verbatim Type Locality: "Moctezuma River, State of San Luis Potosi."

Type Locality: Río Moctezuma (Gulf of México–Pánuco Dr.), S of Tampamolón, San Luis Potosí, México [L=21.5069, -98.7865].

Type Specimens: Lectotype, ANSP 99517 designated by Johnson and Baker (1973:162). Paralectotypes, ANSP 99516 ($n=1$); INHS 82194 ($n=4$).

Remarks: Johnson and Baker (1973:162) noted that "Lectotype here selected, ANSP 99517a, this was the measured specimen, the figured ones were not located." The former statement by Johnson and Baker is puzzling because the lectotype they designated exactly matches the published figure. The specific type locality is based upon Hinkley's collection site on the Moctezuma.

Taxonomic History:

Lampsilis moctezumensis Pilsbry, 1910 — Simpson (1914)
Lampsilis (Actinonaias) moctezumensis Pilsbry, 1910 — Frierson (1927)
Actinonaias moctezumensis (Pilsbry, 1910) — Haas (1969), Graf & Cummings (2007; 2021), Czaja et al. (2023)

Lampsilis (Disconaias) pilsbryana Frierson, 1927. A classification and annotated check list of the North American naiades. Baylor University Press, Waco, Texas. 111 pp. page 83, species 368. Figured in Frierson (1912, pl. 3, figs. 4–5, as *Lampsilis purpuriatus*) and Johnson (1972a:162, pl. 27, fig. 3).

Verbatim Type Locality: "Valles River, Mecos, México."

Type Locality: Río Valles (Gulf of México–Pánuco–Tampaón Dr.), Micos, San Luis Potosí, México [L=22.1004, -99.1529].

Type Specimens: Lectotype, UMMZ 91175 designated by Johnson (1972a:147; Johnson and Baker 1973:166; Johnson 1979:33, all as the holotype). Paralectotypes, ANSP 99551 ($n=14$) (Johnson 1972a:147; Johnson and Baker 1973:166, both as paratypes).

Remarks: Frierson (1927:83) discussed the nomenclatorial history of *Lampsilis pilsbryana* and other species he placed in the genus *Disconaias* and then stated that he thought *L. pilsbryana* was the same as Say's *Unio purpuriata*. How he arrived at that conclusion is a mystery because there are no types of *U. purpuriata* and it was never figured, but he gave it a new name anyway. Frierson (1927:83) also noted that the specimen he figured in 1912 and called *purpuriata* was distributed by A.A. Hinkley as *Nephronaias strebeli* (Lea, 1866) and equated that species with *Nephronaias aztecorum* (Philippi, 1847). The specimens in the MUSSELp database identified as *D. purpuriata* from the Pánuco are *D. fimbriata*. Johnson (1998:434) placed *fimbriata* along with *panacoensis*, *mexicanus*, *lapidosus*, and *connectens* in the synonymy of *D. disca*, but that hypothesis still needs to be tested. Certainly, the shell morphology of *D. fimbriata* is different compared to *panacoensis*, *mexicanus*, *lapidosus*, and *connectens*. *Actinonaias walkeri* H.B. Baker, 1922 from Veracruz was placed in the synonymy of *Disconaias fimbriata* (Graf and Cummings 2007, following Frierson 1927), but we have placed it in *Mesonaias* below.

Taxonomic History:

Lampsilis (Disconaias) purpuriata (Say, 1831) — Haas (1929; 1969)

Order Unionoida
Superfamily Unionoidea Rafinesque, 1820
Family Unionidae Rafinesque, 1820
Subfamily Ambleminae Rafinesque, 1820
Tribe Lampsilini von Ihering, 1901
Genus *Friersonia* Ortmann, 1912

Friersonia Ortmann, 1912. Notes upon the families and genera of the najades. *Annals of the Carnegie Museum* 8(2):222–365+3 plates. Page 318, fig. 19.

Diagnosis: See species account below.

Type Species: *Lampsilis iridella* Pilsbry & Frierson, 1908, by original designation.

Taxonomic History:

Lampsilis Rafinesque, 1820 — Pilsbry (1910)
Friersonia Ortmann, 1912 — Simpson (1914), Thiele (1934), Haas (1969), Contreras-Arquieta (2000), Graf & Cummings (2007; 2021), Pfeiffer et al. (2019b)
Lampsilis (Friersonia) Ortmann, 1912 — Frierson (1927), Haas (1929)

Distribution: Endemic to the Pánuco basin.

Remarks: This is a monotypic genus with no published data regarding its relationship to other members of the tribe Lampsilini.

PLATE 30 ***Friersonia iridella*** **(Pilsbry & Frierson, 1908). Top Left:** *Lampsilis iridella* Pilsbry & Frierson, 1908. Holotype, ANSP 93810a (49 mm L, 26 mm H). Río Valles (Pánuco Dr.), Valles [L = 21.9829, -99.0184]. **Middle Right:** *Lampsilis novileonis* Pilsbry, 1910. Holotype, ANSP 99518 (78 mm L, 43 mm H). Río Los Gatos [= Río Casas Viejas] (Pánuco–Tampaón–Valles Dr.), N of Valles [L = 22.1415, -99.1071]. **Middle Left and Bottom Right:** *Friersonia iridella* (Pilsbry & Frierson, 1908). INHS 90148.1 (58 mm L, 30 mm H). Río Tampaón (Pánuco Dr.), Puente de Dios, 22.1 km SSW of Ciudad Valles [L = 21.8069, -99.0745]. Gravid female collected on 14 December 2018. All, San Luis Potosí, México.

PLATE 31 ***Friersonia iridella* (Pilsbry & Frierson, 1908). Top:** INHS 90148.2 (55 mm L, 28 mm H). Rio Tampaón, Puente de Dios, 22.1 km SSW of Ciudad Valles [21.8069, -99.0745]. 14 December 2018. **Bottom:** INHS 90396.12 (59 mm L, 32 mm H). Rio Valles (Tampaón Dr.), Estacion Micos, 21 km NW of Valles [L=22.1141, -99.1616]. 8 December 2018. Both, Panucó basin, San Luis Potosí, México.

PLATE 32 ***Friersonia* cf. *iridella* (Pilsbry & Frierson, 1908). Top & Bottom:** INHS 90191.1 (45 mm L, 25 mm H) and INHS 90191.2 (41 mm L, 24 mm H). Río Huichihuayán (Panucó Dr.), 8 km NE Xilitla, San Luis Potosí, México [21.4363, -98.9334]. 10 December 2018.

PLATE 33 ***Friersonia iridella*** **(Pilsbry & Frierson, 1908). Top & Bottom:** *Lampsilis novileonis* Pilsbry, 1910. Paralectotypes, INHS 21757 (57 mm L, 27 mm H) and INHS 82193 (36–62 mm L, 19–32 mm H). Río Los Gatos (Pánuco–Tampaón–Valles Dr.), N of Valles, San Luis Potosí, México [L = 22.1415, -99.1071].

Friersonia iridella (Pilsbry & Frierson, 1908)

Plates 30–33, Map 39

Province(s): Pánuco.

Distribution and Status: This species is found in the Río Pánuco and adjacent basins (Carrizal and Cazones) in San Luis Potosí, Tamaulipas, and Veracruz. *Friersonia iridella* was widespread and common at 19 sites nine sites in the Pánuco, Carrizal, and Cazones basins in 1982 and seven in the Pánuco drainage in 2018. Not assessed (IUCN 2024).

Anatomy: Shell thin, oblong, and wide posteriorly. Anterior end rounded. Posterior sloping above, subtruncate at the end. Beak-sculpture double-looped, consisting of six to eight fine bars, of which the later ones are distinctly double-looped, and the latest are interrupted in the middle. Pseudocardinal teeth small, a single stout one in the right valve, two in the left valve. Lateral teeth narrow, double in the left valve and single in the right. Periostracum smooth, obliquely corrugated along the posterior slope, and dull yellow or pale green with narrow green rays. Nacre bluish-silver and very iridescent.

As pointed out by Ortmann, *Friersonia* has a unique anatomy. While similar to that seen in genera like *Lampsilis* where the marsupium is limited to the posterior part of the demibranch, in *Friersonia*, the marsupium extends over the whole ventral margin of the outer demibranch (Plates 30–31 above).

Ortmann's (1912:318–319, pl. 19, fig. 8) description of the soft parts is quoted here. "Inner lamina of inner gills connected with abdominal sac. Edge of mantle in front of branchial slightly lamellate, with fine and distinct crenulations, disappearing gradually in front, but without papillae. A brown streak of pigment along this part of the edge. Gills of gravid female rather long, the outer marsupial gill covering all of the inner gill, except its anterior end. Marsupium consisting of many ovisacs, occupying the larger posterior section of the outer gill. When gravid, 'the ovisacs swell very little, and they are only slightly compressed in the basal part, which is largely enclosed between the laminae of the gill. The ovisacs reach considerably beyond the edge of the gill, and in this region, they are curved backward in a peculiar manner, subcylindrical, and tapering toward a point directed backward at the hind end of the marsupium. The edge of the marsupium has brownish black pigment in spots. The marsupium has also a remarkably sharp edge, its posterior point, and the recurved ovisacs are unique. Placentae not very solid. Glochidia higher than long. Length 0.20; height 0.22 mm. Glochidia lying all through the placental mass, of medium size, and subovate in shape.

Anal and supra-anal separated by a rather short mantle-connection. Anal finely crenulated. Branchial with papillae. Palpi rather large, posterior margins connected about one-half of their length. Diaphragm and anterior attachment of gills normal. Septa and water-tubes as usual. Gills of gravid female rather long, the outer marsupial gill covering all of the inner gill, except its anterior end. Marsupium quite long and formed by a greater part of the outer gill than usual, yet there is a portion at the anterior end and a very small one at the posterior end, which are non-marsupial. Ovisacs over fifty."

Biology: Ortmann (1912:319) reported that three gravid females, one with eggs and two with glochidia, were collected by A.A. Hinkley in December 1906 and January 1907 from the Río Valles, Valles, San Luis Potosí, México [CM 61.4495, as cotypes]. Gravid and slightly gravid females were collected from the Río Tampaón, at Álvaro Obregón on 30 January 1982 (USNM 1282416–17) and at Puente de Dios, 22.1 km SSW of Ciudad Valles on 14 December 2018 (INHS 90148, Plates 30–31); and the Río Valles, Estacion Micos, 21 km NW of Valles on 8 December 2018 (INHS 90396), all in the Panucó basin in San Luis Potosí, México. Gravid females were also collected from Arroyo los Gatos, Nuevo Morelos, in Tamaulipas, on 5 November 1978 (USNM 1282427) and an unnamed creek (Gulf of Mexico–Cazones Dr.), 5.6 km NW of Poza Rica, in Veracruz, México on 22 February 1982 (USNM 1282432). Although both are in the Panucó basin, the gill structure of the gravid individuals collected from the Río Huichihuayán (Moctezuma Dr.), 8 km NE Xilitla on 10 December 2018 (INHS 90191.1–2, Plate 32) is unlike those from the Río Valles and is likely a new species.

Adult Size: Length to 9 cm.

Similar Species: *Actinonaias medellina* (Lea, 1838) and *Disconaias fimbriata* (Frierson, 1907).

MAP 39 Distribution of *Friersonia iridella* (Pilsbry & Frierson, 1908).

Lampsilis iridella Pilsbry & Frierson, 1908. Description of a Mexican *Lampsilis*. *The Nautilus* 22(8):81–82. Page 82, plate 12, two lower, right-hand figures published in December 1907. Also figured in Johnson (1972a:164, pl. 28, fig. 1).

Verbatim Type Locality: "Valles, Mexico."

Type Locality: Río Valles (Pánuco–Tampaón Dr.), Valles, San Luis Potosí, México [L=21.9829, -99.0184].

Type Specimens: Holotype, ANSP 93810a (Johnson 1971:86; 1972:164, pl. 28, fig. 1). Paratypes, ANSP 93810 (n=18½), NHMUK 1908.12.17.5 (n=1) (Johnson 1971:86); CHAS 22285 (n=10); CM 61.4495 (n=5) (Brooks and Brooks 1931:175; Parodiz and Tripp 1988:121); FMNH 21845 (n=5); INHS 21720 (n=34); INHS 82207 (n=8); MCZ 100405 (n=3) (Johnson 1956:120, 1972a:144); MNHN-IM-2000-1811 (n=2). "UMMZ [not located]" (Johnson 1979:32, possibly UMMZ 50883 (n=1) or UMMZ 66445 (n=6), both from Valles).

Remarks: There is confusion regarding name, date, the types, and type locality of *Lampsilis iridella*. The name was first published in December 1907 as "*L. iridella*" on plate 12 (following page 140) of *The Nautilus*, Volume 21 (incidentally erroneously printed as "Vol. XXII" on the first page (p. 85) of that number) referring to a specimen figured in two photographs on the lower right side of the plate (Frierson 1907). That was followed by a formal description by Pilsbry and Frierson (1908). Pilsbry and Frierson described *Lampsilis iridella* from "Valles, México. Type No. 93810 A.N.S.P., collected by Mr. A.A. Hinkley. Cotypes in the collections of Hinkley and Frierson." Johnson

(1971:86; 1972a:144, pl. 28, fig. 1; 1979:32), Johnson and Baker (1973:159), and Parodiz and Tripp (1988:121) all recognized ANSP 93810 (*n*=1) as the holotype and stated that it was from the Río Valles in San Luis Potosí, México. However, the label associated with ANSP 93810 is not from the published type locality, but rather from Tampico, on the Gulf coast of México, which is nowhere mentioned in the original description, but was a locality sampled by Hinkley (Pilsbry and Hinkley 1909:519, fig. 1), and there are two other lots of *F. iridella* from Tampico (FMNH 21846 and UF 93152). It is quite possible that labels associated with the specimens sent by Hinkley to Pilsbry were mixed up as the majority of the specimen lots of *Friersonia iridella* are in rivers upstream from Tampico in streams draining the Sierra Madre de Oriental.

Taxonomic History:

Lampsilis iridella Pilsbry & Frierson, 1908 — Pilsbry (1910), Johnson (1972)
Friersonia iridella (Pilsbry & Frierson, 1908) — Ortmann (1912), Simpson (1914), Thiele (1934), Haas (1969), Fuller (1975), Contreras-Arquieta (2000), Graf & Cummings (2007; 2021), Kiser et al. (2022), Czaja et al. (2023)
Lampsilis (Friersonia) iridella Pilsbry & Frierson, 1908 — Frierson (1927), Haas (1929)

Lampsilis novileonis Pilsbry, 1910. Unionidae of the Panuco River system, Mexico. *Proceedings of the Academy of Natural Sciences of Philadelphia* 61(3):532–539+3 plates. Page 537, pl. 27, figs. 1–2.

Verbatim Type Locality: "Casas Viejas River, State of San Luis Potosi."

Type Locality: Río Los Gatos [= Río Casas Viejas] (Gulf of México–Pánuco–Tampaón–Valles Dr.), N of Valles, San Luis Potosí [L=22.1415, -99.1071].

Type Specimens: Lectotype, ANSP 99518 designated by Johnson and Baker (1971:163, as the figured holotype). Paralectotypes, ANSP 99519 (*n*=4); INHS 21757 (*n*=1); INHS 82193 (*n*=4); INHS 82278 (*n*=5).

Remarks: Pilsbry mentioned a "type" specimen which was presumably the specimen figured.

Taxonomic History:

Lampsilis novileonis Pilsbry, 1910 — Simpson (1914)
Lampsilis (Disconaias) novi-leonensis [sic] Pilsbry, 1910 — Frierson (1927)
Lampsilis (Disconaias) novileonis Pilsbry, 1910 — Haas (1969)
Disconaias novileonis (Pilsbry, 1910) — Graf & Cummings (2007; 2021), Czaja et al. (2023)

Order Unionoida
Superfamily Unionoidea Rafinesque, 1820
Family Unionidae Rafinesque, 1820
Subfamily Ambleminae Rafinesque, 1820
Tribe Lampsilini von Ihering, 1901
Genus *Mesonaias* Crosse & Fischer, 1894

Mesonaias Crosse & Fischer, 1894. *In* P. Fischer and H. Crosse. 1894. Études sur les mollusques terrestres et fluviatiles du Mexique et du Guatemala. *In* M.H.–Milne Edwards (ed.). *Mission Scientifique au Mexique et dans l'Amérique centrale. Recherches Zoologiques pour servir á l'Histoire de la Faune de l'Amérique Centrale et du Mexique.* 15th Livraison. 62–72, pp. 489–576, plates 59–62, Imprimerie Nationale, Paris Volume 2. 731 pp. Page 556. Introduced as subgenus of *Unio* Philipsson, 1788.

Diagnosis: "Coquille transverse, ovale, élevée; sommets peu saillants; surface lisse ou striée concentriquement; dents latérales antérieures obliques, comprimées (*U. explicatus*, Morelet)." [Shell transverse, oval, elevated; slightly protruding beaks; smooth or concentrically striated surface; oblique, compressed anterior lateral teeth (*U. explicatus*, Morelet)].

Type Species: *Unio explicatus* Morelet, 1859, by original designation.

Taxonomic History:

Unio (*Mesonaias*) Crosse & Fischer, 1894 — Fischer & Crosse (1894b)
Unio (*Uniomerus*) Conrad, 1853 — von Martens (1900b) [in part]
Actinonaias Crosse & Fischer, 1894 — H.B. Baker (1922)
Lampsilis (sec. *Mesonaias*) Crosse & Fischer, 1894 — Thiele (1934)
Mesonaias Crosse & Fischer, 1894 — Morrison (1968), Contreras-Arquieta (2000)
Lampsilis (*Mesonaias*) Crosse & Fischer, 1894 — Haas (1969)
Cyrtonaias Crosse & Fischer, 1894 — Nagel (2025)

Sphenonaias Crosse & Fischer, 1894 — Graf & Cummings (2007; 2021), Pfeiffer et al. (2019a), Inoue et al. (2020a) [all in part]

Cyrtonaias Crosse & Fischer, 1894 — Graf & Cummings (2007; 2021) [in part]

Distribution & Remarks: Graf and Cummings (2021:17) recognized four species in the genus *Cyrtonaias. Cyrtonaias berlandierii* (Lea, 1857) of the Río Grande and the southwestern United States is recognized here as a full species separate from *tampicoensis* (Lea, 183). The other species they listed as being in *Cyrtonaias* are now placed in *Mesonaias*. North to South they are: *Mesonaias tampicoensis* (Lea, 183) in the Pánuco and Papaloapán basins, *M. umbrosa* (Lea, 1856) in the Papaloapán drainage, and *M. explicata* (Morelet, 1849) (including *Cyrtonaias sapperi* von Ihering, 1901) in the Usumacinta basin. If the Papaloapán population of *Mesonaias* proves to be distinct, it would be *Mesonaias tecomatensis* (Lea, 1841).

Pfeiffer et al. (2019a:112) found that *Cyrtonaias explicata* was nested deeply within the Lampsilini and was sister to *Sphenonaias microdon*, whereas *Cyrtonaias tampicoensis* [from Texas—here as *Cyrtonaias berlandierii* (Lea, 1857)] was recovered in its customary position as one of the earlier diverging lineages of the Lampsilini and sister to *Glebula rotundata* (Lamarck, 1819). The same sister relationships were found by Inoue et al. (2020a). Pfeiffer et al. further stated that the obvious polyphyly of *Cyrtonaias* required taxonomic revisions, and we have chosen to sink *Sphenonaias microdon* into *Mesonaias explicata* below based upon a re-examination of the types and place the species that occur south of the Río Grande, previously in *Cyrtonaias*, into *Mesonaias*.

MAP 40 Distribution of the genus *Mesonaias* Crosse & Fischer, 1894.

PLATE 34 ***Mesonaias explicata* (Morelet, 1849). Top Left:** *Unio explicatus* Morelet, 1849. Lectotype, NHMUK 1893.2.4.2027 (115 mm L, 62 mm H). Río Usumacinta, Balancán, Tabasco, México [L=17.8000, -91.5333]. **Top Right:** *Lampsilis* (*Proptera*) *lividus* Simpson, 1900. Lectotype, NHMUK 20190695. Río Usumacinta, El Petén Department, Guatemala [W=16.8246, -90.8822]. **Middle Left:** *Unio microdon* von Martens, 1887. Holotype, ZMB/MOLL.39465 (107 mm L, 57 mm H). Río de las Salinas (Usumacinta Dr.), El Petén Department, Guatemala [W=16.4666, -90.5500]. **Middle Right:** *Unio* (*Lampsilis*) *rovirosai* Pilsbry, 1900. Holotype, ANSP 85916 (111 mm L, 71 mm H). Laguna de Atasta (Usumacinta Dr.), San Juan Bautista, Tabasco, México [L=18.0000, -92.9500]. **Bottom Left:** *Lampsilis sapperi* von Ihering, 1901. Lectotype, SMF 3477 (114 mm L, 66 mm H). Río Chixoy (Usumacinta Dr.), El Petén Department, Guatemala [W=16.0693, -90.4409].

Mesonaias explicata (Morelet, 1849)

Plate 34, Map 41

Province(s): Usumacinta.

Distribution and Status: Endemic to the Río Usumacinta basin in Chiapas and Tabasco, México and El Petén Department, Guatemala (Johnson 1998:438, 454). *Mesonaias explicata* was found alive at only one site in the Río Chacamax (Usumacinta Dr.) at La Libertad in Chiapas, México in 2019. Not assessed (IUCN 2024).

Anatomy: Shell large, elongated oval, smooth or concentrically striated, and relatively thick. Beaks slightly elevated. Beak cavity deep. Anterior end rounded. Posterior end rounded or truncated. Ventral margin straight or slightly convex. Left valve with two compressed pseudocardinals, the anterior the higher, and two lateral teeth. Right valve with one pseudocardinal and one lateral. Dorsal ligament elongated. Periostracum smooth, tan to dark brown, or black. Nacre white, pink, or light purple.

Biology: Pfeiffer et al. (2019a:112) noted that *Mesonaias explicata* (as *C. explicata/S. microdon*) exhibited characteristics typical of the Lampsilini including a modified marsupium that extended ventrally in comparison to the rest of the demibranch and (slightly) sexually dimorphic shells. Pfeiffer et al. (2019a:114, figs. 3a–b) figured the ventral expansion of the marsupium in a non-gravid female (UF 507898), and the slightly sexually dimorphic shells collected from the Río Usumacinta, opposite Emiliano Zapata, Tabasco, México on 8 May 2015.

Adult Size: Length to 10 cm.

Similar Species: *Pachynaias spheniopsis* (Morelet, 1849).

MAP 41 Distribution of *Mesonaias explicata* (Morelet, 1849).

Unio explicatus Morelet, 1849. Testacea novissima insulæ Cubanæ, et Americæ Centralis. Librairie de L'Académie Nationale de Medecine, Paris Part 1:1–31. Page 28. Figured in Fischer and Crosse (1894b:594, pl. 61, fig. 1).

Verbatim Type Locality: "flumen Usumasinta, ad pagum Balancan Tabascensium."

Type Locality: Río Usumacinta (Gulf of México–Grijalva Dr.), Balancán, Tabasco, México [L = 17.8000, -91.5333].

Type Specimens: Lectotype, NHMUK 1893.2.4.2027 designated by Johnson (1971:83, as the holotype; Johnson 1998:452; Breure et al. 2018:282, figs. 393–394). Paralectotypes, MNHN-IM-2000-1693 (*n* = 2) (Breure et al. 2018:282); SMF 316727 (*n* = 1), SMF 316726 (*n* = 1) (Nagel 2025:163, the latter as a probable syntype).

Taxonomic History:

Unio (*Mesonaias*) *explicatus* Morelet, 1849 — Fischer & Crosse (1894b)
Unio (*Uniomerus*) *explicatus* Morelet, 1849 — von Martens (1900b)
Unio (*Elliptio*) *explicatus* Morelet, 1849 — von Martens (1901c)
Lampsilis (*Proptera*) *explicatus* (Morelet, 1849) — Simpson (1900)
Lampsilis (*Proptera*) *explicata* (Morelet, 1849) — Simpson (1914)
Actinonaias explicata (Morelet, 1849) — H.B. Baker (1922)
Lampsilis (*Cyrtonaias*) *explicata* (Morelet, 1849) — Frierson (1927), Haas (1929)
Lampsilis (sec. *Mesonaias*) *explicatus* (Morelet, 1849) — Thiele (1934)
Lampsilis (*Mesonaias*) *explicata* (Morelet, 1849) — Haas (1969)
Mesonaias explicatus (Morelet, 1849) — Morrison (1968), Contreras-Arquieta (2000)
Lampsilis explicata (Morelet, 1849) — Johnson (1998), Contreras-Arquieta (2000)
Cyrtonaias explicata (Morelet, 1849) — Graf & Cummings (2007; 2021), Pfeiffer et al. (2019a), Czaja et al. (2023), Nagel (2025)

Unio testudineus Reeve, 1865. Monograph of the genus *Unio*. *In* L. Reeve and G.B. Sowerby (eds.). *Conchologia Iconica* 16:[unpaginated] + plates 19–30. Plate 22, species 101.

Remarks: Misidentified reference to *Unio testudineus* Morelet, 1849 which is *Psoronaias semigranosa* (von dem Busch in Philippi, 1843).

Taxonomic History:

Unio (*Mesonaias*) *explicata* Morelet, 1849 — Fischer & Crosse (1894)
Unio (*Uniomerus*) *explicatus* Morelet, 1849 — von Martens (1900b)
Lampsilis (*Proptera*) *lividus* Simpson, 1900 — Simpson (1900)
Lampsilis lividus Simpson, 1900 — de Leon (1912)
Lampsilis (*Proptera*) *livida* Simpson, 1900 — Simpson (1914)
Lampsilis (*Cyrtonaias*) *livida* Simpson, 1900 — Frierson (1927)
Lampsilis (*Cyrtonaias*) *microdon* (von Martens, 1887) — Haas (1929)

Unio microdon von Martens, 1887. Süsswassermuscheln aus Guatemala. *Sitzungsberichte der Gesellschaft naturforschender Freunde zu Berlin* 7:106–108. Page 107. Figured in von Martens (1900b, pl. 35, figs. 1, 1a–c).

Verbatim Type Locality: "cum praecedente." [Rio de las Salinas, prov. Peten, Guatemalae.]

Type Locality: Río de las Salinas (Gulf of México–Grijalva–Usumacinta Dr.), El Petén Department, Guatemala [W = 16.4666, -90.5500].

Type Specimens: Holotype by monotypy, ZMB/MOLL.39465.

Remarks: Zilch (1967.120) listed SMF 3477 as the holotype and SMF 3478 (*n* = 2) as paratypes. SMF 3477 and 3478 are the lectotype and paralectotypes of *Unio sapperi* von Ihering, 1901, which at the time was synonymous with *Lampsilis* (*Proptera*) *microdon*. von Martens (1887:108) gave size ranges of 93–107 mm L, 50–57 mm H although there appears to be only one specimen extant.

Taxonomic History:

Unio microdon von Martens, 1887 — Fischer & Crosse (1894b)
Unio (*Nephronaias*) *microdon* von Martens, 1887 — von Martens (1900b)
Nephronaias microdon (von Martens, 1887) — Simpson (1914)
Elliptio (*Sphenonaias*) *microdon* (von Martens, 1887) — Frierson (1927)
Lampsilis (*Cyrtonaias*) *microdon* (von Martens, 1887) — Haas (1929), Zilch (1967)
Nephronaias (*Sphenonaias*) *microdon* (von Martens, 1887) — Haas (1969)
Sphenonaias microdon (von Martens, 1887) — Graf & Cummings (2007; 2021), Pfeiffer et al. (2019a), Inoue et al. (2020a), Ramírez et al. (2022), Czaja et al. (2023)

Unio (*Lampsilis*) *rovirosai* Pilsbry, 1900. Notes on some southern Mexican shells. *The Nautilus* 13(12):139–141. Page 140. Figured in Pilsbry (1904:788, pl. 54, figs. 1–1a).

Verbatim Type Locality: "Laguna de Atasta, nr. San Juan Bautista, Tabasco, Mexico."

Type Locality: Laguna de Atasta (Gulf of México–Grijalva–Carrizal Dr.), San Juan Bautista, Tabasco, México [L = 18.0037, -92.9443].

Type specimen: Holotype by monotypy, ANSP 85916 (Johnson and Baker 1973:169, incorrectly as ANSP 58916).

Remarks: Pilsbry mentioned the type as being a female. Pilsbry (1900:141) stated that it was "allied to *U. umbrosus*" [= *Mesonaias umbrosa*] which is endemic to Papaloapán Province. An attempt to locate specimens at the type locality in December 2019 was unsuccessful.

Taxonomic History:

Lampsilis (*Proptera*) *rovirosai* (Pilsbry, 1900) — Simpson (1900; 1914)
Unio (*Lampsilis*) *rovirosai* Pilsbry, 1900 — von Martens (1901c)
Lampsilis rovirosai (Pilsbry, 1900) — Pilsbry (1904), de Leon (1912)
Lampsilis rovirosai rovirosai (Pilsbry, 1900) — H.B. Baker (1922)
Proptera rovirosai (Pilsbry, 1900) — Frierson (1927), Haas (1969)
Lampsilis revirosai [sic] (Pilsbry, 1900) — Contreras-Arquieta (2000)
Lampsilis rousrosai [sic] (Pilsbry, 1900) — Contreras-Arquieta (2000)
Potamilus rovirosai (Pilsbry, 1900) — Graf & Cummings (2007), Czaja et al. (2023)
Cyrtonaias umbrosa (Lea, 1856) — Graf & Cummings (2021)

Lampsilis sapperi von Ihering, 1901. The Unionidae of North America. (cont.). *The Nautilus* 15(5):50–53. Page 50. Also figured in Haas (1929:313, figs. 2–3) and Nagel (2025:167. fig. 164).

Verbatim Type Locality: "Chixoy River in Guatemala."

Type Locality: Río Chixoy (Gulf of México–Grijalva–Usumacinta–Salinas Dr.), El Petén Department, Guatemala [W = 16.0693, -90.4409].

Type Specimens: Holotype, SMF 3477 (Haas 1929:313; Zilch 1967:120). Paratypes, SMF 3478 (*n* = 2) (Haas 1929:313; Zilch 1967:120).

Remarks: von Ihering (1901:51) noted that "I believe the *L. sapperi* more allied to *L. umbrosus* Lea than to *explicatus*, being, however, of a more elongate form with produced posterior extremity and larger size." We restrict Mesonaias umbrosa to Gulf coastal drainages (Coatzacoalcos, Jamapa, and Papaloapán basins) in Veracruz, México in Papaloapán Province. Baker (1922:19) postulated that "From the original comparison, *L. sapperi* von Ihering (1901) appears to resemble an old specimen of *A. explicata...*" but he still recognized *Lampsilis sapperi* (von Ihering, 1901) as valid. The Río Chixoy is called Río Negro from its sources in the highlands of Huehuetenango until it reaches the Chixoy hydroelectric dam. After the dam, the river is called Río Chixoy and flows north along the border of Alta Verapaz and Quiché Guatemala until it reaches the Mexican border. It continues along the border for another 113 km as the Río Salinas until it converges with the Río la Pasión to form Río Usumacinta.

Taxonomic History:

Lampsilis (Proptera) sapperi von Ihering, 1901 — Simpson (1914)
Lampsilis sapperi von Ihering, 1901 — H.B. Baker (1922)
Lampsilis (Cyrtonaias) sapperi von Ihering, 1901 — Frierson (1927), Haas (1969)
Lampsilis (Cyrtonaias) microdon (von Martens, 1887) — Haas (1929), Zilch (1967)
Cyrtonaias sapperi (von Ihering, 1901) — Graf & Cummings (2007; 2021), Nagel (2025)

Lampsilis (Proptera) lividus Simpson, 1900. Synopsis of the naiades, or pearly fresh-water mussels. *Proceedings of the United States National Museum* 22(1205):501–1044. Page 571.

Verbatim Type Locality: "Rio Usumasinta, Mexico, and Guatemala."

Type Locality: Río Usumacinta (Gulf of México–Grijalva Dr.), El Petén Department, Guatemala [W=16.8246, -90.8822].

Type Specimens: Lectotype, NHMUK 20190695, here designated.

Remarks: This name is based upon Reeve's 1865 *Unio testudineus*, Plate 22, species 101. Simpson (1900:571) described it as a new species noting that "Reeve described and figured this form in the *Conchologia*, supposing it to be the *testudineus* of Morelet, which is a very different thing. Fischer and Crosse in *Mission Scientifique*, believing it to be equal to Morelet's *explicatus*, placed it in the synonymy of that species. I have never seen an absolutely authentic specimen of *explicatus*, but if Fischer and Crosse have accurately figured and described it, I am sure it is different from Reeve's species, the latter being evidently more inflated, smoother, differently shaped, and having a great deal higher beaks." Simpson (1900:571) also noted that Reeve listed the locality as "Rio Usumasinto, Portugal."

Taxonomic History:

Lampsilis lividus Simpson, 1900 — de Leon (1912), H.B. Baker (1922)
Lampsilis (Proptera) livida Simpson, 1900 — Simpson (1914)
Lampsilis (Cyrtonaias) livida Simpson, 1900 — Frierson (1927), Haas (1969)
Lampsilis (Cyrtonaias) microdon (von Martens, 1887) — Haas (1929)
Lampsilis explicata (Morelet, 1849) — Johnson (1998)
Cyrtonaias livida (Simpson, 1900) — Graf & Cummings (2007), Czaja et al. (2023)
Cyrtonaias explicata (Morelet, 1849) — Graf & Cummings (2021)

PLATE 35 ***Mesonaias tampicoensis* (Lea, 1838). Top Left:** *Unio tampicoensis* Lea, 1838. Lectotype, USNM 84843 (112 mm L, 66 mm H). Río Pánuco, above Tampico, Tamaulipas, México [L = 22.2078, -97.8482]. **Middle Right:** *Unio tecomatensis* Lea, 1841. Lectotype, USNM 85338 (89 mm L, 63 mm H). Río Tecomate (Papaloapán Dr.), Tlacotalpan, Veracruz, México [L = 18.6083, -95.6583]. **Middle Left:** *Actinonaias (Disconaias) walkeri* H.B. Baker, 1922. Holotype, UMMZ 31844 (105 mm L, 60 mm H). Río San Juan (Papaloapán Dr.), Cuatotolapam, Veracruz, México [L = 18.1241, -95.2998]. **Bottom Right:** *Lampsilis rovirosai sanjuanensis* H.B. Baker, 1922. Holotype, UMMZ 34890 (82 mm L, 53 mm H). Río San Juan (Papaloapán Dr.), Cuatotolapam, Veracruz, México [L = 18.1241, -95.2998].

PLATE 36 ***Mesonaias tampicoensis* (Lea, 1838).** INHS 90091.2 (115 mm L, 75 mm H). Rio Valles (Pánuco–Tampaón Dr.), Álvaro Obregón, at the confluence of the ríos Valles and Tampaón, San Luis Potosí, México [L=21.8461, -98.9438]. Gravid female collected on 9 December 2018.

Mesonaias tampicoensis (Lea, 1838)

Plates 35–36, Map 42

Province(s): Pánuco; Papaloapán.

Distribution and Status: Disjunct populations are present in the ríos Pánuco and Papaloapán. If the Papaloapán population proves to be distinct, it would be *Mesonaias tecomatensis* (Lea, 1841). Simpson (1914:173) gave Honduras as part of the range, and there are two specimens (UMMZ 121213) labeled as coming from Honduras without additional locality data, but we regard those as spurious. *Mesonaias tampicoensis* was found widespread and common in the Pánuco and Papaloapán basins in San Luis Potosí and Veracruz, México in 1982 and 2018. *Cyrtonaias tampicoensis tecomatensis* (Lea, 1841) is considered endangered by the Mexican Secretariat of Environment and Natural Resources (Semarnat 2019). Not assessed (IUCN 2024).

Anatomy: Shell elliptically rounded and smooth. Ventral margin straight. Beaks moderately full, with faint sculpture. Beak cavity deep. Left valve with two pseudocardinals and two slightly curved laterals. Right valve with two pseudocardinals and one lateral. Lateral teeth stout and rose-colored in some specimens. Periostracum somewhat shiny, tan, dark chestnut, or black. Nacre white, pink, or purple.

Biology: Gravid females were collected from the Río Valles (Pánuco Dr.), Álvaro Obregón, at confluence of the ríos Valles and Tampaón, San Luis Potosí, México on 9 December 2018 (INHS 90091, Plate 36). Also, from the Río San Juan (Papaloapán Dr.), 33 km S of Santiago Tuxtla Veracruz, México on 15 February 1982 (USNM 1282378–79).

Adult Size: Length to 13 cm.

Similar Species: *Mesonaias umbrosa* (Lea, 1856).

MAP 42 Distribution of *Mesonaias tampicoensis* (Lea, 1838).

Unio tampicoensis Lea, 1838. Description of new freshwater and land shells. *Transactions of the American Philosophical Society* 6(N.S.):1–154+plates 1–24. Page 24, pl. 7, fig. 18. Also figured in Küster (1862:275, pl. 93, fig. 1), Sowerby (1867, pl. 57, sp. 291), and von Martens (1900b, pl. 33, figs. 1, 1a–b, ZMB/MOLL.305834).

Verbatim Type Locality: "River Tampico, México. River Medellin, Mexico."

Type Locality: Río Pánuco (Gulf of Mexico Dr.), above Tampico, Tamaulipas, México [W=22.2078, -97.8482].

Type Specimens: Lectotype, USNM 84843 designated and type locality fixed by Johnson (1999a:26; Johnson and Baker 1973:172; and Johnson 1974a:143, both as the holotype).

Remarks: Regarding the type locality, Lea (1838b:24) noted that "The specimen figured here, I owe to the kindness of Richard Ronaldson, Esq., who procured it from the commander of a vessel trading to Tampico. It was brought some distance above that city. Dr. Burrough subsequently procured the same species from the river Medellin, ten miles south of Vera Cruz." The second specimen collected by Burrough and referred to by Lea was later described by him as *Unio umbrosus* in 1856.

Taxonomic History:

Margarita (Unio) tampicoensis (Lea, 1838) — Lea (1836, *nomen nudem*)
Unio tampicoensis Lea, 1838 — Hanley (1843; 1856), Conrad (1853), Sowerby (1867), Wright (1888)
Margaron (Unio) tampicoensis (Lea, 1838) — Lea (1852; 1870)
Unio (Hyridella) tampicoensis (Lea, 1838) — Adams & Adams (1857)
Unio tampicoënsis Lea, 1838 — Küster (1862)
Unio (Cyrtonaias) tampicoensis Lea, 1838 — Fischer & Crosse (1894b)
Lampsilis (Proptera) tampecoensis [sic] (Lea, 1838) — Simpson (1900) [in part]
Unio (Lampsilis) tampicoensis Lea, 1838 — von Martens (1900b) [in part]
Lampsilis tampicoensis (Lea, 1838) — Pilsbry (1910), de Leon (1912)
Lampsilis (Proptera) tampicoensis (Lea, 1838) — Simpson (1914) [in part]
Lampsilis (Cyrtonaias) tampicoensis (Lea, 1838) — Frierson (1927), Haas (1929)
Cyrtonaias tampicoensis (Lea, 1838) — Morrison (1968), Burch (1975) [in part], Williams et al. (1993; 2017) [in part], Howells et al. (1996) [in part], Johnson (1999a) [in part], Contreras-Arquieta (2000), Graf & Cummings (2007) [in part], Cummings & Graf (2010) [in part], Graf & Cummings (2021) [in part], Kiser et al. (2022), Czaja et al. (2023) [in part]
Lampsilis (Cyrtonaias) tampicoensis tampicoensis (Lea, 1838) — Haas (1969) [in part]

Unio tecomatensis Lea, 1841. On fresh water and land shells (cont.). Proceedings of the American Philosophical Society 2(17):30–34. Page 30. Figured in Lea (1842a TAPS Vol. 8:234; 1842b Obs. 3:72, pl. 21, fig. 48) and Fischer and Crosse (1894b:589, pl. 65, fig. 6; pl. 66, figs. 4–4a).

Verbatim Type Locality: "Tecomate River, near Tlocatalpam, Mex."

Type Locality: Río Papaloapán (Gulf of Mexico Dr.), Tlacotalpan, Veracruz, México [L=18.6083, -95.6583].

Type Specimens: Lectotype, USNM 85338 designated by Johnson (1999a:26; Johnson and Baker 1973:172; Johnson 1974a:144, both as the figured holotype). Paralectotype, ANSP 129940 ($n=1$) (Johnson and Baker 1973:172, as a paratype and as ANSP 129941). Tlacotalpan is located at the confluence of the ríos Papaloapán and San Juan.

Taxonomic History:

Unio tecomatensis Lea, 1841 — Conrad (1853)
Margaron (Unio) tecomatensis (Lea, 1841) — Lea (1852; 1870)
Unio (Hyridella) tecomatensis Lea, 1841 — Adams & Adams (1857)
Unio (Cyrtonaias) tecomatensis Lea, 1841 — Fischer & Crosse (1894b)
Unio (Lampsilis) tampicoensis var. *tecomatensis* Lea, 1841 — von Martens (1900b)
Lampsilis (Proptera) tecomatensis (Lea, 1841) — Simpson (1900; 1914)
Lampsilis tecomatensis (Lea, 1841) — de Leon (1912)
Unio tecomensis [sic], sp. aff. Lea, 1841 — de Gregorio (1914)
Actinonaias (Leptodea?) tecomatensis (Lea, 1841) — H.B. Baker (1922)
Lampsilis (Cyrtonaias) tampicoensis tecomatensis (Lea, 1841) — Frierson (1927), Haas (1969)
Cyrtonaias tecomatensis (Lea, 1841) — Morrison (1968), Contreras-Arquieta (2000)

Actinonaias (Disconaias) walkeri H.B. Baker, 1922. The Mollusca collected by the University of Michigan-Walker Expedition in Southern Vera Cruz, México. I. Occasional Papers of the Museum of Zoology, University of Michigan No. 106. 95 pp. Page 20, pl. 1, figs. 1–2 (young specimens); pl. 9, fig. 49, as the "type"; pl. 10, figs. 48–50, fig. 49, as the "type"; pl. 11, figs. 48–49 (hinge).

Verbatim Type Locality: "Rio San Juan."

Type Locality: Río San Juan (Gulf of México–Papaloapán Dr.), Cuatotolapam, Veracruz, México [L=18.1241, -95.2998].

Type Specimens: Holotype, UMMZ 31844 by original designation (as the "type" without the catalogue number; Johnson and Baker 1973:174; Johnson 1979:34; Johnson 1998:454, pl. 27, fig. 2, as a paratype). Paratypes, ANSP 133694 (n=2) (Johnson and Baker 1973:174; Johnson 1998:454)

Remarks: H.B. Baker (1922:20) mentioned that he had "Fifteen specimens, including odd valves, from Rio San Juan" and mentioned a "type" but did not assign any catalogue numbers. MCZ 319586 (n=1) is labeled as a paratype (ex. UMMZ 31844) on their web page, but it was never published as such.

Taxonomic History:

Lampsilis (Disconaias) fimbriata Frierson, 1907 — Frierson (1927), Haas (1969), Taylor (1997)
Actinonaias walkeri H.B. Baker, 1922 — Johnson (1998), Contreras-Arquieta (2000)

Lampsilis rovirosai sanjuanensis H.B. Baker, 1922. The Mollusca collected by the University of Michigan-Walker Expedition in Southern Vera Cruz, México. I. Occasional Papers of the Museum of Zoology, University of Michigan No. 106. 95 pp. Page 27, pl. 7 [8], figs. 39–42, fig. 40, as the "type."

Verbatim Type Locality: "Rio San Juan."

Type Locality: Río San Juan (Gulf of México–Papaloapán Dr.), Cuatotolapam, Veracruz, México [L=18.1241, -95.2998].

Type Specimens: Holotype, UMMZ 34890 by original designation (as the "type" without a catalogue number; Johnson 1979:34). Paratypes, ANSP 133691 (n=2½) (Johnson and Baker 1973:169).

Remarks: Baker (1922:27) mentioned that he had "Fourteen specimens, including odd valves from the Rio San Juan."

Taxonomic History:

Lampsilis (Cyrtonaias) umbrosa (Lea, 1856) — Frierson (1927)

Lampsilis (Cyrtonaias) tampicoensis kusteriana Frierson, 1927. A classification and annotated check list of the North American naiades. Baylor University Press, Waco, Texas. 111 pp. Page 85 sp. 383d.

Verbatim Type Locality: "Tampico in Mexiko."

Type Locality: Río Pánuco (Gulf of Mexico Dr.), Tampico, Tamaulipas, México [L=22.2166, -97.8500].

Type Specimens: Unknown. Presumed lost.

Remarks: Based on the figure of *Unio tampicoënsis* in Küster (1862:275, pl. 93, fig. 1). Johnson (1972a:168, pl. 30, fig. 1) reproduced Küster's figure and gave the size as 98 mm L, 67 mm H and stated that "Type in collection of von dem Busch, which is unknown" (Johnson 1972a:145; 1999a:27).

Taxonomic History:

Lampsilis (Cyrtonaias) tampicoensis kusteriana Frierson, 1927 — Haas (1969)
Cyrtonaias tampicoensis (Lea, 1838) — Johnson (1999a)

PLATE 37 ***Mesonaias umbrosa* (Lea, 1856). Top:** *Unio umbrosus* Lea, 1856. Lectotype, USNM 84848 (89 mm L, 48 mm H). Río Medellin (Jamapa Dr.), Veracruz, Veracruz, México [L = 19.0574, -96.1550]. **Bottom:** *Unio alienigenus* Crosse & Fischer, 1894. Lectotype, MNHN-IM-2000-1634 (82 mm L, 51 mm H). Río Coatzacoalcos (Gulf of Mexico Dr.), Veracruz, México [W = 18.1160, -94.4218].

Mesonaias umbrosa (Lea, 1856)

Plate 37, Map 43

Province(s): Papaloapán.

Distribution and Status: Found in Gulf coastal drainages (Coatzacoalcos, Jamapa, and Papaloapán basins) in Veracruz, México. *Mesonaias umbrosa* was found widespread and common in the Papaloapán basin in Veracruz, México in 1982. It was not found in a limited survey of the Papaloapán drainage in 2022. Not assessed (IUCN 2024).

Anatomy: Shell obovate, subsolid, and slightly inflated. Beaks rather low. Beak cavity relatively shallow. Left valve with two pseudocardinals and two curved laterals. Right valve with one pseudocardinal, a vestigial one above it, and one lateral. Periostracum smooth, chestnut brown or black with obscure rays in some individuals. Nacre white, salmon, or rose-colored and iridescent.

Biology: Gravid females were collected from various sites in the Río Papaloapán basin in Veracruz, México from 15–17 February 1982 (USNM 1282346-56 and USNM 1282358).

Adult Size: Length to 10 cm.

Similar Species: *Mesonaias tampicoensis* (Lea, 1838) occurs syntopically in the Papaloapán basin, but it is round in shape with elevated beaks and typically darker in color.

MAP 43 Distribution of *Mesonaias umbrosa* (Lea, 1856).

Unio umbrosus Lea, 1856. Description of twenty-five new species of exotic Uniones. *Proceedings of the Academy of Natural Sciences of Philadelphia* 8(2):92–95. Page 95. Figured in Lea (1858a JANSP Vol. 3:311; 1858b Obs. 6:31, pl. 30, fig. 26), Sowerby (1866, pl. 33, sp. 170), Fischer and Crosse (1894b:587, pl. 67, figs. 1, 1a–b), and von Martens (1900b, pl. 35, figs. 3–5, 5a–b, NHMUK 1901.6.22.1513–1515).

Verbatim Type Locality: "Medellin River, Mexico."

Type Locality: Río Medellin (Gulf of Mexico–Jamapa Dr.), Veracruz, Veracruz, México [L=19.0574, -96.1550].

Type Specimens: Lectotype, USNM 84848 designated by Johnson (1974a:151, as the figured holotype). Paralectotype, MCZ 178872 (*n*=1) (Johnson 1956:140, as a paratype); ANSP A14690 (soft parts only).

Taxonomic History:

Unio umbrosus Lea, 1856 — Sowerby (1866)
Margaron (Unio) umbrosus (Lea, 1856) — Lea (1870)
Unio (Cyrtonaias) umbrosus Lea, 1856 — Fischer & Crosse (1894b)
Unio (Lampsilis) tampicoensis var. *umbrosus* Lea, 1856 — von Martens (1900b)
Lampsilis (Proptera) umbrosus (Lea, 1856) — Simpson (1900)
Lampsilis umbrosus (Lea, 1856) — Pilsbry (1904), de Leon (1912), Contreras-Arquieta (2000)
Lampsilis (Proptera) umbrosa (Lea, 1856) — Simpson (1914)
Actinonaias umbrosa (Lea, 1856) — H.B. Baker (1922)
Lampsilis (Cyrtonaias) umbrosa (Lea, 1856) — Frierson (1927), Haas (1929)
Lampsilis (Cyrtonaias) umbrosa umbrosa (Lea, 1856) — Haas (1969)
Cyrtonaias umbrosa von Martens, 1900 — Contreras-Arquieta (2000)
Cyrtonaias umbrosa (Lea, 1856) — Graf & Cummings (2007; 2021), Czaja et al. (2023)

Unio alienigenus Crosse & Fischer, 1894. Diagnoses molluscorum novorum, reipublicæ Mexicanæ incolarum. *Journal de Conchyliologie* 41(4):293–297. Page 294. Figured in Fischer and Crosse (1894b:590, pl. 65, figs. 5–5a; pl. 66, fig. 3) and von Martens (1900b, pl. 35, figs. 2–2a, ZMB/MOLL.305836).

Verbatim Type Locality: "Goatzalcoalcos, provinciæ Vera Cruz dictæ, in republica Mexicana."

Type Locality: Río Coatzacoalcos (Gulf of México Dr.), Veracruz, México [W=18.1160, -94.4218].

Type Specimens: Lectotype, MNHN-IM-2000-1634, here designated (the figured specimen).

Remarks: The date of publication is often cited as 1893 but is 1894 according to Winckworth (1936:156).

Taxonomic History:

Unio (Cyrtonaias) alienigenus Crosse & Fischer, 1893 — Fischer & Crosse (1894b)
Unio (Lampsilis) alienigenus Crosse & Fischer, 1893 — von Martens (1900b; 1901b)
Lampsilis (Proptera) alienigenus (Crosse & Fischer, 1893) — Simpson (1900)
Lampsilis alienigenus (Crosse & Fischer, 1893) — de Leon (1912)
Lampsilis (Proptera) alienigena (Crosse & Fischer, 1893) — Simpson (1914)
Actinonaias umbrosa (Lea, 1856) — H.B. Baker (1922)
Lampsilis (Cyrtonaias) alienigena (Crosse & Fischer, 1893) — Frierson (1927)
Lampsilis (Cyrtonaias) umbrosa alienigena (Crosse & Fischer, 1893) — Haas (1969)

Order Unionoida
Superfamily Unionoidea Rafinesque, 1820
Family Unionidae Rafinesque, 1820
Subfamily Ambleminae Rafinesque, 1820
Tribe Lampsilini von Ihering, 1901
Genus *Pachynaias* Crosse & Fischer, 1894

Pachynaias Crosse & Fischer, 1894. *In* P. Fischer and H. Crosse. 1894. Études sur les mollusques terrestres et fluviatiles du Mexique et du Guatemala. *In* M.H.–Milne Edwards (ed.). *Mission Scientifique au Mexique et dans l'Amérique centrale. Recherches Zoologiques pour servir á l'Histoire de la Faune de l'Amérique Centrale et du Mexique.* 15th Livraison. 62–72, pp. 489–576, plates 59–62, Imprimerie Nationale, Paris Volume 2. 731 pp. Page 556. Introduced as subgenus of *Unio* Philipsson, 1788.

Diagnosis: "Coquille très-épaisse, renflée, subtrigone, subéquilatérale, ornée seulement de sillons concentriques (*U. spheniopsis.* Morelet)." [Shell very thick, swollen, subtrigonal, subequilateral, decorated only with concentric furrows (*U. spheniopsis.* Morelet)].

Type Species: *Unio spheniopsis* Morelet, 1849, by original designation.

Taxonomic History:

Quadrula (sec. *Pachynaias*) Crosse & Fischer, 1894 — Simpson (1900; 1914)
Unio (*Pleurobema*) Rafinesque, 1819 — von Martens (1900b) [in part]
Nephronaias Crosse & Fischer, 1894 — von Ihering (1901)
Elliptio (*Pachynaias*) Crosse & Fischer, 1894 — Frierson (1927)
Pachynaias Crosse & Fischer, 1894 — Haas (1929), Thiele (1934), Modell (1942), Haas (1969), Lopes-Lima (2017), Pfeiffer et al. (2019a–b), Inoue et al. (2020a), Smith et al. (2020), Graf & Cummings (2021), Nagel (2025)
Ellipsaria (*Pachynaias*) Crosse & Fischer, 1894 —Modell (1964)
Pavhynaias [sic] Crosse & Fischer, 1894 — Contreras-Arquieta (2000)

Distribution & Remarks: von Martens (1900b:517) incorrectly stated that *Unio sallei* Crosse & Fischer, 1894 was the type species of *Pachynaias* Crosse & Fischer, 1894. Graf and Cummings (2021:18) listed two species of *Pachynaias* Crosse & Fischer, 1894 (Ambleminae: Lampsilini). We only recognize one species of *Pachynaias* placing *P. rugososulcatus* (Lea, 1866) under *P. spheniopsis* (Morelet, 1849).

PLATE 38 ***Pachynaias spheniopsis* (Morelet, 1849)**. **Top Left:** *Unio spheniopsis* Morelet, 1849. Lectotype, NHMUK 1893.2.4.2028 (50 mm L, 42 mm H). Río Usumacinta (Grijalva Dr.), El Petén Department, Guatemala [W = 16.8246, -90.8822]. **Middle Right:** *Unio rugososulcatus* Lea, 1866. Lectotype, USNM 84392 (76 mm L, 56 mm H). Central America. **Bottom Left:** *Unio* (*Elliptio*) *sallei* var. *grossus* von Martens, 1900. Holotype, ZMB/MOLL.108850 (62 mm L, 44 mm H). Río de las Salinas (Usumacinta Dr.), El Petén Department, Guatemala [W = 16.4666, -90.5500].

Pachynaias spheniopsis (Morelet, 1849)

Plate 38, Map 44

Province(s): Usumacinta.

Distribution and Status: Endemic to the Río Usumacinta drainage in Tabasco, México and El Petén Department, Guatemala. *Pachynaias spheniopsis* was collected alive from four sites in the Usumacinta drainage: two in 1990–1991 and one in 2015 from Tabasco México and one in the Río de la Pasion, El Petén Department, Guatemala in 2105. Not assessed (IUCN 2024).

Anatomy: Shell triangular, solid, thick, and inflated. Anterior end rounded, dorsal margin descending from the beaks in a nearly straight, slightly convex line. A very obtuse ridge descends from the beaks to the posterior end of the ventral margin and forming a sulcus. Ventral margin slightly convex at the middle. Beaks are not very prominent. Beak cavity deep. Beak sculpture consists of concentric, irregular ridges that covers the surface of the shell. Left valve with one pseudocardinal and two short laterals. Right valve with one strong pseudocardinal and one lateral. Periostracum typically tan or light brown, and rough. Nacre white or bluish white.

Biology: A gravid female was collected from the Río Usumacinta, opposite Emiliano Zapata, Tabasco, México on 8 May 2015 (UF 507900) (Pfeiffer et al. 2019a:114, figs. 3c–d).

Adult Size: Length to 6 cm.

Similar Species: *Mesonaias explicata* (Morelet, 1849).

MAP 44 Distribution of *Pachynaias spheniopsis* (Morelet, 1849).

Unio spheniopsis Morelet, 1849. Testacea novissima insulæ Cubanæ, et Americæ Centralis. Librairie de L'Académie Nationale de Medecine, Paris Part 1:1–31. Page 29. Figured in Fischer and Crosse (1894a:583, pl. 61, figs. 2, 2a–b).

Verbatim Type Locality: "H. regiones superiores fluminis Usumasinta."

Type Locality: Río Usumacinta (Gulf of México–Grijalva Dr.), El Petén Department, Guatemala [W=16.8246, -90.8822].

Type Specimens: Lectotype, NHMUK 1893.2.4.2028 designated by Johnson (1971:94, as the holotype; Breure et al. 2018:438, figs. 1076–1077). Paralectotypes, NHMUK 1893.2.4.2029 (*n*=1) (Johnson 1971:94, as a paratype; Breure et al. 2018:438); MNHN-IM-2000-1791 (*n*=1) (Breure et al. 2018:438).

Taxonomic History:

Unio (*Pachynaias*) *spheniopsis* Morelet, 1849 — Fischer & Crosse (1894b)
Unio (*Pleurobema*) *spheniopsis* Morelet, 1849 — von Martens (1900b)
Quadrula (sec. *Pachynaias*) *spheniopsis* (Morelet, 1849) — Simpson (1900; 1914)
Nephronaias spheniopsis (Morelet, 1849) — von Ihering (1901)
Quadrula spheniopsis (Morelet, 1849) — de Leon (1912)
Elliptio (*Pachynaias*) *spheniopsis* (Morelet, 1849) — Frierson (1927)
Pachynaias spheniopsis (Morelet, 1849) — Haas (1929; 1969), Thiele (1934), Morrison (1968), Graf & Cummings (2007; 2021), Pfeiffer et al. (2019), Inoue et al. (2020a), Smith et al. (2020), Ramírez et al. (2022)
Pavhynaias [sic] *spheniopsis* (Morelet, 1849) — Contreras-Arquieta (2000)
Psoronaias martensi Frierson, 1927 — Ramírez et al. (2022)

Unio rugososulcatus Lea, 1866. Description of twelve new species of Unionidae from South America. *Proceedings of the Academy of Natural Sciences of Philadelphia* 18:33–35. Page 33. Figured in Lea (1868a JANSP Vol. 6:266, 1869a Obs. 12:26, pl. 34, fig. 81).

Verbatim Type Locality: "Central America?"

Type Locality: Central America.

Type Specimens: Lectotype, USNM 84392 designated by Johnson (1974a:127, as the figured holotype). Paralectotype, SMF 3474 (Haas 1929:318, as a paratype; Zilch 1967:121, as a syntype. Both mentioned that additional type specimens may be in Madrid (Nagel 2025:169 as a topotype).

Remarks: It is unclear exactly which species von Martens (1900b:514) thought this name applied to noting in his account of *Unio rubicundus* that it [*Unio rugososulcatus*] is "supposed to be from Central America, is rather near the present species, but it is a little more triangular, without prominences and plaits and has the hinder part distinctly biangulated." He later (1900:522) stated that "Lea supposes it to be from Central America and compares the species with his *Unio newcombianus* but it is much longer. As regards its resemblance to *U. rubicundus*, see *anteà*, p. 514."

Taxonomic History:

Margaron (*Unio*) *rugososulcatus* (Lea, 1866) — Lea (1870)
Unio rugoso–sulcatus Lea, 1866 — von Martens (1900c, *nomen dubium*)
Unio rugosulcatus [sic] Lea, 1866 — von Martens (1901c)
Quadrula (sec. *Pachynaias*) *rugososulcata* (Lea, 1866) — Simpson (1900; 1914)
Elliptio (*Pachynaias*) *rugososulcatus* (Lea, 1866) — Frierson (1927)
Pachynaias rugososulcata (Lea, 1866) — Zilch (1967), Haas (1929 [in part]; 1969), Graf & Cummings (2021), Nagel (2025)
Pachynaias rugosulcata [sic] (Lea, 1866) — Graf & Cummings (2007)

Unio (*Elliptio*) *sallei* var. *grossus* von Martens, 1900. Land and freshwater Mollusca. *In* F.D. Godman and O. Salvin (eds.). *Biologia Centrali-Americana* Part 155, Fascile 63–65:497–520, plates 33–35. Page 517, pl. 34, figs. 2, 2a–b.

Verbatim Type Locality: "Rio Salinas."

Type Locality: Río de las Salinas (Gulf of México–Grijalva–Usumacinta Dr.), El Petén Department, Guatemala [W = 16.4666, -90.5500].

Type specimen: Holotype by monotypy, ZMB/MOLL.108850.

Remarks: Although Frierson (1927:39) and Haas (1929:331; 1969:246) considered *grossus* to be synonymous with *Nephronaias sallei*, it is clearly *Pachynaias* and was determined as such on a label accompanying the type. That is also borne out in that von Martens (1900b:517) incorrectly stated that *Unio sallei* Crosse & Fischer, 1893 was the type species of *Pachynaias* Crosse & Fischer, 1894.

Taxonomic History:

Plagiola (Arotonaias) sallei var. *grossa* (von Martens, 1900) — Simpson (1914)
Elliptio (Barynaias) sallei-grossus (von Martens, 1900) — Frierson (1927)
Elliptio (Barynaias) sallei (Crosse & Fischer, 1894) — Haas (1929)
Barynaias sallei (Crosse & Fischer, 1894) — Haas (1969)

Order Unionoida
Superfamily Unionoidea Rafinesque, 1820
Family Unionidae Rafinesque, 1820
Subfamily Ambleminae Rafinesque, 1820
Tribe Popenaiadini Heard & Guckert, 1970
Genus *Micronaias* Simpson, 1900

Micronaias Simpson, 1900. Synopsis of the naiades, or pearly fresh-water mussels. *Proceedings of the United States National Museum* 22(1205):501–1044. Page 743. Introduced as a subgenus of *Unio* Philipsson, 1788.

Diagnosis: "Shell small, oval, a little produced near the posterior base, with a moderate posterior ridge and pointed near the base behind; beaks rather prominent, the sculpture being apparently rather fine, irregularly concentric ridges, having a tendency to fall into two loops; the whole surface strongly and closely concentrically ridged; pseudocardinals compressed, high, slightly curved upward; laterals compressed, curved; beak cavities moderately deep; nacre whitish; anterior muscle scars deep, rough; posterior well marked" (Simpson 1900:743).

Type Species: *Unio aratus* Lea, 1843, by original designation.

Taxonomic History:

Unio (sec. *Micronaias*) Simpson, 1900 — Simpson (1900; 1914)
Unio (*Arotonaias*) von Martens, 1900 — von Martens (1900b) [in part]
Unio (*Simonaias*) Crosse & Fischer, 1894 — von Martens (1900b) [in part]
Elliptio (*Micronaias*) Simpson, 1900 — Frierson (1927), Haas (1929; 1969)
Elliptio (sec. *Micronaias*) Simpson, 1900 — Thiele (1934)
Micronaias Simpson, 1900 — Modell (1942), Graf & Cummings (2007; 2021), Lopes-Lima et al. (2017), Pfeiffer et al. (2019b)

Distribution & Remarks: All members of the genus *Micronaias* are found in the Río San Juan Basin, Nicaragua, Lake Nicaragua, and adjacent Lake Managua. The members of *Micronaias* have never been tested for monophyly.

PLATE 39 ***Micronaias arata*** **(Lea, 1843). Top Left:** *Unio aratus* Lea, 1843. Holotype, USNM 84395 (17 mm L, 13 mm H). Lake Nicaragua (San Juan Dr.), Río San Juan Department, Nicaragua [W = 11.1350, -84.8367]. **Top Right:** *Unio nuculinus* Philippi, 1849. Lectotype, ZMB/MOLL.117922 (17 mm L, 10 mm H). Nicaragua. **Middle Left:** *Unio globulosus* von Martens, 1900. Holotype, ZMB/MOLL.305844 (25 mm L, 16 mm H). Central America. ***Micronaias fallaciosa*** **(Simpson, 1914). Middle Right:** *Unio* (*Micronaias*) *fallaciosus* Simpson, 1914. Lectotype, USNM 1673591 (32 mm L, 23 mm H). Lake Nicaragua (San Juan Dr.), Río San Juan Department, Nicaragua [W = 11.1350, -84.8367]. ***Micronaias granadensis*** **(Lea, 1868). Bottom Left:** *Unio granadensis* Lea, 1868. Lectotype, USNM 84388 (36 mm L, 23 mm H). Lake Nicaragua (San Juan Dr.), Río San Juan Department, Nicaragua [W = 11.1350, -84.8367].

Micronaias arata (Lea, 1843)

Plate 39, Map 45

Province(s): San Juan.

Distribution and Status: Endemic to the Río San Juan Basin in Nicaragua with all records coming from Lake Nicaragua and adjacent Lake Managua. Not assessed (IUCN 2024).

Anatomy: Shell small, solid, oval, and elongate with a well-defined posterior ridge with two raised curved lines extending from the beak to the posterior margin that ends behind in a point below the median line. Beak cavity shallow. Anterior end rounded. Ventral edge convex. Left valve with two somewhat compressed pseudocardinals and two thin laterals. Right valve with one pseudocardinal and one lateral. Periostracum light brown, tan, or olive. The surface is rough with raised ridges and rayless. Nacre white or occasionally salmon.

Marsupium pad-like, occupying all but the extreme posterior part of the outer gills. Gills branching, elongated, wider behind, inner the larger throughout, free from the abdominal sac or united to it. Palpi large and rounded behind. Mantle with a thickened border. Anal opening crenulate (Simpson 1900:743).

Biology: Unknown.

Adult Size: Length to 3.5 cm.

Similar Species: *Micronaias fallaciosa* (Simpson, 1914) and *Micronaias granadensis* (Lea, 1868) also occur in Lake Nicaragua.

MAP 45 Distribution of *Micronaias arata* (Lea, 1843).

Unio aratus Lea, 1843. Privately published sheet. Redescribed in Lea, 1845. Description of new fresh water and land shells. *Transactions of the American Philosophical Society* 9(N.S):275–282+plates 39–42. Page 282, pl. 42, fig. 12. Also figured in von Martens (1900e, pl. 39, figs. 1, 2, 2a–b, ZMB/MOLL.305843).

Verbatim Type Locality: "Hab......"

Type Locality: Lake Nicaragua (Caribbean–San Juan Dr.), Río San Juan Department, Nicaragua [W=11.1350, -84.8367].

Type Specimens: Holotype by monotypy, USNM 84395 (Johnson 1974:15).

Remarks: No type locality was given in the original description or redescription. In the redescription, Lea stated that "The habitat of this shell is not known. The person from whom I procured it could not even inform me whether it was exotic or not." Lea in the index to his Synopsis of the family Unionidae (1870:95) listed the locality as Lake Nicaragua. The label with the holotype has [Lake Nicaragua, Nicaragua] in brackets suggesting that the locality was added later. How it was determined to be from the lake is unknown.

Taxonomic History:

Unio aratus Lea, 1843 — Sowerby (1868), Tate (1870)
Margaron (*Unio*) *aratus* (Lea, 1843) — Lea (1852; 1870)
Unio (*Hyridella*) *aratus* Lea, 1843 — Adams & Adams (1857)
Unio (sec. *Hyridella*) *aratus* Lea, 1843 — Paetel (1890)
Unio (*Arotonaias*) *aratus* Lea, 1843 — von Martens (1900b)
Unio (sec. *Micronaias*) *aratus* Lea, 1843 — Simpson (1900; 1914)
Elliptio (*Micronaias*) *aratus* (Lea, 1843) — Frierson (1927), Haas (1929; 1969)
Elliptio (sec. *Micronaias*) *aratus* (Lea, 1843) — Thiele (1934)
Micronaias arata (Lea, 1843) — Graf & Cummings (2007; 2021)

Unio nuculinus Philippi, 1849. Centuria tertia Testaceorum novorum. *Zeitschrift für Malakozoologie* 5(11):161–176. Page 176. Figured in von Martens (1900e, pl. 39, figs. 6, 6a–b).

Verbatim Type Locality: "Nicaragua."

Type Locality: Nicaragua.

Type Specimens: Lectotype, ZMB/MOLL.117922, here designated (Coan and Kabat 2017:71, as a figured syntype).

Taxonomic History:

Unio nuculinus Philippi, 1848 — Lea (1852; 1870)
Unio (*Simonaias*) *nuculinus* Philippi, 1848 — von Martens (1900b)
? *Plagiola nuculinus* (Philippi, 1848) — Simpson (1900)
Unio (sec. *Micronaias*) *aratus* Lea, 1843 — Simpson (1914)
Elliptio (*Micronaias*) *aratus* (Lea, 1843) — Haas (1969)
Micronaias arata (Lea, 1843) — Coan & Kabat (2017)

Unio globulosus von Martens, 1900. Land and freshwater Mollusca. *In* F.D. Godman and O. Salvin (eds.). *Biologia Centrali-Americana* Part 155, Fascile 63–65:497–520, plates 33–35. Page 499, and Part 158, Fascile 68–71:537–568, plates 38–39. Pl. 39, figs. 3, 3a–b (Dunker's specimen).

Verbatim Type Locality: "Central America."

Type Locality: Central America.

Type Specimens: Holotype by monotypy, ZMB/MOLL.305844.

Remarks: Introduced as a synonym of *Unio aratus* Lea, 1845 and attributed to Dunker by von Martens.

Micronaias fallaciosa (Simpson, 1914)

Plate 39, Map 46

Province(s): San Juan.

Distribution and Status: Endemic to the Río San Juan Basin in Nicaragua with all records coming from Lake Nicaragua, the adjacent Lake Managua, and the Río San Juan. Not assessed (IUCN 2024).

Anatomy: Shell small, triangular, somewhat inflated, and solid with high beaks. Beak cavity shallow. Posterior ridge sharp. Left valve with two pseudocardinals and two laterals. Right valve with two pseudocardinals and one lateral. Surface sculpture with coarse concentric ridges. Periostracum yellow or tan. Nacre white or salmon.

Biology: Unknown.

Adult Size: Length to 3.2 cm.

Similar Species: *Micronaias arata* (Lea, 1843) and *Micronaias granadensis* (Lea, 1868) also occur in Lake Nicaragua. *Micronaias fallaciosa* is more triangular than *M. arata*. *Micronaias granadensis* is less angular and has a dark periostracum.

MAP 46 Distribution of *Micronaias fallaciosa* (Simpson, 1914).

Unio (*Micronaias*) *fallaciosus* Simpson, 1914. A descriptive catalogue of the naiades, or pearly fresh-water mussels. Parts I–III. Bryant Walker, Detroit, Michigan xii + 1540 pp. Page 715.

Verbatim Type Locality: "Nicaragua."

Type Locality: Lake Nicaragua (Caribbean–San Juan Dr.), Río San Juan Department, Nicaragua [W = 11.1350, -84.8367].

Type Specimens: Lectotype, USNM 1673591 (formerly USNM 73062) (Johnson 1975:12, as the figured holotype).

Remarks: Simpson provided a single set of measurements with no indication of a single primary type. He did not mention any other specimens. However, there are two specimens in the lot that contains the lectotype. The specimen identified as the "holotype" and figured in Johnson (1975:56, pl. 3, fig. 4) appears to be the larger of the two specimens in the lot USNM 1673591 and its length approximates that given by Simpson (32 mm).

Taxonomic History:

Unio (sec. *Micronaias*) *fallaciosus* Simpson, 1914 — Simpson (1914)
Elliptio (*Micronaias*) *fallaciosus* (Simpson, 1914) — Frierson (1927), Haas (1929; 1969)
Micronaias fallaciosa (Simpson, 1914) — Graf & Cummings (2007; 2021)

Micronaias granadensis (Lea, 1868)

Plate 39, Map 47

Province(s): San Juan.

Distribution and Status: Endemic to Lake Nicaragua in the Río San Juan Basin of Nicaragua. Not assessed (IUCN 2024).

Anatomy: Shell rather small, triangular, and somewhat solid with a well-developed posterior ridge. Beak cavity moderately deep. Ventral edge slightly curved or nearly straight ending in a point behind at a level with or just above the base line. Anterior end rounded. Left valve with two pseudocardinals and two short, curved laterals. Right valve with one pseudocardinal, sometimes with a vestigial one above, and one lateral. Periostracum brown or black and rough that sometimes appears silky when fresh. Nacre white or pink.

Biology: Unknown.

Adult Size: Length to 3.7 cm.

Similar Species: *Micronaias arata* (Lea, 1843) and *Micronaias fallaciosa* (Simpson, 1914) also occur in Lake Nicaragua.

MAP 47 Distribution of *Micronaias granadensis* (Lea, 1868).

Unio granadensis Lea, 1868. Description of nine species of Unionidae from Lake Nicaragua, Central America. *Proceedings of the Academy of Natural Sciences of Philadelphia* 20:94–96. Page 95. Figured in Lea (1868a JANSP Vol. 6:293; 1869a Obs. 12:53, pl. 42, fig. 103).

Verbatim Type Locality: "Lake Nicaragua, Cent. Amer."

Type Locality: Lake Nicaragua (Caribbean–San Juan Dr.), Río San Juan Department, Nicaragua [W = 11.1350, -84.8367].

Type Specimens: Lectotype, USNM 84388 designated by Johnson and Baker (1973:156; Johnson 1974:15, both as the holotype). Paralectotypes, ANSP 56543 (n=2) (Johnson and Baker 1973:156, as paratypes).

Taxonomic History:

Margaron (Unio) granadensis (Lea, 1868) — Lea (1870)
Unio granadensis Lea, 1868 — Wright (1888)
Unio (Arotonaias) granadensis Lea, 1868 — von Martens (1900b)
Unio (sec. *Micronaias*) *granadensis* Lea, 1868 — Simpson (1900; 1914)
Elliptio (Micronaias) granadensis (Lea, 1868) — Frierson (1927), Haas (1929; 1969)
Micronaias granadensis (Lea, 1868) — Graf & Cummings (2007; 2021)

Micronaias marshalli Morrison, 1943. Two new Orinoco unionids, with notes on *Unio granadensis* Lea and *U. patulus* Lea. *The Nautilus* 57(1):14–16 + 1 plate. Page 15. Not figured.

Remarks: Morrison (1943:15) stated that "Some years ago Wm. B. Marshall noted in Ms. that *Unio granadensis* Lea (Proc. A.N.S.P. 12: 95, 1868) was a homonym of *Unio granadensis* [sic] Conrad (Proc. A.N.S.P. 7: 256, 1855), but did not publish the fact. In recognition of his discovery of the untenable use of Lea's name, the species is hereby designated *Micronaias marshalli*, new name for *U. granadensis* Lea 1868, from Lake Nicaragua." The presumed homonymy of *Unio granadensis* Lea 1868 with *Unio grandensis* Conrad, 1855 is based on Morrison's misspelling of Conrad's name. Morrison's new name was an unnecessary replacement.

Order Unionoida
Superfamily Unionoidea Rafinesque, 1820
Family Unionidae Rafinesque, 1820
Subfamily Ambleminae Rafinesque, 1820
Tribe Popenaiadini Heard & Guckert, 1970
Genus *Nephritica* Frierson, 1927

Nephritica Frierson, 1927. A classification and annotated check list of the North American naiades. Baylor University Press, Waco, Texas. 111 pp. Page 38. Introduced as a subgenus of *Elliptio* Rafinesque, 1819.

Diagnosis: None. Frierson (1927:38) did not diagnose *Nephritica* but rather described *Elliptio (Nephritica) haricotti* and placed it in *Nephritica* along with *Elliptio (N.) poeyanus.* That diagnosis is as follows: "Shell elliptical; rounded at both ends; dorsum slightly arched; base somewhat arcuate. Beaks low, approximate, bearing faint concentric bars. Epidermis dull, light to dark brown, bearing faint vestigial rays near the beaks. Shell sub-cylindrical, post ridge raised, but rounded; a shallow depression from beak to base produces an arcuate basal margin, giving an elongate kidney shape. Nacre flesh color, rather pale. Cardinals double in the left valve, laterals short, slightly curved. Length 46; Altitude 26; Diameter 17 mm."

Type Species: *Unio poeyanus* Lea, 1857, by original designation.

Taxonomic History:

Unio (sec. *Elliptio*) Rafinesque, 1819 — Simpson (1900; 1914)
Unio (Nephronaias) Crosse & Fischer, 1894 — von Martens (1900a)
Elliptio (Nephritica) Frierson, 1927 — Frierson (1927)
Elliptio Rafinesque, 1819 — Thiele (1934)
Nephronaias (Nephritica) Frierson, 1927 — Haas (1969)
Nephritica Frierson, 1927 — Johnson (1972), Graf & Cummings (2007; 2021), Lopes-Lima et al. (2017), Pfeiffer et al. (2019b)

Distribution & Remarks: The genus *Nephritica* currently contains two species: *Nephritica haricotti* (Frierson, 1927) from Guatemala, without locality, and *Nephritica poeyana* (Lea, 1857) from the Río Balsas (Pacific Dr.) basin in Guerrero and Morelos, México. The members of this genus have never been tested for monophyly. Why Frierson chose to create this genus and describe *Nephritica haricotti* is a mystery. Perhaps because they are both "pinched" in the middle like some members of the genus *Elliptio.*

PLATE 40 ***Nephritica poeyana* (Lea, 1857)**. *Unio poeyanus* Lea, 1857. Lectotype, USNM 85894 (56 mm L, 25 mm H). Río Balsas (Pacific Dr.), Coyuca de Catalán, Guerrero, México [L = 18.3289, -100.6948].

Nephritica poeyana (Lea, 1857)

Plate 40, Map 48

Province(s): Mesa Central and Balsas.

Distribution and Status: Known only from the Río Balsas (Pacific Dr.) and in the Río Amacuzac basin in Guerrero and Morelos, México. No live individuals have ever been collected, and the last shell collection was made in 1979 (ANSP 334456). Not assessed (IUCN 2024).

Anatomy: Shell elongate and equilateral with a low, rather rounded posterior ridge. Dorsal line nearly straight. Ventral edge convex. Beaks low their sculpture consisting of very faint ridges. Beak cavity shallow. Anterior and posterior ends rounded. Left valve with two pseudocardinals and two feeble laterals. Right valve with one pseudocardinal and one lateral. Periostracum straw-colored in young shells, becoming brown in adults, occasionally with fine green rays. Nacre white.

Biology: Unknown.

Adult Size: Length to 5 cm.

Similar Species: *Nephritica poeyana* (Lea, 1857) is the only freshwater mussel species reported from Morelos, México. *Anodontites trapesialis* (Lamarck, 1819) has been collected from the Río Balsas in Guerrero, but it is edentate.

MAP 48 Distribution of *Nephritica poeyana* (Lea, 1857).

Unio poeyanus Lea, 1857. Description of three new species of exotic naïades. *Proceedings of the Academy of Natural Sciences of Philadelphia* 9:85. Page 85. Figured in Lea (1858a JANSP Vol. 3:315; 1858b Obs. 6:35, pl. 32, fig. 30) and Sowerby (1868, pl. 90, sp. 486).

Verbatim Type Locality: "Rio de las Balsas, near Coyucan, México."

Type Locality: Río Balsas (Pacific Dr.), Coyuca de Catalán, Guerrero, México [L = 18.3289, -100.6948].

Type Specimens: Lectotype, USNM 85894 designated by Johnson (1974a:115, as the holotype).

Remarks: Lea (1857:316) mentioned having two specimens. The location of the paralectotype is unknown. The shell morphology of this species looks almost identical to *Popenaias popeii* and similar to *P. berezai*, and it may be a Pacific member of the genus. This warrants further investigation.

Taxonomic History:

Unio poeyanus Lea, 1857 — Sowerby (1868), de Leon (1912)
Margaron (*Unio*) *poeyanus* (Lea, 1857) — Lea (1870)
Unio (*Graphonaias*) *poeyanus* Lea, 1857 — Fischer & Crosse (1894b)
Unio (*Nephronaias*) *poeyanus* Lea, 1857 — von Martens (1900a–b)
Unio (sec. *Elliptio*) *poeyanus* Lea, 1857 — Simpson (1900; 1914)
Elliptio (*Nephritica*) *poeyanus* (Lea, 1857) — Frierson (1927)
Elliptio poeyana (Lea, 1857) — Thiele (1934)
Nephronaias (*Nephritica*) *poeyana* (Lea, 1857) — Haas (1969)
Nephritica poeyana (Lea, 1857) — Graf & Cummings (2007; 2021), Czaja et al. (2023)

PLATE 41 ***Nephritica haricotti*** **(Frierson, 1927)**. **Top:** *Elliptio* (*Nephritica*) *haricotti* Frierson, 1927. Lectotype, UMMZ 97047 (46 mm L, 26 mm H). Guatemala. **Bottom:** Paralectotype, ANSP 145240.

Nephritica haricotti (Frierson, 1927)

Plate 41

Province(s): Unknown.

Distribution and Status: Known only from Guatemala without a specific locality. Not assessed (IUCN 2024).

Anatomy: Shell elliptical, rounded at both ends and kidney shaped. Hinge line slightly arched. Ventral edge somewhat arcuate. Beaks low bearing faint concentric bars. The posterior ridge raised and rounded with a shallow depression from beak to the posterior end. Pseudocardinals double in the left valve, and the laterals are short and slightly curved. Periostracum dull, light to dark brown with faint vestigial rays near the beaks. Nacre light pink.

Biology: Unknown.

Adult Size: Length to 4.6 cm.

Similar Species: None.

Elliptio (*Nephritica*) *haricotti* Frierson, 1927. A classification and annotated check list of the North American naiades. Baylor University Press, Waco, Texas. 111 pp. Page 38, species 150. Figured in Frierson (1928:139, pl. 3, fig. 4) and Johnson (1972a:166, pl. 29, fig. 4).

Verbatim Type Locality: "Guatemala."

Type Locality: Guatemala.

Type Specimens: Lectotype, UMMZ 97047 designated by Johnson (1972a:144, as the holotype and as *harciotti* [sic]). Paralectotype, ANSP 145240 ($n = 1$) (Johnson 1972a:144; Johnson and Baker 1973:157, both as a paratype).

Remarks: Frierson provided only one set of measurements for the two shells mentioned in the original description but did not identify any types. The only two specimens known are the types, and without a clear type locality, it will be extremely difficult to ascertain what this species is and its relationship to other Central American mussels.

Taxonomic History:

Nephronaias (*Nephritica*) *haricotti* (Frierson, 1927) — Haas (1969)
Elliptio harciotti [sic] Frierson, 1927 — Johnson (1972; 1979)
Nephritica haricotti (Frierson, 1927) — Graf & Cummings (2007; 2021)

Order Unionoida
Superfamily Unionoidea Rafinesque, 1820
Family Unionidae Rafinesque, 1820
Subfamily Ambleminae Rafinesque, 1820
Tribe Popenaiadini Heard & Guckert, 1970
Genus *Nephronaias* Crosse & Fischer, 1894

Nephronaias Crosse & Fischer, 1894. *In* P. Fischer and H. Crosse. 1894. Études sur les mollusques terrestres et fluviatiles du Mexique et du Guatemala. *In* M.H.–Milne Edwards (ed.). *Mission Scientifique au Mexique et dans l'Amérique centrale. Recherches Zoologiques pour servir á l'Histoire de la Faune de l'Amérique Centrale et du Mexique.* 15th Livraison. 62–72, pp. 489–576, plates 59–62, Imprimerie Nationale, Paris Volume 2. 731 pp. Page 556. Introduced as subgenus of *Unio* Philipsson, 1788.

Diagnosis: "Coquille petite, ovale, transverse, réniforme, inéquilatérale, acuminée en arrière, peu renflée; bord ventral sinueux; dents latérales antérieures fortes et courtes (*U. plicatulus*, Charpentier)." [Shell small, oval, transverse, kidney-shaped, unequilateral, acuminate posteriorly, slightly swollen; sinuous ventral edge; strong and short anterior lateral teeth (*U. plicatulus*, Charpentier)].

Type Species: *Unio plicatulus* Küster, 1856, by original designation.

Remarks: Unfortunately, *Unio plicatulus* Küster, 1856 is unidentifiable by its short description, figure, the nebulous type locality of "Mexiko," and no known type specimen(s). Baker (1922:7) recognized this and stated that "all of Küster's figures look remarkably alike" and that "the best place for *U. plicatulus*, and the section *Nephronaias* with it, is under the synonomy [sic] of *A. medellina* (Lea)." Baker made the unsubstantiated assumption that the type locality of *Unio plicatulus* Küster, 1856 was from the same river as *A. medellina*. However, the type locality of *U. plicatulus* mentions no specific river. Said by Fischer and Crosse (1894b:613) to belong to the same group as *Nephronaias calamitarum* (Morelet, 1849). We follow von Martens (1900b:502) and consider *U. plicatulus* to be a synonym of *Nephronaias aztecorum* (Philippi, 1847).

Taxonomic History:

Unio (*Nephronaias*) Crosse & Fischer, 1894 — von Martens (1900a)
Nephronaias Crosse & Fischer, 1894 — Simpson (1900; 1914), von Ihering (1901), Pilsbry (1920), Pfeiffer et al. (2019b), Inoue et al. (2020a), Graf & Cummings (2021), Nagel (2025)
Lampsilis Rafinesque, 1820 — Pilsbry (1910) [in part]
Elliptio (*Nephronaias*) Crosse & Fischer, 1894 — Frierson (1927), Haas (1929)
Elliptio (sec. *Nephronaias*) Crosse & Fischer, 1894 — Thiele (1934)
Nephronaias (s.s.) Crosse & Fischer, 1894 — Frierson (1917), Haas (1969), Johnson (1981)
Villosa Frierson, 1927 — Johnson (1981)

Cænonaias Crosse & Fischer, 1894. *In* P. Fischer and H. Crosse. 1894. Études sur les mollusques terrestres et fluviatiles du Mexique et du Guatemala. *In* M.H.–Milne Edwards (ed.). *Mission Scientifique au Mexique et dans l'Amérique centrale. Recherches Zoologiques pour servir á l'Histoire de la Faune de l'Amérique Centrale et du Mexique.* 15th Livraison. 62–72, pp. 489–576, plates 59–62, Imprimerie Nationale, Paris Volume 2. 731 pp. Page 556. Spelled as *Cænonaias* on p. 556, but then *Coenonaias* on p. 596. Introduced as subgenus of *Unio* Philipsson, 1788.

Diagnosis: "Coquille ovalaire ou subrhomboidale, médiocrement haute, peu transverse, striée concentriquement (*U.* æruginosus, Morelet)." [Shell oval or subrhomboidal, moderately high, slightly transverse, concentrically striated (*U. aeruginosus*, Morelet)].

Type Species: *Unio aeruginosus* Morelet, 1849, by original designation.

Taxonomic History:

Unio (*Elliptio*) Rafinesque, 1819 — von Martens (1900b)
Elliptio (*Leptonaias*) Crosse & Fischer, 1894 — H.B. Baker (1922, as *Coenonaias*), Frierson (1927), Haas (1929)
Nephronaias (s.s.) Crosse & Fischer, 1894 — Modell (1964), Haas (1969)
Nephronaias Crosse & Fischer, 1894 — Simpson (1900; 1914), Graf & Cummings (2007; 2021)

Leptonaias Crosse & Fischer, 1894. *In* P. Fischer and H. Crosse. 1894. Études sur les mollusques terrestres et fluviatiles du Mexique et du Guatemala. *In* M.H.–Milne Edwards (ed.). *Mission Scientifique au Mexique et dans l'Amérique centrale. Recherches Zoologiques pour servir á l'Histoire de la Faune de l'Amérique Centrale et du Mexique.* 15th Livraison. Feuilles 62–72, pp. 489–576, plates 59–62, Imprimerie Nationale, Paris Volume 2. 731 pp. Page 557. Introduced as subgenus of *Unio* Philipsson, 1788.

Diagnosis: "Coquille petite, réniforme, un peu déprimée, inéquilatérale; sommets obtus; côté postérieur déclive; bord ventral un peu sinueux; dents latérales antérieures fortes, mais courtes (*U. ravistellus*, Morelet)." [Shell small, kidney-shaped, a little depressed, inequilateral; obtuse beaks; posterior side sloping; slightly sinuous ventral edge; strong but short anterior lateral teeth (*U. ravistellus*, Morelet)].

Type Species: *Unio ravistellus* Morelet, 1849, by original designation.

Taxonomic History:

Unio (*Elliptio*) Rafinesque, 1819 — von Martens (1900a)
Elliptio (*Leptonaias*) Crosse & Fischer, 1894 — H.B. Baker (1922)
Elliptio (sec. *Nephronaias*) Crosse & Fischer, 1894 — Thiele (1934)
Nephronaias Crosse & Fischer, 1894 — Modell (1964)
Nephronaias (s.s.) Crosse & Fischer, 1894 — Haas (1969)

Simonaias Crosse & Fischer, 1894. *In* P. Fischer and H. Crosse. 1894. Études sur les mollusques terrestres et fluviatiles du Mexique et du Guatemala. *In* M.H.–Milne Edwards (ed.). *Mission Scientifique au Mexique et dans l'Amérique centrale. Recherches Zoologiques pour servir á l'Histoire de la Faune de l'Amérique Centrale et du Mexique.* 15th Livraison. 62–72, pp. 489–576, plates 59–62, Imprimerie Nationale, Paris Volume 2. 731 pp. Page 557. Introduced as subgenus of *Unio* Philipsson, 1788.

Diagnosis: "Coquille transverse, subcylindrique; côté antérieur très-court; surface sillonnée concentriquement (*U. Tabascoensis*, Charpentier)." [Shell transverse, subcylindrical; very short anterior side; concentrically furrowed surface (*U. Tabascoensis*, Charpentier)].

Type Species: *Unio tabascoënsis* Küster, 1856, by original designation.

Taxonomic History:

Unio (*Simonaias*) Crosse & Fischer, 1894 — von Martens (1900a)
Elliptio (*Nephronaias*) Crosse & Fischer, 1894 — H.B. Baker (1932, as *Simononaias*)
Elliptio (sec. *Nephronaias*) Crosse & Fischer, 1894 — Thiele (1934)
Nephronaias (s.s.) Crosse & Fischer, 1894 — Haas (1969)

Nepronaias Simpson, 1900. Synopsis of the naiades, or pearly fresh-water mussels. *Proceedings of the United States National Museum* 22(1205):501–1044. Page 591.
Remarks: Error for *Nephronaias* Crosse & Fischer, 1894.

Nephronajas Ortmann, 1912. Notes upon the families and genera of the najades. *Annals of the Carnegie Museum* 8(2):222–365+3 plates. Page 324.
Remarks: Error for *Nephronaias* Crosse & Fischer, 1894.

Simononaias H.B. Baker, 1922. The Mollusca collected by the University of Michigan-Walker Expedition in Southern Vera Cruz, Mexico. I. Occasional Papers of the Museum of Zoology, University of Michigan No. 106. 95 pp. Page 8.
Remarks: Error for *Simonaias* Crosse & Fischer, 1894.

Simononaias Modell, 1964. Das natürliche system der najaden. 3. Archiv für Molluskenkunde 93:71–126. Page 90.
Remarks: Error for *Simonaias* Crosse & Fischer, 1894.

Distribution & Remarks: The speciose genus *Nephronaias* was extensively reworked here. Graf & Cummings (2021:20) listed 24 species in the genus *Nephronaias* Crosse & Fischer, 1894. We recognize 20 species of *Nephronaias,* and they range from Central México south to Panamá and two species in Cuba. *Nephronaias oregonensis* (Lea, 1852), *N. mellea* (Lea, 1859), *N. rugulosa* (Küster, 1856), *N. sphenoryhncha* (Crosse & Fischer, 1894), and *N. tabascoensis* (Küster, 1856) are all considered as *Nomina dubia* here because of the lack of a type, a figure, a poor description, suspect or nebulous type locality, or a combination thereof. *Nephronaias chapalanus* (von Martens, 1900) from Lago de Chapala (Pacific Dr.) was resurrected from the synonymy of the widely disjunct *N. aztecorum* (Philippi, 1847) of the Gulf of México. Other changes include sinking *Nephronaias sajensis* (Frierson, 1927) under *Nephronaias dysonii* (Lea, 1859) and elevating *Nephronaias ortmanni* (Frierson, 1913) from the synonymy of *N. sphenoryhncha* (Crosse & Fischer, 1894), which we consider a *nomen dubium*. We also sunk and *Unio championi* von Martens, 1900, *Unio*

Most of the species of *Nephronaias* are widely distributed along the Caribbean and Pacific coasts from Belize to Panamá (Map 49). There is a considerable gap between southern Costa Rica and northern Panamá where we lack any mussel records at all. There are four species of *Nephronaias* from Gulf of Mexico drainages. *Nephronaias aztecorum* (Philippi, 1847) is known from both the Pánuco and Papaloapán basins. *Nephronaias aeruginosa* (Morelet, 1849) and *Nephronaias sallei* (Crosse & Fischer, 1894) are from the Usumacinta drainage in southern México and Guatemala. *Nephronaias calamitarum* (Morelet, 1849) is shared between the Usumacinta and the Río Dulce (Caribbean–Gulf of Honduras) drainage in Guatemala.

MAP 49 Distribution of the genus *Nephronaias* Crosse & Fischer, 1894.

PLATE 42 ***Nephronaias aeruginosa* (Morelet, 1849).** *Unio aeruginosus* Morelet, 1849. Lectotype, NHMUK 1893.2.4.2019 (50 mm L, 30 mm H). Río Michol (Usumacinta Dr.), near Palenque, Chiapas, México [L = 17.4871, -92.0565].

Nephronaias aeruginosa (Morelet, 1849)

Plate 42, Map 50

Province(s): Usumacinta.

Distribution and Status: Known only from a few specimens from the ríos Michol and Chacamax in the Grijalva–Usumacinta drainage in Chiapas, México. Not assessed (IUCN 2024).

Anatomy: Shell solid and subrhomboid. Ventral edge straight or moderately convex. Beaks moderately elevated and smooth. Left valve with one solid, somewhat ragged pseudocardinal and two laterals. Right valve with two pseudocardinals, the upper the smaller, and one lateral. Periostracum yellow or chestnut with slight concentric sculpture. Nacre white or pink.

Biology: Unknown.

Adult Size: Length to 5 cm.

Similar Species: *Nephronaias calamitarum* (Morelet, 1849) co-occurs in the Usumacinta basin.

MAP 50 Distribution of *Nephronaias aeruginosa* (Morelet, 1849).

Unio æruginosus Morelet, 1849. Testacea novissima insulæ Cubanæ, et Americæ Centralis. Librairie de L'Académie Nationale de Medecine, Paris Part 1:1–31. Page 29. Figured in Fischer and Crosse (1894b:596, pl. 62, figs. 2, 2a–b).

Verbatim Type Locality: "In rivulo Michol, circa Palenqueanum vicum."

Type Locality: Río Michol (Gulf of México–Grijalva–Tulija Dr.), near Palenque, Chiapas, México [L = 17.4871, -92.0565].

Type Specimens: Lectotype, NHMUK 1893.2.4.2019 designated by Johnson (1971:79, as the holotype; Breure et al. 2018:197, figs. 34–35). Paralectotype, NHMUK 1893.2.4.2020 (Johnson 1971:79, as a paratype).

Taxonomic History:

Unio (*Cænonaias*) *æruginosus* Morelet, 1849 — Fischer & Crosse (1894b)
Unio (*Elliptio*) *æruginosus* Morelet, 1849 — von Martens (1900b)
Nephronaias æruginosus (Morelet, 1849) — Simpson (1900), de Leon (1912)
Nephronaias æruginosa (Morelet, 1849) — Simpson (1914)
Elliptio (*Nephronaias*) *aeruginosus* (Morelet, 1849) — Frierson (1927), Haas (1929)
Nephronaias (s.s.) *aeruginosa* (Morelet, 1849) — Haas (1969)
Nephronaias aeruginosa (Morelet, 1849) — Graf & Cummings (2007; 2021), Czaja et al. (2023)

PLATE 43 ***Nephronaias aztecorum* (Philippi, 1847). Top Left:** *Unio strebelii* Lea, 1866. Lectotype, USNM 85661 (74 mm L, 42 mm H). Río Medellin (Jamapa Dr.), Veracruz, Veracruz, México [L = 19.0574, -96.1550]. **Middle Right:** *Unio soledadensis* Crosse & Fischer, 1893. Lectotype, MNHN-IM-2000-1785 (67 mm L, 33 mm H). Río Jamapa (Gulf of Mexico Dr.), Soledad de Doblado, Veracruz, México [L = 19.0415, -96.4253]. **Middle Left:** *Unio aztecorum* "forma" *major* von Martens, 1900. Holotype, ZMB/MOLL.220172. **Bottom Right:** *Lampsilis* (*Disconaias*) *martensi* Frierson, 1927. Holotype, ZMB/MOLL.220183 (94 mm L, 58 mm H). Bottom two both from the Río Misantla, Misantla, Veracruz, México [L = 19.9310, -96.8464].

Nephronaias aztecorum (Philippi, 1847)

Plate 43, Map 51

Province(s): Pánuco; Papaloapán.

Distribution and Status: Widespread along the Gulf coast from the Pánuco basin south to the Jamapa basin in Veracruz, México. Live *Nephronaias aztecorum* were common at nine sites in the Pánuco, Pantapec, and Actopan basins in 1978, 1982, 2017, and 2018. Not assessed (IUCN 2024).

Anatomy: Shell thin to moderately solid, oblong, and slightly inflated. Anterior and posterior ends rounded. Ventral margin straight or slightly concave. Beaks low. Pseudocardinals small, two in the left valve and one in the right. Laterals thin and straight to slightly curved, two in the left valve and one in the right. Periostracum brown, sometimes with indistinct green rays. Nacre purple, occasionally white.

Biology: Gravid females were collected from Rio Pantepec (Pánuco Dr.), Alamo, Veracruz, México on 23 February 1982 (USNM 1282382–84).

Adult Size: Length to 10 cm.

Similar Species: *Sphenonaias plexus* (Conrad, 1838).

MAP 51 Distribution of *Nephronaias aztecorum* (Philippi, 1847).

Unio impressus Menke, 1830. Synopsis methodica Molluscorum generum omnium et specierum earum, quae in Museo Menkeano adservantur; cum synonymia critica et novarum specierum diagnosibus. Georgi Uslar, Pyrmonti 168 pp. Page 107.

Remarks: *Nomen nudem*. Although von Martens (1900b:502) placed this taxon in the synonymy of *Unio aztecorum*, it was named but never described by Menke.

Unio aztecorum Philippi, 1847. Testaceorum novorum centuria. *Zeitschrift für Malakozoologie* 4(6):84–96. Page 95. Figured in Philippi (1849:28, pl. 6, fig. 2) and von Martens (1900b, pl. 36, figs. 2, 2a–b [ZMB/MOLL.9104], 3–3a [ZMB/MOLL.305837], 4–4a [ZMB/MOLL.220182]).

Verbatim Type Locality: "Republica Mexicana."

Type Locality: México.

Type Specimens: Unknown. Presumed lost.

Remarks: There are two specimens in the ANSP (A14693) in alcohol from Veracruz, México that are labeled as paralectotypes and collected by Hermann Strebel (from the Isaac Lea Collection). However, Philippi stated that his specimen was from Frederik Liebmann, and these are likely not types.

Taxonomic History:

Unio aztecorum Philippi, 1847 — Philippi (1849), Küster (1862)
Margaron (*Unio*) *aztecorum* (Philippi, 1847) — Lea (1870)
Unio (sec. *Naïa*) *aztecorum* Philippi, 1847 — Paetel (1890)
Unio (*Graphonaias*) *aztecorum* Philippi, 1847 — Fischer & Crosse (1894b)
Unio (*Nephronaias*) *aztecorum* Philippi, 1847 — von Martens (1900b)
Lampsilis (*Proptera*) *aztecorum* (Philippi, 1847) — Simpson (1900)
Lampsilis aztecorum (Philippi, 1847) — Pilsbry (1910), de Leon (1912)
Nephronaias aztecorum (Philippi, 1847) — Simpson (1914), Contreras-Arquieta (2000), Graf & Cummings (2007; 2021), Coan & Kabat (2017), Kiser et al. (2022), Czaja et al. (2023)
Lampsilis (*Disconaias*) *aztecorum* (Philippi, 1847) — Frierson (1927)
Elliptio (*Nephronaias*) *aztecorum* (Philippi, 1847) — Haas (1929)
Nephronaias (s.s.) *aztecorum* (Philippi, 1847) — Haas (1969)

Unio plicatulus Küster, 1856. Die flusperlmuscheln (*Unio* et *Hyria.*). *In* Abbildungen nach der Natur mit Beschreibungen. Systematisches Conchylien-Cabinet von Martini und Chemnitz. Küster edition. 9(Part 2):137–160, pls. 45–54. Page 154, pl. 44, fig. 4.

Verbatim Type Locality: "Mexico."

Type Locality: México.

Type Specimens: Unknown. Presumed lost.

Remarks: Attributed to "Charpentier" by Küster (1856), Fischer & Crosse (1894b), and Thiele (1934). *Unio plicatulus* Küster, 1856 is the type species of *Nephronaias* (see above). Although regarded as valid by most previous workers, von Martens (1900b:502) thought it was synonymous with *Nephronaias aztecorum* (Philippi, 1847), and Haas (1929:322) thought it was likely *Nephronaias aeruginosus* (Morelet, 1856) stating that "Auch *Unio plicatulus* Charpentier dürfte mit ihr identic sein." ["*Unio plicatulus* Charpentier is also likely to be identical to it"]. Baker (1922:8) considered it as *Actinonaias medellina* (Lea, 1838) under the unsupported supposition that it was from the Río Medellin. Without a type and a nebulous type locality, it is anyone's guess as to what Küster had before him. Based up on the figure and nacre color, we follow von Martens (1900b:502) and consider *U. plicatulus* to be a synonym of *Nephronaias aztecorum* (Philippi, 1847).

Taxonomic History:

Unio (*Nephronaias*) *plicatulus* Küster, 1856 — Fischer & Crosse (1894b)
Unio (*Nephronaias*) *aztecorum* (Philippi, 1847) — von Martens (1900b)
Nephronaias plicatulus (Küster, 1856) — Simpson (1900), de Leon (1912)
Nephronaias plicatula (Küster, 1856) — Simpson (1914), Graf & Cummings (2007; 2021)
Nephronaias plicatulus (Küster, 1856) — Frierson (1917)
Actinonaias medellina (Lea, 1838) — H.B. Baker (1922)

Elliptio (*Nephronaias*) *plicatulus* (Küster, 1856) — Frierson (1927)
Elliptio (*Nephronaias*) *aeruginosus* (Morelet, 1856) — Haas (1929)
Elliptio (sec. *Nephronaias*) *plicatula* (Küster, 1856) — Thiele (1934)
Nephronaias (s.s.) *plicatula* (Küster, 1856) — Haas (1969)

Unio strebelii Lea, 1866. Description of five new species of the genus *Unio*. *Proceedings of the Academy of Natural Sciences of Philadelphia* 18:133. Figured in Lea (1868b JANSP Vol. 6:318; 1869a Obs. 12:78, pl. 51, fig. 131) and von Martens (1900d, pl. 37, figs. 1–1a [ZMB/MOLL.220183], 2–2a [ZMB/MOLL.220173]).

Verbatim Type Locality: "Vera Cruz, México."

Type Locality: Río Medellin (Gulf of México–Jamapa Dr.), Veracruz, Veracruz, México [L=19.0574, -96.1550].

Type Specimens: Lectotype, USNM 85661 designated by Johnson (1974a:138, as the holotype).

Remarks: Johnson added Río Medellin to the type locality. Frierson (1927) later described *Lampsilis* (*Disconaias*) *martensi* based upon von Martens (1900d, pl. 37, figs. 1–1a [ZMB/MOLL.220183]).

Taxonomic History:

Margaron (*Unio*) *strebelii* (Lea, 1866) — Lea (1870)
Unio (*Mesonaias*) *strebeli* Lea, 1866 — Fischer & Crosse (1894b)
Nephronaias medellinus (Lea, 1838) — Simpson (1900)
Unio (*Nephronaias*) *aztecorum* var. *strebeli* Lea, 1866 — von Martens (1900b; 1900d, 1901c)
Unio aztecorum var. *strebeli* Lea, 1866 — von Martens (1901c)
Lampsilis strebeli (Lea, 1866) — Pilsbry (1910)
Lampsilis (*Disconaias*) *aztecorum* (Philippi, 1847) — Frierson (1927)
Lampsilis (*Nephronaias*) *aztecorum* (Philippi, 1847) — Haas (1929)
Nephronaias aztecorum (Philippi, 1847) — Haas (1969)

Unio soledadensis Crosse & Fischer, 1893. Diagnoses molluscorum novorum, Reipublicæ Mexicanæ incolarum. *Journal de Conchyliologie* 41(2):110–111. Page 111. Figured in Fischer and Crosse (1894b:608, pl. 69, figs. 3, 3a–b).

Verbatim Type Locality: "prope Soledad."

Type Locality: Río Jamapa (Gulf of Mexico Dr.), Soledad de Doblado, Veracruz, México [L=19.0415, -96.4253].

Type specimens: Lectotype, MNHN-IM-2000-1785, here designated (the figured specimen). Paralectotype, IRSNB 3736 ($n=1$).

Remarks: Fischer and Crosse (1894b:608) later stated the type locality to be "Habitat prope la Soledad, in provincia Vera Cruz dicta, reipublicæ Mexicanæ." We have placed the type locality at the Río Jamapa, Soledad de Doblado.

Taxonomic History:

Unio (*Graphonaias*) *soledadensis* Crosse & Fischer, 1893 — Fischer & Crosse (1894b)
Unio (*Nephronaias*) *aztecorum* var. *soledadensis* Crosse & Fischer, 1893 — von Martens (1900b)
Nephronaias soledadensis (Crosse & Fischer, 1893) — Simpson (1914)

Unio aztecorum "forma" *major* von Martens, 1900. Land and freshwater Mollusca. *In* F.D. Godman and O. Salvin (eds.). *Biologia Centrali-Americana* Part 157, Fascile 67:529–536, plates 36–37, pl. 37, figs. 3, 3a–b.

Verbatim Type Locality: "Misantla."

Type Locality: Río Misantla (Gulf of Mexico Dr.), Misantla, Veracruz, México [L=19.9310, -96.8464].

Type specimen: Holotype by monotypy, ZMB/MOLL.220172.

Remarks: Not described but figured and named on plate 37, figs. 3, 3a–b. Primary homonym of *Unio* (*Anodontes*) *sirionos* var. *major* d'Orbigny, 1835.

Taxonomic History:

Lampsilis (*Disconaias*) *fimbriata* Frierson, 1907 — Frierson (1927)

Lampsilis (*Disconaias*) *martensi* Frierson, 1927. A classification and annotated check list of the North American naiades. Baylor University Press, Waco, Texas. 111 pp. Page 83, sp. 364. Figured in Johnson (1972:166, pl 29, fig. 2).

Verbatim Type Locality: "Playas of Misantla, Dist. of Misantla, Vera Cruz State, México."

Type Locality: Río Misantla (Gulf of Mexico Dr.), Misantla, Veracruz, México [L=19.9310, -96.8464].

Type Specimens: Holotype by monotypy, ZMB/MOLL.220183.

Remarks: Frierson (1927:83) named this species based on a specimen of *Unio aztecorum* var. *strebeli* Lea, 1866 figured by von Martens (1900d, pl. 37, figs. 1–1a; reproduced in Johnson 1972a:166, pl. 29, fig. 2). Frierson singled out one of the three specimens figured on plate 37 because according to him "The specimen figured by Martens is nothing alike to the genuine '*Unio aztecorum* Philippi.'" Johnson (1972a:146) stated that the "Type in collection of Strebel, Hamburg Mus. Destroyed in World War II." However, ZMB/MOLL.220183 is labeled as the specimen figured by von Martens (1900d, pl. 37, figs. 1–1a) and is the holotype.

Taxonomic History:

Nephronaias (s.s.) *tabascoensis* (Küster, 1856) — Haas (1969)
Lampsilis martensi Frierson, 1927 — Johnson (1972)

PLATE 44 ***Nephronaias calamitarum*** **(Morelet, 1849). Top Left:** *Unio calamitarum* Morelet, 1849. Lectotype, NHMUK 1893.2.4.2010 (51 mm L, 29 mm H). **Top Right:** *Unio calamitarum* var. δ, *nephretica* Fischer & Crosse, 1894. Holotype, NHMUK 1907.12.30.377 (43 mm L, 26 mm H). **Middle Left:** *Unio calamitarum* var. *β, prolongata* Fischer & Crosse, 1894 Holotype, NHMUK 1893.2.4.2008 (65 mm L, 32 mm H). **Middle Right:** *Unio persulcatus* Lea, 1859. Holotype, NHMUK 1900.3.19.22 (51 mm L, 28 mm H). México. **Bottom Left:** *Unio calamitarum* var. γ, *arcuans* Fischer & Crosse, 1894. Holotype, NHMUK 1893.2.4.2009 (46 mm L, 27 mm H). All but *U. persulcatus* Lea, 1859 are from the Río Chacamax (Usumacinta Dr.), Baluntie, near Palenque, Chiapas, México [L = 17.4965, -91.9541].

Nephronaias calamitarum (Morelet, 1849)

Plate 44, Map 52

Province(s): Usumacinta; Gulf of Honduras & Northern Caribbean.

Distribution and Status: Known from the Río Usumacinta drainage in México and Guatemala south to the Río Dulce basin (Gulf of Honduras Dr.) in Guatemala. Found alive at four sites in the Usumacinta basin in Chiapas, México in 2019. Not assessed (IUCN 2024).

Anatomy: Shell rather solid, somewhat elongate, and subrhomboid. Beaks scarcely elevated above the hinge line. The beak sculpture consists of fine, concentric ridges that cover the surface of the shell. There is often a groove on the posterior slope. Left valve with two triangular pseudocardinals and two straight or slightly curved laterals. Right valve with one pseudocardinal, sometimes a vestigial one above it and one lateral. The periostracum is yellowish green in young individuals becoming dark brown in adults. Nacre white, rose, or purple.

Biology: Ortmann (1921a:26) examined specimens (CM 61.6194) collected on 18 February from the Río Blanco, near Livingston, Guatemala and stated that the soft parts had the same structure as those of *Nephronaias ortmanni* (Frierson, 1913). He also reported that the glochidia had the same shape and dimensions: Length 0.23, Height 0.22 mm.

Adult Size: Length to 5 cm.

Similar Species: *Nephronaias aeruginosa* (Morelet, 1849), *Nephronaias dysonii* (Lea, 1859), and *Nephronaias yzabalensis* (Crosse & Fischer, 1893).

Remarks: von Martens (1900b:505–506) grouped *Unio calamitarum* Morelet, 1849, *Unio tabascoensis* Küster, 1856, *Unio coloratus* Küster, 1856, *Unio persulcatus* Lea, 1859, *Unio calamitarum* var. δ, *nephretica* Fischer & Crosse, 1894, *Unio calamitarum* var. γ, *arcuans* Fischer & Crosse, 1894, *Unio calamitarum* var. *β*, *prolongata* Fischer & Crosse, 1894, and a newly described variety, *Unio* (*Simonaias*) *cuprinus* var. *subventralis* von Martens, 1900, all under *Unio* (*Simonaias*) *cuprinus* Lea, 1831. We have tentatively placed *Unio cuprinus*, which has the nebulous type locality of México, under *Popenaias metallica* (Say, 1831) and consider *Unio tabascoensis* Küster, 1856 and *Unio coloratus* Küster, 1856 as *Nomina dubia* (see below).

MAP 52 Distribution of *Nephronaias calamitarum* (Morelet, 1849).

Unio calamitarum Morelet, 1849 Testacea novissima insulæ Cubanæ, et Americæ Centralis. Librairie de L'Académie Nationale de Medecine, Paris Part 1:1–31. Page 30. Figured in Sowerby (1868, pl. 74, sp. 385).

Verbatim Type Locality: "H. rivulum Baluntiè, prope Palenqueanum vicum."

Type Locality: Río Chacamax, Baluntie, near Palenque (Gulf of México–Grijalva–Usumacinta Dr.), Chiapas, México [L = 17.4965, -91.9541].

Type Specimens: Lectotype, NHMUK 1893.2.4.2010 designated and figured in Johnson (1971:80, pl. 2, fig. 5; Breure et al. 2018:228, figs. 159–160). Paralectotype, SMF 316485 (*n* = 2) (Nagel 2025:163). Possible paralectotype NHMUK 1907.12.30.377, from Morelet.

Remarks: MNHN-IM-2000-1653 (*n* = 3) are marked as syntypes on the label and identified as paralectotypes on the MNHN web page. They are not from the type locality but close to it. Also see remarks under *Nephronaias scamnata* (Morelet, 1849) for comments by Johnson (1981:271) on the type locality of *U. calamitarum.*

Taxonomic History:

Unio calamitarum Morelet, 1849 — Sowerby (1868)
Unio (Simonaias) calamitarum Morelet, 1849 — Fischer & Crosse (1894b)
Unio (Simonaias) cuprinus Lea, 1831 — von Martens (1900b)
Nephronaias calimatarum [sic] Morelet, 1849) — Simpson (1900), de Leon (1912)
Nephronaias calamitarum (Morelet, 1849) — von Ihering (1901), Simpson (1914), Hinkley (1920), Goodrich & van der Schalie (1937), Graf & Cummings (2007; 2021), Ramírez et al. (2022), Czaja et al. (2023), Nagel (2025)
Unio dysoni (Lea, 1859) — Frierson (1913)
Unio hjalmarsoni (Dunker, 1858) — Frierson (1913)
Elliptio calamitarum (Morelet, 1849) — Ortmann (1921a)
Elliptio (Nephronaias) calamitarum (Morelet, 1849) — Frierson (1927), Haas (1929)
Nephronaias (s.s.) *calamitarum* (Morelet, 1849) — Haas (1969)
Popenaias calamitarum (Morelet, 1849) — Heard & Guckert (1970)
Nephronaias (s.s.) *scamnata* (Morelet, 1849) — Johnson (1981) [in part]

Unio persulcatus Lea, 1859. Descriptions of twenty-one new species of exotic Unionidae. *Proceedings of the Academy of Natural Sciences of Philadelphia* 11(1859):151–154. Page 153. Figured in Lea (1860a JANSP Vol. 4, Part 3(7):255; 1860c Obs. 7:73, pl. 40, fig. 135).

Verbatim Type Locality: "Mexico."

Type Locality: México.

Type specimen: Holotype by monotypy, NHMUK 1900.3.19.22 (Johnson 1971:91; Johnson 1974a:111).

Taxonomic History:

Unio persulcatus Lea, 1859 — Lea (1860a)
Margaron (Unio) persulcatus (Lea, 1859) — Lea (1870)
Unio (Simonaias) persulcatus Lea, 1859 — Fischer & Crosse (1894b)
Unio (Simonaias) cuprinus var. *persulcatus* Lea, 1859 — von Martens (1900b)
Nephronaias persulcatus (Lea, 1859) — Simpson (1900), de Leon (1912), Frierson (1917)
Unio caprinus [sic] *persulcatus* Lea, 1859 — de Leon (1912)
Nephronaias persulcata (Lea, 1859) — Simpson (1914)
Elliptio (Nephronaias) persulcatus (Lea, 1859) — Frierson (1927)

Unio calamitarum var. *β*, *prolongata* Fischer & Crosse, 1894. Études sur les mollusques terrestres et fluviatiles du Mexique et du Guatemala. 16th Livraison. Feuilles 73–82, pp. 577–656, plates 63–66 in Recherches Zoologiques pour servir À l'Histoire de la Faune de l'Amérique Centrale et du Mexique, Imprimerie Nationale, Paris Volume 2. 731 pp. Page 612, pl. 63, figs. 5–5a.

Verbatim Type Locality: "in flumine Baluntie, prope Palenque, provinciæ Chiapas dictæ, republicæ Mexicanæ."

Type Locality: Río Chacamax (Gulf of México–Grijalva–Usumacinta Dr.), Baluntie, near Palenque, Chiapas, México [L = 17.4965, -91.9541].

Type Specimens: Holotype by monotypy, NHMUK 1893.2.4.2008 (n = ½) (Johnson 1971:91, as NHMUK 1893.2.4.2009 which is the holotype of *Unio calamitarum* var. γ, *arcuans* Fischer & Crosse, 1894, see below). Also see remarks under *Nephronaias scamnata* (Morelet, 1849) for comments on the type locality.

Taxonomic History:

Unio (*Simonaias*) *calamitarum* var. *β*, *prolongata* Fischer & Crosse, 1894 — Fischer & Crosse (1894b)
Unio (*Simonaias*) *cuprinus* var. *tabascoensis* Küster, 1856 — von Martens (1900b)
Elliptio (*Nephronaias*) *calamitarum-prolongus* (Fischer & Crosse, 1894) — Frierson (1927)
Nephronaias (s.s.) *scamnata* (Morelet, 1849) — Johnson (1981) [in part]

Unio calamitarum var. γ, *arcuans* Fischer & Crosse, 1894. Études sur les mollusques terrestres et fluviatiles du Mexique et du Guatemala. 16th Livraison. Feuilles 73–82, pp. 577–656, plates 63–66 in Recherches Zoologiques pour servir À l'Histoire de la Faune de l'Amérique Centrale et du Mexique, Imprimerie Nationale, Paris Volume 2. 731 pp. Page 612, pl. 64, figs. 5–5a.

Verbatim Type Locality: "in flumine Baluntie, prope Palenque, provinciæ Chiapas dictæ, republicæ Mexicanæ."

Type Locality: Río Chacamax (Gulf of México–Grijalva–Usumacinta Dr.), Baluntie, near Palenque, Chiapas, México [L = 17.4965, -91.9541].

Type Specimens: Holotype by monotypy, NHMUK 1893.2.4.2009 (Johnson 1971:79).

Taxonomic History:

Unio (*Simonaias*) *calamitarum* var. γ, *arcuans* Fischer & Crosse, 1894 — Fischer & Crosse (1894b)
Unio (*Simonaias*) *cuprinus* var. *arcuans* Fischer & Crosse, 1894 — von Martens (1900b)
Nephronaias calamitarum var. *arcuans* (Fischer & Crosse, 1894) — Simpson (1914)

Unio calamitarum var. δ, *nephretica* Fischer & Crosse, 1894. Études sur les mollusques terrestres et fluviatiles du Mexique et du Guatemala. 16th Livraison. Feuilles 73–82, pp. 577–656, plates 63–66 in Recherches Zoologiques pour servir À l'Histoire de la Faune de l'Amérique Centrale et du Mexique, Imprimerie Nationale, Paris Volume 2. 731 pp. Page 612. Not figured.

Verbatim Type Locality: "in flumine Baluntie, prope Palenque, provinciæ Chiapas dictæ, republicæ Mexicanæ."

Type Locality: Río Chacamax (Gulf of México–Grijalva–Usumacinta Dr.), Baluntie, near Palenque, Chiapas, México [L = 17.4965, -91.9541].

Type Specimens: Holotype by monotypy, NHMUK 1907.12.30.377.

Remarks: A specimen in the NHMUK 1907.12.30.377 from "Baluntie, nr. Palenque" from the Morelet Collection is labeled as a paralectotype of *Unio nephretica*. However, Fischer and Crosse (1894:613) stated that they had "Un spécimen de la collection S. Hanley. figuré par Sowerby dans le *Conchologia Iconica*, constitue la variété δ." [A specimen from the S. Hanley collection figured by Sowerby in the *Conchologia Iconica*, constitutes the variety δ.]

Taxonomic History:

Unio (*Simonaias*) *calamitarum* var. δ, *nephritica* [sic] Fischer & Crosse, 1894 — von Martens (1900b)

Nephronaias calamitarum var. *nephritica* [sic] (Fischer & Crosse, 1894) — Simpson (1914)

Nephronaias (s.s.) *calamitarum* (Morelet, 1849) — Haas (1969)

PLATE 45 *Nephronaias caldwellii* (Lea, 1858). **Top:** *Unio caldwellii* Lea, 1858. Holotype, USNM 84396 (58 mm L, 38 mm H). Isthmus of Darien, Canal Zone (Caribbean Dr.), Darién Province, Panamá [L=9.0300, -79.6344]. **Bottom:** *Unio averyi* Lea, 1859. Lectotype, USNM 84385 (58 mm L, 41 mm H). Río Chucunaque (Pacific–Turia Dr.), Isthmus of Darien, Darién Province, Panamá [W=8.1583, -77.6901].

Nephronaias caldwellii (Lea, 1858)

Plate 45, Map 53

Province(s): Oaxaca, México to Colombia.

Distribution and Status: Widely distributed in Pacific drainages from Chiapas, México, Guatemala, and Nicaragua southwest to Panamá. Found living in the Lara basin in Darién Province, and the Santa Maria drainage in Veraguas Province, Panamá in 2023. Not assessed (IUCN 2024).

Anatomy: Shell triangular and rounded with a feeble, double posterior ridge. Beaks rather high. Beak sculpture consists of fine, irregular ridges that cover the surface of the shell. Left valve with two ragged pseudocardinals and two laterals. Right valve with one pseudocardinal and one lateral. Periostracum tan to dark brown. Nacre white or pink.

Biology: Inhabits medium-sized streams in a mixed gravel, sand, and mud substrate.

Adult Size: Length to 4.5 cm.

Similar Species: *Nephronaias macnielii* (Lea, 1869) and *Nephronaias rowellii* (Lea, 1859).

Remarks: The generic placement of this species is uncertain. It has most recently been placed in *Barynaias* (now *Psoronaias*), but others have variously placed it in *Arotonaias* (von Martens 1900b:498) or *Nephronaias* (Simpson 1900:593; Haas 1929:320). We have placed it in *Nephronaias* (pending further analysis) as no species of *Psoronaias* or *Arotonaias* have been found in any Pacific drainage. We have also included a few populations in Chiapas, México and southern Guatemala here, but given the separation from the populations in Panamá, they may be a new species. In Lea's discussion of *N. caldwellii* in Observations 7 (1860:83), he noted that "A single specimen was brought by Dr. Caldwell from his perilous expedition on the Isthmus of Darien with Lieut. Strain, and these molluscs formed part of the food on which the party subsisted."

MAP 53 Distribution of *Nephronaias caldwellii* (Lea, 1858).

Unio caldwellii Lea, 1858. Description of four new fresh-water molluscs, from the Isthmus of Darien and Honduras. *Proceedings of the Academy of Natural Sciences of Philadelphia* 10(1858):118. Page 118. Figured in Lea (1860a JANSP Vol. 4, Part 3(7):265; 1860c Obs. 7:83, pl. 43, fig. 145).

Verbatim Type Locality: "Isthmus of Darien."

Type Locality: Isthmus of Darien, Canal Zone (Caribbean Dr.), Darién Province, Panamá [L=9.0300, -79.6344].

Type Specimens: Holotype by monotypy, USNM 84396 (Johnson 1974:26).

Taxonomic History:

Unio caldwellii Lea, 1858 — Lea (1860a), Sowerby (1868), Wright (1888)
Margaron (*Unio*) *caldwellii* (Lea, 1858) — Lea (1870)
Unio (*Arotonaias*) *caldwellii* Lea, 1858 — von Martens (1900b)
Nephronaias caldwellii (Lea, 1858) — Simpson (1900; 1914)
Elliptio (*Barynaias*) *averyi* (Lea, 1859) — Frierson (1927)
Elliptio (*Nephronaias*) *averyi* (Lea, 1859) — Haas (1929)
Barynaias caldwellii (Lea, 1858) — Haas (1969), Graf & Cummings (2007; 2021)

Unio averyi Lea, 1859. Description of a new species of *Unio* from the Isthmus of Darien. *Proceedings of the Academy of Natural Sciences of Philadelphia* 11(1859):281. Figured in Lea (1860a JANSP Vol. 4, Part 3(7):269; 1860c Obs. 7:87, pl. 44, fig. 149).

Verbatim Type Locality: "Isthmus of Darien."

Type Locality: Río Chucunaque (Pacific–Turia Dr.), Isthmus of Darien, Darién Province, Panamá [W=8.1583, -77.6901].

Type Specimens: Lectotype, USNM 84385 designated by Johnson (1974a:18, as the holotype).

Remarks: Locality data from Lea (1860a:269; 1870:96) and the label with the lectotype given as the Chumnagua River, Isthmus of Darien.

Taxonomic History:

Unio averyi Lea, 1859 — Lea (1860a), Wright (1888)
Margaron (*Unio*) *averyi* (Lea, 1859) — Lea (1870)
Unio (*Arotonaias*) *caldwellii* Lea, 1859 — von Martens (1900b)
Nephronaias averyi (Lea, 1859) — Simpson (1900; 1914), von Ihering (1901)
Elliptio (*Barynaias*) *averyi* (Lea, 1859) — Frierson (1927)
Elliptio (*Nephronaias*) *averyi* (Lea, 1859) — Haas (1929)
Barynaias caldwellii (Lea, 1858) — Haas (1969)

PLATE 46 ***Nephronaias chapalanus* (von Martens, 1900). Top:** *Unio aztecorum* var. *chapalanus* von Martens, 1900. Holotype, ZMB/MOLL.9106 (60 mm L, 32 mm H). Lago de Chapala (Pacific–Río Grande de Santiago Dr.), Jalisco, México [W=20.3049, -102.7781]. **Bottom:** *Nephronaias aztecorum tolimanensis* Pilsbry, 1920. Holotype, ANSP 46195 (35 mm L, 19 mm H). Río Armeria (Pacific Dr.), Tolimán, Jalisco, México [L=19.5773, -103.9238].

Nephronaias chapalanus (von Martens, 1900)

Plate 46, Map 54

Province(s): Mesa Central and Balsas.

Distribution and Status: Known only from the type locality, Lago de Chapala and three specimens (the types of *Nephronaias tolimanensis*) from the Río Armeria basin in Jalisco. All in Pacific drainages. No recent collections are known. Not assessed (IUCN 2024).

Anatomy: Shell thin to solid, oblong, and slightly inflated. Rounded both anteriorly and posteriorly. Ventral margin slightly concave. Pseudocardinal teeth small and peg-like, laterals slightly curved. Periostracum brown with faint green rays. Nacre white.

Biology: Unknown.

Adult Size: Length to 6 cm.

Similar Species: None.

Remarks: *Nephronaias chapalanus* (von Martens, 1900) is the only member of the genus in a Pacific coast drainage in northern México. The closest Pacific basin *Nephronaias* in México are in Oaxaca and Chiapas about 500 km to the southeast. Previous workers have considered this to be a subspecies of *Nephronaias aztecorum* (Philippi, 1847), but that species is found along the Gulf Coast and does not occur in any Pacific drainage. USNM 892341 is a relatively thin-shelled, young specimen collected alive from the Río Fuerte in Sinaloa, México that is over 1,000 km north of the next known location of any *Nephronaias* in México (red dot on Map 54). Initially identified as *Lampsilis* sp., it is the only mussel with teeth we have seen from Sinaloa or the surrounding states of Durango, Nayarit, or Sonora, and it may be an undescribed species.

MAP 54 Distribution of *Nephronaias chapalanus* (von Martens, 1900).

Unio (*Nephronaias*) *aztecorum* var. *chapalanus* von Martens, 1900. Land and freshwater Mollusca. *In* F.D. Godman and O. Salvin (eds.). *Biologia Centrali-Americana* Part 155, Fascile 63–65:497–520, plates 33–35. Page 504, and Part 157, Fascile 67:529–536, plates 36–37. Pl. 36, figs. 5–5a.

Verbatim Type Locality: "Laguna de Chapala, a large lake situated on the elevated plain of Central México, discharging its waters by the Río Grande de Santiago into the Pacific."

Type Locality: Lago de Chapala (Pacific–Río Grande de Santiago Dr.), Jalisco, México [W=20.3049, -102.7781].

Type specimen: Holotype by monotypy, ZMB/MOLL.9106.

Remarks: Described in von Martens (1900b) and figured in von Martens (1900d).

Taxonomic History:

Nephronaias aztecorum var. *chapalanus* (von Martens, 1900) — Simpson (1914)
Nephronaias (s.s.) *aztecorum* (Philippi, 1847) — Haas (1969)

Nephronaias aztecorum tolimanensis Pilsbry, 1920. Mollusks from Lake Chapala, State of Jalisco and vicinity. *Proceedings of the Academy of Natural Sciences of Philadelphia* 72(2):192–194. Page 194, fig. 4.

Verbatim Type Locality: "Tolimán, State of Jalisco."

Type Locality: Río Armeria (Pacific Dr.), Tolimán, Jalisco, México [L=19.5773, -103.9238].

Type Specimens: Holotype, ANSP 46195, by original designation.

Remarks: ANSP 46196 (n=1+fragments) from the Río Grande (Pacific–Armeria Dr.), Zapotitlan, Jalisco, México [L=19.5480, -103.8060] are labeled as syntypes in the ANSP, but Pilsbry only identified it as an additional specimen examined. According to Pilsbry (1920a:194) "Differs from *N. aztecorum* and *N. a. chapalensis* by the smaller size and lower form, the shell not so wide posteriorly."

Taxonomic History:

Nephronaias (s.s.) *aztecorum* (Philippi, 1847) — Haas (1969)

PLATE 47 ***Nephronaias copanensis* (von Martens, 1900). Top:** *Unio* (*Sphenonaias*) *tehuantepecensis* var. *copanensis* von Martens, 1900. Lectotype, ZMB/MOLL.220170 (69 mm L, 36 mm H). Río Copan (Caribbean–Gulf of Honduras–Motagua Dr.), Honduras [W = 14.8333, -89.1500]. **Bottom: *Nephronaias flucki* Bartsch, 1906**. Holotype, USNM 186121 (64 mm L, 27 mm H). Río Kukalaya (Caribbean–Layasiksa Dr.), NW of Kukalaya, Región Autónoma del Atlántico Norte, Nicaragua [L = 14.0353, -84.0790].

Nephronaias copanensis (von Martens, 1900)

Plate 47, Map 55

Province(s): Gulf of Honduras & Northern Caribbean.

Distribution and Status: Known only from the Río Copan a tributary of the Río Grande de Zacapa (Gulf of Honduras–Motagua Dr.), in Honduras. No recent collections are known. Not assessed (IUCN 2024).

Anatomy: Shell elongate, subrhomboid, and very solid when old. Anterior end rounded. Posterior end truncated. The beaks low and in all specimens examined very badly eroded. Beak cavity shallow. Left valve with two triangular, ragged pseudocardinals and two laterals. Right valve with one pseudocardinal and one small lateral. Periostracum brown or black and rough. Nacre white, orange, or purple.

Biology: Unknown.

Adult Size: Length to 7 cm.

Similar Species: *Nephronaias ortmanni* (Frierson, 1913) co-occurs with *Nephronaias copanensis* (von Martens, 1900) in the Motagua basin, but it is much thicker shell with fine radiating ridges on the surface of the shell.

MAP 55 Distribution of *Nephronaias copanensis* (von Martens, 1900).

Unio (*Sphenonaias*) *tehuantepecensis* var. *copanensis* von Martens, 1900. Land and freshwater Mollusca. *In* F.D. Godman and O. Salvin (eds.). *Biologia Centrali-Americana* Part 155, Fascile 63– 65:497–520. Page 502, and Part 157, Fascile 67:529–536, plates 36–37. Pl. 36, figs. 6–6a.

Verbatim Type Locality: "Honduras: Rio Copan, an affluent of the Motagua."

Type Locality: Río Copan (Caribbean–Gulf of Honduras–Motagua–Río Grande de Zacapa Dr.), Honduras [W = 14.8333, -89.1500].

Type specimen: Lectotype, ZMB/MOLL.220170, here designated (the figured specimen).

Remarks: ZMB/MOLL.220171 ($n=2$) are labeled as paratypes in the collection at Berlin. However, von Martens mentioned receiving only two specimens from von Ihering. Since we cannot determine which one is actually part of the original two, we will not designate a paralectotype.

Taxonomic History:

Nephronaias copanensis (von Martens, 1900) — Simpson (1914), Graf & Cummings (2007; 2021)

Elliptio (*Nephronaias*) *copanensis* (von Martens, 1900) — Frierson (1927), Haas (1929)

Nephronaias (s.s.) *copanensis* (von Martens, 1900) — Haas (1969)

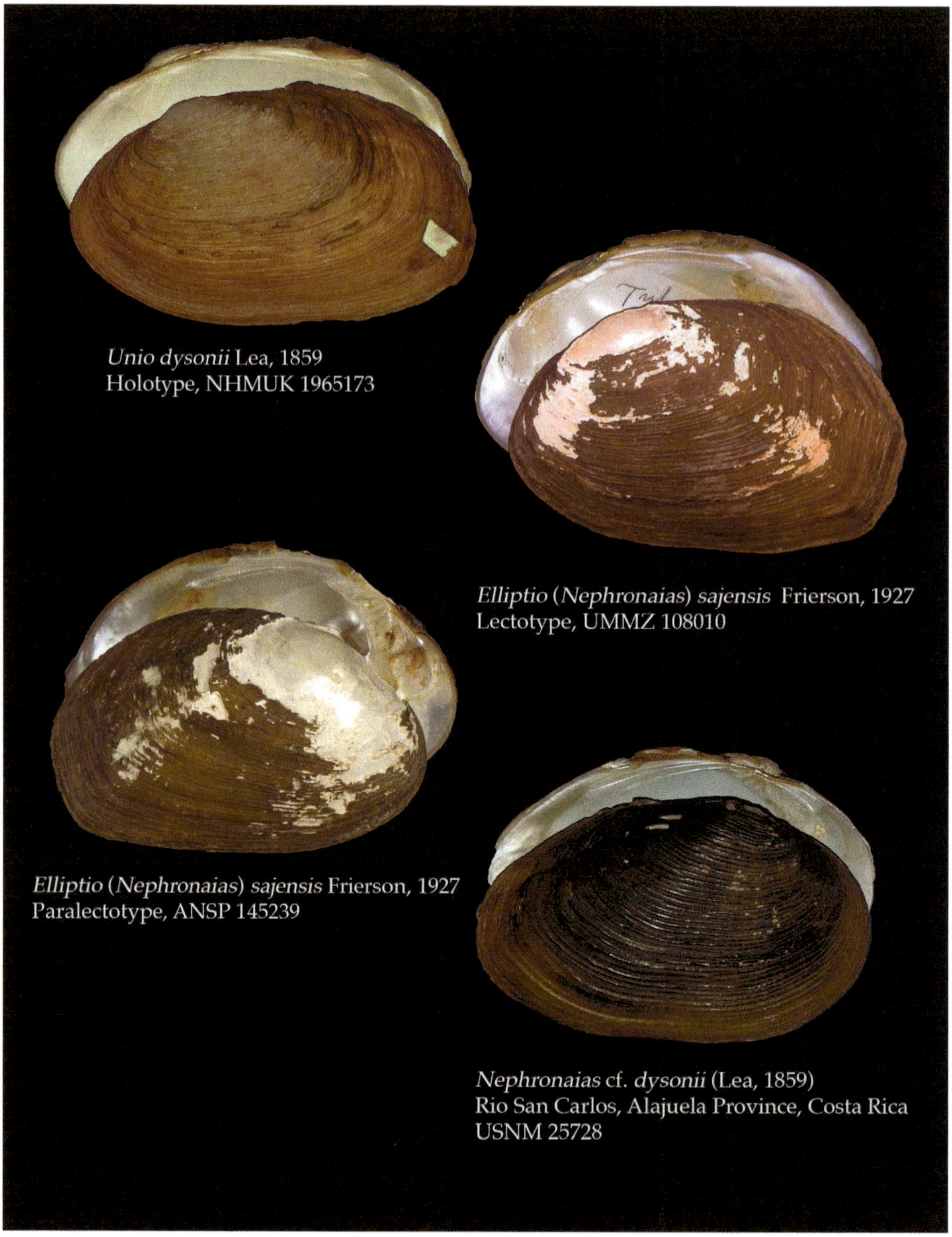

PLATE 48 ***Nephronaias dysonii* (Lea, 1859). Top Left:** *Unio dysonii* Lea, 1859. Holotype, NHMUK 1965173 (45 mm L, 28 mm H). Honduras. **Middle Right:** *Elliptio* (*Nephronaias*) *sajensis* Frierson, 1927. Lectotype, UMMZ 108010 (57 mm L, 37 mm H). Río Sejá (Caribbean–Gulf of Honduras–Dulce Dr.), 8 km W Esmeralda, Izabal Department, Guatemala [L = 15.6872, -88.9828]. **Middle Left:** *Elliptio* (*Nephronaias*) *sajensis* Frierson, 1927. Paralectotype, ANSP 145239. **Bottom Right:** *Nephronaias* cf. *dysoni* (Lea, 1859), USNM 25728. Rio San Carlos (Caribbean–San Juan Dr.), Alajuela Province, Costa Rica [10.7833, -84.2000].

PLATE 49 ***Nephronaias dysonii*** **(Lea, 1859).** *Elliptio* (*Nephronaias*) *sajensis* Frierson, 1927. **Top:** Paralectotypes, INHS 21772. **Bottom:** INHS 21773. Río Sejá (Caribbean–Gulf of Honduras–Dulce Dr.), 8 km W Esmeralda, Izabal Department, Guatemala [L =15.6872, -88.9828]. Specimen size ranges 20–40 mm.

Nephronaias dysonii (Lea, 1859)

Plates 48–49, Map 56

Province(s): Gulf of Honduras & Northern Caribbean.

Distribution and Status: Widespread from northern Belize, south to Guatemala and Honduras. Recent collections of live individuals are known from numerous sites in Belize and Honduras. Not assessed (IUCN 2024).

Anatomy: Shell quadrate and solid, becoming elongate in some specimens. The beaks are low and the beak cavity shallow. Dorsal margin slightly arched. Ventral margin varies from convex to concave. Left valve with two compressed pseudocardinals and two slightly curved laterals. Right valve with two pseudocardinals and one lateral. Periostracum rough, yellow or straw-colored, occasionally with rays on the posterior slope in small individuals, becoming brown in adults. Nacre white, pink, salmon, or purple.

Biology: Unknown.

Adult Size: Length to 6 cm.

Similar Species: *Nephronaias calamitarum* (Morelet, 1849) and *Nephronaias yzabalensis* (Crosse & Fischer, 1893).

Remarks: There are three lots labeled as *Nephronaias dysonii* (Lea, 1859) [ANSP 125622 (n=1), USNM 25018 (n=4), and USNM 25728 (n=1)] from the Río San Carlos (Caribbean–San Juan Dr.) in Costa Rica that may be an undescribed species (Plate 48). An annotation on the back of the label of USNM 25728 by Joseph P.E. Morrison reads "not *dysonii*." The surface of the shell is ribbed in the specimens from the Río San Carlos, whereas it is smooth in all other species of *Nephronaias* except for the elongate *N. ortmanni* (Frierson, 1914) from the Río Motagua basin in Guatemala. The population in the Río Humuya (Caribbean–Ulua Dr.) in Honduras are also elongate and atypical that warrants further study.

MAP 56 Distribution of *Nephronaias dysonii* (Lea, 1859).

There is also an elongate, thick-shelled mussel with a dark, almost black periostracum and purple nacre that co-occurs with the yellow-colored *Nephronaias dysonii* (Lea, 1859) in Yalbac Creek, Cayo District, Belize, which may be a new species (Figure 41).

Unio dysonii Lea, 1859. Descriptions of twenty-one new species of exotic Unionidae. *Proceedings of the Academy of Natural Sciences of Philadelphia* 11(1859):151–154. Page 152. Figured in Lea (1860a JANSP Vol. 4, Part 3(7):252; 1860c Obs. 7:70, pl. 39, fig. 132).

Verbatim Type Locality: "Honduras."

Type Locality: Honduras.

Type Specimens: Holotype by monotypy, NHMUK 1965173 (Johnson 1971:83; Johnson 1974:50).

Taxonomic History:

Unio dysoni Lea, 1859 — Lea (1860a)
Margaron (*Unio*) *dysonii* (Lea, 1859) — Lea (1870)
Unio (*Simonaias*) *dysoni* Lea, 1859 — von Martens (1900b)
Nephronaias dysonii (Lea, 1859) — Simpson (1900; 1914), Hinkley (1920), Graf & Cummings (2007; 2021)
Nephronaias dysoni (Lea, 1859) — Frierson (1917), Pilsbry (1920b)
Elliptio (*Nephronaias*) *dysoni* (Lea, 1859) — Frierson (1927)
Elliptio (*Nephronaias*) *dysonii* (Lea, 1859) — Haas (1929)
Nephronaias goascoranensis (Lea, 1858) — Richards & Boekelman (1937)
Nephronaias ortmanni (Frierson, 1914) — Andrews (1969)
Nephronaias (s.s.) *dysonii* (Lea, 1859) — Haas (1969)

Elliptio (*Nephronaias*) *sajensis* Frierson, 1927. A classification and annotated check list of the North American naiades. Baylor University Press, Waco, Texas. 111 pp. Page 37, species 139. Figured in Frierson (1928:139, pl. 2, fig. 2) and Johnson (1972a:168, pl. 30, fig. 2).

Verbatim Type Locality: "Saja River, Guatemala, at Esmaralda [sic]."

Type Locality: Río Sejá (Caribbean–Gulf of Honduras–Dulce Dr.), 8 km W Esmeralda, Izabal Department, Guatemala [L = 15.6872, -88.9828].

Type Specimens: Lectotype, UMMZ 108010 designated by Johnson (1972a:148; Johnson and Baker 1973:169; Johnson 1979:34, all as the holotype). Paralectotypes, ANSP 145239 (n=2) (Johnson 1972a:148; Johnson and Baker 1973:169, both as paratypes); FMNH 68188 (n=5); INHS 21772 (n=6), INHS 21773 (n=22).

Remarks: Regarding the type locality, Hinkley (1920:45) stated that "This plantation [Esmeralda] is on the left bank of the Rio Dulce some three miles below the old Fort San Felipe, now in ruins. This stream is some five miles west of Esmeralda." Frierson (1927:37) noted that specimens were distributed by A.A. Hinkley as *Unio dysoni* Lea from which it differed in being larger, shorter in proportion, and more biangular behind. The river is spelled as Río Sejá on current maps.

Taxonomic History:

Elliptio sajensis Frierson, 1927 — Johnson (1972)
Nephronaias (s.s.) *sajensis* (Frierson, 1927) — Haas (1969)
Nephronaias sajensis (Frierson, 1927) — Graf & Cummings (2007; 2021)

FIGURE 41 *Nephronaias* sp. Yalbac Creek, Cayo District, Belize, 27 June 2019. 17.3813–89.0482. (Photo by Jan Meerman iNaturalist 2019).

Nephronaias flucki (Bartsch, 1906)

Plate 47, Map 57

Province(s): Gulf of Honduras & Northern Caribbean.

Distribution and Status: Known primarily from the ríos Kukalaya and Coco in Región Autónoma del Atlántico Norte, Nicaragua. No recent collections of live individuals are known. Not assessed (IUCN 2024).

Anatomy: Shell elliptical, moderately inflated, and truncated posteriorly. Ventral margin straight or slightly concave. Beaks low and the beak cavity moderately deep. The pseudocardinal in the right valve is triangular. The pseudocardinals of the left valve are slender. Laterals narrow, straight or slightly curved. Surface of the periostracum is marked by numerous concentric lines and radiating rays that are more pronounced on the anterior end. Periostracum light to dark brown. Nacre white or copper.

Biology: Unknown.

Adult Size: Length to 6.5 cm.

Similar Species: *Arotonaias reticulata* (Simpson, 1900) co-occurs in the Río Coco in Nicaragua, and both have a rugose periostracum but *A. reticulata* is thicker and round in shape.

MAP 57 Distribution of *Nephronaias flucki* Bartsch, 1906.

Nephronaias flucki Bartsch, 1906. Description of two new naiads. *Proceedings of the United States National Museum* 30:393–395 + 3 plates. Page 393, pls. 27–29, fig. 2 (holotype), fig. 3 (paratype).

Verbatim Type Locality: "Wounta River, northwest of Kukallaya, Nicaragua."

Type Locality: Río Kukalaya [= Wounta River] (Caribbean–Layasiksa Dr.), NW of Kukalaya, Región Autónoma del Atlántico Norte, Nicaragua [L = 14.0353, -84.0790].

Type Specimens: Holotype, USNM 186121, by original designation (published as USNM 106121). Paratypes, CM 61.3953 ($n=2$) (Brooks and Brooks 1931:175, as part of the type lot); ANSP 97585 ($n=2$) (Johnson and Baker 1973:155); MCZ 167991 ($n=2$) (Johnson 1956:115, incorrectly as MCZ 167491).

Remarks: Specimens in the MCZ (167986, $n=4$) are labeled as paratypes, but they are from Wasla, Nicaragua, which is not the type locality. Bartsch (1906:393) mentioned having only four specimens before him so three of the above specimens published as paratypes are topotypes. *Nephronaias flucki* Bartsch, 1906 was placed in *Sphenonaias* by Frierson (1927:38) noting that "The figure of this species was by an error reduced in size just half and being unnoted, leaves the impression that the shell is allied to the '*Unio calamitarum*, etc.' a resemblance not borne by the specimen." This opinion by Frierson is not consistent with our examination of the type. As noted in the original description by Bartsch (1906:394), "*N. flucki* can be readily separated from the other species of *Nephronaias* of the region by its rugose sculpture, in which respect it resembles *N. dysoni* Lea, from Costa Rica, and by its wavy radiating lines and dark nacre." No Central American species formerly placed in *Sphenonaias* have surface sculpture like that present on *N. flucki*, and we return it to *Nephronaias* as described.

Taxonomic History:

Nephronaias flucki Bartsch, 1906 — Simpson (1914)
Elliptio (*Sphenonaias*) *flucki* (Bartsch, 1906) — Frierson (1927)
Nephronaias (*Sphenonaias*) *flucki* Bartsch, 1906 — Haas (1969)
Sphenonaias flucki (Bartsch, 1906) — Graf & Cummings (2007; 2021)

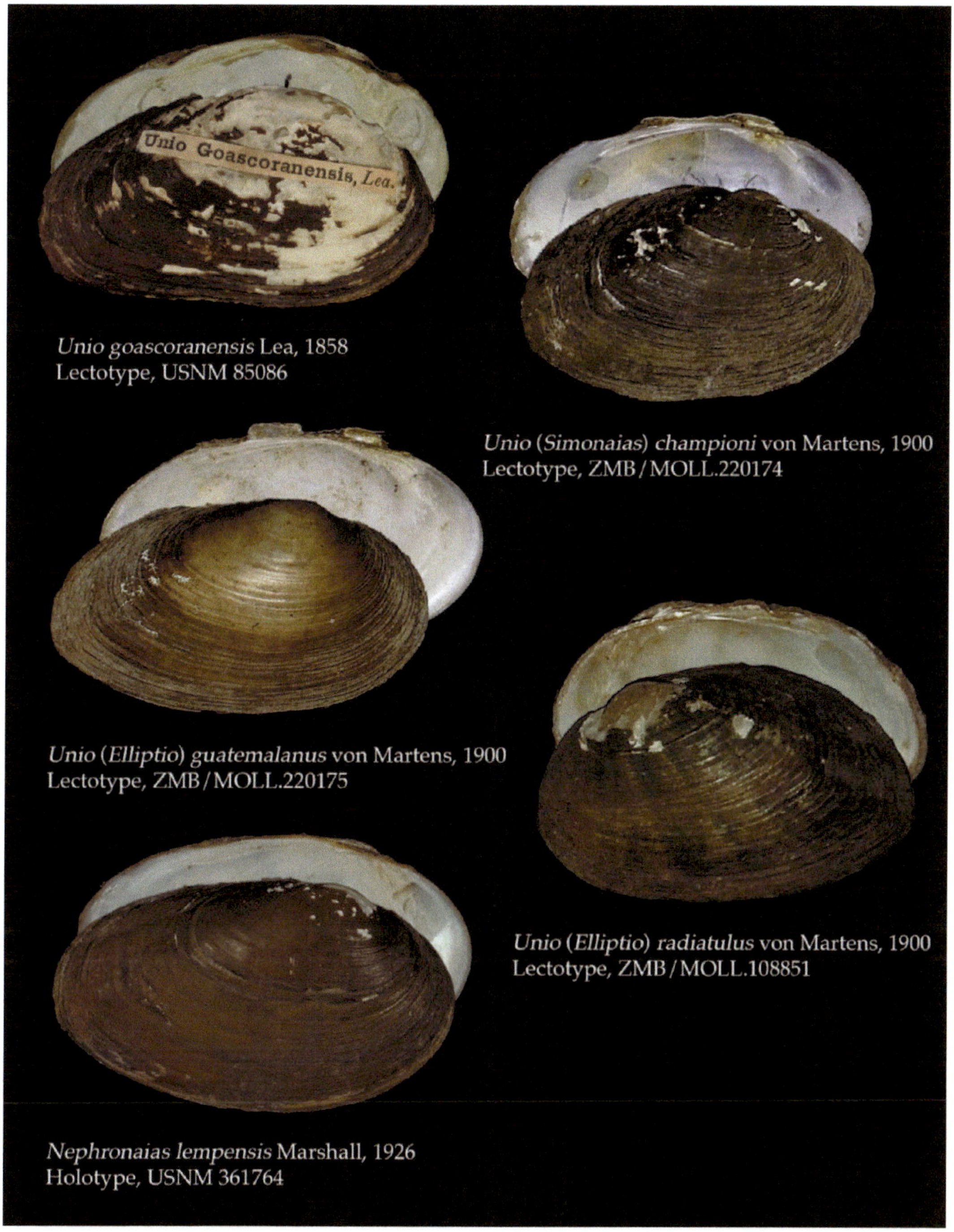

PLATE 50 ***Nephronaias goascoranensis* (Lea, 1858). Top Left:** *Unio. goascoranensis* Lea, 1858. Lectotype, USNM 85086 (58 mm L, 32 mm H). Río Goascorán (Pacific Dr.), Honduras [W = 13.6075, -87.7554]. **Top Right:** *Unio* (*Simonaias*) *championi* von Martens, 1900. Lectotype, ZMB/MOLL.220174 (30 mm L, 17 mm H). **Middle Left:** *Unio* (*Elliptio*) *guatemalanus* von Martens, 1900. Lectotype, ZMB/MOLL.220175 (57 mm L, 30 mm H). **Middle Right:** *Unio* (*Elliptio*) *radiatulus* von Martens, 1900. Lectotype, ZMB/MOLL.108851 (44 mm L, 24 mm H). The last three, Río Michatoya (Pacific Dr.), Paso San Antonio, Escuintla Department, Guatemala [W = 14.0617, -90.6809]. **Bottom Left:** *Nephronaias lempensis* Marshall, 1926. Holotype, USNM 361764 (53 mm L, 30 mm H). Río Lempa (Pacific Dr.), at R.R. bridge, San Vicente Department, El Salvador [L = 13.4270, -88.6992].

Nephronaias goascoranensis (Lea, 1858)

Plate 50, Map 58

Province(s): Oaxaca, México to Colombia; San Juan.

Distribution and Status: Found in both Caribbean and Pacific drainages from the Río Michatoya in Guatemala southeast in the ríos Goascorán and Río Grande de Choluteca basins (all Pacific Dr.) in Honduras to the San Juan Basin (Caribbean Dr.) in Nicaragua. Recent collections made in the Río Goascorán, La Unión Department, El Salvador and the Río Grande de San Miguel and Río Lempa basins 2018–2024. Listed as endangered by El Salvador (Ministério de Medio Ambiente y Recursos Naturales 2023:39, as *Nephronaias lempensis*). Not assessed (IUCN 2024).

Anatomy: Shell elongate, subelliptical to subtriangular, and slightly inflated. Dorsal margin curved gradually at the hinge and more sharply behind. Ventral margin nearly straight in males, curved outward at the posterior base in females. Beaks moderately high, with concentric fine ridged sculpture. Beak cavity shallow. Left valve with two small pseudocardinals and two curved laterals. Right valve with one pseudocardinal with a very small tooth above and one lateral. Periostracum olive in the younger individuals, dark brown in adults, and scarcely rayed when young. The periostracum is often eroded. Nacre white and slightly iridescent.

Lea (1860a:258) described the soft parts and noted that no ova were found in the specimen he received. "Branchiae rather large, very thin, slightly curved below, inner one rather the larger, free nearly the whole length of abdominal sack. Palpi rather large, suboval, open nearly the whole length of posterior edge. Mantle thin, thicker along the edge, and bordered with black. Branchial opening rather large, with numerous, rather small, brownish papillae. Anal opening large, with numerous small, brownish papillae. Super-anal opening large, bordered with black and united below. Color of the mass whitish."

Biology: Unknown.

Adult Size: Length to 6 cm.

Similar Species: We have united 4 taxa under *Nephronaias goascoranensis* (Lea, 1858) pending further analysis. von Martens described three species: *Unio* (*Elliptio*) *radiatulus* von Martens, 1900, *Unio* (*Simonaias*) *championi* von Martens, 1900, and *Unio* (*Elliptio*) *guatemalanus* von Martens, 1900 from the same type locality: Río Michatoya (Pacific Dr.), Paso San Antonio, Escuintla Department, Guatemala. Various authors have considered these to be distinct, as *Nephronaias goascoranensis* (Lea, 1858) or as *N. rowellii* (Lea, 1859). *Nephronaias rowellii* (Lea, 1859) is found only in the Chagres basin (Caribbean Dr.) much further south. There are also numerous specimens in collections labeled as *N. rowellii* from Nicaragua and Guatemala that we have tentatively identified as *N. goascoranensis.*

MAP 58 Distribution of *Nephronaias goascoranensis* (Lea, 1858).

Unio goascoranensis Lea, 1858. Description of four new fresh-water molluscs, from the Isthmus of Darien and Honduras. *Proceedings of the Academy of Natural Sciences of Philadelphia* 10(1858):118. Figured in Lea (1860a JANSP Vol. 4, Part 3(7):258; 1860c Obs. 7:76, pl. 41, fig. 139).

Verbatim Type Locality: "River Goascoran, Honduras, Pacific Slope."

Type Locality: Río Goascorán (Pacific Dr.), Honduras [W = 13.6075, -87.7554].

Type Specimens: Lectotype, USNM 85086 designated by Johnson and Baker (1973:156; Johnson 1974:64, both as the holotype). Paralectotypes, ANSP 129957 (n = 5½) (Johnson and Baker 1973:156, as paratypes).

Taxonomic History:

Unio goascoranensis Lea, 1858 — Lea (1860a)
Margaron (*Unio*) *goascoranensis* (Lea, 1858) — Lea (1870)
Unio goascoraensis [sic] Lea, 1858 — Wright (1888)
Unio (*Sphenonaias*) *goascoranensis* Lea, 1858 — von Martens (1900b)
Nephronaias goascoranensis (Lea, 1858) — Simpson (1900; 1914), Zilch (1967), Graf & Cummings (2007; 2021), Czaja et al. (2023)
Nephronaias calamitarum (Morelet, 1849) — von Ihering (1901)
Elliptio (*Nephronaias*) *goascoranensis* (Lea, 1858) — Frierson (1927), Haas (1929)
Nephronaias (s.s.) *goascoranensis* (Lea, 1858) — Haas (1969)

Unio (*Simonaias*) *championi* von Martens, 1900. Land and freshwater Mollusca. *In* F.D. Godman and O. Salvin (eds.). *Biologia Centrali-Americana* Part 155, Fascile 63–65:497–520, plates 33–35. Page 508, and Part 158, Fascile 68–71:537–568, plates 38–39. Pl. 38, figs. 9–9a.

Verbatim Type Locality: "Paso Antonio, in the Pacific coast–region."

Type Locality: Río Michatoya (Pacific Dr.), Paso San Antonio, Escuintla Department, Guatemala [L = 14.0617, -90.6809].

Type Specimens: Lectotype, ZMB/MOLL.220174, here designated (the figured specimen). Paralectotypes, ZMB/MOLL.305841 (*n*=2½), NHMUK 1901.6.22.1538–39 (*n*=2) (Johnson 1971:80, as paratypes).

Remarks: von Martens also figured a specimen on pl. 38, figs. 10, 10a–b (ZMB/MOLL.305842) from the Río Michatoya at Paso San Antonio.

Taxonomic History:

Nephronaias championi (von Martens, 1900) — Simpson (1914), Graf & Cummings (2007; 2021)
Elliptio (*Nephronaias*) *championi* (von Martens, 1900) — Frierson (1927)
Elliptio (*Nephronaias*) *macnielii* (Lea, 1869) — Haas (1929)
Nephronaias (s.s.) *championi* (von Martens, 1900) — Haas (1969)

Unio (*Elliptio*) *guatemalanus* von Martens, 1900. Land and freshwater Mollusca. *In* F.D. Godman and O. Salvin (eds.). *Biologia Centrali-Americana* Part 155, Fascile 63–65:497–520, plates 33–35. Page 519, and Part 158, Fascile 68–71:537–568, plates 38–39. Pl. 38, figs. 7, 7a–b; 8–8a.

Verbatim Type Locality: "W. Guatemala: Paso Antonio, Rio Michatoya."

Type Locality: Río Michatoya (Pacific Dr.), Paso San Antonio, Escuintla Department, Guatemala [L=14.0617, -90.6809].

Type Specimens: Lectotype, ZMB/MOLL.220175, here designated (pl. 38, figs. 7, 7a–b). Paralectotypes, ZMB/MOLL.305840 (*n*=3), SMF 3486 (*n*=1, pl. 38, figs. 8–8a; Haas 1929:321; Nagel 2025:164, both as a paratype). Zilch 1967:102, as a syntype).

Remarks: von Martens (1900b:519) noted that "Two perfect specimens of this species are before me: one of them (b) agrees rather well in its general features with *U. rowelli*, Lea, except in the fore part of the shell."

Taxonomic History:

Unio (*Elliptio*) *guatemalanus* von Martens, 1900 — von Martens (1901c)
Nephronaias goascoranensis (Lea, 1858) — Simpson (1914), Zilch (1967), Nagel (2025)
Nephronaias guatemalanus (von Martens, 1900) — Hinkley (1920)
Elliptio (*Nephronaias*) *guatemalensis* [sic] (von Martens, 1900) — Frierson (1927)
Elliptio (*Nephronaias*) *rowellii* (Lea, 1859) — Haas (1929)
Nephronaias (s.s.) *goascoranensis* (Lea, 1858) — Haas (1969)

Unio (*Elliptio*) *radiatulus* von Martens, 1900. Land and freshwater Mollusca. *In* F.D. Godman and O. Salvin (eds.). *Biologia Centrali-Americana* Part 155, Fascile 63–65:497–520, plates 33–35. Page 518, and Part 158, Fascile 68–71:537–568, plates 38–39. Pl. 39, figs. 8, 8a–b.

Verbatim Type Locality: "W. Guatemala: Rio Naranjo, a confluent of the Rio Michatoya, at the station Naranjo, and Paso Antonio in the lower part of the Rio Michatoya."

Type Locality: Río Michatoya (Pacific Dr.), Paso San Antonio, Escuintla Department, Guatemala [L=14.0617, -90.6809].

Type Specimens: Lectotype, ZMB/MOLL.108851, here designated (the figured specimen) and type locality fixed. Paralectotype, ZMB/MOLL.220176 (*n*=1) from the Río Naranjo (Pacific–Michatoya Dr.), Escuintla Department, Guatemala [W=14.07674, -90.7771].

Remarks: von Martens (1900b:518) recognized this taxon as valid, but Simpson (1914:261) and others that followed placed it under *Unio oregonensis* Lea, 1852—a taxon we regard as a *nomen dubium* below.

Taxonomic History:

Unio (*Sphenonaias*) *goascoranensis* Lea, 1858 — von Martens (1900b) [in part]
Nephronaias oregonensis (Lea, 1852) — Simpson (1914)
Elliptio (*Nephronaias*) *oregonensis* (Lea, 1852) — Frierson (1927)
Nephronaias (s.s.) *oregonensis* (Lea, 1852) — Haas (1969)

Nephronaias lempensis Marshall, 1926. New land and fresh-water mollusks from Central and South America. *Proceedings of the United States National Museum* 69(2638):1–12+3 plates. Page 4, pl. 2, figs. 4, 6; pl. 3, fig. 4.

Verbatim Type Locality: "Rio Lempe [sic], Salvador, at the railroad bridge."

Type Locality: Río Lempa (Pacific Dr.), at R.R. bridge, San Vicente Department, El Salvador [L=13.4270, -88.6992].

Type Specimens: Holotype by monotypy, USNM 361764 (Johnson 1974b:11).

Remarks: Marshall noted that there were "five other specimens Cat. No. 360388 U.S.N.M., collected at the same time and same place," but he did not designate them as types.

Taxonomic History:

Nephronaias (s.s.) *lempensis* Marshall, 1926 — Haas (1969)
Nephronaias lempensis Marshall, 1926 — Graf & Cummings (2007; 2021), Agudo-Padrón (2019)

PLATE 51 ***Nephronaias gundlachi*** **(Dunker, 1858). Top Left:** *Unio gundlachi* Dunker, 1858. Lectotype, ZMB/MOLL.108720 (70 mm L, 39 mm H). Río Taco Taco (Caribbean Dr.), Artemisa Province, Cuba [W = 16.8246, -90.8822]. **Middle Right:** Paralectotype, ZMB/MOLL.108854. ***Nephronaias hjalmarsoni*** **(Dunker, 1858). Bottom Left:** *Unio hjalmarsoni* Dunker, 1858. Holotype, ZMB/MOLL.108719 (36 mm L, 18 mm H). Río Chamelecón (Caribbean Dr.), Honduras [W = 15.7713, -87.8443].

Nephronaias gundlachi (Dunker, 1858)

Plate 51, Map 59

Province(s): Cuba.

Distribution and Status: Known only from western Cuba. No known recent collections. Not assessed (IUCN 2024).

Anatomy: Shell elliptical and moderately solid. Most conspicuously inflated at or just behind the middle. Posterior ridge scarcely developed and feebly biangulate. Beaks low with fine corrugated ridges that cover the surface of the shell. Left valve with two subcompressed pseudocardinals and two thin laterals. Right valve with one subcompressed pseudocardinal, a vestigial one above, and one lateral. Periostracum olive green to dark brown or black and occasionally with indistinct rays. Shells sexually dimorphic with a pronounced marsupial expansion in females.

Biology: Unknown.

Adult Size: Length to 9 cm.

Similar Species: *Nephronaias scamnata* (Morelet, 1849) is much more elongate with a smooth periostracum.

Remarks: Johnson synonymized this taxon under *Leunio lienosus* (Conrad, 1834) (as *Villosa*) of eastern North America based upon it being sexually dimorphic; a hypothesis that has never been tested.

MAP 59 Distribution of *Nephronaias gundlachi* (Dunker, 1858).

Unio gundlachi Dunker, 1858. Einige neue species der naiaden. *Malakozoologische Blätter* 5:225–229. Page 228. Figured in Haas (1929:328, figs. 6–7, as the typus, figs. 8–11, as paratypen) and Johnson (1981:275, pl. 38, figs. 2, MCZ 94849, fig. 4; MCZ 20862).

Verbatim Type Locality: "flumine Taco Taco dicto insulae Cubanae."

Type Locality: Río Taco Taco (Caribbean Dr.), Artemisa Province, Cuba [W=16.8246, -90.8822].

Type Specimens: Lectotype, ZMB/MOLL.108720 designated by Haas (1929:328, as the type; Johnson 1981:275, as ZMB uncatalogued and as the holotype). Paralectotypes, ZMB/MOLL.108854 (*n*=8); MCZ 94849 (*n*=1) (Haas 1929:328; Johnson 1981:276, both as paratypes).

Taxonomic History:

Unio gundlachi Dunker, 1858 — Arango (1865), Sowerby (1866)
Nephronaias gundlachi (Dunker, 1858) — Simpson (1900; 1914), Graf & Cummings (2007; 2021), Serrand & Cummings (2014)
Actinonaias gundlachi (Dunker, 1858) — Frierson (1917)
Lampsilis (*Actinonaias*) *gundlachi* (Dunker, 1858) — Frierson (1927)
Elliptio (*Nephronaias*) *gundlachi* (Dunker, 1858) — Haas (1929)
Nephronaias (s.s.) *gundlachi* (Dunker, 1858) — Haas (1969)
Villosa gundlachi (Dunker, 1858) — Johnson (1981)

Nephronaias hjalmarsoni (Dunker, 1858)

Plate 51, Map 60

Province(s): Gulf of Honduras & Northern Caribbean.

Distribution and Status: Known from the Caribbean (Chamelecón and Aguan) drainages in Honduras. No known recent collections. Not assessed (IUCN 2024).

Anatomy: Shell small, elliptical, and moderately compressed. Anterior end narrowly rounded. Posterior end wider and somewhat truncated. Beaks slightly elevated and located about one-third back from the anterior end. Pseudocardinals somewhat thick, oblique, and deeply furrowed but not crenulated. Periostracum yellow and covered with fine, concentric ridges. Nacre white.

Biology: Unknown.

Adult Size: Length to 3.5 cm.

Similar Species: *Nephronaias hjalmarsoni* is the only known mussel from the Chamelecón and Aguan basins.

MAP 60 Distribution of *Nephronaias hjalmarsoni* (Dunker, 1858).

Unio hjalmarsoni Dunker, 1858. Einige neue species der naiaden. *Malakozoologische Blätter* 5:225–229. Page 227. Holotype figured in von Martens (1900e, pl. 38, figs. 11–11a, from Dunker).

Verbatim Type Locality: "Rio Chamelicon [sic], Honduras."

Type Locality: Río Chamelecón (Caribbean Dr.), Honduras [W = 15.7713, -87.8443].

Type Specimens: Holotype by monotypy, ZMB/MOLL.108719.

Remarks: Frierson (1927:37) noted that "A rather doubtfully valid species—the type being evidently quite a young shell, having no marked characters, and is quite likely nothing more than *Unio goascoranensis* Lea." *Nephronaias goascoranensis* is a Pacific basin species, and we provisionally regard *N. hjalmarsoni* as valid pending further analysis.

Taxonomic History:

Unio (*Simonaias*) *hjalmarsoni* Dunker, 1858 — von Martens (1900b)

Nephronaias hjalmarsoni (Dunker, 1858) — Simpson (1914), Graf & Cummings (2007; 2021)

Elliptio (*Nephronaias*) *hjalmarsoni* (Dunker, 1858) — Frierson (1927)

Elliptio (*Nephronaias*) *goascoranensis* (Lea, 1858) — Haas (1929)

Nephronaias (s.s.) *hjalmarsoni* (Dunker, 1858) — Haas (1969)

PLATE 52 ***Nephronaias macnielii* (Lea, 1869). Top Left and Middle Right:** *Unio macnielii* Lea, 1869. Lectotype, USNM 84414 (43 mm L, 25 mm H). Río Telica (Pacific Dr.), Corcuera, Leon Department, Nicaragua [L = 12.4721, -86.9954]. **Middle Left and Bottom Right:** Paralectotype, ANSP 129943.

Nephronaias macnielii (Lea, 1869)

Plate 52, Map 61

Province(s): San Juan; Oaxaca, México to Colombia.

Distribution and Status: Found in Lake Nicaragua (Río San Juan Dr.) and in a few small Pacific coast drainages in Nicaragua and Costa Rica. No known recent collections. Not assessed (IUCN 2024).

Anatomy: Shell slightly elongate or quadrate with a well-developed, double posterior ridge that ends in a distinct truncation on the posterior end. Anterior end rounded. Ventral margin straight. The beaks are not elevated with sculpture consisting of fine, double loops. Left valve with two pseudocardinals and two laterals. Right valve with one pseudocardinal and one lateral. Surface with fine, concentric striations. Periostracum brown or olive green, occasionally with a few faint rays. Nacre white or bluish white.

Biology: Unknown.

Adult Size: Length to 6 cm.

Similar Species: *Nephronaias dysonii* (Lea, 1859) is quadrate in shape and similar to *N. macnielii*, but they do not co-occur.

MAP 61 Distribution of *Nephronaias macnielii* (Lea, 1869).

Unio macnielii Lea, 1869. Descriptions of six new species of freshwater shells. *Proceedings of the Academy of Natural Sciences of Philadelphia* 21:124–125. Page 124. Figured in Lea (1874a JANSP Vol. 8:25; 1874b Obs. 13:29, pl. 8, fig. 22).

Verbatim Type Locality: "Rio Gigillillo, Corcuera, Nicaragua, Cen. Am."

Type Locality: Río Telica [= Gigillillo] (Pacific Dr.), Corcuera, Leon Department, Nicaragua [L=12.4721, -86.9954].

Type Specimens: Lectotype, USNM 84414 designated by Johnson and Baker (1973:161, incorrectly as *macneili*; Johnson 1974:88, both as the holotype). Paralectotype, ANSP 129943 (*n* = 1) (Johnson and Baker 1973:161, as a paratype).

Remarks: UF 175164 (*n* = 1) is labeled as "cotype," and while it was collected or sent by J.A. MacNiel, it lacks specific locality data.

Taxonomic History:

Unio (*Simonaias*) *macnieli* Lea, 1869 — von Martens (1900b)
Nephronaias macnielii (Lea, 1869) — Simpson (1900; 1914), Graf & Cummings (2007; 2021), Czaja et al. (2023)
Nephronaias macneilii [sic] (Lea, 1869) — Pilsbry (1920b)
Elliptio (*Nephronaias*) *macnielii* (Lea, 1869) — Haas (1929)
Nephronaias (s.s.) *macnielii* (Lea, 1869) — Haas (1969)

PLATE 53 ***Nephronaias ortmanni* (Frierson, 1913). Top:** *Unio* (*Nephronaias*) *ortmanni* Frierson, 1913. Lectotype, ANSP 117544 (82 mm L, 44 mm H). Río Motagua (Caribbean–Gulf of Honduras Dr.), Quirigua, Izabal Department, Guatemala [L = 15.2520, -89.0348]. **Bottom:** Paralectotype, FMNH 21863.

PLATE 54 ***Nephronaias ortmanni* (Frierson, 1913).** *Unio* (*Nephronaias*) *ortmanni* Frierson, 1913. Paralectotypes, INHS 21716. Río Motagua (Caribbean–Gulf of Honduras Dr.), Quirigua, Izabal Department, Guatemala [L=15.2520, -89.0348].

Nephronaias ortmanni (Frierson, 1913)

Plates 53–54, Map 62

Province(s): Gulf of Honduras & Northern Caribbean.

Distribution and Status: Known only from the type locality in the Río Motagua (Gulf of Honduras Dr.) in Guatemala. No known recent collections. Not assessed (IUCN 2024).

Anatomy: Shell large, thick, and elliptical, to subtrapezoid in shape. Beaks small, incurved, pointed, low, and without sculpture. The beak cavity is rather shallow. The pseudocardinals are double in both valves and the laterals single in the right valve. The whole surface of the shell is densely covered with thin concentric ridges. The periostracum is greenish yellow and finely rayed in young individuals. Adults are rayless and dark brown. Nacre white, salmon, or light purple.

The inner lamina of the inner gills is free from the abdominal sac, except at its anterior end. Marsupium moderately swollen, its edge remaining sharp, when charged. Marsupium in the outer gills, placentae are present, sublanceolate, not very solid. Glochidia subcircular: 0.23 mm L, 0.22 mm H. The anal opening with crenulations. Not sexually dimorphic. Color of soft parts (in alcohol) pale (Ortmann 1921a:25).

Biology: Gravid in February (Frierson 1913:15).

Adult Size: Length to 8.5 cm.

Similar Species: *Nephronaias calamitarum* (Morelet, 1849) co-occurs with *Nephronaias ortmanni* in the Río Motagua, but it is not nearly as elongate or solid and has a truncated posterior end.

MAP 62 Distribution of *Nephronaias ortmanni* (Frierson, 1913).

Unio (*Nephronaias*) *ortmanni* Frierson, 1913. *Unio* (*Nephronaias*) *ortmanni*, N. sp. *The Nautilus* 27(2):14–15. Page 14, pl. 2. [Plate 2 appears in Vol. 27(3)]. Also figured in Johnson (1972a:168, pl. 30, fig. 3).

Verbatim Type Locality: "Conchins River, near Quirigua, Guatemala (Atlantic drainage)."

Type Locality: Río Motagua [= Conchins] (Caribbean–Gulf of Honduras Dr.), Quirigua, Izabal Department, Guatemala [L = 15.2520, -89.0348].

Type Specimens: Lectotype, ANSP 117544 designated by Johnson (1971:90; as the figured holotype; Johnson 1972a:146,168 as ANSP 977544a; Johnson and Baker 1973:164; Johnson 1979:33; Parodiz and Tripp 1988:123). Paralectotypes, ANSP 476876 (*n* = 2); CM 61.6197 (*n* = 16), CM 61.6196 (*n* = 8) (Brooks and Brooks 1931:171, as cotypes; Parodiz 1967:27, as syntypes; Parodiz and Tripp 1988:123); FMNH 21863 (*n* = 1); INHS 21715 (*n* = 4), INHS 21716 (*n* = 13); NHMUK 1914.1.7.305–307 (*n* = 3) (Johnson 1971:90, as paratypes); MCZ 20916 (*n* = 3) (Johnson 1956:127, as cotypes; Johnson 1972a:146; Parodiz and Tripp 1988:123), MCZ 23330 (*n* = 2), MCZ 36476 (*n* = 2); MCZ 63011 (*n* = 6); UF 229382 (*n* = 2); UMMZ 108005 (*n* = 9) (Johnson 1972a:146); USNM 252118 (*n* = 4½).

Remarks: Frierson (1913, 1917) initially considered *N. ortmanni* distinct but similar to *Nephronaias sphenorhyncha* (Crosse & Fischer, 1894), but later (Frierson 1927:36, followed by Haas 1929:327) synonymized it with *N. sphenorhyncha*, which was described from somewhere in México. There are no specimens referable to either taxon that have been collected between México and the Río Motagua in Guatemala. Other than the overall shape of the two, there is no compelling reason to unite *Nephronaias sphenorhyncha* with *N. ortmanni*. Fischer and Crosse (1894b:618) noted that the thickness and outline of the shell of *N. sphenorhyncha* was different than anything they had seen from México, and although *N. ortmanni* is a thick-shelled species, it has numerous fine ridges on the surface of the shell that combined with its elongate shape is unique to this species. Without any locality data, it is impossible to tell what *N. sphenorhyncha* is and we have placed it among the *Nomina dubia* below.

Taxonomic History:

Nephronaias ortmanni (Frierson, 1913) — Frierson (1917), Hinkley (1920)
Elliptio ortmanni (Frierson, 1913) — Ortmann (1921a)
Elliptio (*Nephronaias*) *sphenorhynchus* (Crosse & Fischer, 1893) — Frierson (1927), Haas (1929), Johnson (1972)
Popenaias ortmanni (Frierson, 1913) — Heard & Guckert (1970)

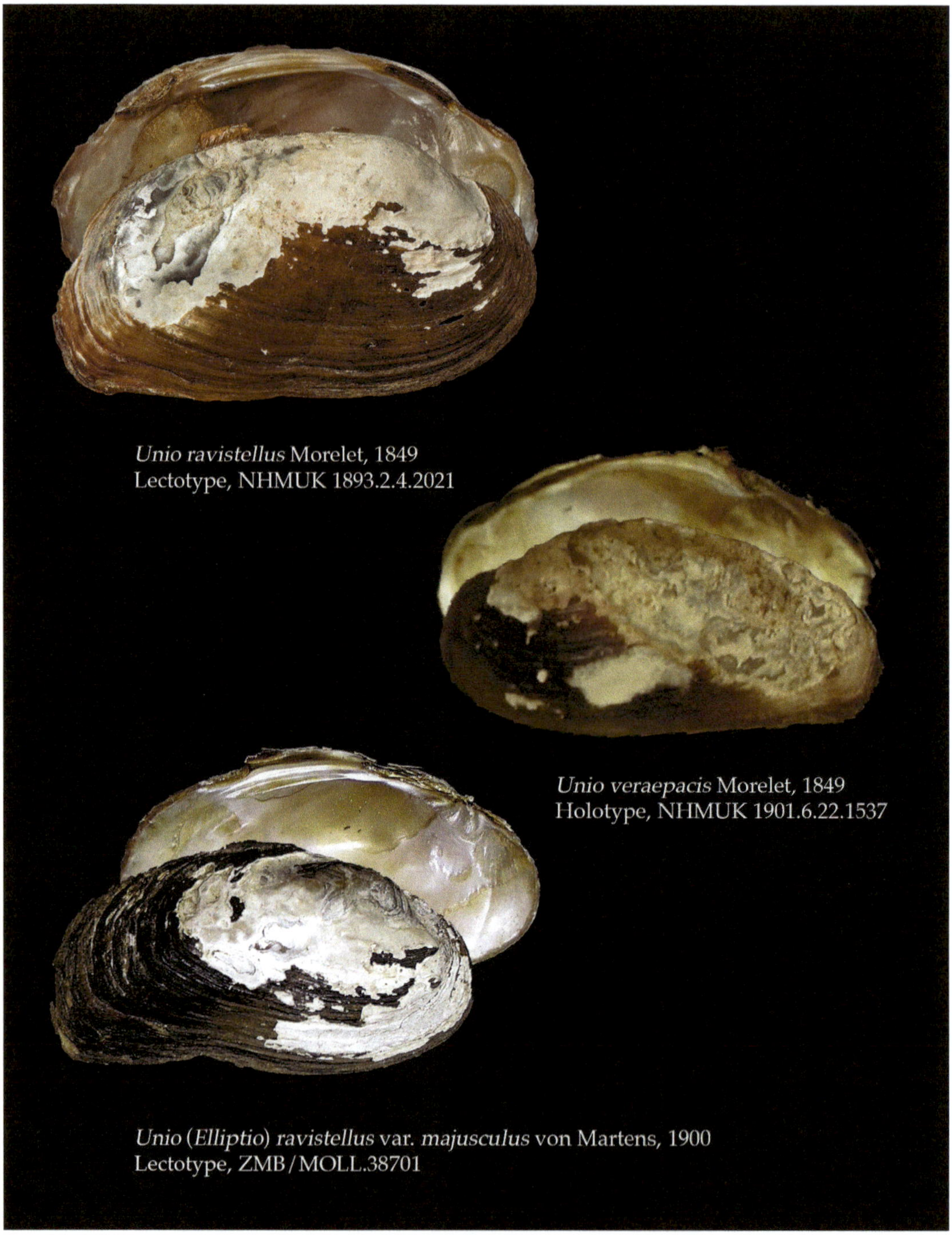

PLATE 55 ***Nephronaias ravistella* (Morelet, 1849). Top Left:** *Unio ravistellus* Morelet, 1849. Lectotype. NHMUK 1893.2.4.2021 (40 mm L, 22 mm H). Lago de Izabal (Caribbean–Gulf of Honduras Dr.), Izabal Department, Guatemala [W = 15.6166, -89.0500]. **Middle Right:** *Unio veraepacis* Tristram, 1863. Holotype, NHMUK 1901.6.22.1537 (45 mm L, 25 mm H). Alta Verapaz Department, Guatemala. **Bottom Left:** *Unio* (*Elliptio*) *ravistellus* var. *majusculus* von Martens, 1900. Lectotype, ZMB/MOLL.38701 (59 mm L, 28 mm H). Lago de Izabal (Caribbean–Gulf of Honduras Dr.), Izabal Department, Guatemala [W = 15.6166, -89.0500].

Nephronaias ravistella (Morelet, 1849)

Plate 55, Map 63

Province(s): Gulf of Honduras & Northern Caribbean.

Distribution and Status: Known primarily from Lago de Izabal in and around Jocolo, Izabal Department, Guatemala. There is a lot of four specimens (NHMUK 20240217) labeled as *Unio verapazanus* [sic] from "Peten, Guatemala" and UF 270358 ($n = 1$) from "Tabasco, México" would place this species in the Usumacinta drainage 500 km northwest of all other known occurrences. These are considered dubious and not plotted on Map 63 below. Not assessed (IUCN 2024).

Anatomy: Shell elongate, small, rather compressed, and strongly concentrically striated. Beaks low, often eroded and sculpture not observed. Ventral margin straight or concave. Left valve with two solid pseudocardinals and two laterals Right valve with two pseudocardinals and a single lateral. Periostracum yellow with green rays in smaller individuals, brown in adults. Nacre white, pink, or purple.

Ortmann (1921a:27) reported that the anatomy resembled that of *Nephronaias ortmanni*, but that the posterior margins of the palpi were connected at their bases only, and that the inner lamina of the inner gills was free from the abdominal sac only for about one-half of the length of the latter (or very slightly more), while it was connected with it in the anterior half (or slightly less). He stated that this was a rather unusual condition in North American Unionidae but that this character cannot be regarded as essential, before a larger number of Central American mussels can be investigated. Glochidia like those of *Nephronaias ortmanni* and *N. calamitarum*: Length 0.23, Height 0.22 mm. Color of soft parts (in alcohol) pale, with black pigment in the region of the branchial, anal, and supra-anal openings.

Biology: Ortmann (1921a:27) examined one male and three females (CM 61.7069) collected by A.A. Hinkley in January 1914 from Lago de Izabal, at Jocolo, Guatemala. He noted that one of the females was gravid only with eggs. He also examined seven others (one male and six gravid females) from the same locality (CM 61.9211) collected on 9 January 1917 and that three of them had glochidia. Ortmann noted that the breeding season seemed to be similar to that of *Nephronaias calamitarum* and *N. yzabalensis* with eggs and glochidia in January.

MAP 63 Distribution of *Nephronaias ravistella (*Morelet, 1849).

Adult Size: Length to 5 cm.

Similar Species: *Nephronaias yzabalensis* (Crosse & Fischer, 1893) is syntopic with *N. ravistella* in Lago de Izabal, Guatemala, but *N. yzabalensis* is quadrate in shape and truncated posteriorly.

Unio ravistellus Morelet, 1849. Testacea novissima insulæ Cubanæ, et Americæ Centralis. Librairie de L'Académie Nationale de Medecine, Paris Part 1:1–31. Page 29. Figured in Fischer and Crosse (1894a, pl. 61, figs. 4, 5–5a), von Martens (1900e, pl. 38. fig. 5, juv., ZMB/MOLL.305839), and Breure et al. (2018:407, fig. 943).

Verbatim Type Locality: "H. in lacu Yzabal, reipublicæ Guatemalensis."

Type Locality: Lago de Izabal (Caribbean–Gulf of Honduras Dr.), Izabal Department, Guatemala [W=15.6166, -89.0500].

Type Specimens: Lectotype, NHMUK 1893.2.4.2021 designated by Johnson (1971:92, as the holotype; Breure et al. 2018:407, figs. 943–944). Paralectotypes, NHMUK 18932.4.2022–2023 (*n*=2) (Johnson 1971:92, as paratypes), MNHN-IM-2000-1767 (*n*=3), MNHN-IM-2000-1768 (*n*=5) (Breure et al. 2018:407).

Taxonomic History:

Unio raristellus [sic] Morelet, 1849 — Lea (1852)
Unio ravistellus Morelet, 1849 — Küster (1856), Sowerby (1868)
Margaron (*Unio*) *ravistellus* (Morelet, 1849) — Lea (1870)
Unio (*Leptonaias*) *ravistellus* Morelet, 1849 — Fischer & Crosse (1894b)
Unio (*Elliptio*) *ravistellus* Morelet, 1849 — von Martens (1900b)
Nephronaias ravistellus (Morelet, 1849) — Simpson (1900), Frierson (1917), Hinkley (1920)
Nephronaias ravistella (Morelet, 1849) — Simpson (1914), Graf & Cummings (2007; 2021)
Elliptio ravistellus (Morelet, 1849) — Ortmann (1921a)
Elliptio (*Leptonaias*) *ravistellus* (Morelet, 1849) — H.B. Baker (1922)
Elliptio (*Nephronaias*) *ravistellus* (Morelet, 1849) — Frierson (1927), Haas (1929)
Nephronaias (s.s.) *ravistella* (Morelet, 1849) — Haas (1969)
Popenaias ravistellus (Morelet, 1849) — Heard & Guckert (1970)

Unio veræ pacis Tristram, 1863. Supplemental catalogue of terrestrial and fluviatile mollusks collected in Guatemala by O. Salvin, Esq., M.A., F.Z.S. *Proceedings of the Zoological Society of London* 1863(3):411–414. Page 414. Figured in von Martens (1900e, pl. 38, figs. 1–1a).

Verbatim Type Locality: Not given.

Type Locality: Alta Verapaz Department, Guatemala.

Type Specimens: Holotype by monotypy, NHMUK 1901.6.22.1537.

Remarks: The holotype figured in von Martens (1900e, pl. 38, figs. 1–1a) was referred to as "Tristram's type." No type locality was given by Tristram, but it was inferred from the title and introductory text of his paper where he stated that "Mr. Salvin's later and much smaller collection, formed chiefly in the district of Vera Paz and the Lake Peten. This region has supplied all the novelties of the present catalogue." Possibly from the Lago de Izabal basin, Guatemala. According to Simpson (1900:738), "Only a Latin description was given by Tristram, and the species is not identifiable. It is probably close to *U. popeii* and may possibly be that." Simpson later (1914:701) placed it under *Unio popeii* Lea, 1857 and was predictively followed by Frierson (1927:38) and later Haas (1969:201). However, Haas previously (1929:328) mentioned receiving a specimen from Frierson labeled as *ravistellus*. Given that none of the records for *P. popeii* (or the genus *Popenaias*) extend south of Veracruz into Guatemala and that Tristram's "type" of *veraepacis* looks nothing like *Unio popeii* Lea, 1857, it was recently suggested to be a synonym of *Nephronaias ravistella* (Morelet, 1849) (Inoue et al. 2020a:470), and we have placed it here.

Taxonomic History:

Unio veræ pacis Tristram, 1863 — Fischer & Crosse (1894b)
Unio (Elliptio) ravistellus var. *veræpacis* Tristram, 1863 — von Martens (1900b)
Unio (sec. *Elliptio*) *veræ-pacis* Tristram, 1863 — Simpson (1900)
Elliptio (Nephronaias) verae-pacis (Tristram, 1863) — Haas (1929)
Unio (sec. *Elliptio*) *popeii* Lea, 1857 — Simpson (1914)
Elliptio (Popenaias) popei (Lea, 1857) — Frierson (1927)
Nephronaias (Popenaias) popeii (Lea, 1857) — Haas (1969)
Popenaias popei (Lea, 1857) — Howells et al. (1996)
Nephronaias ravistella (Morelet, 1849) — Inoue et al. (2020a)

? Unio vellicatus Reeve, 1865. Monograph of the genus *Unio*. *In* L. Reeve and G.B. Sowerby (eds.). *Conchologia Iconica* 16:[unpaginated]+plates 19–30. Plate 22, species 103.

Verbatim Type Locality: "Guatemala."

Type Locality: Guatemala.

Type Specimens: Unknown. Presumed lost (Johnson 1971:96).

Remarks: Considered by some to be synonymous with *Nephronaias veraepacis* (Tristram, 1863), but without a type or locality data other than Guatemala, it is impossible to say for sure.

Taxonomic History:

Unio vellicatus Reeve, 1865 — Lea (1870)
Unio (Leptonaias) vellicatus Reeve, 1865 — Fischer & Crosse (1894b)
Unio (Elliptio) ravistellus var. *veræpacis* Tristram, 1863 — von Martens (1900b)
Nephronaias vellicatus (Reeve, 1865) — Simpson (1900)
Nephronaias vellicata (Reeve, 1865) — Simpson (1914)
Elliptio (Nephronaias) ravistellus-vellicatus (Reeve, 1865) — Frierson (1927)
Elliptio (Nephronaias) verae-pacis (Tristram, 1863) — Haas (1929)

Unio (Elliptio) ravistellus var. *majusculus* von Martens, 1900. Land and freshwater Mollusca. *In* F.D. Godman and O. Salvin (eds.). *Biologia Centrali-Americana* Part 155, Fascile 63–65:497–520, plates 33–35. Page 516, and Part 158, Fascile 68–71:537–568, plates 38–39. Pl. 38. figs. 2–2a, 3–3a, 4–4a, and 6 (from Otto Stoll).

Verbatim Type Locality: "E. Guatemala: Lake of Yzabal."

Type Locality: Lago de Izabal (Caribbean–Gulf of Honduras Dr.), Izabal Department, Guatemala [W=15.6166, -89.0500].

Type Specimens: Lectotype, ZMB/MOLL.38701, here designated (pl.38. figs. 2–2a). Paralectotypes, ZMB/MOLL.305838 (*n*=3, pl. 38. figs. 3–4, 6).

Remarks: Described and named on page 516, but labeled as "*Unio ravistellus* v. *veraepacis*" on plate 38.

Taxonomic History:

Nephronaias ravistella var. *majuscula* (von Martens, 1900) — Simpson (1914)

PLATE 56 ***Nephronaias rowellii* (Lea, 1859). Top Left:** *Unio rowellii* Lea, 1859. Lectotype, USNM 84398 (43 mm L, 25 mm H). Río Chagres (Caribbean Dr.), Panamá Province, Panamá [W = 9.2291, -79.5361]. **Middle Right & Bottom Left: *Nephronaias tehuantepecensis* (Crosse & Fischer, 1894).** *Unio tehuantepecensis* Crosse & Fischer, 1894. Syntypes, MNHN-IM-2000-1799 (68 mm L, 38 mm H). Río Novillero (Pacific Dr.), near Santa Efigenia, Isthmus of Tehuantepec, Oaxaca, México [L = 16.4038, -94.2161]. Figured in Fischer and Crosse 1894b:618, pl. 65, fig. 3 (middle), fig. 4 (bottom).

Nephronaias rowellii (Lea, 1859)

Plate 56, Map 64

Province(s): Gulf of Honduras & Northern Caribbean; San Juan; Oaxaca, México to Colombia.

Distribution and Status: Known only from the Chagres Basin in Panamá. Not assessed (IUCN 2024). Not found in a 2023 survey of Panamá. Listed as endangered by El Salvador (Ministério de Medio Ambiente y Recursos Naturales 2023:39).

Anatomy: Shell elongate or subelliptical, rounded anteriorly and bluntly pointed posteriorly. Ventral margin convex. Beaks moderately high and full, with fine, double-looped sculpture. Beak cavity moderately deep. Left valve with two somewhat irregular pseudocardinals and two curved laterals. Right valve with one pseudocardinal and a faint tooth above it and one lateral. Surface concentrically radiated. Periostracum tan, olive, or dark green and rayed in some individuals, often dark brown in adults. Nacre usually white but occasionally copper colored or pink.

Biology: Unknown.

Adult Size: Length to 7.5 cm.

Similar Species: None.

Remarks: *Nephronaias rowellii* (Lea, 1859) is the only native mussel in the Río Chagres (Caribbean Dr.), Panamá Province, Panamá.

MAP 64 Distribution of *Nephronaias rowellii* (Lea, 1859).

Unio rowellii Lea, 1859. Descriptions of twenty-one new species of exotic Unionidae. *Proceedings of the Academy of Natural Sciences of Philadelphia* 11(1859):151–154. Page 153. Figured in Lea (1860a JANSP Vol. 4, Part 3(7):256; 1860c Obs. 7:74, pl. 40, fig. 136).

Verbatim Type Locality: "Charges River, New Granada."

Type Locality: Río Chagres (Caribbean Dr.), Panamá Province, Panamá [W=9.2291, -79.5361].

Type Specimens: Lectotype, USNM 84398 designated by Johnson (1974a:125, as the figured holotype).

Remarks: The Republic of New Granada was a centralist unitary republic consisting primarily of present-day Colombia and Panamá with parts of Costa Rica, Ecuador, Venezuela, Peru, and Brazil.

Taxonomic History:

Unio rowellii Lea, 1859 — Lea (1860a), Sowerby (1868), Singley (1893)
Unio rowelli [sic] Lea, 1859 — Tate (1870)
Margaron (*Unio*) *rowellii* (Lea, 1859) — Lea (1870)
Unio (*Elliptio*) *rowelli* [sic] Lea, 1859 — von Martens (1900b)
Nephronaias rowellii (Lea, 1859) — Simpson (1900; 1914), Graf & Cummings (2007; 2021), Agudo-Padrón (2019)
Elliptio (*Nephronaias*) *rowellii* (Lea, 1859) — Frierson (1927), Haas (1929)
Nephronaias (s.s.) *rowellii* (Lea, 1859) — Haas (1969)

PLATE 57 ***Nephronaias sallei* (Crosse & Fischer, 1894)**. **Top:** *Unio sallei* Crosse & Fischer, 1894. Figured Syntype, MNHN-IM-2000-1776 (65 mm L, 43 mm H). México. **Bottom:** Syntypes, IRSNB 3737 (right valve), IRSNB 3738 (left valve). México.

Nephronaias sallei (Crosse & Fischer, 1894)

Plate 57, Map 65

Province(s): Usumacinta.

Distribution and Status: Endemic to the Río Usumacinta basin. Although described from México (without a specific locality) and cited by Czaja et al. (2023:86) as occurring in Tabasco, other than the syntypes there are no other specimens known from México. All specimens with locality data are from the Río Usumacinta in Guatemala. Goodrich and van der Schalie (1937:46) noted that it was one of the most common species in the Rio de la Pasión in Guatemala, but that it appeared to be restricted to the central part of the river. Not assessed (IUCN 2024).

Anatomy: Shell triangular to ovate, very solid, and somewhat inflated. Beaks low and beak cavity deep. Left valve with two pseudocardinals and two short laterals. Right valve with one large, triangular pseudocardinal, and one lateral. Periostracum rough and typically dark brown. Nacre white or pink.

Biology: Unknown.

Adult Size: Length to 7.5 cm.

Similar Species: *Nephronaias aeruginosa* (Morelet, 1849) and *Nephronaias calamitarum* (Morelet, 1849).

Remarks: Incorrectly cited as the type of *Pachynaias* Crosse & Fischer, 1894 by von Martens (1900b:517).

MAP 65 Distribution of *Nephronaias sallei* (Crosse & Fischer, 1894).

Unio sallei Crosse & Fischer, 1894. Diagnosis Mollusci novi, Reipublicæ Mexicanæ incolæ. *Journal de Conchyliologie* 41(3):179. Figured in Fischer and Crosse (1894b:619, pl. 68, figs. 3–3a).

Verbatim Type Locality: "Republica Mexicana."

Type Locality: México.

Type Specimens: Syntypes, MNHN-IM-2000-1776 (*n*=3); IRSNB 3737 (*n*=½), IRSNB 3738 (*n*=1), IRSNB 3739 (*n*=4).

Remarks: The date of publication is often cited as 1893 but is 1894 according to Winckworth (1936:156).

Taxonomic History:

Unio (*Barynaias*) *sallei* Crosse & Fischer, 1893 — Fischer & Crosse (1894b)
Unio (*Elliptio*) *sallei* Crosse & Fischer, 1893 — von Martens (1900b)
Plagiola (*Arotonaias*) *sallei* (Crosse & Fischer, 1893) — Simpson (1914)
Plagiola (*Artonaias*) [sic] *sallei* (Crosse & Fischer, 1893) — H.B. Baker (1922)
Elliptio (*Barynaias*) *sallei* (Crosse & Fischer, 1893) — Frierson (1927), Haas (1929)
Plagiola sallei (Crosse & Fischer, 1893) — Goodrich & van der Schalie (1937)
Barynaias sallei (Crosse & Fischer, 1893) — Haas (1969), Graf & Cummings (2007; 2021), Czaja et al. (2023)

PLATE 58 ***Nephronaias scamnata*** **(Morelet, 1849). Top Left:** *Unio scamnatus* Morelet, 1849. Lectotype. NHMUK 1893.2.4.1976 (52 mm L, 30 mm H). **Middle Right:** Paralectotype. NHMUK 1893.2.4.1977 (female) (52 mm L, 32 mm H). **Bottom Left:** *Unio* (*Simonaias*) *proclinatus* von Martens, 1900. Syntype, ZMB/MOLL.33793 (60 mm L, 30 mm H). All, Río Taco Taco (Caribbean Dr.), Artemisa Province, Cuba [W = 16.8246, -90.8822].

Nephronaias scamnata (Morelet, 1849)

Plate 58, Map 66

Province(s): Cuba.

Distribution and Status: Although all museum records are from western Cuba, von Martens (1900b:508) reported *Unio* (*Simonaias*) *proclinatus* von Martens, 1900 from Gibara, Province of Holguín, in the far eastern part of the island. DD—Data Deficient (IUCN 2024); Not assessed (NatureServe 2024).

Anatomy: Shell elongate rhomboid or elliptical, and compressed, with a low, double posterior ridge. Beaks low with fine double-looped sculpture. Left valve with two subtriangular pseudocardinals and two slender laterals. Right valve with one strong pseudocardinal, a vestige of a second above, and one lateral. Surface with concentric ridges that covers the entire shell. Periostracum brown or olive, occasionally with a few faint rays. Nacre white, pink, or purple.

Biology: Unknown.

Adult Size: Length to 8 cm.

Similar Species: *Nephronaias gundlachi* (Dunker, 1858).

MAP 66 Distribution of *Nephronaias scamnata* (Morelet, 1849).

Unio scamnatus Morelet, 1849. Testacea novissima insulæ Cubanæ, et Americæ Centralis. Librairie de L'Académie Nationale de Medecine, Paris Part 1:1–31. Page 30. Figured in Johnson (1971:92, pl. 2, fig. 3; 1981:271, pl. 37, fig. 1) and Breure et al. (2018:420, fig. 1002).

Verbatim Type Locality: "H. amnem Cacajajicara, in insula Cuba."

Type Locality: Río Taco Taco (Caribbean Dr.), Artemisa Province, Cuba [W = 16.8246, -90.8822].

Type Specimens: Lectotype, NHMUK 1893.2.4.1976 designated by Johnson (1971:92; Johnson 1981:271; Breure et al. 2018:420). Paralectotypes, NHMUK 1893.2.4.1977–78 (*n*=2) (Johnson 1971:92); MNHN-IM-2000-1777 (*n*=1).

Remarks: Described from the Río Cacajajicara, Johnson (1971:92) noted that was=Río Las Pazas, and in 1981:271, he spelled it as Río Las Pozas, which drains to the Gulf of México. Johnson (1971:92; 1981:271) also stated that the type locality was in Pinar del Rio Province, Cuba, but the Río Taco-Taco is in Artemisa Province as is the Río Las Pozas. Although the Río Taco-Taco is not mentioned in the original description, it is on a label with the lectotype in the NHMUK but may have been added by Johnson as the catalogue simply reads "Cuba." Johnson (1981:271) also included *Unio calamitarum* Morelet, 1849 and *Unio calamitarum* var. *β*, *prolongata* Fischer & Crosse, 1894 in the synonymy of *U. scamnatus* thinking the locality data with the types of those taxa were erroneous. Johnson stated that "The lectotype of *Unio calamitarum* and the figure of *U. c. prolongata* are certainly *N. scamnata*, and as no other specimens have been seen from Central America, it is assumed that both taxa are based on Cuban shells from the Morelet collection." There are no data to support that assumption. Johnson (1981:273) also figured two specimens: MCZ 6585, from Pinar del Rio Province, Cuba (fig. 4), and MCZ 126800, from Rio Caiguanabo [= Ariguanabo], San Diego de los Baños, Pinar del Rio Province, Cuba (fig. 5).

Taxonomic History:

Unio scamnatus Morelet, 1849 — Arango (1865)
Margaron (*Unio*) *scamnatus* (Morelet, 1849) — Lea (1870)
Unio (*Simonaias*) *scamnatus* Morelet, 1849 — von Martens (1900b)
Unio (*Simonaias*) *scamnatus* Morelet, 1849 — von Martens (1900c, *nomen dubium*)
Nephronaias scamnatus (Morelet, 1849) — Simpson (1900), von Ihering (1901)
Nephronaias scamnata (Morelet, 1849) — Simpson (1914), Graf & Cummings (2007; 2021), Serrand & Cummings (2014)
Elliptio (*Nephronaias*) *scamnatus* (Morelet, 1849) — Frierson (1927), Haas (1929)
Nephronaias (s.s.) *scamnata* (Morelet, 1849) — Haas (1969), Johnson (1981) [in part]

Unio (*Simonaias*) *proclinatus* von Martens, 1900. Land and freshwater Mollusca. *In* F.D. Godman and O. Salvin (eds.). *Biologia Centrali-Americana* Part 155, Fascile 63–65:497–520, plates 33–35. Page 508. Not figured.

Verbatim Type Locality: "Cuba: Rio Taco and Gibara, in the eastern part of the island, and Pinar del Rio, farther west."

Type Locality: Río Taco Taco (Caribbean Dr.), Artemisa Province, Cuba [W=16.8246, -90.8822].

Type Specimens: Syntypes, ZMB/MOLL.33793 (*n*=2) (Johnson 1981:271, as "presumed to be in the ZMB"); ZMB/MOLL.305853 (*n*=5), both from the Río Taco Taco.

Taxonomic History:

Nephronaias scamnata (Morelet, 1849) — Simpson (1914)
Elliptio (*Nephronaias*) *scamnatus* (Morelet, 1849) — Frierson (1927), Haas (1929)
Nephronaias (s.s.) *scamnata* (Morelet, 1849) — Haas (1969), Johnson (1981)

Nephronaias tehuantepecensis (Crosse & Fischer, 1894)

Plate 56, Map 67

Province(s): Oaxaca, México to Colombia.

Distribution and Status: Known from the Río Tehuantepec, the adjacent ríos Perros and Novillero and Laguna Dolores (All Pacific Dr.) in Oaxaca, México. Not assessed (IUCN 2024).

Anatomy: Shell oblong and elongate, the length exceeding twice the height. Beaks are low and located at the anterior third. The dorsal and ventral margins are weakly convex. The anterior end rounded and the posterior end obliquely subtruncate. Two compressed pseudocardinal teeth in the left valve and a single pseudocardinal in the right valve. The lateral teeth are smooth and not very long. Periostracum light brown to black and occasionally slightly glossy. Nacre white.

Biology: Unknown.

Adult Size: Length to 3 cm.

Similar Species: *Nephronaias tehuantepecensis* (Crosse & Fischer, 1894) is the only species with lateral and pseudocardinal teeth in a Pacific Coast drainage in the Isthmus of Tehuantepec, Oaxaca.

MAP 67 Distribution of *Nephronaias tehuantepecensis* (Crosse & Fischer, 1894).

Unio tehuantepecensis Crosse & Fischer, 1894. Diagnoses molluscorum novorum, reipublicæ Mexicanæ incolarum. *Journal de Conchyliologie* 41(4):293–297. Page 296. Figured in Fischer and Crosse (1894b:618, pl. 65, figs. 3–3a, 4).

Verbatim Type Locality: "prope Santa Efigenia, in isthmo Tehuantepec dicto, provinciæ Oajaca, reipublicæ Mexicanæ."

Type Locality: Río Novillero (Pacific Dr.), near Santa Efigenia, Isthmus of Tehuantepec, Oaxaca, México [L=16.4038, -94.2161].

Type Specimens: Syntypes, MNHN-IM-2000-1799 (*n*=2, the figured specimens).

Remarks: Frierson (1927:37) synonymized *Unio tehuantepecensis* Crosse & Fischer, 1894 with *Elliptio (Sphenonaias) callosus* (Lea, 1841), a taxon of dubious identity originally described from near Columbus, Ohio, but thought to be a Mexican species by Simpson (1900). However, there are no data to support their assertion that *Unio callosus* Lea, 1841 was collected in México. We consider *U. callosus* to be a synonym of *Unio ligamentina* Lamarck, 1819.

Haas (1929:321) synonymized *Unio tehuantepecensis* with *Elliptio (Nephronaias) goascoranensis* (Lea, 1858), which was described from the Río Goascorán (Pacific Dr.), Honduras. Haas later (1969:202) considered *U. tehuantepecensis* as valid and placed it in *Popenaias* as *Nephronaias (Popenaias) tehuantepecensis* (Crosse & Fischer, 1894) along with *P. popeii* and *P. metallica.* Why Haas placed it in *Popenaias* is unknown. *Nephronaias goascoranensis* (Lea, 1858) is also found in Pacific basin further south in Guatemala and Nicaragua.

Based upon an examination of its shell morphology, *Unio tehuantepecensis* could be a member of numerous Central American genera. It had not been treated since Haas (1969) and was recognized as valid in the most recent global checklist of freshwater mussels (Graf & Cummings 2021:19). However, due to its disjunct distribution and the limited number of specimens in museum collections, the taxonomic validity (including its placement in *Popenaias*) and distribution of *Popenaias tehuantepecensis* is uncertain. An examination of the syntypes and other specimens in the UMMZ and ZMB identified as *Popenaias tehuantepecensis* indicate that it might be synonymous with *Nephronaias rowellii* (Lea, 1859) as suggested by Inoue et al. (2020a:476), but we restrict that species to the Río Chagres of Panamá. The date of publication is often cited as 1893 but is 1894 according to Winckworth (1936:156).

Taxonomic History:

Unio (Sphenonaias) tehuantepecensis Crosse & Fischer, 1893 — Fischer & Crosse (1894b), von Martens (1900b)
Nephronaias tehuantepecensis (Crosse & Fischer, 1894) — Simpson (1900; 1914), de Leon (1912)
Glabaris tehuantepecensis (Crosse & Fischer, 1893) — de Leon (1912)
Unio tehuantepecensis Crosse & Fischer, 1893 — H.B. Baker (1922)
Unio (Sphenonaias) callosus Lea, 1841 — Frierson (1927)
Elliptio (Nephronaias) goascoranensis (Lea, 1858) — Haas (1929)
Nephronaias (Popenaias) tehuantepecensis (Crosse & Fischer, 1893) — Haas (1969)
Popenaias tehuantepecensis (Crosse & Fischer, 1893) — Graf & Cummings (2007; 2021), Inoue et al. (2020a), Czaja et al. (2023)

PLATE 59 ***Nephronaias tempisquensis*** **Pilsbry, 1920. Top:** Holotype, ANSP 105225 (28 mm L, 15 mm H). Río Tempisque (Pacific Dr.), Filadelfia de Guanacaste, Guanacaste, Costa Rica [L = 10.4484, -85.5500]. **Bottom:** ***Nephronaias yzabalensis*** **(Crosse & Fischer, 1893).** *Unio yzabalensis* Crosse & Fischer, 1893. Syntype, MNHN-IM-2000-1810 (45 mm L, 28 mm H). Lago de Izabal (Caribbean–Gulf of Honduras Dr.), Izabal Department, Guatemala [W = 15.6166, -89.0500]. Figured in Fischer and Crosse (1894b:597, pl. 64, fig. 4).

Nephronaias tempisquensis (Pilsbry, 1920)

Plate 59, Map 68

Province(s): Oaxaca, México to Colombia.

Distribution and Status: Known only from the Río Tempisque basin (Pacific Dr.) in Costa Rica. Not assessed (IUCN 2024).

Anatomy: Shell oblong and elongate, the length exceeding twice the height. Beaks are low and located at the anterior third. The dorsal and ventral margins are weakly convex. The anterior end rounded and the posterior end obliquely subtruncate. Two compressed pseudocardinal teeth in the left valve and a single pseudocardinal in the right valve. The lateral teeth are smooth and not very long. Surface somewhat glossy, with very fine, hair-like threads prominent on the posterior end. Periostracum yellow with numerous green rays. Nacre white.

Biology: Unknown.

Adult Size: Length to 3 cm.

Similar Species: The only other species with lateral and pseudocardinal teeth in a Pacific Coast drainage in Costa Rica is a single specimen of the quadrate-shaped *Nephronaias macnielii* (Lea, 1869) from the Rio Higuerón (Pacific–Tempisque Dr.), Taboga, Guanacaste Province, Costa Rica.

MAP 68 Distribution of *Nephronaias tempisquensis* Pilsbry, 1920.

Nephronaias tempisquensis Pilsbry, 1920. Costa Rican land and freshwater mollusks. *Proceedings of the Academy of Natural Sciences of Philadelphia* 72(1):2–10. Page 9, fig. 6.

Verbatim Type Locality: "Rio Tempisque, Filadelfia."

Type Locality: Río Tempisque (Pacific Dr.), Filadelfia de Guanacaste, Guanacaste, Costa Rica [L = 10.4484, -85.5500].

Type Specimens: Holotype, ANSP 105225 (*n* = ½, the figured specimen) (Johnson 1980:278). Paratypes, ANSP 454373 (*n* = $2^3/_2$, ex. 105225) (Johnson 1980:278).

Remarks: Pilsbry indicated that the type and paratypes had the same number, and he gave two sets of measurements and figured one. However, since he did not explicitly tie his type designation to a particular specimen, it might not be reasonable to call it a holotype by original designation. The paratypes have since been recatalogued as ANSP 454373.

Taxonomic History:

Nephronaias (s.s.) *tempisquensis* Pilsbry, 1920 — Haas (1969)

Nephronaias tempisquensis Pilsbry, 1920 — Taylor (1993), Graf & Cummings (2007; 2021)

Nephronaias yzabalensis (Crosse & Fischer, 1893)

Plate 59, Map 69

Province(s): Gulf of Honduras & Northern Caribbean.

Distribution and Status: Known from the Dulce–Saja and Lago de Izabal basins (Gulf of Honduras Dr.), Izabal Department, Guatemala. No recent collections are known. Not assessed (IUCN 2024).

Anatomy: Shell solid, rhomboid elliptical and somewhat compressed, with a low, double posterior ridge. Beaks low, with sculpture consisting of concentric ridges. Left valve with two pseudocardinals and two laterals. Right valve with two compressed pseudocardinals, the upper one smaller, and one lateral. Surface covered with concentric ridges. Periostracum brown or olive with green rays. Nacre typically white, rarely salmon, pink, or purple.

Biology: Ortmann (1921a:26) examined two specimens (CM 61.9212) collected by A.A. Hinkley on 6 January 1917 from the Río Saja (trib. to the Río Dulce, below Lago de Izabal), Guatemala. He noted that both were females and gravid but did not have developed glochidia. He also noted that one had a white nacre and the other purple, and that "the anatomy is identical with that of *E. ortmanni* and *calamitarum* in every particular. Of course, the glochidia have not been observed."

Adult Size: Length to 4.5 cm.

Similar Species: This species occurs syntopically with *Nephronaias calamitarum* (Morelet, 1849) in the Río Blanco near Livingston and with *Nephronaias dysonii* (Lea, 1859) in the Río Saja, both Izabal Department, Guatemala: a complex that warrants further investigation. It also occurs with *Nephronaias ravistella* (Morelet, 1849) in Lago de Izabal, but *N. ravistella* is more elongate and not truncated on the posterior end.

MAP 69 Distribution of *Nephronaias yzabalensis* (Crosse & Fischer, 1893).

Unio yzabalensis Crosse & Fischer, 1893. Diagnoses Molluscorum Reipublicæ Mexicanæ et Guatemalæ incolarum. *Journal de Conchyliologie* 40(3):294–296. Page 294. Figured in Fischer and Crosse (1894b:597, pl. 64, figs. 4, 4a–b) and von Martens (1900e, pl. 39, figs. 9–11, ZMB/MOLL.305848).

Verbatim Type Locality: "lacu Yzabal, Guatemalæ."

Type Locality: Lago de Izabal (Caribbean–Gulf of Honduras Dr.), Izabal Department, Guatemala [W=15.6166, -89.0500].

Type Specimens: Syntypes, MNHN-IM-2000-1810 (*n*=2).

Remarks: The date of publication is often cited as 1892 but is 1893 according to Winckworth (1936:156).

Taxonomic History:

Unio (*Cænonaias*) *yzabalensis* Crosse & Fischer, 1892— Fischer & Crosse (1894b)
Unio (*Simonaias*) *yzabalensis* Crosse & Fischer, 1892— von Martens (1900b)
Nephronaias yzabalensis (Crosse & Fischer, 1892) — Simpson (1900; 1914), Graf & Cummings (2007; 2021)
Elliptio yzabalensis (Crosse & Fischer, 1892) — Ortmann (1921a)
Elliptio (*Nephronaias*) *ravistellus-yzabalensis* (Crosse & Fischer, 1892) — Frierson (1927)
Elliptio (*Nephronaias*) *yzabalensis* (Crosse & Fischer, 1892) — Haas (1929)
Nephronaias (s.s.) *yzabalensis* (Crosse & Fischer, 1892) — Haas (1969)
Popenaias yzabalensis (Crosse & Fischer, 1892) — Heard & Guckert (1970)

Order Unionoida
Superfamily Unionoidea Rafinesque, 1820
Family Unionidae Rafinesque, 1820
Subfamily Ambleminae Rafinesque, 1820
Tribe Popenaiadini Heard & Guckert, 1970
Genus *Popenaias* Frierson, 1927

Popenaias Frierson, 1927. A classification and annotated check list of the North American naiades. Baylor University Press, Waco, Texas. 111 pp. Page 38. Introduced as a subgenus of *Elliptio* Rafinesque, 1819.

Type Species: *Unio popeii* Lea, 1857, by original designation.

Diagnosis: As with *Nephritica*, Frierson named the genus but did not provide a diagnosis for it. He did designate a type species and the following is the diagnosis of *Unio popeii* Lea, 1857: "Testâ lævi, transversâ, compresso-cylindraceâ, ad basim sub-emerginatâ, valdè inæquilaterali, ad latere planulatâ, posticè truncatâ; valvulis subtenuibus, anticè crassioribus; natibus parvis, prominulis, ad apices granalatis; epidermide vel tenebroso-olivâ vel fuscâ, obsoletè radiatâ; dentibus cardinalibus compressis, erectis, acuminatis crenulatisque; lateralibus prælongis, lamellatis subrectisque; margaritâ vel albâ salmonis tinctê et iridescente." [Test smooth, transverse, compressed-cylindrical, sub-emergent at the base, very unequal, flat at the side, truncate at the back; subthin valves, previously thicker; small, prominent, granulated at the tips; epidermis either dark-olive or brown, obsolete and radiant; cardinal teeth compressed, erect, pointed and crenulate; with lateral prolongations, lamellated and substraight; pearl or white salmon dyed and iridescent.] (Lea 1857:102).

Remarks: This genus was found to be monophyletic with *Popenaias popeii* sister to *P. berezai*/*P. semirasa* (Inoue et al. 2020a).

Taxonomic History:

Elliptio (*Popenaias*) Frierson, 1927 — Haas (1929)
Elliptio (sec. *Popenaias*) Frierson, 1927 — Thiele (1934)
Popenaias Frierson, 1927 — Modell (1942), Johnson (1999a), Graf & Cummings (2007; 2021), Bouchet & Ricroi (2010), Williams et al. (2017), Pfeiffer et al. (2019b), Inoue et al. (2020a)
Nephronaias (*Popenaias*) Frierson, 1927 — Haas (1969)

Distribution and Status: This genus of four species is widespread from the Upper Río Grande basin in Texas and New Mexico south to the Jamapa basin in Veracruz, México (Map 70). *Popenaias popeii* (Lea, 1857) has disappeared from much of the Río Grande and was listed as a Federally Endangered Species in the USA in 2018. *Popenaias metallica* (Say, 1831) was described and was apparently endemic to Lago de Chalco, an endorheic lake near Ciudad de México. It is unlikely that the population is still extant since the lake is heavily polluted and has been periodically drained.

MAP 70 Distribution of the genus *Popenaias* Frierson, 1927.

PLATE 60 ***Popenaias berezai*** **Inoue et al., 2020**. Holotype, CNMO 8037 (ex. INHS 90397.5) (73 mm L, 36 mm H). Río Valles (Río Pánuco–Río Tampaón), Estación Micos, 20 km NW of Ciudad Valles, San Luis Potosí, México [L=22.1141, -99.1616]. 8 December 2018.

Popenaias berezai (Inoue et al., 2020)

Plate 60, Map 71

Province(s): Pánuco; Papaloapán?

Distribution and Status: Found in the Soto la Marina, Pánuco, Carrizal, and Cazones drainages in San Luis Potosí, Tamaulipas, and Veracruz, México. However, the full extent of its distribution is currently unknown. Widespread and common in the Pánuco basin. One questionable record (ANSP 142600) from the Río Jamapa in Veracruz in Papaloapán Province. Not assessed (IUCN 2024).

Anatomy: Shell thin to thick and elongate subtrapezoidal. Posterior ridge biangulate. Dorsal and ventral margins almost parallel and nearly straight. Beaks broad and even with hinge line or elevated slightly above. Beak cavity shallow. Faint irregular corrugations sometimes present on posterior-central disk. Left valve with two thin lateral teeth, straight to slightly curved and two triangular, compress pseudocardinal teeth. Right valve with single thin and relatively long lateral tooth, and one compressed triangular, serrated pseudocardinal tooth. Periostracum glossy or clothlike, brown to dark gray or black, with faint thin rays. Nacre typically fine metallic pink or copper colored and iridescent posteriorly.

Mantle tan to off-white, sometimes lighter outside of pallial line; mantle wall usually with black, brown, and white mottling adjacent to incurrent and excurrent apertures. Apertures with dark shades of tan and gray, or black distally, often multicolor. Supra-anal aperture pursed with mantle bridge separating supra-anal and excurrent apertures. Margin of supra-anal aperture smooth and shorter than excurrent aperture. Excurrent aperture with reduced, off-white to dark conical, single stalked papillae. Incurrent aperture larger than excurrent aperture and bearing larger papillae, particularly in gravid females. Papillae are conical, elongate and vary from simple, bifid, and trifid within a single individual; in gravid females, elongation may be extreme such that papillae take on an arborescent appearance. Gills creamy white to tan may show streaks of dark pigmentation posteriorly in non-gravid and partially gravid individuals. Female outer gill marsupial, entire gill is utilized to brood glochidia except the extreme posterior portion and is distended or pad-like when gravid. Labial palps creamy-white to off-white and often translucent, pointed distally and almost spade-like or boot-shaped in appearance. Visceral mass usually pearly white, often with tan areas dorsally. Foot off-white to tan (Inoue et al. 2020a:467).

Biology: Ortmann reported two females gravid with eggs only in late December and early January from the Río Valles San Luis Potosí, México. Two of four specimens collected on 25 March 1986 from the Río Moctezuma at Tamazunchale, San Luis Potosí, México also were gravid (Smith et al. 2003:335).

Gravid females were collected from the Río Tampaón, Álvaro Obregón, at confluence of the ríos Valles and Tampaón (INHS 90183), the Río Claro, 5.5 km SE of Tamazunchale (INHS 90134), the Río Amajac, 5.1 km SE of Tamazunchale (INHS 90152), and the Río Tancuilín, 5.5 km S of Axtla de Terrazas (INHS 90175), all San Luis Potosí, México from 8 to 11 December 2018.

Adult Size: Length to 8 cm.

Similar Species: *Popenaias semirasa* (Pilsbry, 1910).

MAP 71 Distribution of *Popenaias berezai* Inoue et al., 2020.

Popenaias berezai Inoue, Cummings, Tiemann, Miller, Johnson, Smith & Randklev, 2020. A new species of *Popenaias* Frierson, 1927, from the Gulf coastal rivers of central México (Bivalvia: Unionida: Unionidae) with comments on the genus. *Zootaxa* 4816(4):457–490. Page 464, figs. 5a, 6a, 7a, and 8a.

Type Locality: Río Valles (Río Pánuco–Río Tampaón), Estación Micos, 20km NW of Ciudad Valles, San Luis Potosí, México [L=22.1141, -99.1616].

Type Specimens: Holotype, CNMO 8037 (ex. INHS 90397.5). Paratypes, CNMO 8038 (3), FMNH 344800 (3), INHS 90397 (9), NHMUK 20191101 (3), NRI 9603 (3), UF 548816 (3), USNM 1617023 (3), all from the type locality; Río Claro, 5.5km SE of Tamazunchale, 21.2215, -98.7557, INHS 90134 (3); Río Amajac, 5.1 km SE of Tamazunchale, 21.2272, -98.7543, CNMO 8039 (2), FMNH 344801 (2), INHS 90152 (10), NHMUK 20191102 (2), NRI 9604 (2), UF 548817 (2), USNM 1617024 (2); Río Amajac, Temamatla, 21.2366, -98.7618, CNMO 8041 (5), FMNH 344803 (5), INHS 90199 (33), NHMUK 20191103 (5), NRI 9606 (5), UF 548819 (5), USNM 1617026 (5); Río Huichihuayán, Ejido Tazaquil, 21.4726, -98.9576, INHS 90167 (1); Río Huichihuayán, 8km NE Xilitla, 21.4363, -98.9334, INHS 90192 (2); Río Tancuilín, 5.5km S of Axtla de Terrazas, 21.3900, -98.8754, CNMO 8040 (2), FMNH 344802 (2), INHS 90175 (9), NHMUK 20191103 (2), NRI 9605 (2), UF 548818 (2), USNM 1617025 (2); all Río Moctezuma Dr; Río Tampaón (Río Pánuco Dr.), Álvaro Obregón, at confluence of the ríos Valles and Tampaón, 21.8448, -98.9437, INHS 90183 (12). All the above from San Luis Potosí, México.

Synonymy:

Unio (*sec. Elliptio*) *popeii* Lea, 1857 — Simpson (1900; 1914) [in part]
Lampsilis popei (Lea, 1857) — Pilsbry (1910) [in part]
Elliptio popei (Lea, 1857) — Ortmann (1912)
Nephronaias (*Popenaias*) *popei* (Lea, 1857) — Haas (1969) [in part]
Popenaias popeii (Lea, 1857) — Smith et al. (2003), Graf & Cummings (2007; 2021); Pfeiffer et al. (2019a) [all in part]
Popenaias berezai Inoue et al., 2020 — Kiser et al. (2022), Czaja et al. (2023)

PLATE 61 ***Popenaias metallica* (Say, 1831). Top:** *Unio cuprinus* Lea, 1831. Lectotype, USNM 86121 (53 mm L, 28 mm H). México. **Bottom:** *Unio metallica* Say, 1831. Topotype, AMNH 31036. Lago de Chalco (Endorheic Dr.), Ciudad de México, México [W = 19.2635, -98.9701].

Popenaias metallica (Say, 1831)

Plate 61, Map 72

Province(s): Mesa Central and Balsas.

Distribution and Status: Known only from the type locality, Lago de Chalco, an endorheic lake near Ciudad de México, México. Not assessed (IUCN 2024). Likely extinct as Lago de Chalco has been heavily modified and polluted.

Anatomy: Shell small, thin, elongate, and slightly inflated, with a small post-dorsal wing. Ventral margin slightly curved and convex. Beaks low, with a few broken, double-looped ridges. Beak cavity shallow. Left valve with two pseudocardinals and two thin laterals. Right valve with one pseudocardinal, a vestigal one above it, and one lateral. Shell surface smooth. Periostracum chestnut brown. Nacre pink or copper colored and iridescent posteriorly.

Biology: Unknown.

Adult Size: Length to 4.8 cm.

Similar Species: None. *Popenaias metallica* (Say, 1831) is the only species with pseudocardinal and lateral teeth ever reported from Lago de Chalco.

MAP 72 Distribution of *Popenaias metallica* (Say, 1831).

Unio metallicus Say, 1831. New terrestrial and fluviatile shells of North America. (cont.). The Disseminator. [2nd Series]. New Harmony, Indiana 1(29):unpaginated, 15 January 1831.

Verbatim Type Locality: "Mexico....Lake Chalco?"

Type Locality: Lago de Chalco (Endorheic Dr.), Ciudad de México, México [W = 19.2635, -98.9701].

Type Specimens: Unknown. Presumed lost.

Remarks: Although no type specimens are known to be extant, there are four lots of freshwater mussels from Lago de Chalco. One of which is the lectotype of *Anodonta chalcoensis* Crosse & Fischer, 1893 (MNHN-IM-2000-1656). The other three, AMNH 29105 (n=2), AMNH 31036 (n=1), and UMMZ 91173 (n=1), are specimens identified as *Popenaias metallica* (Say, 1831). They are the only freshwater mussels we know of from Lago de Chalco, the type locality of *Unio metallicus* Say, 1831. It is unlikely that mussels are still extant since the lake is heavily polluted and has been periodically drained. There are five other lots variously labeled as *Unio metallicus* or *Unio cuprinus* (see below) without specific locality data that may be this species. Regarding its generic placement, given that all the known species of the genus *Popenaias* are from Gulf of Mexico tributaries, it is unknown if *Unio metallicus* Say, 1831 is a member of the group.

Taxonomic History:

Unio metallicus Say, 1831 — Ferussac (1835), Say (1840), Conrad (1853)
Margarita (*Unio*) *cuprinus* (Lea, 1831) — Lea (1836)
Unio cuprinus Lea, 1831 — Hanley (1843; 1856)
Margaron (*Unio*) *cuprinus* (Lea, 1831) — Lea (1852; 1870)
Unio (*Simonaias*) *cuprinus* Lea, 1831 — Fischer & Crosse (1894b)
Unio metallicus Say, 1831 — von Martens (1900c, *nomen dubium*, 1901c)
Unio (*Simonaias*) *cuprinus* var. *persulcatus* Lea, 1859 — von Martens (1900b)
Lampsilis (*Proptera*) *metallicus* (Say, 1831) — Simpson (1900)
Lampsilis metallicus (Say, 1831) — de Leon (1912)
Lampsilis (*Proptera*) *metallica* (Say, 1831) — Simpson (1914)
Elliptio (*Popenaias*) *metallicus* (Say, 1831) — Frierson (1927)
Nephronaias (*Popenaias*) *metallica* (Say, 1831) — Haas (1969)
Popenaias metallica (Say, 1831) — Graf & Cummings (2007; 2021), Inoue et al. (2020a), Czaja et al. (2023)

Unio cuprinus Lea, 1831. Observations on the naïades, and descriptions of new species of that and other families. *Transactions of the American Philosophical Society* 4(N.S.):63–121+plates 3–18. Page 94, pl. 12, fig. 24. Also figured in Lea (1834, Obs. 1:94, pl. 12, fig. 24) and Sowerby (1867, pl. 66, sp. 336).

Verbatim Type Locality: "Mexico."

Type Locality: México.

Type Specimens: Lectotype, USNM 86121 designated by Johnson (1974a:40, as the holotype).

Remarks: It could be argued that given the lack of a specific type locality that *Unio cuprinus* Lea, 1831 could be one of the Central American mussels, we recognize as a *nomin dubium* below. We provisionally place it here following Simpson (1900; 1914), Frierson (1927), and Haas (1969).

Taxonomic History:

Unio cuprinus Lea, 1831 — Hanley (1843; 1856), Sowerby (1868), Inoue et al. (2020a)
Margarita (*Unio*) *cuprinus* (Lea, 1831) — Lea (1836)
Margaron (*Unio*) *cuprinus* (Lea, 1831) — Lea (1852; 1870)
Unio (*Naidea*) *cuprinus* Lea, 1831 — Adams & Adams (1857)
Unio (*Simonaias*) *cuprinus* Lea, 1831 — Fischer & Crosse (1894b), von Martens (1900b; 1901c)
Lampsilis (*Proptera*) *metallicus* (Say, 1831) — Simpson (1900)
Lampsilis metallicus (Say, 1831) — de Leon (1912)

Lampsilis (*Proptera*) *metallica* (Say, 1831) — Simpson (1914)
Elliptio (*Popenaias*) *metallicus* (Say, 1831) — Frierson (1927)
Nephronaias (*Popenaias*) *metallica* (Say, 1831) — Haas (1969)

Unio aereus Reeve, 1865. Monograph of the genus *Unio*. *In* L. Reeve and G.B. Sowerby (eds.). *Conchologia Iconica* 16: [unpaginated]+plates 19–30. Plate 30, species 160.

Verbatim Type Locality: "Hab. —?"

Type Locality: Unknown.

Type Specimens: NHMUK, lost (Johnson 1971:78).

Remarks: Reeve (1865) described and figured *Unio aereus* in his monograph of the genus *Unio*. Although he provided no locality and no type is known to exist, Lea (1870:61) decided (presumably based upon the color of the nacre) that it was synonymous with *Unio cuprinus* Lea, 1831. Others (see below) considered it synonymous with *Unio metallicus* Say, 1831, which also lacks a type specimen but does have a type locality (Lago de Chalco, México). We have seen no specimens in collections we examined identified as *Unio aereus* Reeve, 1865, and without a type or locality, *Unio aereus* Reeve, 1865 is unrecognizable, but we place it here following previous workers.

Taxonomic History:

Margaron (*Unio*) *cuprinus* (Lea, 1831) — Lea (1870)
Unio aereus Reeve, 1865 — Paetel (1890)
Lampsilis (*Proptera*) *metallicus* (Say, 1831) — Simpson (1900)
Lampsilis metallicus (Say, 1831) — de Leon (1912)
Lampsilis (*Proptera*) *metallica* (Say, 1831) — Simpson (1914)
Elliptio (*Popenaias*) *metallicus* (Say, 1831) — Frierson (1927)
Nephronaias (*Popenaias*) *metallica* (Say, 1831) — Haas (1969)

PLATE 62 ***Popenaias semirasa* (Pilsbry, 1910). Top:** *Lampsilis semirasa* Pilsbry, 1910. Lectotype, ANSP 99542 (47 mm L, 26 mm H). **Bottom:** *Lampsilis metallica ganina* Pilsbry, 1910. Lectotype, ANSP 99523 (47 mm L, 25 mm H). Both, Río Gallinas (Pánuco–Tampaón Dr.), 4.8 km SW San Dieguito, San Luis Potosí, México [L=21.9869, -99.2548].

Popenaias semirasa (Pilsbry, 1910)

Plate 62, Map 73

Province(s): Pánuco.

Distribution and Status: Endemic and fairly common in the ríos Gallinas and Valles in the Pánuco basin of San Luis Potosí, México. Not assessed (IUCN 2024).

Anatomy: Shell elongate and oblong. Dorsal margin arched. Ventral margin straight or slightly concave. Posterior end is wide in females, less so in males. The beaks are low and located at the anterior fifth of the shell with double-looped, concentric sculpture. Pseudocardinal teeth compressed, rather small and the lateral teeth short, about one-third the shell's length. Shell surface smooth. Periostracum dark brown, with indistinct green rays, most evident on the posterior half. Nacre white, pink, or purple.

Biology: Found in small to medium-sized rivers in sand and gravel.

Adult Size: Length to 6.5 cm.

Similar Species: *Popenaias berezai* Inoue et al., 2020.

MAP 73 Distribution of *Popenaias semirasa* (Pilsbry, 1910).

Lampsilis semirasa Pilsbry, 1910. Unionidae of the Panuco River system, Mexico. *Proceedings of the Academy of Natural Sciences of Philadelphia* 61(3):532–539 + 3 plates. Page 534, pl. 26, figs. 7–9.

Verbatim Type Locality: "Ganina River three miles southwest of San Dieguito, in mud and gravel in clear water about knee–deep, where there was no current. Also, in the Valles River two miles above Mecos."

Type Locality: Río Gallinas (Gulf of México–Pánuco–Tampaón Dr.), 4.8 km SW San Dieguito, San Luis Potosí, México [L = 21.9869, -99.2548].

Type Specimens: Lectotype, ANSP 99542 designated and type locality fixed by Johnson and Baker (1973:170). Paralectotypes, ANSP 454002 (*n*=18, ex. ANSP 99542), ANSP 99541 (*n*=9); CHAS 25165 (*n*=9); INHS 82199 (*n*=10), INHS 82202 (*n*=3), INHS 82203 (*n*=2); MCZ 189370 (*n*=25) (Johnson 1956:134, as cotypes); UMMZ 39172 (Johnson 1979:34, now UMMZ 91174), all from the type locality. ANSP 99540 (*n*=14) from Río Valles, 3.2 km above Micos, San Luis Potosí, México [L=22.1163, -99.1663].

Remarks: Pilsbry gave three sets of measurements and figured two without identifying one as a type. Haas (1969:450) placed this species in synonymy with *Friersonia iridella*, which it clearly is not.

Taxonomic History:

Lampsilis semirasa Pilsbry, 1910 — Simpson (1914), Fuller (1975)
Lampsilis (Friersonia) semirasa Pilsbry, 1910 — Frierson (1927)
Friersonia iridella (Pilsbry & Frierson, 1908) — Haas (1969)
Popenaias ganina (Pilsbry, 1910) — Inoue et al. (2020a)
Popenaias semirasa (Pilsbry, 1910) — Kiser et al. (2022), Czaja et al. (2023)

Lampsilis metallica ganina Pilsbry, 1910. Unionidae of the Panuco River system, Mexico. *Proceedings of the Academy of Natural Sciences of Philadelphia* 61(3):532–539+3 plates. Page 539, pl. 27, fig. 6.

Verbatim Type Locality: "Ganina River, three miles southwest of San Dieguito."

Type Locality: Río Gallinas (Gulf of México–Pánuco–Tampaón Dr.), 4.8 km SW San Dieguito, San Luis Potosí, México [L= 21.9869, -99.2548].

Type Specimens: Lectotype, ANSP 99523 designated by Johnson and Baker (1973:156, as the holotype). Paralectotypes, ANSP 99524 (*n*=4); CHAS 25176 (*n*=21); INHS 82200 (*n*=3); MCZ 189371 (*n*=15) (Johnson 1956:116, as cotypes).

Taxonomic History:

Lampsilis (Proptera) metallicus var. *ganinus* Pilsbry, 1910 — Simpson (1914)
Elliptio (Popenaias) metallicus-ganinus (Pilsbry, 1910) — Frierson (1927)
Elliptio (Popenaias) metallicus var. *ganinus* (Pilsbry, 1910) — Haas (1929)
Nephronaias (Popenaias) metallica (Pilsbry, 1910) — Haas (1969)
Popenaias ganina (Pilsbry, 1910) — Inoue et al. (2020a), Czaja et al. (2023)

Remarks: After the publication by Inoue et al. (2020a), it was brought to their attention by Philippe Bouchet (Muséum national d'Histoire naturelle) that they treated *Popenaias ganina* (Pilsbry, 1910) as a valid species, basionym *Lampsilis metallica ganina* Pilsbry, 1910; and they treated *Lampsilis semirasa* Pilsbry, 1910 as a synonym. However, this is in contradiction to ICZN Art. 24.1, which states that, of two names established simultaneously, the one established at higher rank (*semirasa*) has precedence.

Unio (Popenaias) popei regalis "Pilsbry" Haas, 1929. Bemerkungen über mittelamerikanische Najaden. *Senckenbergiana* 11(5/6):310–344. Page 330.

Remarks: Not described or figured. *Nomen nudem.*

Taxonomic History:

Popenaias ganina (Pilsbry, 1910) — Inoue et al. (2020a)

Order Unionoida
Superfamily Unionoidea Rafinesque, 1820
Family Unionidae Rafinesque, 1820
Subfamily Ambleminae Rafinesque, 1820
Tribe Popenaiadini Heard & Guckert, 1970
Genus *Psoronaias* Crosse & Fischer, 1894

Psoronaias Crosse & Fischer, 1894. *In* P. Fischer and H. Crosse. 1894. Études sur les mollusques terrestres et fluviatiles du Mexique et du Guatemala. M.H.–Milne Edwards (ed.). *Mission Scientifique au Mexique et dans l'Amérique centrale. Recherches Zoologiques pour servir á l'Histoire de la Faune de l'Amérique Centrale et du Mexique* 15th Livraison. 62–72, pp. 489–576, plates 59–62, Imprimerie Nationale, Paris Volume 2. 731 pp. Page 556. Introduced as a subgenus of *Unio* Philipsson, 1788.

Diagnosis: "Forme très-variable, ovale, subtrigone, cordiforme ou rostrée, plus ou moins inéquilatérale; surface chargée de petits tubercules et portant parfois en arrière plusieurs plis; dents latérales antérieures de la valve droite inégales: la supérieure faible, comprimée; l'inférieure épaisse et sillonnée (*U. psoricus*, Morelet)." [Very variable shape, oval, subtrigonal, heart-shaped or rostrate, more or less unequal; surface covered with small tubercles and sometimes bearing several folds behind; anterior lateral teeth of the right valve unequal: the upper weak, compressed; the lower thick and furrowed (*U. psoricus*, Morelet)].

Type Species: *Unio psoricus* Morelet, 1851, by original designation.

Taxonomic History:

Unio (*Psoronaias*) Crosse & Fischer, 1894 — von Martens (1900a)
Unio (*Pleurobema*) Rafinesque, 1819 — von Martens (1900b) [in part]
Unio (*Elliptio*) Rafinesque, 1819 — Simpson (1900; 1914) (in part)
Quadrula (*Rotundaria*) — Simpson (1900; 1914) [in part]
Elliptio Rafinesque, 1819 — Pilsbry (1910) [in part], Contreras-Arquieta (2000) [in part]
Psoronaias Crosse & Fischer, 1894 — Frierson (1917, 1927), Haas (1929; 1930; 1969), Thiele (1934), Goodrich & van der Schalie (1937), van der Schalie (1940), Modell (1942; 1964), Heard & Guckert (1970), Graf & Cummings (2007; 2021), Lopes-Lima et al. (2017), Pfeiffer et al. (2019a–b), Inoue et al. (2020a), Smith et al. (2020), Nagel (2025)
Psorionaias [sic] Crosse & Fischer, 1894 — Pfeiffer et al. (2019a)

Barynaias Crosse & Fischer, 1894. *In* P. Fischer and H. Crosse. 1894. Études sur les mollusques terrestres et fluviatiles du Mexique et du Guatemala. *In* M.H.–Milne Edwards (ed.). *Mission Scientifique au Mexique et dans l'Amérique centrale. Recherches Zoologiques pour servir á l'Histoire de la Faune de l'Amérique Centrale et du Mexique.* 15th Livraison. 62–72, pp. 489–576, plates 59–62, Imprimerie Nationale, Paris Volume 2. 731 pp. Page 557. Introduced as subgenus of *Unio* Philipsson, 1788.

Diagnosis: "Coquille très-épaisse, relativement courte, striée concentriquement; dents latérales antérieures très-épaisses (*U. pigerrimus*, Crosse et Fischer)." [Shell very thick, relatively short, concentrically striated; very thick anterior lateral teeth (*U. pigerrimus*, Crosse and Fischer)].

Type Species: *Unio pigerrimus* Crosse & Fischer, 1894, by original designation.

Taxonomic History:

Unio (*Barynaias*) Crosse & Fischer, 1894 — Fischer & Crosse (1894b)
Unio (*Arotonaias*) von Martens, 1900 — von Martens (1900b) [in part]
Unio (*Elliptio*) Rafinesque, 1819 — von Martens (1900b) [in part]

Unio Philipsson, 1788 — Pilsbry (1904), Contreras-Arquieta (2000)
Elliptio (*Barynaias*) Crosse & Fischer, 1894 — Frierson (1927), Haas (1929; 1930)
Psoronaias Crosse & Fischer, 1894 — Frierson (1927) [in part]
Elliptio (*Nephronaias*) Crosse & Fischer, 1894 — Haas (1929) [in part]
Elliptio (sec. *Barynaias*) Crosse & Fischer, 1894 — Thiele (1934)
Plagiola Rafinesque, 1819 — Goodrich & van der Schalie (1937)
Barynaias Crosse & Fischer, 1894 — Modell (1942), Haas (1969), Graf & Cummings (2007; 2021), Lopes-Lima et al. (2017), Pfeiffer et al. (2019b)
Elliptio — Contreras-Arquieta (2000)

Remarks: Crosse and Fischer (1894:557) listed *U. pigerrimus* as the type of *Barynaias*, but later (1894:582), listed *U. pigerrimus* under *Psoronaias* [as *Unio* (*Psoronaias*) *pigerrimus* Crosse & Fischer, 1894)], and *U. sallei* was listed as the sole example of *Barynaias* (Crosse and Fischer (1894:619). After examining the lectotype of *U. pigerrimus* Crosse & Fischer, 1894 (MNHN-IM-2000-1761), we have placed it here in *Psoronaias* as a synonym of *Psoronaias crocodilorum* (Morelet, 1849).

Martensnaias Frierson, 1927. A classification and annotated check list of the North American naiades. Baylor University Press, Waco, Texas. 111 pp. Page 38. Introduced as a subgenus of *Elliptio* Rafinesque, 1819.

Diagnosis: As with *Popenaias*, Frierson provided no diagnosis for his genus *Martensnaias*. The following is the diagnosis of *Unio rubicundus* von Martens, 1900. "Testa oblonga, trapezoidea, concentrice grosse striata, strüs nonnullis subnodulosis, in declivitate posteriore pliculis numerosis obliquis sursum arcuatis sculpta, rufo-fusca, parte antica longiuscule rotundata, parte postica declivi-rostrata et subperpendiculariter truncata, linea angulari a verticibus descendente, margine ventrali antice valde ascendente, medio et postico recto; dentes cardinales validi, triangulares, sulcati et crenulati; facies interna purpurea, linea palliali antice a margine valde remota." [The shell is oblong, trapezoidal, concentrically coarsely striated, with some subnodular striae, the posterior slope with numerous folds obliquely arching upward, reddish-brown, the anterior part rounded, the posterior part sloping-rostrated and subperpendicularly truncated, the angular line descending from the vertices, the ventral margin anteriorly very ascending, middle and straight behind; cardinal teeth strong, triangular, furrowed and crenulated; internal nacre purple, the pallial line anteriorly very distant from the margin.]

"Although a single valve only of this species has been obtained, I am compelled to propose a new name for it. In size and outlines *U. rubicundus* so nearly resembles *U. corium*, Sowerby, that, at first sight, it seemed to be nothing more than a smooth variety of that species; the sculpture, however, is different: this is due to the presence of coarse concentric striae, some of those on the umbonal region rising here and there into flat prominences but the summits themselves (which, fortunately, are well preserved) are simply concentrically striated; there is no trace of granulated structure. On the hinder slope there are small plaits, bent upwards as in *U. pliciferus*. The cardinal teeth rather strong. The sculpture is unlike "that of" the species of the subgenus *Psoronaias*, and, as placed here, near *U. pliciferus*, in the subgenus *Elliptio. U. rubicundus* differs from all its allies in the large size and thickness of the shell and in the stout teeth" (von Martens 1900b:514).

Type species: *Unio rubicundus* von Martens 1900, by original designation.

Taxonomic History:

Unio (*Elliptio*) Rafinesque, 1819 — von Martens (1900b)
Unio (sec. *Elliptio*) Rafinesque, 1819 — Simpson (1914)
Elliptio (*Martensnaias*) Frierson, 1927 — Frierson (1927)
Elliptio Rafinesque, 1819 — Thiele (1934)
Nephronaias (*Martensnaias*) Frierson, 1927 — Haas (1969)

Martensnaias Frierson, 1927 — Modell (1942), Johnson (1972), Graf & Cummings (2007; 2021), Pfeiffer et al. (2019b)
Martinsnaias [sic] Frierson, 1927 — Lopes-Lima et al. (2017)

Distribution & Remarks: Lopes-Lima et al. (2017) considered *Martensnaias* as *incertae cedis* Ambleminae. This monotypic genus was described based upon a single valve and compared to *Psoronaias* by von Martens (1900b) who thought it might be a smooth form of *U. corium* Reeve, 1864 and thought it near *U. pliciferus*. We have placed it with the other members of the morphologically variable genus *Psoronaias* pending further analysis.

Psorula Haas, 1930. Über nord- und mittelamerikanische Najaden. *Senckenbergiana* 12(6):317–330. Page 319. Introduced as a subgenus of *Amblema* Rafinesque, 1820.

Diagnosis: "Schale rundlich bis gerundet viereckig, bauchig bis ganz zusammengedrückt: Wirbel vorregend, mit tiefen, oft ganz zusammengpreßten Wirbelhöhlen, ihre Skulptur aus vielen, engstehenden Runzeln geht in die warzig oder körnig schuppige der Schalenaußenfläche über, die sich außerdem noch meist durch den Besitz stark wulstiger Zuwachsstreifen auszeichnet; Perlmutter weiß bis violet.

Zu dieser neuen Untergattung rechne ich außer den vorhin aufgezählten Arten noch meine (Haas 1929, S. 339) *Rotundaria salinarum*. Mit meiner Einteilung der von Crosse & Fischer und später von Frierson zu *Psoronaias* gerechneten Arten komme ich fast ganz zum gleichen Ergebnis wie Simpson, mit der einzigen Ausnahme, daß ich *Unio ostreatus* Mor. nicht wie dieser zu der hier von mir als *Psorula* zusammengefaßten Gruppe der Quadrulinen, sondern als Unionine zu *Psoronaias* im berichtigten Sinne stelle."

[Shell rounded to rounded quadrangular, bulbous to completely compressed: beaks protruding, with deep, often completely compressed beak cavities, their sculpture of many narrow folds, merges into the warty or granular surface of the shell, which is also usually characterized by the possession of strongly bulging growth lines; Nacre, mother-of-pearl white to violet.

In addition to the species listed above, I also include my (Haas 1929, p. 339) *Rotundaria salinarum* in this new subgenus. With my classification of the species recognized by Crosse & Fischer and later by Frierson as *Psoronaias*, I arrive at almost exactly the same conclusion as Simpson, with the only exception that *Unio ostreatus* Mor. is not like this group of quadrulines I have summarized here as *Psorula*, but as Unionine to *Psoronaias* in the corrected sense].

Type Species: *Quadrula rudis* Simpson, 1900, by original designation.

Taxonomic History:
Amblema (sec. *Psorula*) Haas, 1930 — Thiele (1934)
Psorula Haas, 1930 — Haas (1969), Modell (1942), Graf & Cummings (2007), Lopes-Lima et al. (2017), Pfeiffer et al. (2019a), Inoue et al. (2020a)
Psoronaias Crosse & Fischer, 1894 — Modell (1964)

Distribution & Remarks: *Psoronaias* as a widespread genus endemic to Central America in Gulf of Mexico tributaries and occurs from the Río San Fernando in Nuevo León, in northern México (Pánuco Province) south to the Usumacinta basin of México and Guatemala, where most of the nominal species were described (Map 74). Graf and Cummings (2021:20) moved five species formerly in the genus *Psorula* into *Psoronaias*: *Psoronaias percompressa* (von Martens, 1887), *Psoronaias profunda* (Simpson, 1914), *Psoronaias quadrata* (Simpson, 1914), *Psoronaias salinarum* (Haas, 1929), and *Psoronaias usumasintae* (Crosse & Fischer, 1893). We sunk all five of the preceding into *Psoronaias crocodilorum* (Morelet, 1849), which we elevated from *Psoronaias semigranosa* (von dem Busch in Philippi, 1843).

Based upon geography and recent genetic analysis (Pfeiffer et al. 2019a:112), we recognize four species in genus. *Psoronaias semigranosa* (von dem Busch in Philippi, 1843) is endemic to the Pánuco Basin and *Psoronaias crocodilorum* (Morelet, 1849) is found only in the Usumacinta. *Psoronaias distincta* is found in the Papaloapán and Coatzacoalcos basins in Veracruz. *Psoronaias kuxensis* is small member of the genus and a narrow endemic in the Usumacinta drainage found primarily in Kux Creek, Alta Verapaz Department, Guatemala.

Psoronaias shares several conchological similarities with the Nearctic genera *Quadrula* Rafinesque, 1820 and *Cyclonaias* Pilsbry in Ortmann & Walker, 1922, including a circular shell outline and strongly sculptured shells. Because of these similarities, *Psoronaias* has historically been considered a member of the tribe Quadrulini (Haas 1969; Modell 1942, 1964; Starobogatov 1970), although *Psoronaias* has also been attributed to the Pleurobemini (Heard & Guckert 1970; Starobogatov 1970). Despite these conchological similarities, *Psoronaias* was not resolved as part of the Quadrulini—they were recovered in a distantly related clade, along with *Popenaias popeii*, which represents the sister group to the Lampsilini.

Despite the morphological disparity of the 15 taxa examined (most of which were described from the Usumacinta basin) and previously recognized as valid species of *Psorula* or *Psoronaias*, all are grouped here under *Psoronaias crocodilorum* (Morelet, 1849). This is based upon genetic analysis of six morphological species: *Psoronaias semigranosa* (Philippi, 1843) (UF 507899)-–Río Usumacinta opposite of Emiliano Zapata, Tabasco, México; *Psoronaias guatemalensis* (Simpson,1900) (UF 507901-–stream (Usumacinta Dr.) near La Cascada, Chiapas, México; *Psoronaias percompressa* (von Martens,1887) (UF 507905); *Psoronaias profunda* (Simpson, 1914) (UF 507908); *Psoronaias usumasintae* (Crosse & Fischer, 1892) (UF 507907); and *Psorula rudis* (Simpson, 1900) (UF 507909), the last four all collected from the Rio de la Pasi6n (Usumacinta Dr.), 8.8 km E of Sayaxché, El Petén Department, Guatemala, that were, for all intents and purposes, genetically indistinguishable. The average pairwise COI uncorrected p-distance among those terminals, 0.163%, was less than expectations for interspecific variation in amblemines (Pfeiffer et al. 2016; 2018) and below the general DNA barcode cut-off to delimit species (Hebert et al. 2003; Pfeiffer et al. 2019a:112).

MAP 74 Distribution of the genus *Psoronaias* Crosse & Fischer, 1894.

PLATE 63 ***Psoronaias crocodilorum*** **(Morelet, 1849). Top Left:** *Unio crocodilorum* Morelet, 1849. Lectotype, NHMUK 1893.2.4.2030 (68 mm L, 40 mm H). **Top Right:** *Unio testudineus* Morelet, 1849, Lectotype, NHMUK 1894.3.22.1 (90 mm L, 70 mm H). **Middle Left:** *Unio ostreatus* Morelet, 1849. Lectotype, NHMUK 1894.3.22.2 (64 mm L, 60 mm H). **Middle Right:** *Unio morini* Morelet, 1851. Lectotype, NHMUK 1893.2.4.2026 (73 mm L, 47 mm H). All four from the Río Usumacinta, El Petén Department, Guatemala [D=16.8246, -90.8822]. **Bottom Left:** *Unio psoricus* Morelet, 1851. Lectotype, NHMUK 1893.2.4.2033 (83 mm L, 63 mm H). Río Usumacinta, Tabasco, México [D=17.8833, -91.8208]. **Bottom Right:** *Unio corium* Reeve, 1864. Holotype, NHMUK 1965209 (103 mm L, 63 mm H). Chiapas (Grijalva Dr.), México [D=16.4100, -92.4086].

PLATE 64 ***Psoronaias crocodilorum* (Morelet, 1849). Top Left:** *Unio percompressus* von Martens, 1887. Figured Syntype, ZMB/MOLL.39463 (94 mm L, 64 mm H). Río de las Salinas, El Petén Department, Guatemala [D = 16.4666, -90.5500]. **Top Right:** *Unio pigerrimus* Crosse & Fischer, 1894. Lectotype, MNHN-IM-2000-1761 (59 mm L, 38 mm H). México. **Middle Left:** *Unio* (*Psoronaias*) *crocodilorum* var. *β, semipustulata* Fischer & Crosse, 1894. Holotype, NHMUK 1893.2.4.2032 (70 mm L, 45 mm H). **Middle Right:** *Quadrula rudis* Simpson, 1900. Holotype, USNM 84210 (103 mm L, 65 mm H). Río Yaxchilán, El Petén Department, Guatemala [D = 16.8666, -90.9166]. **Bottom Left:** *Unio* (*Psoronaias*) *crocodilorum* var. γ, *praestricta* Fischer & Crosse, 1894. Holotype, NHMUK 1893.2.4.2031 (60 mm L, 40 mm H). **Bottom Right:** *Quadrula* (*Rotundata*) *quadrata* Simpson, 1914. Lectotype, USNM 128996 (67 mm L, 53 mm H). The last two are from the Río Usumacinta, El Petén Department, Guatemala [D = 16.8246, -90.8822].

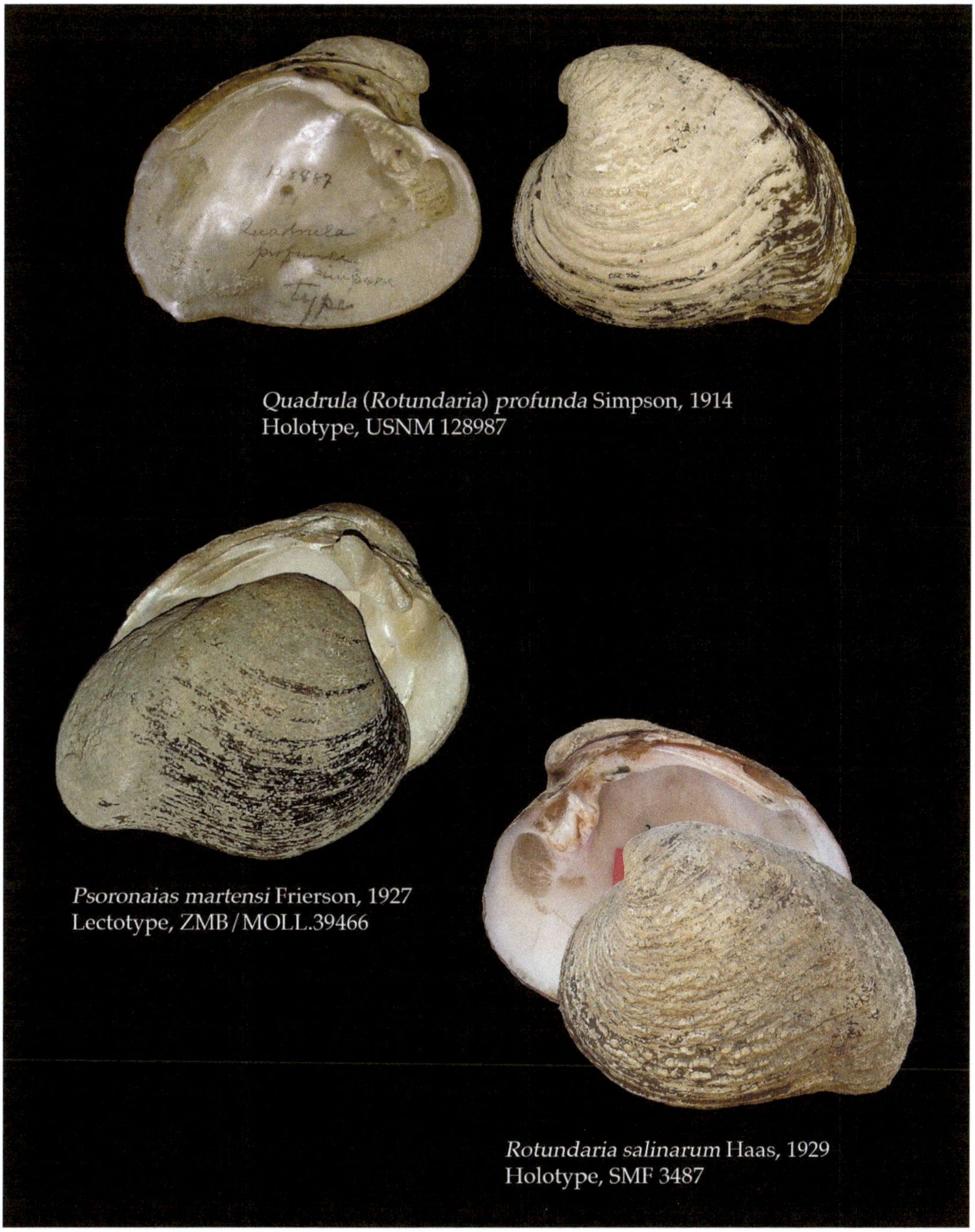

PLATE 65 ***Psoronaias crocodilorum*** **(Morelet, 1849). Top:** *Quadrula* (*Rotundaria*) *profunda* Simpson, 1914. Holotype, USNM 128987 (77 mm L, 64 mm H). Río Usumacinta, El Petén Department, Guatemala [D = 16.8246, -90.8822]. **Middle:** *Psoronaias martensi* Frierson, 1927. Lectotype, ZMB/MOLL.39466 (68 mm L, 54 mm H). **Bottom:** *Rotundaria salinarum* Haas, 1929. Holotype, SMF 3487 (76 mm L, 61 mm H). Bottom two, Río de las Salinas (Usumacinta Dr.), El Petén Department, Guatemala [D = 16.4666, -90.5500].

Psoronaias crocodilorum (Morelet, 1849)

Plates 63–65, Map 75

Province(s): Usumacinta.

Distribution and Status: Endemic to the Usumacinta basin, México and Guatemala. Not assessed (IUCN 2024). Numerous weathered shells found in the ríos Chajul, Lacanja, Lacantún, and Tzendales (Usumacinta Dr.) in collections made from 2010 to 2020 (Ramírez et al. 2022:23). Collected alive in the Rio de la Pasion (Usumacinta Dr.), 9 km E of Sayaxche, El Petén Department, Guatemala in 2015 (Pfeiffer et al. 2019:113).

Anatomy: Shell solid, inflated, triangular or ovate. Posterior ridge well developed, sometimes slightly double and ending in a point posteriorly. Beaks high and often heavily sculptured with numerous pustules that cover most or all the shell. Beak cavity deep. Left valve with two ragged pseudocardinals and two laterals. Right valve with a single pseudocardinal and one lateral. Periostracum green in young shells, becoming brown or black in adults. Nacre purple, copper colored, or white.

Biology: Unknown.

Adult Size: Length to 9.5 cm.

Similar Species: *Psoronaias kuxensis* Frierson, 1917 also occurs in the Usumacinta basin, but it is smaller in size and typically inhabits smaller streams.

Remarks: There are numerous shell morphologies and described species from the Usumacinta, but all are synonymized here under *Psoronaias crocodilorum* (Morelet, 1849).

MAP 75 Distribution of *Psoronaias crocodilorum* (Morelet, 1849).

Unio crocodilorum Morelet, 1849. Testacea novissima insulæ Cubanæ, et Americæ Centralis. Librairie de L'Académie Nationale de Medecine, Paris Part 1:1–31. Page 28. Figured in Fischer and Crosse (1874b:577, pl. 60, fig. 3; pl. 67, fig. 3) and von Martens (1900a:495, pl. 31, figs. 3, 3a–c).

Verbatim Type Locality: "H. flumen Usumasinta."

Type Locality: Río Usumacinta (Gulf of México–Grijalva Dr.), El Petén Department, Guatemala [D=16.8246, -90.8822].

Type Specimens: Lectotype, NHMUK 1893.2.4.2030 designated by Johnson (1971:81, as the holotype; Breure et al. 2018:258, figs. 293–294). Paralectotypes, NHMUK 1965161–62 (*n*=2); NHMUK 1908.12.15.4 (*n*=1); MNHN-IM-2000-1667 (*n*=1) (Breure et al. 2018:228).

Remarks: Johnson (1971:81) also stated that "2 paratypes NHMUK 1893.2.4.2031–32, were also figured, pl. 60, figs. 45 and were named varieties semipustulata and praestricta respectively." Breure et al. (2018:258) also considered NHMUK 1893.2.4.2031–32 as paralectotypes of *Unio crocodilorum* Morelet, 1849. However, NHMUK 1893.2.4.2031 is the holotype of *Unio crocodilorum* var. γ, *praestricta* Fischer & Crosse, 1894 and NHMUK 1893.2.4.2032 the holotype of *Unio crocodilorum* var. *β*, *semipustulata* Fischer & Crosse, 1894 (see below).

Taxonomic History:

Unio crocodilorum Morelet, 1849 — Reeve (1864)
Margaron (*Unio*) *pliciferus* (Lea, 1838) — Lea (1870)
Unio (*Psoronaias*) *crocodilorum* Morelet, 1849 — Fischer & Crosse (1894b), von Martens (1900a, 1901b)
Unio (sec. *Elliptio*) *crocodilarum* [sic] Morelet, 1849 — Simpson (1900; 1914), de Leon (1912)
Psoronaias crocodilarum [sic] (Morelet, 1849) — Frierson (1917), Haas (1929)
Elliptio plexus crocodilarum [sic] (Morelet, 1849) — H.B. Baker (1922)
Psoronaias crocodilorum (Morelet, 1849) — Frierson (1927), Haas (1930; 1969), Graf & Cummings (2007), Breure et al. (2018), Ramírez et al. (2022)
Psoronaias semigranosus (von dem Busch in Philippi, 1843) — Goodrich & van der Schalie (1937)
Psoronaias semigranosa (von dem Busch in Philippi, 1843) — Pfeiffer et al. (2019b), Smith et al. (2020), Graf & Cummings (2021) [in part], Czaja et al. (2023) [in part]

Unio testudineus Morelet, 1849. Testacea novissima insulæ Cubanæ, et Americæ Centralis. Librairie de L'Académie Nationale de Medecine, Paris Part 1:1–31. Page 28. Figured in Fischer and Crosse (1874a:571, pl. 62, fig. 3; pl. 70, fig. 3).

Verbatim Type Locality: "H. cum præcedente." [= flumen Usumasinta].

Type Locality: Río Usumacinta (Gulf of México–Grijalva Dr.), El Petén Department, Guatemala [D=16.8246, -90.8822].

Type Specimens: Lectotype, NHMUK 1894.3.22.1 designated by Johnson (1971:96, as the holotype; Breure et al. 2018:452, figs. 1140–1141).

Remarks: Fischer and Crosse (1894:572) gave the type locality as "Guatemala, dans le Rio Usumacinta, department du Peten (A. Morelet)."

Taxonomic History:

Unio testudineus Morelet, 1849 — Reeve (1865), de Leon (1912)
Unio (*Psoronaias*) *testudineus* Morelet, 1849 — Fischer & Crosse (1894b)
Unio (*Psoronaias*) *semigranosus* var. *testudineus* Morelet, 1849 — von Martens (1900a)
Elliptio plexus crocodilarum [sic] (Morelet, 1849) — H.B. Baker (1922)
Unio (sec. *Elliptio*) *testudineus* Morelet, 1849 — Simpson (1900; 1914)
Psoronaias testudinea (Morelet, 1849) — Haas (1929; 1969)
Psoronaias testudineus (Morelet, 1849) — Haas (1930)
Psoronaias semigranosa (von dem Busch in Philippi, 1843) — Breure et al. (2018), Ramírez et al. (2022), Nagel (2025)

Unio ostreatus Morelet, 1849. Testacea novissima insulæ Cubanæ, et Americæ Centralis. Librairie de L'Académie Nationale de Medecine, Paris Part 1:1–31. Page 29. Figured in Fischer and Crosse (1874a:573, pl. 63, fig. 3; pl. 70, fig. 4).

Verbatim Type Locality: "H. flumen Usumasinta."

Type Locality: Río Usumacinta (Gulf of México–Grijalva Dr.), El Petén Department, Guatemala [D=16.8246, -90.8822].

Type Specimens: Lectotype, NHMUK 1894.3.22.2 designated by Johnson (1971:90, as the holotype; Breure et al. 2018:377, figs. 806–807). Paralectotypes, MNHN-IM-2000–2102 (n=2) (Breure et al. 2018:337, as MNHN-IM-2000–2101).

Taxonomic History:

Unio (*Psoronaias*) *ostreatus* Morelet, 1849 — Fischer & Crosse (1894b), von Martens (1900a)
Quadrula (*Rotundaria*) *ostreata* (Morelet, 1849) — Simpson (1900; 1914)
Quadrula ostreata (Morelet, 1849) — de Leon (1912)
Psoronaias ostreatus (Morelet, 1849) — Frierson (1927), Haas (1930), Breure et al. (2018)
Psoronaias ostreata (Morelet, 1849) — Haas (1929; 1969), Graf & Cummings (2007; 2021), Ramírez et al. (2022), Czaja et al. (2023)

Unio psoricus Morelet, 1851. Testacea novissima insulae Cubanae, et Amer. Centralis. Librairie de L'Académie Nationale de Medecine, Paris Part 2:1–30. Page 25. Figured in Fischer and Crosse (1874a:572, pl. 61) and von Martens (1900a, pl. 31, figs. 1–1a; ZMB/MOLL.39466, fig. 2. ZMB/MOLL.305830).

Verbatim Type Locality: "H. flumen Usumasinta."

Type Locality: Río Usumacinta (Gulf of México–Grijalva Dr.), Tabasco, México [D=17.8833, -91.8208].

Type Specimens: Lectotype, NHMUK 1893.2.4.2033 designated by Johnson (1971:91, as the holotype; Breure et al. 2018:397, figs. 897–898). Paralectotype, NHMUK 1893.2.4.2034 (Johnson 1971:91, as a paratype; Breure et al. 2018:397).

Remarks: Johnson (1975:35) incorrectly listed USNM 84210 as a paratype of *Unio psoricus* Morelet, 1851. That specimen is the holotype of *Quadrula rudis* Simpson, 1900 (see below).

Taxonomic History:

Unio (*Psoronaias*) *psoricus* Morelet, 1851 — Fischer & Crosse (1894b), von Martens (1900a)
Unio (sec. *Elliptio*) *psoricus* Morelet, 1851 — Simpson (1900; 1914)
Unio psoricus Morelet, 1851 — de Leon (1912)
Psoronaias psoricus (Morelet, 1851) — Frierson (1917), Haas (1930), Thiele (1934)
Elliptio psoricus (Morelet, 1851) — H.B. Baker (1922)
Psoronaias semigranosus-psoricus (Morelet, 1851) — Frierson (1927)
Psoronaias psorica (Morelet, 1851) — Haas (1929; 1969)
Psoronaias semigranosus (von dem Busch in Philippi, 1843) — Goodrich & van der Schalie (1937)
Psoronaias semigranosa (von dem Busch in Philippi, 1843) — Breure et al. (2018), Graf & Cummings (2021), Ramírez et al. (2022)

Unio morini Morelet, 1851. Testacea novissima insulae Cubanae, et Amer. Centralis. Librairie de L'Académie Nationale de Medecine, Paris Part 2:1–30. Page 25. Figured in Fischer and Crosse (1874a:576, pl. 60, fig. 2; pl. 67, fig. 4) and von Martens (1900a, pl. 31, figs. 3, 3a–c, ZMB/MOLL.305831, as *Unio crocodilorum*).

Verbatim Type Locality: "H. cum præcedente [H. flumen Usumasinta]."

Type Locality: Río Usumacinta (Gulf of México–Grijalva Dr.), El Petén Department, Guatemala [D=16.8246, -90.8822].

Type Specimens: Lectotype, NHMUK 1893.2.4.2026 designated by Johnson (1971:88, as the holotype; Breure et al. 2018:359, figs. 725–726).

Taxonomic History:

Unio (*Psoronaias*) *morini* Morelet, 1851 — Fischer & Crosse (1894b), von Martens (1900a)
Unio (sec. *Elliptio*) *morini* Morelet, 1851 — Simpson (1900; 1914)
Elliptio plexus crocodilarum [sic] (Morelet, 1849) — H.B. Baker (1922)
Psoronaias morini (Morelet, 1851) — Frierson (1927), Haas (1929, 1930; 1969), Graf & Cummings (2007), Breure et al. (2018), Ramírez et al. (2022)
Psoronaias semigranosus (von dem Busch in Philippi, 1843) — Goodrich & van der Schalie (1937)
Psoronaias semigranosa (von dem Busch in Philippi, 1843) — Graf & Cummings (2021)

Unio corium Reeve, 1864. Monograph of the genus *Unio. In* L. Reeve and G.B. Sowerby (eds.). *Conchologia Iconica* 16: [unpaginated]+plates 2–18. Plate 10, species 39. Also figured in von Martens (1900a, pl. 31, figs. 4, 4a–b, ZMB/MOLL.305832).

Verbatim Type Locality: "Chiapa [sic], México."

Type Locality: Chiapas (Gulf of México–Grijalva Dr.), México [D=16.4100, -92.4086].

Type Specimens: Holotype by monotypy, NHMUK 1965209 (Johnson 1971:81).

Taxonomic History:

Margaron (*Unio*) *semigranosus* (von dem Busch in Philippi, 1843) — Lea (1870)
Unio (*Psoronaias*) *corium* Reeve, 1864 — Fischer & Crosse (1894b), von Martens (1900a)
Unio (sec. *Elliptio*) *corium* Reeve, 1864 — Simpson (1900; 1914)
Unio corium Reeve, 1864 — Pilsbry (1910), de Leon (1912), H.B. Baker (1922)
Psoronaias semigranosus-corium (Reeve, 1864) — Frierson (1927)
Psoronaias semigranosus (von dem Busch in Philippi, 1843) — Haas (1930), Goodrich & van der Schalie (1937)
Psoronaias semigranosa corium (Reeve, 1864) — Haas (1969)

Unio percompressus von Martens, 1887. Süsswassermuscheln aus Guatemala. *Sitzungsberichte der Gesellschaft naturforschender Freunde zu Berlin* 7:106–108. Page 107. Figured in von Martens (1900a, pl. 32, figs. 1–1a, 2–2a, 3, 3a–b) and Pfeiffer et al. (2019a:113, fig. 2e).

Verbatim Type Locality: "Rio de Salinas, prov. Peten, Guatemalae."

Type Locality: Río de las Salinas (Gulf of México–Grijalva–Usumacinta Dr.), El Petén Department, Guatemala [D=16.4666, -90.5500].

Type Specimens: Lectotype, ZMB/MOLL.39463 (fig. 1), here designated. Paralectotypes, ZMB/MOLL.305833 (*n*=2, figs. 2–3), ZMB/MOLL.108849 (*n*=1).

Taxonomic History:

Unio (*Psoronaias*) *percompressus* von Martens, 1887 — Fischer & Crosse (1894b), von Martens (1900a)
Quadrula (*Rotundaria*) *percompressa* (von Martens, 1887) — Simpson (1900; 1914)
Psoronaias percompressus (von Martens, 1887) — Frierson (1927)
Rotundaria percompressa (von Martens, 1887) — Haas (1929)

Amblema (Psorula) percompressa (von Martens, 1887) — Haas (1930)
Psorula percompressa (von Martens, 1887) — Haas (1969), Graf & Cummings (2007), Pfeiffer et al. (2019a), Inoue et al. (2020a)
Psoronaias percompressa (von Martens, 1887) — Graf & Cummings (2021), Ramírez et al. (2022), Czaja et al. (2023)

Unio usumasintæ Crosse & Fischer, 1893. Diagnoses Molluscorum Reipublicæ Mexicanæ et Guatemalæ incolarum. *Journal de Conchyliologie* 40(3):294–296. Page 294. Figured in Fischer and Crosse (1894a:574, pl. 63, fig. 4; pl. 67, fig. 5).

Verbatim Type Locality: "in flumine Usumasinta, Provinciæ Peten dictæ, Guatemalæ."

Type Locality: Río Usumacinta (Gulf of México–Grijalva Dr.), El Petén Department, Guatemala [D=16.8246, -90.8822].

Type Specimens: Unknown. Presumed lost. Perhaps in the MNHN.

Remarks: von Martens (1900b:500) spelling "*usumacintæ*" is an unjustified emendation to match his spelling of the type locality. The date of publication is often cited as 1892 but is 1893 according to Winckworth (1936:156). Size given as 47 mm L, 42 mm H.

Taxonomic History:
Unio (Psoronaias) usumasintæ Crosse & Fischer, 1892— Fischer & Crosse (1894b)
Unio (Pleurobema) usumacintæ [sic] Crosse & Fischer, 1892— von Martens (1900b)
Quadrula (Rotundaria) usumasintæ (Crosse & Fischer, 1892) — Simpson (1900; 1914)
Quadrula usumacintæ [sic] (Crosse & Fischer, 1892) — de Leon (1912)
Psoronaias usumacintae [sic] (Crosse & Fischer, 1893) — Frierson (1927)
Rotundaria usumacintae [sic] (Crosse & Fischer, 1892) — Haas (1929)
Amblema (Psorula) usumasintae (Crosse & Fischer, 1892) — Haas (1930)
Psorula usumacintae [sic] (Crosse & Fischer, 1892) — Haas (1969)
Psorula usumasintae (Crosse & Fischer, 1892) — Graf & Cummings (2007), Pfeiffer et al. (2019a)
Psoronaias usumasintae (Crosse & Fischer, 1892) — Graf & Cummings (2021), Czaja et al. (2023)

Unio pigerrimus Crosse & Fischer, 1894. Diagnoses molluscorum novorum, reipublicæ Mexicanæ incolarum. *Journal de Conchyliologie* 41(4):293–297. Page 293. Figured in Fischer and Crosse (1894b:582, pl. 65, figs. 1–1a).

Verbatim Type Locality: "republica Mexicana."

Type Locality: México.

Type Specimens: Lectotype, MNHN-IM-2000-1761, here designated (the figured specimen). Paralectotype, IRSNB 3735 (*n*=1).

Remarks: "San Gabriel" is given as the type locality on the MNHN web page (March 2023) but that locality is not mentioned in the original description or by Crosse & Fischer, 1894. There is a San Gabriel on the Río Grijalva basin near Tuxtla Gutiérrez, Chiapas. The date of publication is often cited as 1893 but is 1894 according to Winckworth (1936:156).

Although Fischer and Crosse (1894b:557) designated *Unio pigerrimus* Crosse & Fischer, 1894 as the type of *Barynaias*, they subsequently (Fischer and Crosse 1894b:582) placed it under *Psoronaias*. The surface sculpture and shell shape are both like *Psoronaias* as pointed out by von Martens (1900a:494). H.B. Baker (1922:9) came to the same conclusion stating the "The section *Barynaias* C. and F. (1894) is described in Part VII of the 'Moll. terr. et fluv. de Mex.,' with *U. pigerrimus* as the single example, while in Part VIII that species is put under *Psoronaias* and *U. sallei* is listed as the sole example of the former section. *Barynaias* is thus a synonym of *Psoronaias*." We provisionally place *Unio pigerrimus* Crosse & Fischer, 1894 here.

Taxonomic History:

Unio (*Barynaias*) *pigerrimus* Crosse & Fischer, 1893 — Fischer & Crosse (1894b)
Unio (*Psoronaias*) *pigerrimus* Crosse & Fischer, 1893 — Fischer & Crosse (1894b)
Unio (*Psoronaias*) *psoricus* var. *pigerrimus* Crosse & Fischer, 1893 — von Martens (1900a)
Unio (sec. *Elliptio*) *pigerrimus* Crosse & Fischer, 1893 — Simpson (1900; 1914)
Unio pigerrimus Crosse & Fischer, 1893 — de Leon (1912)
Elliptio (*Sphenonaias*) *plexus* (Conrad, 1838) — H.B. Baker (1922)
Elliptio (*Barynaias*) *pigerrimus* (Crosse & Fischer, 1893) — Frierson (1927), Haas (1929)
Elliptio (sec. *Barynaias*) *pigerrima* (Crosse & Fischer, 1893) — Thiele (1934)
Barynaias pigerrimus (Crosse & Fischer, 1893) — Haas (1969)
Barynaias pigerrima (Crosse & Fischer, 1893) — Graf & Cummings (2007; 2021), Czaja et al. (2023)

Unio (*Psoronaias*) *crocodilorum* var. *β*, *semipustulata* Fischer & Crosse. 1894. Études sur les mollusques terrestres et fluviatiles du Mexique et du Guatemala. 16th Livraison. Feuilles 73–82, pp. 577–656, plates 63–66 in Recherches Zoologiques pour servir À l'Histoire de la Faune de l'Amérique Centrale et du Mexique, Imprimerie Nationale, Paris Volume 2. 731 pp. Page 578, pl. 60, fig. 4.

Verbatim Type Locality: Not given, presumed to be the same as *Unio crocodilorum.* "in flumen Usumasinta, Provinciæ Peten dictæ, Guatemalæ."

Type Locality: Río Usumacinta (Gulf of México–Grijalva Dr.), El Petén Department, Guatemala [D = 16.8246, -90.8822].

Type Specimens: Holotype by monotypy, NHMUK 1893.2.4.2032 (Johnson 1971:81, as a paratype; Breure et al. 2018:258, as a paralectotype of *Unio crocodilorum* Morelet, 1849).

Remarks: Plate 60 was published in Septiéme Partie. 15th Livraison. 62–72.

Taxonomic History:

Unio (*Psoronaias*) *crocodilorum* var. *β*, *semipustulata* Fischer & Crosse, 1894 — Fischer & Crosse (1894b)
Unio (*Psoronaias*) *crocodilorum* Morelet, 1849 — von Martens (1900b)
Psoronaias crocodilorum (Morelet, 1849) — Frierson (1927), Haas (1969), Breure et al. (2018)
Psoronaias crocodilarum [sic] (Morelet, 1849) — Haas (1929)

Unio (*Psoronaias*) *crocodilorum* var. *γ*, *praestricta* Fischer & Crosse, 1894. Études sur les mollusques terrestres et fluviatiles du Mexique et du Guatemala. 16th Livraison. Feuilles 73–82, pp. 577–656, plates 63–66 in Recherches Zoologiques pour servir À l'Histoire de la Faune de l'Amérique Centrale et du Mexique, Imprimerie Nationale, Paris Volume 2. 731 pp. Page 578, pl. 60, fig. 5.

Verbatim Type Locality: Not given, presumed to be the same as *Unio crocodilorum.* "in flumine Usumasinta, provinciæ Peten dictæ, Guatemalæ."

Type Locality: Río Usumacinta (Gulf of México–Grijalva Dr.), El Petén Department, Guatemala [D = 16.8246, -90.8822].

Type Specimens: Holotype by monotypy, NHMUK 1893.2.4.2031 (Johnson 1971:81; as a paratype; Breure et al. 2018:258, as a paralectotype of *Unio crocodilorum* Morelet, 1849).

Remarks: Plate 60 was published in Septiéme Partie. 15th Livraison. Feuilles 62–72.

Taxonomic History:

Unio (*Psoronaias*) *crocodilorum* var. γ, *praestricta* Fischer & Crosse, 1894 — Fischer & Crosse (1894b)
Unio (*Psoronaias*) *crocodilorum* Morelet, 1849 — von Martens (1900b)
Psoronaias crocodilorum (Morelet, 1849) — Frierson (1927), Haas (1969), Breure et al. (2018)

Quadrula rudis Simpson, 1900. New and unfigured Unionidae. *Proceedings of the Academy of Natural Sciences of Philadelphia* 52:74–86+5 plates. Page 82, pl. 3, fig. 2.

Verbatim Type Locality: "Rio Taxtunilha, Guatemala."

Type Locality: Río Yaxchilán [= Taxtunilha] (Gulf of México–Grijalva–Usumacinta Dr.), El Petén Department, Guatemala [D=16.8666, -90.9166].

Type specimen: Holotype by monotypy, USNM 84210 (Johnson 1975:19; Pfeiffer et al. 2019a:113, fig. 2c).

Taxonomic History:

Psoronaias crocodilorum var. *rudis* Simpson, 1900 — von Martens (1901b).
Quadrula (*Rotundaria*) *rudis* Simpson, 1900 — Simpson (1914)
Psoronaias rudis (Simpson, 1900) — Frierson (1927)
Rotundaria rudis (Simpson, 1900) — Haas (1929)
Amblema (*Psorula*) *rudis* (Simpson, 1900) — Haas (1930)
Amblema (sec. *Psorula*) *rudis* (Simpson, 1900) — Thiele (1934)
Psoronaias semigranosus (von dem Busch in Philippi, 1843) — Goodrich & van der Schalie (1937)
Psorula rudis (Simpson, 1900) — Haas (1969), Graf & Cummings (2007), Pfeiffer et al. (2019a), Inoue et al. (2020a)
Psoronaias semigranosa (von dem Busch in Philippi, 1843) — Graf & Cummings (2021)

Quadrula (*Rotundata*) *quadrata* Simpson, 1914. A descriptive catalogue of the naiades, or pearly fresh-water mussels. Parts I–III. Bryant Walker, Detroit, Michigan xii+1540 pp. Page 911. Figured in Johnson (1975:18, pl. 3, fig. 8).

Verbatim Type Locality: "Usumacinta River, Guatemala."

Type Locality: Río Usumacinta (Gulf of México–Grijalva Dr.), El Petén Department, Guatemala [D=16.8246, -90.8822].

Type Specimens: Lectotype, USNM 128996 designated by Johnson (1975:18). Topotypes, SMF 3475 (*n*=4) (Haas 1929:334, figs. 14–15, as paratypes; Zilch 1967:109, as syntypes (Nagel 2025:169, see remarks).

Remarks: Haas (1929:334) and Zilch (1967:109) both stated that the four specimens in SMF 3475 were paratypes. However, Simpson (1914:911) stated that "The National Museum possesses three specimens sent by Dr. von Ihering from the above locality..." So, one of the four could be a paralectotype.

Taxonomic History:

Psoronaias quadrata (Simpson, 1914) — Frierson (1927), Andrews (1969), Graf & Cummings (2021)
Rotundaria quadrata (Simpson, 1914) — Haas (1929)
Amblema (*Psorula*) *quadrata* (Simpson, 1914) — Haas (1930)
Psorula quadrata (Simpson, 1914) — Zilch (1967), Haas (1969), Graf & Cummings (2007)

Quadrula (*Rotundaria*) *profunda* Simpson, 1914. A descriptive catalogue of the naiades, or pearly fresh-water mussels. Parts I–III. Bryant Walker, Detroit, Michigan xii+1540 pp. Page 907. Figured in Johnson (1975:17, pl. 3, fig. 6) and Pfeiffer et al. (2019a:113, fig. 2a, as USNM 12987)

Verbatim Type Locality: "Usumacinta River, Guatemala."

Type Locality: Río Usumacinta (Gulf of México–Grijalva Dr.), El Petén Department, Guatemala [D=16.8246, -90.8822].

Type specimen: Holotype by monotypy, USNM 128987.

Remarks: Haas (1929:339, figs. 16–19) figured two specimens as paratypes and Zilch (1967:108) stated that SMF 3476 (*n*=2) were syntypes, but Simpson (1914:908) stated that "The National Museum possesses a single valve of this robust species presented by Dr. von Ihering."

Taxonomic History:

Psoronaias profunda (Simpson, 1914) — Frierson (1927), Graf & Cummings (2021), Czaja et al. (2023), Nagel (2025)
Rotundaria profunda (Simpson, 1914) — Haas (1929)
Amblema (*Psorula*) *profunda* (Simpson, 1914) — Haas (1930)
Psorula profunda (Simpson, 1914) — Zilch (1967), Haas (1969), Graf & Cummings (2007), Pfeiffer et al. (2019a), Inoue et al. (2020a)

Psoronaias martensi Frierson, 1927. A classification and annotated check list of the North American naiades. Baylor University Press, Waco, Texas. 111 pp. page 64, sp. 279.

Verbatim Type Locality: "N. Guatemala: Rio de las Salinas."

Type Locality: Río de las Salinas (Gulf of México–Grijalva–Usumacinta Dr.), El Petén Department, Guatemala [D=16.4666, -90.5500].

Type Specimens: Lectotype, ZMB/MOLL.39466. Paralectotype, ZMB/MOLL.305830

Remarks: Based upon the figure of *Unio psoricus* von Martens (1900a:494, pl. 31, figs. 1–1a, 2). Johnson (1972a:170, pl. 31, fig. 1) reproduced von Martens figure and stated that the holotype was lost and that the "Type figure, here restricted to [Pl. 31] figs. 1, la (Río de las Salinas, Guatemala). Collection of Stoll [location unknown]." The figured specimen from the Stoll Collection is still extant (ZMB/MOLL.39466) and is the lectotype.

Taxonomic History:

Unio psoricus Morelet, 1851 — von Martens (1900a)
Psoronaias martensi Frierson, 1927 — Haas (1930; 1969), Graf & Cummings (2007; 2021)

Rotundaria salinarum Haas, 1929. Bemerkungen über mittelamerikanische Najaden. *Senckenbergiana* 11(5/6):310–344. Page 339, figs. 20–21. Also figured in Nagel (2025:167, fig. 163).

Verbatim Type Locality: "Rio de las Salinas, N. Guatemala."

Type Locality: Río de las Salinas (Gulf of México–Grijalva–Usumacinta Dr.), El Petén Department, Guatemala [D=16.4666, -90.5500].

Type specimen: Holotype by monotypy and original designation, SMF 3487.

Remarks: The type locality is likely the Río Chixoy (Nagel 2025:167).

Taxonomic History:

Amblema (*Psorula*) *salinarum* (Haas, 1929) — Haas (1930)
Psorula salinarum (Haas, 1929) — Zilch (1967), Haas (1969), Graf & Cummings (2007)
Psoronaias salinarum (Haas, 1929) — Graf & Cummings (2021), Nagel (2025)

PLATE 66 ***Psoronaias distincta*** **(Crosse & Fischer, 1893). Top:** *Unio distinctus* Crosse & Fischer, 1893. Holotype, MNHN-IM-2000-1676 (79 mm L, 50 mm H). Río Papaloapán, Chacaltianguis, Veracruz, México [L = 18.3333, -95.8333]. **Bottom:** ***Psoronaias rubicunda*** **(von Martens, 1900)**. *Unio* (*Elliptio*) *rubicundus* von Martens, 1900. Holotype, ZMB/MOLL.108853 (89 mm L, 57 mm H) (n = ½). Río Cahabon (Caribbean–Gulf of Honduras–Lake Izabal Dr.), Coban, Alta Verapaz Department, Guatemala [L = 15.4742, -90.3573].

Psoronaias distincta (Crosse & Fischer, 1893)

Plate 66, Map 76

Province(s): Papaloapán.

Distribution and Status: Known from the Río Papaloapán basin and the adjacent Jamapa (N) and Coatzacoalcos (S) basins (Gulf of Mexico Dr.) in Veracruz, México. Not assessed (IUCN 2024). No recent collections are known.

Anatomy: Shell triangular, slightly inflated, thick, and heavy. Ventral margin curved and slightly convex. Beaks slightly elevated and often eroded. Right valve with two pseudocardinals and one lateral tooth. Left valve with two pseudocardinals, the anterior thick and larger, the posterior shorter, granular and two lateral teeth. Periostracum brown with concentric striae and numerous pustules all over the surface in juveniles and toward the beaks in adults. Nacre purple.

Biology: Unknown.

Adult Size: Length to 8 cm.

Similar Species: *Sphenonaias liebmanni* (Philippi, 1847) also has pustules on the surface of the shell, but it is typically more elongate with a black periostracum. See Baker (1922:10–11) for a discussion on the considerable morphological variability of *Psoronaias distincta*.

MAP 76 Distribution of *Psoronaias distincta* (Crosse & Fischer, 1893).

Unio distinctus Crosse & Fischer, 1893. Diagnoses molluscorum novorum, Reipublicæ Mexicanæ incolarum. *Journal de Conchyliologie* 41(2):110–111. Page 110. Figured in Fischer and Crosse (1894b:579, pl. 69, figs. 2, 2a–b).

Verbatim Type Locality: "in flumine Cosamaloapam, prope Chacaltianguez, provinciæ Vera–Cruz."

Type Locality: Río Papaloapán (Gulf of Mexico Dr.), Chacaltianguis, Veracruz, México [L = 18.3333, -95.8333].

Type Specimens: Holotype by monotypy, MNHN-IM-2000-1676.

Remarks: The type locality was given as the Río Cosamaloapam [sic], near Chacaltianguez. The town of Chacaltianguis is on the Río Papaloapán, 4.5 km SW of the city of Cosamaloapan.

Taxonomic History:

Unio (*Psoronaias*) *distinctus* Crosse & Fischer, 1893 — Fischer & Crosse (1894b)
Unio (*Psoronaias*) *semigranosus* var. *distinctus* Crosse & Fischer, 1893 — von Martens (1900a)
Unio (sec. *Elliptio*) *distinctus* Crosse & Fischer, 1893 — Simpson (1914)
Psoronaias distinctus (Crosse & Fischer, 1893) — Frierson (1917, 1927), Haas (1930)
Elliptio (*Sphenonaias*) *plexus distinctus* (Crosse & Fischer, 1893) — H.B. Baker (1922)
Psoronaias distincta (Crosse & Fischer, 1893) — Haas (1969), Graf & Cummings (2007; 2021), Czaja et al. (2023) [in part]
Psoronaias semigranosa (von dem Busch in Philippi, 1843) — Czaja et al. (2023) [in part]

PLATE 67 ***Psoronaias kuxensis*** **Frierson, 1917. Top Left:** Lectotype, ANSP 117094 (50 mm L, 30 mm H). Kux Creek (Usumacinta–Chixoy Dr.), Chama, Alta Verapaz Department, Guatemala [L = 15.6166, -90.5500]. **Middle Right:** *Quadrula guatemalensis* Simpson, 1900. Holotype, USNM 128991 (45 mm L, 27 mm H). Río Usumacinta, El Petén Department, Guatemala [D = 16.8246, -90.8822]. **Middle Left:** *Elliptio divaricatus* Marshall, 1926. Holotype, USNM 362398 (32 mm L, 20 mm H). Finca de Providencia (Usumacinta–San Pedro Dr.), El Petén Department, Guatemala [L = 17.2277, -90.7828]. **Bottom Right:** *Elliptio* (*Nephronaias*) *hermanni* Haas, 1929. Holotype, SMF 3488 (46 mm L, 25 mm H). Arroyo Limón (Usumacinta Dr.), Tabasco, México [W = 17.8875, -92.5166].

Psoronaias kuxensis (Frierson, 1917)

Plate 67, Map 77

Province(s): Usumacinta.

Distribution and Status: Endemic to the Usumacinta basin, with most specimens coming from Kux Creek in the Río Chixoy drainage in Alta Verapaz Department, Guatemala. A disjunct population occurs in Arroyo Limón in Tabasco, México. Not assessed (IUCN 2024). Collected alive from the Río Seco, 1 km SW La Cascada, Cascadas Welib Ha, Chiapas, México in 2015 and 2019.

Anatomy: Shell small, subrhomboid, compressed, and rough. Shell round anteriorly. Dorsal margin arched. Posterior slope low, prominent, and biangular. Ventral margin straight. Beaks low and located far forward, often eroded, and pustulose in juveniles. Beak cavity deep. Pseudocardinal teeth double in the left valve, single in the right. Lateral teeth slightly curved or nearly straight, separated by an interdentum. Periostracum yellow, olive, or dark brown with fine pustules. Nacre pink or purple.

Biology: Unknown.

Adult Size: Length to 5.5 cm.

Similar Species: *Psoronaias crocodilorum* (Morelet, 1849) is also found in the Usumacinta drainage, but it is larger in size and has not been found in the Río Chixoy basin.

MAP 77 Distribution of *Psoronaias kuxensis* Frierson, 1917.

Psoronaias kuxensis Frierson, 1917. New genera and species of Central American naiades. *The Nautilus* 31(2):47–49 + plate 7. Page 48, pl. 7, figs. 1–2. Also figured in Johnson (1972a:170, pl. 31, fig. 3).

Verbatim Type Locality: "Kux Creek, Chama, Guatemala."

Type Locality: Kux Creek (Gulf of México–Grijalva–Usumacinta–Chixoy Dr.), Chama, Alta Verapaz Department, Guatemala [L = 15.6166, -90.5500].

Type Specimens: Holotype ANSP 117094 by original designation (without a catalogue number; Johnson 1972a:145, as 117094a in the text and as 117044a in the plate legend, and as the lectotype). Paratypes, ANSP 467442 (*n*=1) (Johnson 1972a:145 as ANSP 117094, as a paralectotype); FMNH 68336 (*n*=2); MCZ 45465 (*n*=2) Johnson (1972a:145, incorrectly as MCZ 108080 and as paralectotypes); INHS 82263 (*n*=2), INHS 82277 (*n*=10).

Remarks: Frierson (1917:49) also mentions that "A few dead specimens were obtained on the bank of the Isaibha River (Chama) of which the Kux Creek is a tributary." We can find no river of that name anywhere.

Taxonomic History:

Psoronaias kuxensis Frierson, 1917 — H.B. Baker (1922), Frierson (1927), Haas (1929; 1930; 1969), van der Schalie (1940), Johnson (1972), Graf & Cummings (2007; 2021)

Quadrula guatemalensis Simpson, 1900. New and unfigured Unionidae. *Proceedings of the Academy of Natural Sciences of Philadelphia* 52:74–86+5 plates. Page 83, pl. 2, fig. 4. Also figured in Pfeiffer et al. (2019a:113, fig. 2k).

Verbatim Type Locality: "Rio Usumasintæ, Guatemala."

Type Locality: Río Usumacinta (Gulf of México–Grijalva Dr.), El Petén Department, Guatemala [D=16.8246, -90.8822].

Type specimen: Holotype by monotypy, USNM 128991 (Johnson 1975:13; Pfeiffer et al. 2019a:113, fig. 2k).

Taxonomic History:

Quadrula (Rotundaria) guatemalensis Simpson, 1900 — Simpson (1900; 1914)
Quadrula guatemalensis Simpson, 1900 — de Leon (1912)
Psoronaias guatemalensis (Simpson, 1900) — Frierson (1927), Graf & Cummings (2021), Czaja et al. (2023)
Amblema (Psorula) guatemalensis (Simpson, 1900) — Haas (1930)
Psorula guatemalensis (Simpson, 1900) — Haas (1969), Graf & Cummings (2007), Pfeiffer et al. (2019a), Inoue et al. (2020a)

Elliptio divaricatus Marshall, 1926. New land and fresh-water mollusks from Central and South America. *Proceedings of the United States National Museum* 69(2638):1–12+3 plates. Page 5, pl. 1, figs. 1–2; pl. 3, fig. 3.

Verbatim Type Locality: "Finca de Providencia, Guatemala."

Type Locality: Finca de Providencia (Gulf of México–Grijalva–Usumacinta–San Pedro Dr.), El Petén Department, Guatemala [L=17.2277, -90.7828].

Type Specimens: Holotype, USNM 362398. Paratype, USNM 362399 (*n*=1); both by original designation (Johnson 1974b:9).

Remarks: The placement of this species is uncertain. As stated by Marshall (1926:6), "The nearest relative to this species is *Quadrula guatemalensis* Simpson. The genus *Quadrula* does not seem to be the proper allocation for *guatemalensis* nor for the present species, which is here placed doubtfully in the genus *Elliptio*, to remain until the true generic positions of the Mexican and Central American granulous Maiads [sic] are finally determined." It is placed here in *Psoronaias* following

Pfeiffer et al. (2019a:111) who looked at *Psoronaias guatemalensis* (Simpson, 1900) genetically and found it was sister to all the other *Psoronaias* he examined. There are numerous Finca de Providencia's in Guatemala, and the precise type locality of *E. divaricatus* is uncertain.

Elliptio (*Nephronaias*) *hermanni* Haas, 1929. Bemerkungen über mittelamerikanische Najaden. *Senckenbergiana* 11(5/6):310–344. Page 323, figs. 12–13. Also figured in Nagel (225:164, fig. 159).

Verbatim Type Locality: "Arroyo Limon, Staat Tabasco, Mexiko."

Type Locality: Arroyo Limón (Gulf of México–Grijalva–Usumacinta Dr.), Tabasco, México [W=17.8875, -92.5166].

Type Specimens: Holotype, SMF 3488, by original designation (Solem 1967a:114; Zilch 1967:102).

Remarks: All of the other material mentioned by Haas (1929:324) has been destroyed (Zilch 1967:102; Nagel 2025:164). All previous workers have placed this taxon under *Nephronaias*. The holotype, unlike other *Nephronaias*, is extremely pustulose and it is provisionally placed here. Unfortunately, a search for specimens in 2019 at the type locality (Arroyo Limón) to gather genetic data was unsuccessful.

Taxonomic History:

Nephronaias (s.s.) *hermanni* (Haas, 1929) — Haas (1969)

Nephronaias hermanni (Haas, 1929) — Zilch (1967), Graf & Cummings (2007; 2021), Czaja et al. (2023), Nagel (2025)

Psoronaias rubicunda (von Martens, 1900)

Plate 66, Map 78

Province(s): Gulf of Honduras & Northern Caribbean.

Distribution and Status: Known only from the type locality in the Río Cahabon (Gulf of Honduras Dr.), upstream of Lago de Izabal in Alta Verapaz Department, Guatemala. Not assessed (IUCN 2024).

Anatomy: Shell oblong, triangular or subrhomboid and solid with a well-developed posterior ridge. Beaks moderately high and positioned about two-fifths of the way from the anterior end of the shell. Posterior end truncated. Ventral edge slightly convex. Left valve with two strong pseudocardinals and two laterals. Surface with rather strongly concentrically striae. Periostracum light brown. Nacre white or light purple.

Biology: Unknown.

Adult Size: Length to 9cm.

Similar Species: *Nephronaias calamitarum* (Morelet, 1849), *Nephronaias ravistella* (Morelet, 1849), and *Nephronaias yzabalensis* (Crosse & Fischer, 1893) all occur in the Lago de Izabal basin.

Remarks: Described from a single valve there are two other specimens from the Río Cahabon, Coban, Alta Verapaz Department, Guatemala that may also be this species.

MAP 78 Distribution of *Psoronaias rubicunda* (von Martens, 1900).

Unio (*Elliptio*) *rubicundus* von Martens, 1900. Land and freshwater Mollusca. *In* F.D. Godman and O. Salvin (eds.). *Biologia Centrali-Americana* Part 155, Fascile 63–65:497–520, plates 33–35. Page 514, pl. 34, figs. 3–3a.

Verbatim Type Locality: "N. Guatemala: Coban."

Type Locality: Río Cahabon (Caribbean–Gulf of Honduras–Lake Izabal Dr.), Coban, Alta Verapaz Department, Guatemala [L = 15.4742, -90.3573].

Type Specimens: Holotype by monotypy, ZMB/MOLL.108853 ($n = ½$).

Taxonomic History:

Unio (sec. *Elliptio*) *rubicundus* von Martens, 1900 — Simpson (1914)
Elliptio (*Martensnaias*) *rubicundus* (von Martens, 1900) — Frierson (1927)
Elliptio rubicunda (von Martens, 1900) — Thiele (1934)
Nephronaias (*Martensnaias*) *rubicunda* (von Martens, 1900) — Haas (1969)
Martensnaias rubicunda (von Martens, 1900) — Graf & Cummings (2007; 2021)

PLATE 68 ***Psoronaias semigranosa* (von dem Busch in Philippi, 1843). Top Left:** *Unio semigranosus* von dem Busch in Philippi, 1843. Lectotype, MCZ 152861. **Middle Right:** Paralectotype, MCZ 409059. Both, Río Pánuco, near Tampico, Tamaulipas, México [L=22.2166, -97.8500]. **Bottom Left:** *Elliptio herrerae* Marshall, 1923. Holotype, USNM 347183 (71 mm L, 45 mm H). Estera del Avileño (Río San Fernando Dr.), Nuevo León, México [L=24.8444, -99.2583].

PLATE 69 ***Psoronaias semigranosa*** **(von dem Busch in Philippi, 1843).** INHS 90189 (38 mm L, 25 mm H). Río Huichihuayán (Pánuco–Moctezuma–Axtla Dr.), 8 km NE Xilitla, San Luis Potosí, México [21.4363, -98.9334]. Partially gravid female collected on 10 December 2018.

Psoronaias semigranosa (von dem Busch in Philippi, 1843)

Plates 68–69, Map 79

Province(s): Pánuco.

Distribution and Status: Known from the San Fernando, Carrizal, and Pánuco basins in Nuevo León, Querétaro, San Luis Potosí, and Tamaulipas. Not assessed (IUCN 2024). Collected alive at various sites in the Pánuco and San Fernando since 1990.

Anatomy: Shell triangular, elongate, and solid with a low double posterior ridge. Beaks low and their sculpture consisting of irregular, ridges. Beak cavity moderately deep. Ventral margin slightly curved and convex. Left valve with two pseudocardinals and two laterals. Right valve with two pseudocardinals, the upper smaller and one lateral. Surface covered with numerous small pustules. Periostracum yellow, olive, or brown. Nacre copper, pink, or purple. Rarely white.

Biology: Partially gravid (developing eggs, no glochidia) females were collected from the Río Tampaón (Río Pánuco Dr.), Álvaro Obregón, San Luis Potosí on 30 January 1982 (USNM 1282607) and from the Río Huichihuayán (Pánuco–Moctezuma–Axtla Dr.), 8 km NE Xilitla, San Luis Potosí, México on 10 December 2018 (INHS 90189, Plate 69).

Adult Size: Length to 11 cm.

Similar Species: None.

MAP 79 Distribution of *Psoronaias semigranosa* (von dem Busch in Philippi, 1843).

Unio semigranosus von dem Busch in Philippi, 1843. *Unio* Tab I. Abbildungen und Beschriebungen neuer oder wenig gekannter Conchylien. Cassel. Theodor Fischer 1:19–20. Page 19, pl. 1, figs. 1–3. Also figured in Reeve (1864, pl. 10, species 36, NHMUK 196496), von Martens (1900a, pl. 30, figs. 1–4, ZMB/MOLL.305826–305829), and Pfeiffer et al. (2019a:113, fig. 2i, MCZ 152861).

Verbatim Type Locality: "Flumen prope Tampico, Mexico."

Type Locality: Río Pánuco (Gulf of Mexico Dr.), near Tampico, Tamaulipas, México [L=22.2166, -97.8500].

Type Specimens: Lectotype, MCZ 152861, here designated (Johnson 1956:134, as paratypes; Pfeiffer et al. 2019a:113, fig. 2i). Paralectotype, MCZ 409059.

Taxonomic History:

Unio semigranosus von dem Busch in Philippi, 1843 — Conrad (1853), Hanley (1856), Küster (1861), Reeve (1864), de Leon (1912), Strecker (1931)

Margaron (*Unio*) *semigranosus* (von dem Busch in Philippi, 1843) — Lea (1852; 1870)

Unio (*Diplodon*) *semigranosus* von dem Busch in Philippi, 1843 — Adams & Adams (1857)

Unio semigranosa von dem Busch in Philippi, 1843 — Singley (1893)

Unio (*Psoronaias*) *semigranosus* von dem Busch in Philippi, 1843 — Fischer & Crosse (1894b), von Martens (1900a)

Unio (sec. *Elliptio*) *semigranosus* von dem Busch in Philippi, 1843 — Simpson (1900; 1914)

Quadrula semigranosa (von dem Busch in Philippi, 1843) — Pilsbry (1910)

Elliptio semigranosus (von dem Busch in Philippi, 1843) — H.B. Baker (1922)

Psoronaias semigranosus (von dem Busch in Philippi, 1843) — Frierson (1917; 1927), Haas (1930), Goodrich & van der Schalie (1937)

Psoronaias semigranosa (von dem Busch in Philippi, 1843) — Haas (1929), Graf & Cummings (2007; 2021) [in part], Pfeiffer et al. (2019a), Inoue et al. (2020a), Smith et al. (2020), Kiser et al. (2022), Czaja et al. (2023) [in part]

Psoronaias semigranosa semigranosa (von dem Busch in Philippi, 1843) — Haas (1969)

Elliptio herrerae Marshall, 1923. New pearly freshwater mussels from México and Uruguay. *Proceedings of the United States National Museum* 63(2485):1–4+3 plates. Page 1, pl. 1, figs. 6–8, pl. 2, figs. 4–5, pl. 3.

Verbatim Type Locality: "Estera del Avileno, in the State of Tamaulipas, México."

Type Locality: Estera del Avileño (Río San Fernando Dr.), Nuevo León, México [L=24.8444, -99.2583].

Type Specimens: Holotype, USNM 347183. Paratype, USNM 347184 ($n=1$); both by original designation (Johnson 1974b:10).

Remarks: Until recently, this species was known only from the three specimens from Río San Fernando basin (Gulf of Mexico drainage) in Nuevo León (not Tamaulipas as stated in the original description). Three additional lots, INHS 89433 ($n=11$), INHS 89427 ($n=1$, both ex. OSUM), and Bell Museum 4054 ($n=½$), extend the known range north to the Río San Fernando.

Taxonomic History:

Psoronaias herrerae (Marshall, 1923) — Frierson (1927), Haas (1929; 1930; 1969), Graf & Cummings (2007; 2021), Czaja et al. (2023)

Elliptio herrerae Marshall, 1924 — Contreras-Arquieta (2000)

Order Unionoida
Superfamily Unionoidea Rafinesque, 1820
Family Unionidae Rafinesque, 1820
Subfamily Ambleminae Rafinesque, 1820
Tribe Popenaiadini Heard & Guckert, 1970
Genus *Sphenonaias* Crosse & Fischer, 1894

Sphenonaias Crosse & Fischer, 1894. *In* P. Fischer and H. Crosse. 1894. Études sur les mollusques terrestres et fluviatiles du Mexique et du Guatemala. M.H.–Milne Edwards (ed.). *Mission Scientifique au Mexique et dans l'Amérique centrale. Recherches Zoologiques pour servir á l'Histoire de la Faune de l'Amérique Centrale et du Mexique.* 15th Livraison. 62–72, pp. 489–576, plates 59–62, Imprimerie Nationale, Paris Volume 2. 731 pp. Page 557. Introduced as subgenus of *Unio* Philipsson, 1788.

Diagnosis: "Coquille oblongue, allongée, ren flée en avant, rostrée en arrière; dents latérales antérieures épaisses (*U. Liebmanni*, Philippi)." [Shell oblong, elongated, rounded in front, rostrate behind; thick anterior lateral teeth (*U. Liebmanni*, Philippi)].

Type Species: *Unio liebmanni* Philippi, 1847, by original designation.

Taxonomic History:

Unio (sec. *Elliptio*) Rafinesque, 1819 — Simpson (1900; 1914)
Unio (*Sphenonaias*) Crosse & Fischer, 1894 — von Martens (1900b)
Elliptio (*Sphenonaias*) Crosse & Fischer, 1894 — Frierson (1927)
Elliptio (*Barynaias*) Crosse & Fischer, 1894 — Haas (1929), Thiele (1934)
Nephronaias (*Sphenonaias*) Crosse & Fischer, 1894 — Haas (1969)
Sphenonaias Crosse & Fischer, 1894 — Contreras-Arquieta (2000), Graf & Cummings (2007; 2021), Pfeiffer et al. (2019a–b), Inoue et al. (2020a)

Distribution & Remarks: The genus *Sphenonaias* is endemic to Papaloapán Province in México (Map 80). Graf and Cummings (2021:20) recognized six species of *Sphenonaias*: *Sphenonaias liebmanni* (Philippi, 1847), *S. callosa* (Lea, 1841), *S. colorata* (Küster, 1856), *S. flucki* (Bartsch, 1906), *S. mexicana* (Philippi, 1847), and *S. microdon* (von Martens, 1887). We consider *Sphenonaias callosa, S. colorata,* and *S. mexicana* as *Nomina dubia* below. We recognize three species in *Sphenonaias*: *S. liebmanni* (Philippi, 1847), *Sphenonaias opacata* (Crosse & Fischer, 1894), and *Sphenonaias plexus* (Conrad, 1838) here. The members of *Sphenonaias* are all solid, black, and triangular to elongate.

Unio pigerrimus Crosse & Fischer, 1894 has been cited as the type species for both *Barynaias* and *Psoronaias*, and we have placed it in the synonymy of *Psoronaias semigranosa* (von dem Busch in Philippi, 1843) and two former members of *Barynaias*: *B. opacata* (Crosse & Fischer, 1894) and *B. plexus* (Conrad, 1838) are now placed in *Sphenonaias* (Crosse & Fischer, 1894) (see comments in the account for *P. semigranosa* below for additional discussion).

Nephronaias flucki Bartsch, 1906 from Nicaragua was placed in *Sphenonaias* by Frierson (1927:38) noting that "The figure of this species was by an error reduced in size just half and being unnoted, leaves the impression that the shell is allied to the '*Unio calamitarum*, etc.' a resemblance not borne by the specimen." This opinion by Frierson is not consistent with our examination of the lectotype of *Unio calamitarum*, and we have returned it to *Nephronaias* where it was originally described.

Sphenonaias microdon (von Martens, 1887) was recovered as sister to *Mesonaias explicata* (Morelet, 1849) by Pfeiffer et al. (2019a:112), and we have placed it in *Mesonaias* as noted above. We also moved the southern former members of *Barynaias* (*Unio caldwellii* Lea, 1858 and *Unio sallei* Crosse & Fischer, 1894) back into *Nephronaias*.

Unfortunately, the type species of *Sphenonaias* (*Unio liebmanni* Philippi, 1847) has the ambiguous type locality of "Mexico." That and the fact that the type is lost makes it difficult to place this genus with other Central American mussels. The placement of *Sphenonaias liebmanni* has not been tested and attempts to secure live individuals of *Elliptio* (*Sphenonaias*) *liebmanni cuatotolapamensis* H.B. Baker, 1922 in the Papaloapán basin in 2022 were unsuccessful. It should be noted that the shells obtained in 2022 were extremely difficult to separate from *Psoronaias distincta.*

MAP 80 Distribution of the genus *Sphenonaias* Crosse & Fischer, 1894.

PLATE 70 ***Sphenonaias liebmanni*** **(Philippi, 1847). Top:** *Elliptio* (*Sphenonaias*) *liebmanni cuatotolapamensis* H.B. Baker, 1922. Holotype, UMMZ 31904 (81 mm L, 53 mm H). Río San Juan (Papaloapán Dr.), Cuatotolapam, Veracruz, México [L = 18.1241, -95.2998]. **Bottom:** ***Sphenonaias opacata*** **(Crosse & Fischer, 1894)**. *Unio opacatus* Crosse & Fischer, 1894, Holotype, MNHN-IM-2000-1755 (53 mm L, 36 mm H). México.

Sphenonaias liebmanni (Philippi, 1847)

Plate 70, Map 81

Province(s): Papaloapán.

Distribution and Status: Known from Gulf of Mexico drainages (primarily the Jamapa and Papaloapán) in Veracruz, México. Not assessed (IUCN 2024).

Anatomy: Shell solid, large, elongate, and subelliptical to subrhomboid with a well-developed posterior ridge. Ventral margin varies from convex to concave. Beaks not greatly elevated. Beak cavity moderately shallow. Anterior end rounded. Posterior end pointed. Dorsal margin curved behind the beaks, sometimes biangulate behind. Left valve with two pseudocardinals and two laterals. Right valve with one pseudocardinal and one lateral. Periostracum brown to black with green rays in some individuals. Nacre white, pink, salmon, or purple.

Biology: A slightly gravid female was collected from the Río Cotaxtla (Jamapa Dr.), Cotaxtla, Veracruz, México on 12 February 1982 (USNM 1282644–47).

Adult Size: Length to 9 cm.

Similar Species: *Psoronaias distincta* (Crosse & Fischer, 1893).

Remarks: H.B. Baker (1920:13) described the subspecies *Elliptio* (*Sphenonaias*) *liebmanni cuatotolapamensis*, from the Río San Juan in Veracruz, México. These specimens possess both a rugose or smooth periostracum, dark brown to black in color, with fine, radiating striations or pustules on the surface of shell that start high on the beaks and extend about a third of the way posteriorly. Individuals are difficult to separate from *Psoronaias distincta* (Crosse & Fischer, 1893) as pointed out by Baker (1922:10–14). Although we could list this with the other unidentifiable species from "Mexico," or perhaps under *Psoronaias distincta*, we will retain it here pending further analysis.

MAP 81 Distribution of *Sphenonaias liebmanni* (Philippi, 1847).

Unio liebmanni Philippi, 1847. Testaceorum novorum centuria. *Zeitschrift für Malakozoologie* 4(6):84–96. Page 96. Figured in Philippi (1849:27, pl. 6, fig. 1) and von Martens (1900b, pl. 34, figs. 1, 1a–b, ZMB/MOLL.11912).

Verbatim Type Locality: "Mexico."

Type Locality: México.

Type Specimens: Unknown. Presumed lost.

Remarks: This is an enigmatic species. The type locality is "México" so it could be any number of things. von Martens (1900b:500) noted that "Prof. F. Liebmann, of Copenhagen, travelled in México, in 1840–1843, not only on the eastern slope (Vera Cruz, Papantla, &c.), but on the western also (Acapulco, &c.)."

Taxonomic History:

Unio liebmanni Philippi, 1847 — Philippi (1849), Küster (1862), de Leon (1912)
Margaron (*Unio*) *liebmanni* (Philippi, 1847) — Lea (1870)
Unio (*Sphenonaias*) *liebmanni* Philippi, 1847 — Fischer & Crosse (1894b), von Martens (1900b)
Unio (sec. *Elliptio*) *liebmanni* Philippi, 1847 — Simpson (1900; 1914)
Elliptio (*Sphenonaias*) *liebmanni* (Philippi, 1847) — H.B. Baker (1922), Frierson (1927)
Elliptio (*Barynaias*) *liebmanni* (Philippi, 1847) — Haas (1929)
Nephronaias (*Sphenonaias*) *liebmanni* (Philippi, 1847) — Haas (1969)
Sphenonaias liebmanni (Philippi, 1847) — Graf & Cummings (2007; 2021), Coan & Kabat (2017), Pfeiffer et al. (2019a), Czaja et al. (2023)

Elliptio (*Sphenonaias*) *liebmanni cuatotolapamensis* H.B. Baker, 1922. The Mollusca collected by the University of Michigan-Walker Expedition in Southern Vera Cruz, México. I. Occasional Papers of the Museum of Zoology, University of Michigan No. 106. 95 pp. + 17 plates. Page 13, pl. 1, figs. 6–7 (young); pl. 3, fig. 22 (hinge); pl. 4., figs. 20–25, fig. 22, as the type specimen.

Verbatim Type Locality: "Rio San Juan."

Type Locality: Río San Juan (Gulf of México–Papaloapán Dr.), Cuatotolapam, Veracruz, México [L = 18.1241, -95.2998].

Type Specimens: Holotype, UMMZ 31904. Paratypes, ANSP 133689 (*n* = 2) (Johnson and Baker 1973:152).

Remarks: H.B. Baker (1922:13) identified and figured a "type" (without a catalogue number) and noted that there were "Twenty specimens, including odd valves." Frierson (1927:37) stated that "We list the last variety, following its author, but his figures would appear to ally his species to the '*Lampsilis signata* Pilsbry' rather than to either Marten's or Küster's figures." The two "young" specimens figured in Baker (plate 1, figs. 6–7) are very pustulose, and this taxon might be a junior synonym of *Psoronaias distincta* (Crosse & Fischer, 1893), which was also described from the Papaloapán basin.

Taxonomic History:

Elliptio (*Sphenonaias*) *liebmanni-cuatotolapamensis* H.B. Baker, 1922 — Frierson (1927)
Nephronaias (*Sphenonaias*) *liebmanni* (Philippi, 1847) — Haas (1969)

Sphenonaias opacata (Crosse & Fischer, 1894)

Plate 70, Map 82

Province(s): Papaloapán.

Distribution and Status: Known from a few localities in Veracruz, with almost all of the specimens coming from Laguna de Catemaco in the Papaloapán–Río Grande de Catemaco drainage where it was alive and common in 2022. Not assessed (IUCN 2024).

Anatomy: Shell triangular, solid, inflated, with a high, pronounced, curved posterior ridge and high beaks. Beak cavity rather deep. Anterior end rounded. Ventral margin straight or slightly convex. Left valve with two ragged pseudocardinals, that are partly joined together and two short, curved laterals. Right valve with two pseudocardinals, the upper generally smaller but sometimes equal to the lower in size, with one lateral and often a vestigial second. Periostracum dark brown or black and slightly rayed in the young individuals. Nacre white, pink, or light purple.

Biology: Found in mud and gravel along the lake shore. Females with glochidia collected in December (Morrison 1968:51).

Adult Size: Length to 5 cm.

Similar Species: *Sphenonaias plexus* (Conrad, 1838).

MAP 82 Distribution of *Sphenonaias opacata* (Crosse & Fischer, 1894).

Unio opacatus Crosse & Fischer, 1894. Diagnoses molluscorum novorum, reipublicæ Mexicanæ incolarum. *Journal de Conchyliologie* 41(4):293–297. Page 295. Figured in Fischer and Crosse (1894b:592, pl. 66, figs. 1–1a).

Verbatim Type Locality: "republica Mexicana."

Type Locality: México.

Type Specimens: Holotype by monotypy, MNHN-IM-2000-1755.

Remarks: Crosse and Fischer gave only a single set of measurements, and there are no other specimens in collections we examined with any type designations. The date of publication is often cited as 1893 but is 1894 according to Winckworth (1936:156).

Taxonomic History:

Unio (*Uniomerus*) *opacatus* Crosse & Fischer, 1893 — Fischer & Crosse (1894b)
Unio (*Elliptio*) *opacatus* Crosse & Fischer, 1893 — von Martens (1900b)
Unio (sec. *Elliptio*) *opacatus* Crosse & Fischer, 1893 — Simpson (1900; 1914)
Unio opacatus Crosse & Fischer, 1893 — de Leon (1912)
Plagiola (*Artonaias*) [sic] *opacata* (Crosse & Fischer, 1893) — H.B. Baker (1922)
Elliptio (*Barynaias*) *opacatus* (Crosse & Fischer, 1893) — Frierson (1927)
Elliptio opacata (Crosse & Fischer, 1893) — Morrison (1968)
Barynaias opacatus (Crosse & Fischer, 1893) — Haas (1969)
"Elliptio" opacatus (Crosse & Fischer, 1893) — Heard & Guckert (1970)
Barynaias opacata (Crosse & Fischer, 1893) — Contreras-Arquieta (2000), Graf & Cummings (2007; 2021), Czaja et al. (2023)

PLATE 71 ***Sphenonaias plexus*** **(Conrad, 1838). Top Left:** *Unio plexus* Conrad, 1838. Lectotype, ANSP 129923 (36 mm L, 19 mm H). Veracruz (Jamapa Dr.), Veracruz, México [L = 19.0944, -96.1370]. **Middle Right:** *Unio carbonarius* Lea, 1838. Lectotype, USNM 84080a (61 mm L, 36 mm H). Río Medellin (Jamapa Dr.), Veracruz, Veracruz, México [L = 19.0574, -96.1550]. **Middle Left:** *Unio pliciferus* Lea, 1838. Syntype, USNM 84080 (60 mm L, 38 mm H). México. **Bottom Right:** *Unio pliciferus* var. *β*, *minor* Fischer & Crosse, 1894. Holotype, MNHN-IM-2000-1812 (33 mm L, 23 mm H). Río Medellin (Jamapa Dr.), Veracruz, Veracruz, México [L = 19.0574, -96.1550].

Sphenonaias plexus (Conrad, 1838)

Plate 71, Map 83

Province(s): Papaloapán.

Distribution and Status: Endemic to the Río Jamapa basin in Veracruz, México. Not assessed (IUCN 2024).

Anatomy: Shell elongate, subrhomboid, somewhat inflated, and solid. Beaks low and beak cavity shallow. Anterior end rounded, ventral edge straight to slightly concave. Left valve with two stumpy pseudocardinals and two laterals. Right valve with one ragged pseudocardinal and one, often partly double lateral. Surface usually smooth but may be finely sculptured with radiating corrugations. Periostracum black or dark brown. Nacre white, pink, or light purple.

Biology: A gravid female was collected from the Rio Cotaxtla (Jamapa drainage), Cotaxtla, Veracruz, México on 12 February 1982 (USNM 1282318).

Adult Size: Length to 5.5 cm.

Similar Species: *Sphenonaias liebmanni* (Philippi, 1847) is also known from the Río Jamapa, and they may be conspecific.

MAP 83 Distribution of *Sphenonaias plexus* (Conrad, 1838).

Unio plexus Conrad, 1838. Monography of the family Unionidae, or naiades of Lamarck, (fresh water bivalve shells) or North America, illustrated by figures drawn on stone from nature. J. Dobson, 108 Chestnut Street, Philadelphia, Pennsylvania. 10:81–94 + plates 46–51. Page 89, pl. 49, figs. 1–2.

Verbatim Type Locality: "Vera Cruz."

Type Locality: Veracruz (Jamapa Dr.), Veracruz, México [L = 19.0944, -96.1370].

Type Specimens: Lectotype, ANSP 129923 designated by Johnson and Baker (1973:166). Paralectotypes, ANSP 412414 ($n = 2$) (ex. ANSP 129923).

Taxonomic History:

Margaron (Unio) pliciferus (Lea, 1838) — Lea (1852; 1870)
Unio plexus Conrad, 1838 — Conrad (1853), Pilsbry (1904), de Leon (1912), Contreras-Arquieta (2000)
Unio pliciferus Lea, 1838 — Sowerby (1868)
Unio (Psoronaias) pliciferus Lea, 1838 — Fischer & Crosse (1894b)
Unio (Elliptio) pliciferus Lea, 1838 — von Martens (1900b)
Unio (sec. *Elliptio) plexus* Conrad, 1838 — Simpson (1900; 1914)
Elliptio (Sphenonaias) plexus (Conrad, 1838) — H.B. Baker (1922)
Psoronaias plexus (Conrad, 1838) — Frierson (1927)
Elliptio (Barynaias) plexus (Conrad, 1838) — Haas (1929; 1930)
Barynaias plexus (Conrad, 1838) — Haas (1969), Graf & Cummings (2007; 2021), Czaja et al. (2023)

Unio carbonarius Lea, 1838. Description of new freshwater and land shells. *Transactions of the American Philosophical Society* 6(N.S.):1–154+plates 1–24. Page 37, pl. 11, fig. 32.

Verbatim Type Locality: "River Medellin, México."

Type Locality: Río Medellin (Gulf of México–Jamapa Dr.), Veracruz, Veracruz, México [L=19.0574, -96.1550].

Type Specimens: Lectotype, USNM 84080a designated by Johnson and Baker (1973:150; Johnson 1974a:27, both as the holotype). Paralectotype, ANSP lost (Johnson and Baker 1973:150).

Taxonomic History:

Margarita (Unio) carbonarius — Lea (1836) *nomen nudem*
Unio carbonarius Lea, 1838 — Hanley (1843; 1856)
Margaron (Unio) pliciferus (Lea, 1838) — Lea (1852; 1870)
Unio plexus Conrad, 1838 — Conrad (1853), de Leon (1912)
Unio pliciferus Lea, 1838 — Sowerby (1868)
Unio (Psoronaias) pliciferus Lea, 1838 — Fischer & Crosse (1894b)
Unio (Elliptio) pliciferus var. *carbonarius* Lea, 1838 — von Martens (1900b)
Unio (sec. *Elliptio) plexus* Conrad, 1838 — Simpson (1900; 1914)
Psoronaias plexus (Conrad, 1838) — Frierson (1927)
Barynaias plexus (Conrad, 1838) — Haas (1969)

Unio pliciferus Lea, 1838. Description of new freshwater and land shells. *Transactions of the American Philosophical Society* 6(N.S.):1–154+plates 1–24. Page 61, pl. 17, fig. 53. Also figured in Fischer and Crosse (1894b:580, pl. 65, fig. 2).

Verbatim Type Locality: "México."

Type Locality: México.

Type Specimens: Figured "type" lost (Johnson and Baker 1973:167). Syntypes, USNM 84080 (*n*=3½) (Johnson and Baker 1973:167; Johnson 1974a:115).

Remarks: Johnson (1956:130) also mentioned MCZ 6584 (*n*=7) as ideotypes.

Taxonomic History:

Margarita (Unio) pliciferus — Lea (1836) *nomen nudem*
Unio pliciferus Lea, 1838 — Hanley (1843; 1856), Küster (1856), Sowerby (1868), Wright (1888)
Margaron (Unio) pliciferus (Lea, 1838) — Lea (1852; 1870)

Unio plexus Conrad, 1838 — Conrad (1853)
Unio (*Diplodon*) *pliciferus* Lea, 1838 — Adams & Adams (1857)
Unio (*Psoronaias*) *pliciferus* Lea, 1838 — Fischer & Crosse (1894b)
Unio (sec. *Elliptio*) *plexus* Conrad, 1838 — Simpson (1900; 1914)
Unio (*Elliptio*) *pliciferus* Lea, 1838 — von Martens (1900b; 1901c)
Unio plexus Conrad, 1838 — de Leon (1912)
Psoronaias plexus (Conrad, 1838) — Frierson (1927)
Barynaias plexus (Conrad, 1838) — Haas (1969)

Unio pliciferus var. *β, minor* Fischer & Crosse, 1894. Études sur les mollusques terrestres et fluviatiles du Mexique et du Guatemala. 16th Livraison. Feuilles 73–82, pp. 577–656, plates 63–66 in Recherches Zoologiques pour servir À l'Histoire de la Faune de l'Amérique Centrale et du Mexique, Imprimerie Nationale, Paris Volume 2. 731 pp. Page 581, Plate 69, fig. 4. Also figured in von Martens (1900e, pl. 39, figs. 4–4a, ZMB/MOLL.305845).

Verbatim Type Locality: "Rio Medellin, État de Vera Cruz, Mexique."

Type Locality: Río Medellin (Gulf of México–Jamapa Dr.), Veracruz, Veracruz, México [L=19.0574, -96.1550].

Type Specimens: Holotype by monotypy, MNHN-IM-2000-1812.

Taxonomic History:

Unio (*Psoronaias*) *pliciferus* var. *β, minor* Fischer & Crosse, 1894 — Fischer & Crosse (1894b)
Unio (*Elliptio*) *pliciferus* var. *minor* Fischer & Crosse, 1894 — von Martens (1900b)
Unio (sec. *Elliptio*) *plexus* var. *minor* Fischer & Crosse, 1894 — Simpson (1914)

Order Unionoida
Superfamily Unionoidea Rafinesque, 1820
Family Unionidae Rafinesque, 1820
Subfamily Ambleminae Rafinesque, 1820
Tribe Quadrulini von Ihering, 1901
Genus *Megalonaias* Utterback, 1915

Megalonaias Utterback, 1915. The naiades of Missouri. II. *American Midland Naturalist* 4(4):97–152. Page 123.

Diagnosis: "Shell large, ponderous, broadly rhomboid, moderately inflated, post-dorsal ridge alated, sculptured with regular upcurved undulations; post-umbonal ridge broken with coarse plications running more or less parallel with it; beaks rather low, sculptured with coarse double looped-corrugations which extend out as nodules at base of post-ridge and as zigzag ridges all over umbonal region to upper part of disk; epidermis black; cardinals heavy; laterals long and straight; interdentum short; beak cavities narrowly deep; scars very deeply impressed—especially anterior retractor cicatrix; nacre white to pink.

Branchial opening very large with short papillae; anal and supra-anal also large, almost smooth, separated by short but distinct mantle connection; inner laminae of inner gills partly free from visceral mass or sometimes almost entirely connected; palpi long, enormous; color of soft parts, tan colored, with gills brownish.

Marsupia occupying all four gills, when gravid enormous, pad-like, not so distended at ventral edge; conglutinates, sole-shaped, brown, rather solid; glochidia large, ventral margin obliquely rounded, hinge line long" (Utterback 1915:123–124).

Type Species: *Unio heros* Say, 1829, by original designation.

Taxonomic History:
Unio (*Plectomerus*) Conrad, 1853 — Fischer & Crosse (1894a)
Quadrula (sec. *Crenodonta*) Schlüter, 1838 — Simpson (1900; 1914), Burch (1975)
Unio (*Crenodonta*) Schlüter, 1838 — von Martens (1900a)
Crenodonta Schlüter, 1838 — Ortmann (1912), Haas (1929)
Amblema (s.s.) Rafinesque, 1820 — Frierson (1927)
Amblema (*Megalonaias*) Utterback, 1915 — Frierson (1927), Haas (1930)
Amblema (sec. *Megalonaias*) Utterback, 1915 — Thiele (1934)
Megalonaias Utterback, 1915 — Haas (1969), Lopes-Lima et al. (2019), Contreras-Arquieta (2000), Graf & Cummings (2007; 2021), Pfeiffer et al. (2019b), Nagel (2025)

Magnonaias Utterback, 1915, Utterback, 1915. The naiades of Missouri. II. *American Midland Naturalist* 4(3):41–53. Page 47.
Remarks: Error for *Megalonaias* Utterback, 1915.

Distribution & Remarks: The tribe Quadrulini is represented in Central America by only one geographically widespread genus, *Megalonaias* (Map 84, Pfeiffer et al. 2019a:112). *Megalonaias* is distributed from eastern North America south and west to the Usumacinta basin in Guatemala. Pfeiffer et al. (2018) sank all the described species of *Megalonaias* into *M. nervosa* (Rafinesque, 1820). However, the clade for central and southern México was distinct and came out as sister to all other *Megalonaias* examined (Pfeiffer et al. 2018:144, fig. 2). Regarding the validity of the two taxa, von Martens (1900a:492) summed it up thusly "The descriptions and figures of Fischer and Crosse [referring to *nickliniana* of the Río Moctezuma] and Lea [referring to *eightsii* of the Rio Grande] evidently refer to one and the same species. But as regards the localities, it is scarcely credible that such a characteristic and large-sized *Unio* should live in two rivers so very far distant one from the other, and not have been found in the intermediate ones, as, for instance, in the Pánuco or the Medellin. As to the locality given by Morelet there is no reason to doubt it, as he was himself in that country; but that given by Lea, 'Rio Sabinas,' may be a mistake for Rio de las Salinas, the name of the upper course of the Usumacinta. There is, however, in N.E. México, a town of Sabinas as well as a Rio Sabinas, on the confines of the States of Coahuila and Nuevo Leon, this river being a confluent of the Rio Salado, which discharges itself into the Rio Grande del Norte; there is also a Rio Salinas in the State of Nuevo Leon, a confluent of the Rio San Juan. I must therefore leave this difficulty for future elucidation." It should be noted that von Martens was mistaken in saying that "not have been found in the intermediate ones, as, for instance, in the Pánuco or the Medellin as the Moctezuma is in the Pánuco basin." There are two lots without precise locality data: MCZ 30507, from Honduras and FMNH 50540, from Nicaragua. These records are regarded as spurious until new data can verify its presence in those countries.

MAP 84 Distribution of the genus *Megalonaias* Utterback, 1915.

PLATE 72 ***Megalonaias nickliniana* (Lea, 1834). Top Left:** *Unio nicklinianus* Lea, 1834. Holotype, USNM 83898 (130 mm L, 125 mm H). Río Moctezuma (Río Pánuco Dr.), near Chintón, San Luis Potosí, México [L=21.9672, -98.5618]. **Middle Right:** *Unio digitatus* Morelet, 1851. Lectotype, NHMUK 1893.2.4.2035 (125 mm L, 87 mm H). Río Usumacinta, El Petén Department, Guatemala [W=16.8246, -90.8822]. **Bottom Left:** *Unio* (*Crenodonta*) *stolli* von Martens, 1900. Holotype, ZMB/MOLL.39464 (128 mm L, 93 mm H). Río de las Salinas (Usumacinta Dr.), El Petén Department, Guatemala [W=16.4811, -90.5441].

Megalonaias nickliniana (Lea, 1834)

Plate 72, Map 85

Province(s): Pánuco; Papaloapán; Usumacinta.

Distribution and Status: *Megalonaias nickliniana* (Lea, 1834) is known from the three large Central American basins draining to the Gulf of México: Pánuco, Papaloapán, and the Usumacinta. Considered endangered by the Méxican Secretariat of Environment and Natural Resources (Semarnat 2019). Not assessed (IUCN 2024). It appears to be rare. It was not found in recent surveys of the Río Panucó in 2017–18. Collected in the Río San Juan in the Papaloapán basin in 1910, it was not found in a limited survey in that basin in 2022. The only recent live specimens we are aware of were collected from the Río de la Passion (Usumacinta Dr.), El Petén Department, Guatemala in 2015. It was not found in the Usumacinta in a survey conducted in 2019. There is a specimen (FMNH 50540) labeled as coming from Nicaragua without additional locality data that we regard as spurious.

Anatomy: Shell large, trapezoid, and solid. Beaks somewhat elevated and slightly inflated. Beak cavity very deep. Anterior end rounded. Ventral margin curved and convex. A large dorsal wing is typically present, which is somewhat obliquely truncated posteriorly. The posterior end has a few strongly developed, radial folds. The beak is sculptured with broken corrugations, which change into irregular, radial wrinkles and pustules below. Pseudocardinals rough. Laterals short. Surface of the shell is heavily sculptured with strong concentric ridges. Periostracum brown to black and rayless. Nacre white to light purple.

Biology: Unknown. In North America, various fish hosts for *Megalonaias nervosa* (Rafinesque, 1820) have been reported and include members of the families Anguillidae, Centrarchidae, Clupeidae, Cyprinidae, Ictaluridae, Lepisosteidae, Moronidae, Percidae, and Sciaenidae (O'Dee and Watters. 2000).

Adult Size: Length to 20 cm.

Similar Species: None.

MAP 85 Distribution of *Megalonaias nickliniana* (Lea, 1834).

Unio nicklinianus Lea, 1834. Observations on the naiades; and descriptions of new species of that and other families. *Transactions of the American Philosophical Society* 5(N.S.):23–119+19 plates. Page 28, pl. 1, fig. 1. Also figured in Lea (1834, Obs. 1:140, pl. 1, fig. 1).

Verbatim Type Locality: "China."

Type Locality: Río Moctezuma (Río Pánuco Dr.), near Chintón, San Luis Potosí, México [L=21.9672, -98.5618].

Type specimen: Holotype by monotypy, USNM 83898 (*n*=½) (Johnson 1974a:99).

Remarks: There is also a topotypic series of 8½, specimens (USNM 83898a). *Unio nicklinianus* Lea, 1834 was initially described from "China," which is surely an erroneous interpretation of the writing on the label with the holotype "Canton? Moctezuma R., Central Am." Noted by von Martens (1900a:493) as Rio Moctezuma, an affluent of the southern Rio Panucó. Corrected to Chintón, San Luis Potosí by Pfeiffer et al. (2018:142).

Taxonomic History:

Margarita (*Unio*) *nicklinianus* (Lea, 1834) — Lea (1836)
Unio nicklinianus Lea, 1834 — Hanley (1843; 1856), Sowerby (1866)
Margaron (*Unio*) *nicklinianus* (Lea, 1834) — Lea (1852; 1870)
Plectomerus nicklinianus (Lea, 1834) — Conrad (1853)
Unio (*Metaptera*) *nicklinianus* Lea, 1834 — Adams & Adams (1857)
Unio nicklianus [sic] Lea, 1834 — Küster (1861)
Unio (*Plectomerus*) *nicklinianus* Lea, 1834 — Fischer & Crosse (1894b)
Unio (*Crenodonta*) *nicklinianus* Lea, 1834 — von Martens (1900a)
Quadrula (sec. *Crenodonta*) *nickliniana* (Lea, 1834) — Simpson (1900; 1914)
Quadrula nickliniana (Lea, 1834) — Pilsbry (1910), de Leon (1912)
Amblema (*Megalonaias*) *nickliniana* (Lea, 1834) — H.B. Baker (1922), Frierson (1927)
Crenodonta nickliniama [sic] (Lea, 1834) — Haas (1929)
Megalonaias nickliniana (Lea, 1834) — Morrison (1968), Contreras-Arquieta (2000), Graf & Cummings (2007; 2021), Kiser et al. (2022), Ramírez et al. (2022), Czaja et al. (2023)
Megalonaias nickliniana nickliniana (Lea, 1834) — Haas (1969)
Megalonaias nervosa (Rafinesque, 1820) — Pfeiffer et al. (2018), Ramírez et al. (2022)

Unio digitatus Morelet, 1851. Testacea novissima insulae Cubanae, et Amer. Centralis. Librairie de L'Académie Nationale de Medecine, Paris Part 2:1–30. Page 24. Figured in Fischer and Crosse (1894a:563, pl. 60, fig. 1).

Verbatim Type Locality: "flumen Usumasinta."

Type Locality: Río Usumacinta (Gulf of México–Grijalva Dr.), El Petén Department, Guatemala [W=16.8246, -90.8822].

Type Specimens: Lectotype, NHMUK 1893.2.4.2035 designated by Johnson (1971:82, as the holotype; Breure et al. 2018:271, figs. 342–343, as a syntype).

Remarks: Breure et al. (2018:271) pointed out that Morelet did not state on how many specimens his description was based upon and called NHMUK 1893.2.4.2035 a syntype not realizing that a lectotype had already been designated. Morelet provided only one set of measurements "Longit. 125. — Altit. 92.," and the catalogue entry in the NHMUK indicates that there is only one specimen in the lot. It is probably a holotype by monotypy.

Unio digitatus Morelet, 1851 was incorrectly stated to be the type species of *Plectomerus* Conrad, 1853 by Fischer and Crosse (1894b:556). Regarding *Unio digitatus* Morelet, 1851 and *Unio eightsii* Lea, 1860, von Martens (1900a:492) stated that "The descriptions and figures of Fischer and Crosse and Lea refer to one and the same species. But as regards the localities, it is scarcely credible that such a characteristic and large-sized *Unio* should live in two rivers so very far distant from the other, and not have been found in the intermediate ones, as, for instance, in the Panuco or the Medellin. As to the locality given by Morelet, there is no reason to doubt it, as he was himself in that country; but that given by Lea 'Rio Sabinas,' may be a mistake for Rio de las Salinas, the name of the upper course of the Usumacinta. There is, however, in N.E. Mexico, a town of Sabinas as well Rio Sabinas, on the confines of States of Coahuila and Nuevo Leon, this river being a confluent of the Rio Salado, which discharges itself into the Rio Grande del Norte; there is also a Rio Salinas in the State Nuevo Leon, a confluent of the Rio San Juan. I must therefore leave this difficulty for future elucidation."

Taxonomic History:

Unio digitatus Morelet, 1851 — Wright (1888)
Unio (*Plectomerus*) *digitatus* Morelet, 1851 — Fischer & Crosse (1894b)
Unio (*Crenodonta*) *digitatus* Morelet, 1851 — von Martens (1900a)
Quadrula (sec. *Crenodonta*) *digitata* (Morelet, 1851) — Simpson (1900; 1914)
Quadrula digitata (Morelet, 1851) — de Leon (1912)
Amblema (*Megalonaias*) *nickliniana-digitata* (Morelet, 1851) — Frierson (1927)
Crenodonta digitata (Morelet, 1851) — Haas (1929)
Amblema digitata (Morelet, 1851) — Goodrich & van der Schalie (1937)
Megalonaias digitata (Morelet, 1851) — Morrison (1968), Contreras-Arquieta (2000)
Megalonaias nickliniana digitata (Morelet, 1851) — Haas (1969)
Megalonaias nervosa (Rafinesque, 1820) — Pfeiffer et al. (2018)

Unio (*Crenodonta*) *stolli* von Martens, 1900. Land and freshwater Mollusca. *In* F.D. Godman and O. Salvin (eds.). *Biologia Centrali-Americana* Part 154, Fascile 61–62:481–496, plates 29–32. Page 492, pl. 29, fig. 2.

Verbatim Type Locality: "Rio de las Salinas."

Type Locality: Río de las Salinas (Gulf of México–Grijalva–Usumacinta Dr.), El Petén Department, Guatemala [W=16.4811, -90.5441].

Type specimen: Holotype by monotypy, ZMB/MOLL.39464.

Taxonomic History:

Quadrula (sec. *Crenodonta*) *stolli* (von Martens, 1900) — Simpson (1900; 1914)
Quadrula stolli (von Martens, 1900) — de Leon (1912)
Amblema (*Megalonaias*) *nickliniana-digitata* (Morelet, 1851) — Frierson (1927)
Crenodonta stolli (von Martens, 1900) — Haas (1929)
Megalonaias nickliniana digitata (Morelet, 1851) — Haas (1969)
Megalonaias nervosa (Rafinesque, 1820) — Pfeiffer et al. (2018)

Order Unionoida
Superfamily Etherioidea Deshayes, 1832
Family Mycetopodidae Gray, 1840
Subfamily Mycetopodinae Gray, 1840
Genus *Mycetopoda* d'Orbigny, 1835

Mycetopoda d'Orbigny, 1835. Synopsis terrestrium et fluviatilium molluscorum, in suo per Americam meridionalem itinere, ab. A. d'Orbigny, collectorum. *Revue et Magasin de Zoologie Pure et Appliquée (Paris)* 5(61–62): 1–44. Page 41.

Diagnosis: "Testa elongatissima, soleniformi, æquivalvi, inæquilitera, anterius hiante, muscularibus impressionibus anterius magnopere divisis. Corpore pallio simplici, ad totam longitudineum aperto, nullo siphone proeminente; apatura anali distincta, pede longissimo cylindraceo, ad extremam partem inflato, in testa non contractili, bucca grande, appendicibus labialibus elongatis; branchiis elongatissimis. –Perforat, sicut pholade." [Shell very elongate, soleniform, equivalve, unequal, gaping anteriorly, muscular impressions greatly divided anteriorly. The body is covered with a simple mantle, open to the whole length, with no protruding siphon; distinct anal opening, with a very long cylindrical foot, inflated to the extreme, in a non-contractile shell, large mouth, elongated labial appendages; with the most elongated gills. –Perforate, like a Pholade].

Type Species: *Mycetopoda soleniformis* d'Orbigny, 1835.

Remarks: *Mycetopoda soleniformis* d'Orbigny is the first species listed, but he did not explicitly designate a type species. Herrmannsen (1847b:77), by subsequent designation according to Johnson (1981:282), as *Anodontis siliquosi* [sic] Spix, 1827.

Taxonomic History:

Mycetopoda d'Orbigny, 1835 — Sowerby (1839), Gray (1847), Herrmannsen (1847b), Simpson (1900; 1914), von Ihering (1910), Ortmann (1921b), Frierson (1922; 1927), H.B. Baker (1930), Haas (1931; 1969), Thiele (1934), Modell (1942; 1964), Johnson (1981), Graf (2000), Simone (2006), Graf & Cummings (2007; 2021), Lasso et al. (2009), Bouchet & Rocroi (2010), Lopes-Lima et al. (2017; 2020), Breure et al. (2022), Nagel (2025)

Spatha Lea, 1838 — Herrmannsen (1848)

Mycetopus d'Orbigny, 1840. Voyage dans l'Amérique Méridionale [...] exécuté pendant les années 1826, 1827, 1828, 1829, 1830, 1831, 1832 et 1833. Mollusques 1835–1843. Paris, Chez P. Bertrand, Éditeur, Libraire de la Société Géologique de France. Strasbourg, Chez V. Levrault. Tome V, part 3, Mollusques 758 pp. Plate 66.

Remarks: Error for *Mycetopoda* d'Orbigny, 1835. Fischer (1886:1003) referenced "*Mycetopus* d'Orbigny, 1840" and "*M. soleniformis*, d'Orbigny, Pl. XVIII, fig. 6." d'Orbigny's use of the name *Mycetopus* was the illustration of *Mycetopus soleniformis* on plate 66, figs. 1–3 of the Voyage dans l'*Amérique Méridionale* published in 1840, and page 600 of the text published in 1846 (d'Orbigny 1846a) (see Coan and Kabat 2024a:41–42 for the publication dates of the various parts).

Taxonomic History:

Mycetopus d'Orbigny, 1840 — Gray (1840; 1847), Swainson (1840), Reeve (1842), Hanley (1843), Herrmannsen (1847b), Conrad (1853), Adams & Adams (1857), Hupé (1857), Fischer (1886), von Ihering (1890), von Martens (1900e)

Spatha Lea, 1838 — Herrmannsen (1848)

Platris (*Mycetopus*) d'Orbigny, 1840 — Lea (1852; 1870)

Mycetopoda d'Orbigny, 1835 — Simpson (1900; 1914), Thiele (1934), Modell (1964), Haas (1931; 1969), Simone (2006), Graf & Cummings (2006)

Mycetopus d'Orbigny, 1846. Voyage dans l'Amérique Méridionale (le Brésil, la république Orientale de l'Uruguay, la République Argentine, la Patagonie, la république du Chili, la république de Bolivia, la république du Pérou), exécuté pendant les années 1826, 1827, 1828, 1829, 1830, 1831, 1832 et 1833. Pp. 529–600 *In* Tome V, part 3, Mollusques. Paris, Chez P. Bertrand, Éditeur, Libraire de la Société Géologique de France. Strasbourg, Chez V. Levrault. Page 600.
Remarks: Error for *Mycetopoda* d'Orbigny, 1835.

Mycetopodus d'Orbigny. Mousson, A. 1869. Notiz über einige von Herrn Gustav Wallis aus dem nördlichen Südamerika zurückgebrachte Mollusken. *Malakozoologische Blätter* 16:170–189. Page 189.
Remarks: Error for *Mycetopoda* d'Orbigny, 1835.

Micetopus Heude, R.P. 1877. Conchyologie fluviatile de la province de Nanking et de la Chine centrale. Troisiéme Fascicule. Librairie F. Savy, Paris, France Part III. plates 1724. Plate 21.
Remarks: Error for *Mycetopus* d'Orbigny, 1840.

Mycetodopa Sterki, V. 1901. [Review] Synopsis of the naiades, or pearly fresh water mussels, by Charles Torrey Simpson. *Nachrichtsblatt der Deutschen Malakozoologischen Gesellschaft* 33:145–148. Page 146.
Remarks: Error for *Mycetopoda* d'Orbigny, 1835.

Distribution & Remarks: The genus *Mycetopoda* is wide ranging and occurs from the Río Motagua basin in Guatemala, Central America south to the Río de La Plata basin in Argentina (Map 94). We have elevated *Mycetopoda subsinuata* (Sowerby, 1868) from the synonymy of *M. siliquosa* (Spix in Wagner, 1827), and it is the lone member of the genus *Mycetopoda* in Central America.

Mycetopoda is so named because of its large mushroom-shaped foot. *Mycet*=mushroom, *poda* = foot (Figures 42–43). They use the foot to anchor into clay banks and will quickly retract it

FIGURE 42 *Mycetopoda soleniformis* d'Orbigny, 1835. Río Guayabero, Raudal de Angosturas I, 2 km NW Miraflores, Meta Department, Colombia. 2.2927, -73.8791. (Photo by Mark Sabaj, ANSP).

when touched. Frierson (1927:98) noted that "A.A. Hinkley found shells in the Conchens [sic] River, and when these were pulled out of their clayey beds, often the mushroomed 'foot' was left behind."

FIGURE 43 Specimens of *Mycetopoda subsinuata* (Sowerby, 1868) collected Río Titihuapa (Pacific–Lempa Dr.), Cabañas Department, El Salvador. (Photo by Kentaro Inoue).

MAP 86 Distribution of *Mycetopoda* in Central America and Northwestern South America.

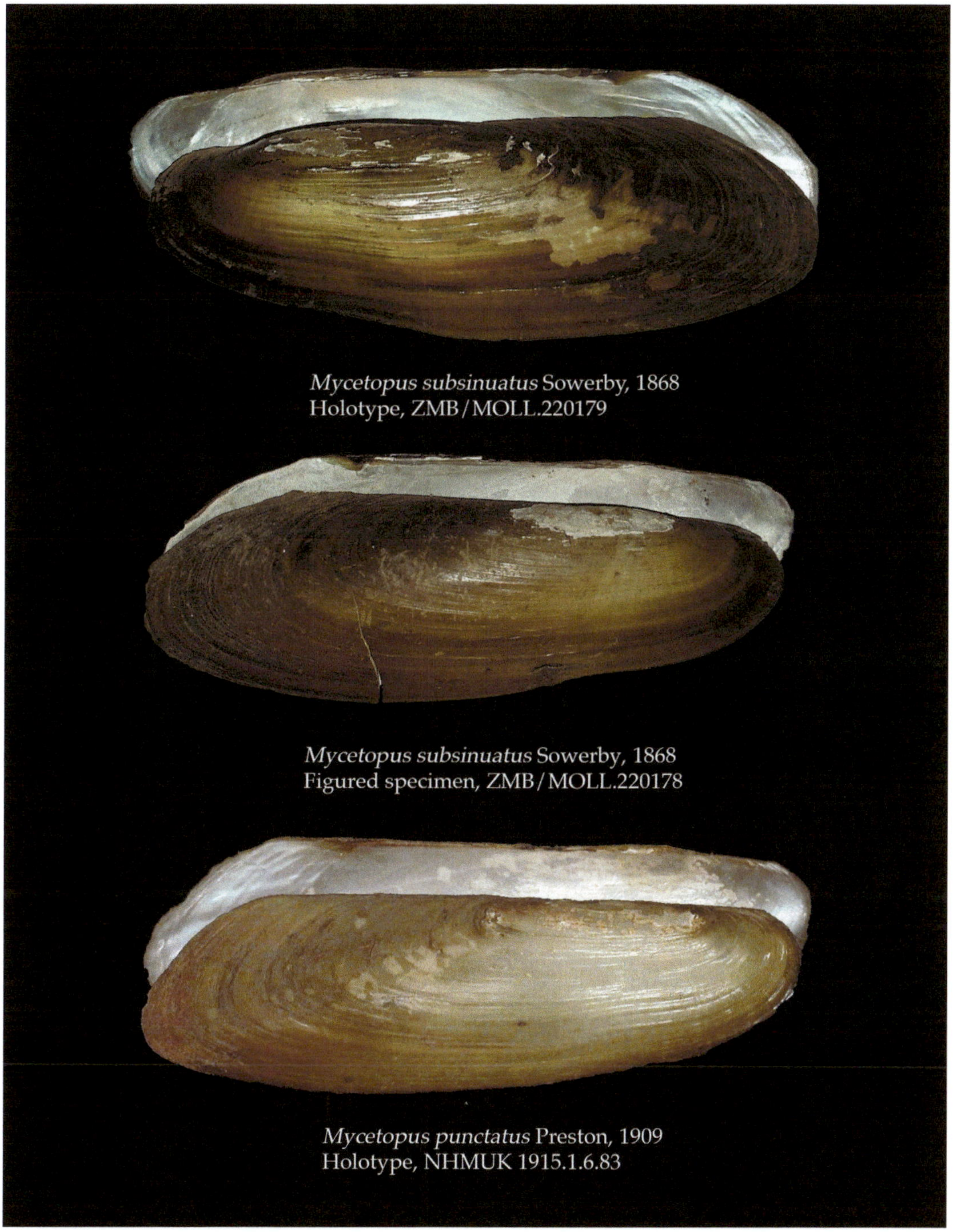

PLATE 73 ***Mycetopoda subsinuata* (Sowerby, 1868).** **Top:** *Mycetopus subsinuatus* Sowerby, 1868. Holotype, ZMB/ MOLL.220179 (118 mm L, 42 mm H). **Middle:** ZMB/MOLL. 220178 (132 mm L, 43 mm H). Río Michatoya (Pacific Dr.), Paso San Antonio, Escuintla Department, Guatemala [L=14.0617, -90.6809]. Figured in von Martens 1900:508, Pl. 41, figs. 5–5a. **Bottom:** *Mycetopus punctatus* Preston, 1909. Holotype, NHMUK 1915.1.6.83 (72 mm L, 22 mm H). Top and bottom both, Río Bogotá (Magdalena Dr.), Bogotá, Distrito Capital, Colombia [L=4.6000, -74.0833].

Mycetopoda subsinuata (Sowerby, 1868)

Plate 73, Map 87

Province(s): Gulf of Honduras & Northern Caribbean; San Juan; Oaxaca, México to Colombia.

Distribution and Status: The only member of the genus *Mycetopoda* in Central America. Known from Lake Izabal in the Río Motagua and Río San Juan–Lake Nicaragua drainage (both Caribbean Dr.) in Nicaragua, and the Río Michatoya in Guatemala, the Río Grande de San Miguel, and Río Lempa basins in El Salvador (all Pacific Dr.) southeast to Colombia where it occurs in the Río Magdalena basin and at Cartagena, Bolívar Department (Caribbean drainage). Although no records exist from Panamá, it will likely be found there. Listed as endangered by El Salvador (Ministério de Medio Ambiente y Recursos Naturales 2023:39). Not assessed (IUCN 2024).

Anatomy: Shell very thin, elongate, slightly inflated, and edentate. Beaks only moderately elevated. Hinge line straight behind the beaks, slightly incurved in front. Anterior end evenly rounded. Posterior end slightly pointed. Periostracum smooth yellow or olive and rayless. Nacre bluish white.

Ortmann (1921b:636) described the soft parts of one male and two females (one gravid with eggs) (CM 61.6199) from the Río Motagua Guatemala, collected by A.A. Hinkley on 6 February 1913 as follows: "The anatomy is that of the South American *Mutelinae*, [Mycetopodidae] but it resembles that of the African *Mutelinae*, [Iridinidae] in having the anal opening closed above for about half of its length. The opening is therefore comparatively short, being only a little longer than the branchial opening, and reaching upward only to about the middle of the adductor muscles. In other South American *Mutelinae* it reaches upward beyond the posterior retractor-muscles. The anal is separated from the branchial opening by a connection of the mantle; its inner edge is smooth. Branchial opening with fine papillae on the inner edge. Although Simpson says that the branchial opening is closed below, I cannot find any trace of a mantle-connection at the lower (anterior) end. There is also no indication that such a connection has been torn during life, or in preservation. Palpi long and low, their lower margins curved, not drawn out into a posterior point, posteriorly with a short truncation, forming the posterior margins, which are not connected.

Foot very large, subcylindrical and subcompressed, at the distal end swollen into a button-like knob. This structure is likewise not seen in other *Muteline* shells. Gills long and narrow, the inner the wider, the anterior ends as usual. Inner lamina of inner gill entirely connected with abdominal sac. Distinct, continuous septa of the *Muteline* type are present. The inner gill of the female is marsupial, with thicker, but not more crowded, septa; the water-tubes are again divided by vertical ridges into an inner and outer compartment, the inner of which serves as ovisac, the outer as secondary water-canal. Eggs small, loosely hanging together. I have not been able to find mature larvae in my gravid female."

Ortmann noted that the inner gill of the female was marsupial, and that the species had a closed anal opening, in which it differed from all South American Mycetopodidae and resembled some African members of the family Iridinidae.

Biology: Members of the genus *Mycetopoda* are typically found embedded in muddy stream banks. Two females (one gravid with eggs) (CM 61.6199) were collected from the Rio Motagua (Caribbean–Gulf of Honduras Dr.), Quirigua, Izabal Department, Guatemala on 6 February 1913. "Eggs small, loosely hanging together. I have not been able to find mature larvae in my gravid female" (Ortmann 1921b:637).

Adult Size: Length to 12 cm.

Similar Species: Nothing.

Remarks: Considered valid by Ortmann (1921b:635) and Haas (1931:104). Ortmann noted "that there is no question about the identity of the Central America specimens with the form from Northern South America."

MAP 87 Distribution of *Mycetopoda subsinuata* (Sowerby, 1868).

Mycetopus subsinuatus Sowerby, 1868. Monograph of the genus *Mycetopus In* L. Reeve and G.B. Sowerby (eds.). *Conchologia Iconica* 16: [unpaginated]+Plates 1–4. Plate 4, species 10. Also figured in von Martens (1900f, pl. 41, figs. 5–5a, ZMB/MOLL.220178, from Stoll).

Verbatim Type Locality: "Bogota, New Granada."

Type Locality: Río Bogotá (Magdalena Dr.), Bogotá, Bogotá, Distrito Capital, Colombia [L=4.6000, -74.0833].

Type Specimens: Holotype by monotypy, ZMB/MOLL.220179 (Sowerby's figured specimen).

Remarks: Although Sowerby's figured specimen (ZMB/MOLL.220179) is in the ZMB, von Martens figured a specimen (ZMB/MOLL.220178) from the Rio Michatoya (Pacific Dr.), Paso San Antonio, Escuintla Department, Guatemala stating "I figure one of Stoll's specimens in order to show its close resemblance to Sowerby figure." von Martens (1900e:540) also recognized *Mycetopus weddellii* Hupé, 1857 as valid and occurring in "Central Nicaragua: on the muddy margins of the Rio Tipitapa" [Caribbean–San Juan–Lake Nicaragua Dr.] citing Tate (1870:160). He further stated that "It is not very probable that the Nicaraguan and Chiquitos [Bolivia] shells belong to the same species, but I can give no more information beyond what is mentioned in the comparative table (*anteà*, p. 539)." von Martens later (1901c:655) stated that "Shells collected at Putumayo, Ecuador, by Hopke and Dübrich pertaining evidently to *M. weddelli*, help to fill the gap in the geographical distribution between Nicaragua and Chiquitos." We consider *Mycetopoda weddelli* as a synonym of *Mycetopoda siliquosa* (Spix in Wagner, 1827) and distinct from *Mycetopoda subsinuata* (Sowerby, 1868).

Taxonomic History:

Mycetopus weddelli Hupé, 1857 — Tate (1870), von Martens (1900e; 1901c)
Platiris (Mycetopus) siliquosus (Spix in Wagner, 1827) — Lea (1870)
Mycetopus longinus (Spix in Wagner, 1827) — von Ihering (1890)
Mycetopus subsinuatus Sowerby, 1868 — von Martens (1900e), Agudo-Padrón (2019)
Mycetopus weddelli Hupé, 1857 — von Martens (1901c)
Mycetopoda subsinuata (Sowerby, 1868) — Simpson (1900; 1914), von Ihering (1910), Ortmann (1921b) [in part], Haas (1931)
Mycetopoda sp. — Hinkley (1920)
Mycetopoda longina (Spix in Wagner, 1827) — Frierson (1922)
Mycetopoda weddellii (Hupé, 1857) — Frierson (1927)
Mycetopoda siliquosa (Spix in Wagner, 1827) — Bonetto (1962), Figueiras (1965), Haas (1969), Johnson (1981), Fechter (1983), Ramírez et al. (2003), Simone (2006), Mansur & dos Santos (2008)

Mycetopus punctatus Preston, 1909. New land, freshwater, and marine shells from South America. *Annals and Magazine of Natural History 8th Series* 3(18):507–513+2 plates. Page 513, pl. 10, fig. 8.

Verbatim Type Locality: "Rio Chenchi, U.S. Colombia."

Type Locality: Río Chenche (Magdalena Dr.), Tolima Department, Colombia [W=3.8617, -74.9538].

Type Specimens: Holotype by monotypy, NHMUK 1915.1.6.83 (Johnson 1971:91).

Taxonomic History:

Mycetopoda punctata (Preston, 1909) — von Ihering (1910), Simpson (1914)
Mycetopoda siliquosa (Spix in Wagner, 1827) — Bonetto (1962), Figueiras (1965), Haas (1969), Johnson (1981), Fechter (1983), Ramírez et al. (2003), Simone (2006), Mansur & dos Santos (2008)

Order Unionoida
Superfamily Etherioidea Deshayes, 1832
Family Mycetopodidae Gray, 1840
Subfamily Anodontitinae Modell, 1942
Genus *Anodontites* Bruguière, 1792

Anodontites Bruguière, 1792. Sur une nouvelle coquille du genre de l'Anodontite. *Journal d'Histoire Naturelle* 1:131–136+plate 8. Page 134 (Map 88).

Diagnosis: "Hauteur, 10 lignes; largeur, 1 pouce 7 lignes; profondeur, 6 lignes et demi. Forme, elargie en avant, et marquée d'un angle peu saillant, oblique, se terminant au bord, arrondie en arrière. Valves, minces, garnies à leur superficie de stries longitudinales creuses, distantes, peu marquees près des bords et à leur face antérieure, croisées par d'autres stries transverses plus serrées, élevées, onduleuses, et lègérement lamelleuses près des bords. Attaches musculaires, trois dans chaque valve; celle du bord antérieur grande et superficielle, les deux du bord postérieur inégales, rénuies, peu profondes. Sommets, cariés, peu saillans, et situés en arrière, au quart de la largeur des valves."

[Height, 10 lines; width, 1 inch 7 lines; depth, 6 and a half lines. Shape, widened in front, and marked with a slightly projecting angle, oblique, ending at the edge, rounded behind. Valves, slender, furnished on the superficial with hollow longitudinal striae, distant, slightly marked near the edges and on the anterior surface, crossed by other transverse striae that are tighter, raised, undulating and slightly lamellar near the edges. Muscle attachments, three in each valve; that of the anterior edge large and superficial, the two of the posterior edge unequal, rounded, shallow. Vertices, cavities, slightly protruding, and located behind, at a quarter of the width of the valves].

Type species: *Anodontites crispata* Bruguière, 1792, by monotypy.

Taxonomic History:

Mutela Scopoli, 1777 — Gray (1847), Adams & Adams (1857)
Anodonta Lamarck, 1799 — Fischer (1886), von Martens (1900c)
Anodontites Bruguière, 1792 — Simpson (1900), Frierson (1927), Haas (1929), H.B. Baker (1930), Modell (1942; 1964), Simone (2006), Graf & Cummings (2007; 2021), Bouchet & Rocroi (2010), Lopes-Lima et al. (2017), Pfeiffer et al. (2019b), Bieler (2021), Nagel (2025)
Anodontites (s.s.) Bruguière, 1792 — Simpson (1914), Ortmann (1921b), Haas (1930; 1931; 1969), Thiele (1934), Johnson (1981)

Patularia Swainson, 1840. *A Treatise on Malacology: or the Natural Classification of Shells and Shell-fish*. London. Printed for Longman, Orme, Brown, Green, & Longman, Peternoster Row 419 pp. Pages 286–287, 381.

Diagnosis: "Has the umbones remarkably swollen, or ventricose; and the shell almost round.... Shell nearly equilateral, round or cordate; no teeth." "*T. ovata* Sw. Ex. Conch, pl. 36. *rotundatus*, Ib. pl. 137."

Type Species: *Patularia ovata* Swainson, 1840, by original designation. As *Anodon rotundatus* on Plate 36 of Swainson, 1841.

Remarks: *Patularia* was erected in 1840 with the type species as "*T. ovata* Sw. Ex. Conch, pl. 36." He designated the type species BEFORE the species was actually described. He was also referencing a publication (*Exotic Conchology*) that had not yet been published. In *Exotic Conchology* (1841), Swainson described the type species as *Patularia ovata* in the text and figured it as *Anodon ovata* Sw. He also noted that Lamarck's "*Anodonta trapezialis* [sic] had priority over *ovata* but still used *ovata* and included *Anodonta trapezialis* Lam." [sic] as a synonym.

Taxonomic History:

Anodonta Lamarck, 1799 — Herrmannsen (1846a), Gray (1847)
Patularia Swainson, 1840 — Herrmannsen (1847c)
Anodonta (*Patularia*) Swainson, 1840 — Adams & Adams (1857), von Martens (1900d)
Anodontites (s.s.) Bruguière, 1792 — Simpson (1914), Ortmann (1921b), Haas (1969)
Anodontites Bruguière, 1792 — Thiele (1934), Simone (2006)
Glabaris Gray, 1847 — Modell (1942)

Anadontites Gray, 1847. A list of the genera of recent Mollusca, their synonyma and types. *Proceedings of the Zoological Society of London* 1847(15):129–219. Page 197.
Remarks: Error for *Anodontites* Bruguière, 1792.

Glabaris Gray, 1847. A list of the genera of recent Mollusca, their synonyma and types. *Proceedings of the Zoological Society of London* 1847(15):129–219. Page 206.

Diagnosis: "Oblong, dorsal margin forming a straight line before and behind the summits, subangulated at its fore and hinder ends. Inside very iridescent: *A. nicaraguæ* and *jewetti*, from Central America; *A. exotica*, Lam., *A. anserina*, Spix, *A. forbesiana*, Lea, &c., from South America" (von Martens, 1900c:525).

Type Species: *Anodonta exotica* Lamarck, 1819, by subsequent designation (Herrmannsen 1852:57).

Taxonomic History:

Anodonta (*Patularia*) Swainson, 1840 — Adams & Adams (1857)
Anodonta (*Glabaris*) Gray, 1847 — von Martens (1900c–d)
Glabaris (s.s.) Gray, 1847 — Simpson (1900)
Glabaris Gray, 1847 — Pilsbry (1910), von Ihering (1910), Modell (1942; 1964), Graf & Cummings (2006), Bouchet & Rocroi (2010)
Anodontites (s.s.) Bruguière, 1792 — Simpson (1914), Ortmann (1921b), Haas (1969)
Anodontites Bruguière, 1792 — Thiele (1934), Simone (2006)

Mesanodon Crosse & Fischer, 1894. *In* P. Fischer and H. Crosse. 1894. Études sur les mollusques terrestres et fluviatiles du Mexique et du Guatemala. *In* M.H.–Milne Edwards (ed.). *Mission Scientifique au Mexique et dans l'Amérique centrale. Recherches Zoologiques pour servir á l'Histoire de la Faune de l'Amérique Centrale et du Mexique.* 15th Livraison. Feuilles 62–72, pp. 489–576, plates 59–62, Imprimerie Nationale, Paris Volume 2. 731 pp. Page 517. Introduced as a subgenus of *Anodonta* Lamarck, 1799.

Diagnosis: "Sommet peu saillant; forme ovale; bord dorsal peu ascendant en arrière des sommets (*A. lurulenta*, Morelet; *A. Strebeli*, Lea; *A. Henryana*, Lea; *A. Tehuantepecensis*, Crosse et Fischer)." [Top slightly protruding; Oval shape; dorsal edge slightly ascending behind the apices (*A. lurulenta*, Morelet; *A. Strebeli*, Lea; *A. Henryana*, Lea; *A. Tehuantepecensis*, Crosse and Fischer)].

Type Species: *Anodon lurulentus* Morelet, 1849, by subsequent designation (Thiele 1934:832).

Remarks: Previous workers have considered *Mesanodon* as a member of the Anodontini upon the assumption that the type species (*Anodon lurulentus* Morelet, 1849) was a member of the genus *Anodonta*. We have moved it into *Anodontites* as a synonym of *Anodontites* Bruguière, 1792.

Taxonomic History:

Anodonta (s.s.) Lamarck, 1799 — von Martens (1900c)
Anodonta Lamarck, 1799 — Thiele (1934), Modell (1964)
Anodonta (*Mesanodon*) Crosse & Fischer, 1894 — Haas (1969)

Euryanodon Crosse & Fischer, 1894. *In* P. Fischer and H. Crosse. 1894. Études sur les mollusques terrestres et fluviatiles du Mexique et du Guatemala. *In* M.H.–Milne Edwards (ed.). *Mission Scientifique au Mexique et dans l'Amérique centrale. Recherches Zoologiques pour servir á l'Histoire de la Faune de l'Amérique Centrale et du Mexique.* 15th Livraison. 62–72, pp. 489–576, plates 59–62, Imprimerie Nationale, Paris Volume 2. 731 pp. Page 518. Introduced as a subgenus of *Anodonta* Lamarck, 1799.

Diagnosis: "Sommet peu saillant; forme allongée; bord dorsal non-ascendant (*A. Bambousearum*, Morelet; *A. cylindracea*, Lea)." [Top slightly protruding; elongated shape; non-ascending dorsal edge (*A. Bambousearum*, Morelet; *A. cylindracea*, Lea)].

Type Species: *Anodon bambousearum* Morelet, 1851, by subsequent designation (Haas 1969).

Taxonomic History:

Anodonta (*Euryanodon*) Lamarck, 1799 — von Martens (1900c, 1900d)
Anodontites (s.s.) Bruguière, 1792 — Thiele (1934), Haas (1969)
Glabaris Gray, 1847 — Modell (1964)
Anodontites Bruguière, 1792 — Simone (2006)

Pseudoleila Crosse & Fischer, 1894. *In* P. Fischer and H. Crosse. 1894. Études sur les mollusques terrestres et fluviatiles du Mexique et du Guatemala. *In* M.H.–Milne Edwards (ed.). *Mission Scientifique au Mexique et dans l'Amérique centrale. Recherches Zoologiques pour servir á l'Histoire de la Faune de l'Amérique Centrale et du Mexique.* 15th Livraison. 62–72, pp. 489–576, plates 59–62, Imprimerie Nationale, Paris Volume 2. 731 pp. Page 518. Introduced as subgenus of *Anodonta* Lamarck, 1799.

Diagnosis: "Coquille très-bâillante; ligne palléale large (*A. ciconia*, Gould)." [Very gaping shell; broad pallial line (*A. ciconia*, Gould)].

Type Species: *Anodon ciconia* Gould, 1851, by monotypy.

Taxonomic History:

Anodonta (*Pseudoleila*) Crosse & Fischer, 1894 — von Martens (1900c, 1900d)
Leila (*Pseudoleila*) Crosse & Fischer, 1894 — Frierson (1927)
Anodontites (s.s.) Bruguière, 1792 — Thiele (1934), Haas (1969)
Glabaris Gray, 1847 — Modell (1964)

Pachyanodon von Martens, 1900, Land and freshwater Mollusca. *In* F.D. Godman and O. Salvin (eds.). *Biologia Centrali-Americana* Part 156, Fascile 66:521–528. Page 525. Introduced as subgenus of *Anodonta* Lamarck, 1799.

Diagnosis: "*Anodontæ latomarginatæ*, v. Martens, in Malak. Blätt. 1868, p. 211; Die dickschaligen südamerikamischen, Anodonten, Clessin, in Martini & Chemnitz, Syst. Conch.-Cab. Ed. 2, *Anodonta* p. 77. —Rounded, solid; the dull lustreless zone inside the ventral and posterior margins very broad: *A. latomarginata* Lea, and other species from South America; *A. luteola*, from the Isthmus of Darien."

Type Species: *Anodonta latomarginata* Lea, 1834, by subsequent designation (Haas 1969:558).

Taxonomic History:

Anodontites (s.s.) Bruguière, 1792 — Haas (1969)
Anodontites Bruguière, 1792 — Simone (2006)
Scolianodon von Martens, 1900. Land and freshwater Mollusca. *In* F.D. Godman and O. Salvin (eds.). *Biologia Centrali-Americana* Part 156, Fascile 66:521–528. Page 525. Introduced as subgenus of *Anodonta* Lamarck, 1799.

Diagnosis: "Oval, summits somewhat unequal. Only known from the Lake of Nicaragua. – *A. inæquivalvis, lenticularis, granadensis*."

Type Species: *Anodonta inaequivalva* Lea, 1868, here designated.

Remarks: Although von Martens (1900d) listed three species under the genus *Scolianodon* on page 525, he later (von Martens 1900d:534) only gave an account for a single species (*Anodonta* (*Scolianodon*) *inæquivalvis*), with two named varieties (*lenticularis* and *granadensis*), all of which were eligible for type status.

Taxonomic History:

Anodontites Bruguière, 1792 — Simone (2006)

Styganodon von Martens, 1900. Land and freshwater Mollusca. In F.D. Godman and O. Salvin (eds.). *Biologia Centrali-Americana* Part 156, Fascile 66:521–528. Page 525. Introduced as subgenus of *Anodonta* Lamarck, 1799.

Diagnosis: "Shell more or less reniform, inside more or less dark and dull, with a broad lusterless marginal zone as in *Pachyanodon*: type *A. tenebricosa*, Lea, from South America. Not yet known from Central America."

Type Species: *Anodonta tenebricosa* Lea, 1834, by original designation.

Taxonomic History:

Glabaris (sec. *Styganodon*) von Martens, 1900 — Simpson (1900)
Anodontites (sec. *Styganodon*) von Martens, 1900 — Simpson (1914)
Anodontites (s.s.) Bruguière, 1792 — Ortmann (1921b), Haas (1969)
Anodontites Bruguière, 1792 — Thiele (1934), Modell (1964), Simone (2006)

Anondontites Marshall, 1925. Microscopic sculpture of pearly fresh-water mussel shells. *Proceedings of the United States National Museum* 67(2576):1–14+4 plates. Page 13.
Remarks: Error for *Anodontites* Bruguière, 1792.

Anodonites Marshall, 1928. New fresh-water and marine bivalve shells from Brazil and Uruguay. *Proceedings of the United States National Museum* 74(2762):1–7+4 plates. Page 4
Remarks: Error for *Anodontites* Bruguière, 1792.

Anodotites Marshall, 1930. New land and fresh-water mollusks from South America. *Proceedings of the United States National Museum* 77(2825):1–7+2 plates. Page 6.
Remarks: Error for *Anodontites* Bruguière, 1792.

Ruganodontites Marshall, 1931. *Ruganodontites*, a new subgenus of South American pearly, fresh-water mussels. *The Nautilus* 45(1):16–20. Page 16.

Diagnosis: "Shell elongate, thick, usually somewhat falcate, with a broad, shallow depression running from the beaks to the middle of the ventral margin. Periostracum thick, microscopically radiately striate, much puckered and wrinkled as if from shriveling, the wrinkling occurring also in the calcareous portion of the shell. The wrinkles are obscurely divided into narrow, gradually widening rays extending from the beaks to the margin. Nacre greenish-livid."

Type Species: *Anodontites colombiensis* Marshall, 1922, by original designation.

Taxonomic History:

Anodontites (sec. *Ruganodontites*) Marshall, 1931 — Thiele (1934)
Anodontites Bruguière, 1792 — Modell (1964), Simone (2006)
Anodontites (*Ruganodontites*) Marshall, 1931 — Haas (1969)

Glabris Prashad, 1932. Notes on some little known South Asiatic Unionidæ. *Proceedings of the Malacological Society, London* 20(3):174–178+plate 14. Page 174.
Remarks: Error for *Glabaris* Gray, 1847.

Andontites Kabat, 1997. Correct family names for the freshwater "muteloid" bivalves (Unionoida: Etherioidea). Occasional Papers on Mollusks, Museum of Comparative Zoology, Harvard University 5(72):379–392. Page 383.
Remarks: Error for *Anodontites* Bruguière, 1792.

Regarding the gender of *Anodontites*, mentions of the binomen in the literature sometimes employ the masculine *crispatus* in combination with *Anodontites,* but the facts are clear. Ortmann (1921b:589, as a footnote) recognized this and stated that "Bruguière uses *Anodontites* as *feminini generis*."

1. *crispatus, -a, -um* is unambiguously adjectival (the past participle of the regular first conjugation verb *crispo, crispare, crispavi, crispatus*; meaning curled or wrinkled in Latin) and *crispata* could not be construed as a noun in apposition.
2. Article 30.1.4.4 of the Code states: "A compound genus-group name ending in the suffix *-ites*, *-oides*, *-ides*, *-odes*, or *-istes* is to be treated as masculine unless its author, when establishing the name, stated that it had another gender or treated it as such by combining it with an adjectival species-group name in another gender form" (Harry G. Lee, pers. comm., 2019).

Distribution & Remarks: The genus *Anodontites* is widespread in Central and South America and ranges from Sinaloa, México (~26° N) south to Patagonia, Argentina (~39° S). Graf and Cummings (2021:15) recognized six species in the genus *Anodontites* as occurring in Central America. We now consider seven as valid. We elevated *Anodontites jaliscoensis* Pilsbry, 1920 from *Anodontites trapesialis* (Lamarck, 1819) and sunk *Anodontites depexa* (von Martens, 1900) into *Anodontites tehuantepecensis* (Crosse & Fischer, 1893), *Anodontites carinata* (Dunker, 1858) into *Anodontites tortilis* (Lea, 1852), and moved *Anodonta lurulenta* Morelet, 1849 from *Anodonta* into *Anodontites. Anodontites cylindracea* (Lea, 1838), *Anodontites montezuma* (Lea, 1841), and *Anodontites trapesialis* (Lamarck, 1819) remain unchanged. Of the seven species of *Anodontites* in Central America, four are endemic (Table 3).

Two species of *Anodontites* are found in Gulf of Mexico drainages: *Anodontites cylindracea* (Lea, 1838)—which is found in all three Central American Gulf of Mexico provinces. *Anodontites lurulenta* (Morelet, 1849) is endemic to the Usumacinta and is known only from the type locality. Interestingly, the wide-ranging mycetopodid *Anodontites trapesialis* (Lamarck, 1819) has moved north from South America into Central America along the Pacific Coast as far north the Río Fuerte in Sinaloa, México and does not inhabit any Gulf of Mexico tributaries. It does however occur in a few Caribbean tributaries as far north as Honduras (Map 86).

MAP 88 Distribution of the genus *Anodontites* Bruguière, 1792 in Central America and northwestern South America.

PLATE 74 ***Anodontites cylindracea* (Lea, 1838). Top Left:** *Anodonta cylindracea* Lea, 1838. Lectotype, USNM 86712 (58 mm L, 33 mm H). Río Medellin (Jamapa Dr.), near Veracruz, Veracruz, México [L = 19.0436, -96.1403]. **Middle Right:** *Anodon bambousearum* Morelet, 1851. Holotype, NHMUK 1965146 (73 mm L, 38 mm H). Río Chacamax or Río Michol (Usumacinta Dr.), near the ruins of Palenque, Chiapas, México [L = 17.4823, -91.9742]. **Bottom Left:** *Anodonta strebelii* Lea, 1868. Holotype, USNM 25433 (39 mm L, 23 mm H). Río Medellin (Jamapa Dr.), Veracruz, Veracruz, México [L = 19.0574, -96.1550].

Anodontites cylindracea (Lea, 1838)

Plate 74, Map 89

Province(s): Pánuco; Papaloapán; Usumacinta.

Distribution and Status: Found in Gulf of Mexico tributaries from the Pánuco basin in San Luis Potosí, south to the Usumacinta drainage in Chiapas, México. There are two disjunct populations in México. If determined to be a distinct species, the Usumacinta population would be *Anodontites bambousearum* (Morelet, 1851). Not assessed (IUCN 2024). Surveys in the Pánuco (2017 and 2018), Papaloapán (2022), and Usumacinta (2019) did not result in locating live individuals, but recent shells were found, and it is likely still extant in those basins.

Anatomy: Shell small, thin, and elongate. Dorsal margin slightly curved. Ventral margin straight to slightly concave. Anterior and posterior ends round. Beaks low and the beak cavity shallow. Lateral and pseudocardinal absent. Periostracum yellow with green rays. Nacre bluish white, violet-tinted, and highly iridescent.

Biology: Unknown.

Adult Size: Length to 8 cm.

Similar Species: The edentate *Pyganodon globosa* (Lea, 1841) and *Utterbackia imbecillis* (Say, 1829) co-occur with *Anodontites cylindracea* (Lea, 1838) in the Papaloapán and Usumacinta basins. *Anodontites cylindracea* has a ligamental sinus that is lacking in both *P. globosa* and *U. imbecillis* (Figure 2).

MAP 89 Distribution of *Anodontites cylindracea* (Lea, 1838).

Anodonta cylindracea Lea, 1838. Description of new freshwater and land shells. *Transactions of the American Philosophical Society* 6(N.S.):1–154+plates 1–24. Page 45, pl. 13, fig. 40.

Verbatim Type Locality: "River Medellin, near Vera Cruz."

Type Locality: Río Medellin (Gulf of México–Jamapa Dr.), near Veracruz, Veracruz, México, Central America [L = 19.0436, -96.1403].

Type Specimens: Lectotype, USNM 86712, here designated.

Remarks: Johnson and Baker (1973:152) and Johnson (1974a:42) stated that the type was not found in ANSP or USNM and identified USNM 86712 as an ideotype. Lea (1838b:45) stated that "Two specimens only of this were sent by Dr. Burrough." USNM 86712 matches the figure in Lea and is here designated as the lectotype. The location of the second specimen is unknown and presumed lost.

Taxonomic History:

Anodonta cylindracea — Lea (1836) *nomen nudem*
Anodonta cylindracea Lea, 1838 — Hanley (1843), Conrad (1853), Clessin (1874c), Cruz & Quesada (2017)
Margarita (*Anodonta*) *cylindracea* (Lea, 1838) — Lea (1836)
Margaron (*Anodonta*) *cylindracea* (Lea, 1838) — Lea (1852; 1870)
Anodonta (*Lamproscapha*) *cylindracea* Lea, 1838 — Adams & Adams (1857)
Anodon cylindracea (Lea, 1838) — Sowerby (1869)
Anodonta (*Euryanodon*) *cylindracea* Lea, 1838 — Fischer & Crosse (1894a), von Martens (1900d)
Glabaris (s.s.) *cylindraceus* (Lea, 1838) — Simpson (1900)
Glabaris cylindracea (Lea, 1838) — von Martens (1901c), Pilsbry (1910)
Glabaris cylindraceus (Lea, 1838) — de Leon (1912)
Anodontites (s.s.) *cylindraceus* (Lea, 1838) — Simpson (1914)
Anodontites cylindracea (Lea, 1838) — Frierson (1927), Graf & Cummings (2021), Kiser et al. (2022), Ramírez et al. (2022), Czaja et al. (2023)
Anodontites (*Euryanodon*) *cylindracea* (Lea, 1838) — Morrison (1968)
Anodontoides [sic] *cylindracea* (Lea, 1838) — Contreras-Arquieta (2000)
Anodontites cylindraceus (Lea, 1838) — Haas (1929), Graf & Cummings (2007), Mansur et al. (2020)

Anodon bambousearum Morelet, 1851. Testacea novissima insulae Cubanae, et Amer. Centralis. Librairie de L'Académie Nationale de Medecine, Paris Part 2:1–30. Page 24. Figured in Fischer and Crosse (1894a:527, pl. 63, figs. 6–6a).

Verbatim Type Locality: "rivulos prope vicum Palenqueanum prov. Chiapa."

Type Locality: Río Chacamax or Río Michol (Gulf of México–Grijalva–Usumacinta Dr.), near the ruins of Palenque, Chiapas, México [L = 17.4823, -91.9742] or [L = 17.4871, -92.0565].

Type specimen: Holotype by monotypy, NHMUK 1965146 (Johnson 1971:80; Breure et al. 2018:212, figs. 99–100, as the lectotype).

Taxonomic History:

Anodonta (*Euryanodon*) *bambousearum* (Morelet, 1851) — Fischer & Crosse (1894a)
Anodonta (*Euryanodon*) *bambusearum* [sic] (Morelet, 1851) — von Martens (1900d)
Glabaris (sec. *Styganodon*) *bambousearum* (Morelet, 1851) — Simpson (1900)
Glabaris bambousearum (Morelet, 1851) — de Leon (1912)
Anodontites (sec. *Styganodon*) *bambousearum* (Morelet, 1851) — Simpson (1914)
Anodontites cylindracea (Lea, 1838) — Frierson (1927)
Anodontites (*Euryanodon*) *bambousearum* (Morelet, 1851) — Morrison (1968)
Anodontites (s.s.) *soleniformis* (Lea, 1838) — Haas (1969)

Anodontoides bambousearum (Morelet, 1851) — Contreras-Arquieta (2000)
Anodonta banbusearum [sic] (Morelet, 1851) — Cruz & Quesada (2017)

Anodonta strebelii Lea, 1868. Descriptions of four new species of exotic Unionidae. *Proceedings of the Academy of Natural Sciences of Philadelphia* 20:150. Page 150. Figured in Lea (1868b JANSP Vol. 6:322;1869a Obs. 12:82, pl. 52, fig. 135).

Verbatim Type Locality: "Vera Cruz, Mexico."

Type Locality: Río Medellin (Gulf of México–Jamapa Dr.), Veracruz, Veracruz, México [L = 19.0574, -96.1550].

Type specimen: Holotype by monotypy, USNM 25433 (Johnson 1974a:137).

Taxonomic History:

Margaron (Anodonta) strebellii [sic] (Lea, 1868) — Lea (1870)
Anodonta strebelii Lea, 1868 — Clessin (1874b)
Anodonta (Mesanodon) strebeli [sic] Lea, 1868 — Fischer & Crosse (1894a)
Anodonta (s.s.) *lurulenta* (Morelet, 1849) — von Martens (1900d)
Glabaris (s.s.) *strebelii* (Lea, 1868) — Simpson (1900)
Glabaris strebeli (Lea, 1868) — Pilsbry (1910), de Leon (1912)
Anodontites (s.s.) *strebelii* (Lea, 1868) — Simpson (1914)
Anodontites cylindracea (Lea, 1838) — Frierson (1927)
Anodontites cylindraceus (Lea, 1838) — Haas (1929)
Anodontites (s.s.) *trapesialis glaucus* (Valenciennes in Humbolt & Bonpland, 1833) — Haas (1969)

Anodonta (Euryanodon) cylindracea var. *β, subarcuatula* Fischer & Crosse, 1894. Études sur les mollusques terrestres et fluviatiles du Mexique et du Guatemala. *In* M.H.–Milne Edwards (ed.). *Mission Scientifique au Mexique et dans l'Amérique centrale. Recherches Zoologiques pour servir á l'Histoire de la Faune de l'Amérique Centrale et du Mexique.* Volume 2. Page 528. Not figured.

Verbatim Type Locality: "Mexique, dans le Rio Medellin, près Vera Cruz, État de Vera Cruz."

Type Locality: Río Medellin (Gulf of México–Jamapa Dr.), near Veracruz, Veracruz, México, Central America [L = 19.0574, -96.1550].

Type specimen: Unknown. Presumed lost.

Taxonomic History:

Anodonta (Euryanodon) cylindracea Lea, 1838 — von Martens (1900d)

PLATE 75 **Top: *Anodontites jaliscoensis* Pilsbry, 1920**. Holotype, ANSP 46197 (46 mm L, 25 mm H). Río Armeria (Pacific Dr.), Tolimán, Jalisco, México [L = 19.5773, -103.9238]. **Bottom: *Anodontites lurulenta* (Morelet, 1849).** *Anodon lurulenta* Morelet, 1849. Lectotype, NHMUK 1893.2.4.2003 (45 mm L, 27 mm H). Río Usumacinta, Boca de San Gerónimo, Tabasco, México [L = 17.9016, -91.9131].

Anodontites jaliscoensis (Pilsbry, 1920)

Plate 75, Map 90

Province(s): Mesa Central and Balsas.

Distribution and Status: Known only from the holotype, a topotype, and a single specimen (AMS 140804) from "muddy creek, 2 km from Singayta, approximately 15 km from San Blas on Tepic Rd., México" collected in 1983, which we have provisionally identified as *Anodontites jaliscoensis* Pilsbry, 1920. As noted by Pilsbry (1920a:192), further work on the Pacific coast drainages of México is badly needed to determine what mussels occur there. It is possibly a juvenile *Anodontites trapesialis*. Not assessed (IUCN 2024).

Anatomy: Shell oblong and elongate, and thin with weak growth lines and minute radial bands of striae in the middle part. No lateral or pseudocardinal teeth. A conspicuous ligamental sinus at the posterior end of the hinge. Periostracum light brown, pale toward the beaks with no evident sculpture. Nacre white, stained with copper in the beak cavity.

Biology: Unknown.

Adult Size: Length to 4.6 cm.

Similar Species: None. *Anodontites trapesialis* (Lamarck, 1819) also is found in Jalisco, but it has not been reported to co-occur with *Anodontites jaliscoensis* Pilsbry, 1920 in the Río Armeria drainage.

MAP 90 Distribution of *Anodontites jaliscoensis* Pilsbry, 1920.

Anodontites jaliscoensis Pilsbry, 1920. Mollusks from Lake Chapala, State of Jalisco and vicinity. *Proceedings of the Academy of Natural Sciences of Philadelphia* 72(2):192–194. Page 193, fig. 3.

Verbatim Type Locality: "Tolimán, State of Jalisco."

Type Locality: Río Armeria (Pacific Dr.), Tolimán, Jalisco, México [L = 19.5773, -103.9238].

Type Specimens: Holotype by monotypy and original designation, ANSP 46197.

Remarks: ANSP 454376 ($n=1$ valve) is labeled as a paralectotype in the ANSP Collection, but Pilsbry mentioned having only one specimen before him.

Taxonomic History:

Anodontites jaliscoensis Pilsbry, 1920 — Frierson (1927)

Anodontites (s.s.) *trapesialis glaucus* (Valenciennes in Humbolt & Bonpland, 1833) — Haas (1969)

Anodontoides [sic] *jaliscoensis* Pilsbry, 1920 — Contreras-Arquieta (2000)

Anodontites lurulenta (Morelet, 1849)

Plate 75, Map 91

Province(s): Usumacinta.

Distribution and Status: Known only from the type locality in the Río Usumacinta in México. Not assessed (IUCN 2024). A visit to the type locality during a survey of the Usumacinta basin in 2019 found the site heavily polluted, and it is possibly extirpated from that locality. It may occur elsewhere the drainage. von Martens (1900d:531) included *Anodonta strebeli* Lea, 1868 from Veracruz, México as a synonym of *Anodonta lurulenta*, but we have placed that taxon under *Anodontites cylindracea* (Lea, 1838) above.

Anatomy: Shell oblong, moderately inflated, thin and edentate. Beaks low and without sculpture. No lateral or pseudocardinal teeth. Periostracum green with numerous narrow rays, shiny on the anterior end. Nacre bluish white, violet-tinted and highly iridescent.

Biology: Unknown.

Adult Size: Length to 5 cm.

Similar Species: *Anodontites cylindracea* (Lea, 1838) has also been found in the Usumacinta basin, but they have not been found together.

MAP 91 Distribution of *Anodontites lurulenta* (Morelet, 1849).

Anodon lurulentus Morelet, 1849. Testacea novissima insulæ Cubanæ, et Americæ Centralis. Librairie de L'Académie Nationale de Medecine, Paris Part 1:1–31. Page 28. Figured in Fischer and Crosse (1894a:523, pl. 64, figs. 6–6a) and Breure et al. (2018:332, figs. 615–616).

Verbatim Type Locality: "Vivit in paludibus fluvii Usumasinta."

Type Locality: Río Usumacinta (Gulf of México–Grijalva Dr.), Boca de San Gerónimo, Tabasco, México [L = 17.9016, -91.9131].

Type Specimens: Lectotype, NHMUK 1893.2.4.2003 designated by Johnson (1971:87, as the holotype; Breure et al. 2018:332, figs. 615–616). Paralectotypes, NHMUK 1893.2.4.2004–2005 ($n=2$) (Johnson 1971:87, as paratypes; Breure et al. 2018:332).

Remarks: The label with the type reads "Marais de S. Geronimo. Yuc." The catalogue entry for NHMUK 1893.2.4.2003 is Yucatán. Fischer and Crosse (1894a:524) noted that "We saw three specimens of this species, from the collection A. Morelet; they are stuck on a cardboard with this mention: Marais de San Geronimo. San Geronimo is a locality in Yucatán. On the other hand, the original description gives for habitat the marshes of the Usumacinta. So, there is an attribution error which unfortunately we cannot rectify." We have concluded that all the specimens labeled as San Geronimo are from Boca de San Gerónimo (a small town located on the Usumacinta) or possibly San Gerónimo, about 6.4 km NE of Boca de San Gerónimo, Tabasco, México. von Martens (1901c:654) stated that "Simpson (loc. cit. p. 630) separates *A. strebeli*, Lea, far apart from *A. lurulenta* Morelet, and places it in the genus *Glabaris*. I dare not oppose this course, for on a fresh examination of Lea's figure I find that the sinulus is rather deep and pointed, as in *Glabaris*, though the general outline is very like that of *A. lurulenta*."

Taxonomic History:

Anodonta (*Mesanodon*) *lurulenta* (Morelet, 1849) — Fischer & Crosse (1894a), Haas (1969)

Glabaris lurulenta (Morelet, 1849) — von Martens (1900d, 1901c)

Anodonta (s.s.) *lurulenta* (Morelet, 1849) — Frierson (1927)

Anodonta lurulenta (Morelet, 1849) — Simpson (1900; 1914), de Leon (1912), Thiele (1934), Graf & Cummings (2007; 2021), Breure et al. (2018), Cruz & Quesada (2017), Ramírez et al. (2022), Czaja et al. (2023)

Anodonta (*Pyganodon*) *globosa lurulenta* (Morelet, 1849) — Morrison (1968)

PLATE 76 ***Anodontites montezuma* (Lea, 1841). Top Left:** *Anodonta montezuma* Lea, 1841. Holotype, AMNH 30871 (38 mm L, 25 mm H). Central America. **Top Right:** *Anodonta granadensis* Lea, 1868. Lectotype, USNM 86655 (35 mm L, 23 mm H). **Middle Left:** *Anodonta lenticularis* Lea, 1868. Holotype, USNM 86704 (35 mm L, 24 mm H). **Middle Right:** *Anodonta inaequivalva* Lea, 1868. Lectotype, USNM 86767 (41 mm L, 28 mm H). **Bottom Left:** *Anodonta atrovirens* Philippi, 1849. Paralectotype, SMF 5164. **Bottom Right:** *Anodonta cornea* Philippi, 1849. Paralectotype, SMF 5163. Last five all, Lake Nicaragua (Caribbean–San Juan Dr.), Río San Juan Department, Nicaragua [W = 11.1350, -84.8367].

Anodontites montezuma (Lea, 1841)

Plate 76, Map 92

Province(s): San Juan.

Distribution and Status: Endemic to Lake Nicaragua and the Río Malacatoya in the Río San Juan basin, Nicaragua. Not assessed (IUCN 2024). No recent collections are known.

Anatomy: Shell round to elliptical, convex, thin, and edendate. Anterior and posterior ends rounded. Dorsal and ventral margins also rounded. Beaks full but not greatly elevated. Periostracum dark green or yellowish green to light brown and rayed. Nacre bluish white and iridescent.

Biology: Unknown.

Adult Size: Length to 4 cm.

Similar Species: *Anodontites trapesialis* (Lamarck, 1819) and *Anodontites tehuantepecensis* (Crosse & Fischer, 1893) co-occur with *Anodontites montezuma* (Lea, 1841) in the Lake Nicaragua/San Juan basin.

MAP 92 Distribution of *Anodontites montezuma* (Lea, 1841).

Anodonta montezuma Lea, 1841. On fresh water and land shells (cont.). *Proceedings of the American Philosophical Society* 2(17):30–34. Page 31. Figured in Lea (1842a TAPS Vol. 8:240; 1842b Obs. 3:78, pl. 23, fig. 55) and Clessin (1876, pl. 72, figs. 7–8).

Verbatim Type Locality: "Central America."

Type Locality: Nicaragua.

Type Specimens: Holotype by monotypy, AMNH 30871 (Boyko and Sage 1996:11, ex. Jay Collection 2057, figured by Lea).

Remarks: Although the type locality was unspecified by Lea, a green label with the holotype indicates that it was collected in Nicaragua.

Taxonomic History:

Margaron (*Anodonta*) *montezuma* (Lea, 1841) — Lea (1852; 1870)
Anodonta montezumæ Lea, 1841 — Hanley (1856)
Anodonta (*Patularia*) *montezuma* Lea, 1841 — Adams & Adams (1857)
Anodon montezianus [sic] (Lea, 1841) — Sowerby (1870)
Anodonta montezuma Lea, 1841 — Clessin (1876), Cruz & Quesada (2017)
Anodonta (s.s.) *montezuma* Lea, 1841 — von Martens (1900d)
Glabaris (s.s.) *montezuma* (Lea, 1841) — Simpson (1900)
Glabaris cylindraceus (Lea, 1838) — de Leon (1912, as *montezianus*, Sowerby)
Anodontites (s.s.) *montezuma* (Lea, 1841) — Simpson (1914)
Anodontites montezuma (Lea, 1841) — Frierson (1927), Mansur et al. (2020), Graf & Cummings (2021)
Anodontites (s.s.) *trigonus montezuma* (Lea, 1841) — Haas (1930; 1969), Zilch (1967)

Anodonta atrovirens Philippi, 1849. Centuria altera Testaceorum novorum. *Zeitschrift für Malakozoologie* 5(9):129–144. Page 130. Not figured.

Verbatim Type Locality: "Nicaragua."

Type Locality: Lake Nicaragua (Caribbean–San Juan Dr.), Río San Juan Department, Nicaragua [W=11.1350, -84.8367].

Type Specimens: Lectotype, Muséum d'Histoire Naturelle de Rouen, France MALA.2015.0.617. MHN designated by Haas (1930:321, fig. 2; Johnson 1951b:32; Nagel 2025:2019). Paralectotype, SMF 5164 (Zilch 1967:132, as a syntype; Nagel 2025:219).

Remarks: Water body from the label with the paralectotype (Haas 1930:320). Corrected date from Kabat and Coan (2016:20).

Taxonomic History:

Anodonta atrovirens Philippi, 1849 — Adams & Adams (1857), Haas (1930)
Anodonta (*Scolianodon*) *inæquivalvis* [sic] Lea, 1868 — von Martens (1900d)
Glabaris atrovireus [sic] (Philippi, 1849) — Simpson (1900)
Anodontites atrovirens (Philippi, 1849) — Simpson (1914)
Anodontites (s.s.) *trigonus montezuma* (Lea, 1841) — Haas (1930), Zilch (1967), Coan & Kabat (2017)
Anodontites cornea (Philippi, 1849) — Johnson (1951a)
Anodontites montezuma (Lea, 1841) — Nagel (2025)

Anodonta cornea Philippi, 1849. Centuria altera Testaceorum novorum. *Zeitschrift für Malakozoologie* 5(9):129–144. Page 130.

Verbatim Type Locality: "Nicaragua."

Type Locality: Lake Nicaragua (Caribbean–San Juan Dr.), Río San Juan Department, Nicaragua [W=11.1350, -84.8367].

Type Specimens: Lectotype Muséum d'Histoire Naturelle de Rouen, France, MALA. 2015.0.618.0.MHN designated by Haas (1930:321, fig. 3; Johnson 1951b:32; Nagel 2025:220). Paralectotype, SMF 5163 (Zilch 1967:132, as a syntype; Coan and Kabat 2017:70; Nagel 2025:220, ex. Museum Rouen). Possible paralectotype, MNHNS 50893 (n=½) (Coan and Kabat 2017:70).

Remarks: Corrected date from Kabat and Coan (2016:20).

Taxonomic History:

Anodonta cornea Philippi, 1849 — Adams & Adams (1857), Lea (1870), Simpson (1900; 1914), von Martens (1900e, *nomen dubium*), Haas (1930; 1969)
Glabaris cornea (Philippi, 1849) — Simpson (1900)
Anodontites cornea (Philippi, 1849) — Simpson (1914), Johnson (1951a)
Anodontites (s.s.) *trigonus montezuma* (Lea, 1841) — Haas (1930), Coan & Kabat (2017), Zilch (1967)
Anodontites montezuma (Lea, 1841) — Nagel (2025)

Anodonta inæquivalva Lea, 1868. Description of nine species of Unionidae from Lake Nicaragua, Central America. *Proceedings of the Academy of Natural Sciences of Philadelphia* 20:94–96. Page 95. Figured in Lea (1868a JANSP Vol. 6:292; 1869a Obs. 12:52, pl. 43, fig. 108).

Verbatim Type Locality: "Lake Nicaragua, Cent. Amer."

Type Locality: Lake Nicaragua (Caribbean–San Juan Dr.), Río San Juan Department, Nicaragua [W=11.1350, -84.8367].

Type Specimens: Lectotype, USNM 86767 designated by Johnson (1951a:83 as *A. inaequivalvis*; Johnson and Baker 1973:158; Johnson 1974:73, all as the holotype). Paralectotypes, ANSP 41836 (n=2) (Johnson and Baker 1973:158, as paratypes); USNM 86767a (n=2), USNM 59873 (n=1).

Taxonomic History:

Anodonta inæqualis [sic] Lea, 1868 — Tate (1870)
Margaron (*Anodonta*) *inæquivalva* (Lea, 1868) — Lea (1870)
Anodonta inaequivalva Lea, 1868 — Clessin (1874b), Cruz & Quesada (2017)
Anodonta (*Scolianodon*) *inæquivalvis* [sic] Lea, 1868 — von Martens (1900d)
Glabaris (s.s.) *inæquivalvis* [sic] (Lea, 1868) — Simpson (1900)
Glabaris cylindracea (Lea, 1838) — de Leon (1912, as *inæquivalvis*)
Anodontites (s.s.) *inæquivalvis* [sic] (Lea, 1868) — Simpson (1914)
Anodontites montezuma (Lea, 1841) — Frierson (1927)
Anodontites (s.s.) *trigonus montezuma* (Lea, 1841) — Haas (1930)
Anodontites cornea (Philippi, 1849) — Johnson (1951a)
Anodontites (s.s.) *inaequivalva* (Lea, 1868) — Haas (1969)
Anodontites inaequivalva (Lea, 1868) — Graf & Cummings (2007)
Anodontites inaequivalvis [sic] (Lea, 1868) — Mansur et al. (2020)

Anodonta lenticularis Lea, 1868. Description of nine species of Unionidae from Lake Nicaragua, Central America. *Proceedings of the Academy of Natural Sciences of Philadelphia* 20:94–96. Page 95. Figured in Lea (1868a JANSP Vol. 6:290; 1869a Obs. 12:50, pl. 41, fig. 102).

Verbatim Type Locality: "Lake Nicaragua, Cent. Amer."

Type Locality: Lake Nicaragua (Caribbean–San Juan Dr.), Río San Juan Department, Nicaragua [W=11.1350, -84.8367].

Type Specimens: Holotype by monotypy, USNM 86704 (Johnson 1951a:83; Johnson 1974:83).

Taxonomic History:

Anodonta lenticularis Lea, 1868 — Sowerby (1870), Clessin (1874b)
Margaron (*Anodonta*) *lenticularis* (Lea, 1868) — Lea (1870)
Anodonta (*Scolianodon*) *inæquivalvis* [sic] var. *lenticularis* Lea, 1868 — von Martens (1900d)

Glabaris (s.s.) *inæquivalvis* [sic] (Lea, 1868) — Simpson (1900)
Anodontites (s.s.) *inæquivalvis* [sic] (Lea, 1868) — Simpson (1914)
Anodontites montezuma (Lea, 1841) — Frierson (1927)
Anodontites (s.s.) *trigonus montezuma* (Lea, 1841) — Haas (1930)
Anodontites cornea (Philippi, 1849) — Johnson (1951a)
Anodontites (s.s.) *inaequivalva* (Lea, 1868) — Haas (1969)
Anodontites inaequivalvis [sic] (Lea, 1868) — Mansur et al. (2020)

Anodonta granadensis Lea, 1868. Description of nine species of Unionidae from Lake Nicaragua, Central America. *Proceedings of the Academy of Natural Sciences of Philadelphia* 20:94–96. Page 96. Figured in Lea (1868a JANSP Vol. 6:288; 1869a Obs. 12:53, pl. 41, fig. 100).

Verbatim Type Locality: "Lake Nicaragua, Cent. Amer."

Type Locality: Lake Nicaragua (Caribbean–San Juan Dr.), Río San Juan Department, Nicaragua [W = 11.1350, -84.8367].

Type Specimens: Lectotype, USNM 86655 designated by Johnson (1951a:83; Johnson and Baker 1973:156; Johnson 1974:65, all as the holotype). Paralectotype, ANSP 41835 (*n* = 1) from Nicaragua (Johnson and Baker 1973:156, as paratypes).

Taxonomic History:

Margaron (*Anodonta*) *granadensis* (Lea, 1868) — Lea (1870)
Anodonta granadensis Lea, 1868 — Clessin (1874b)
Anodonta (*Scolianodon*) *inæquivalvis* var. *granadensis* Lea, 1868 — von Martens (1900d)
Glabaris (s.s.) *granadensis* (Lea, 1868) — Simpson (1900)
Anodontites (s.s.) *granadensis* (Lea, 1868) — Simpson (1914)
Anodontites (s.s.) *trigonus montezuma* (Lea, 1841) — Haas (1930)
Anodontites cornea (Philippi, 1849) — Johnson (1951a)

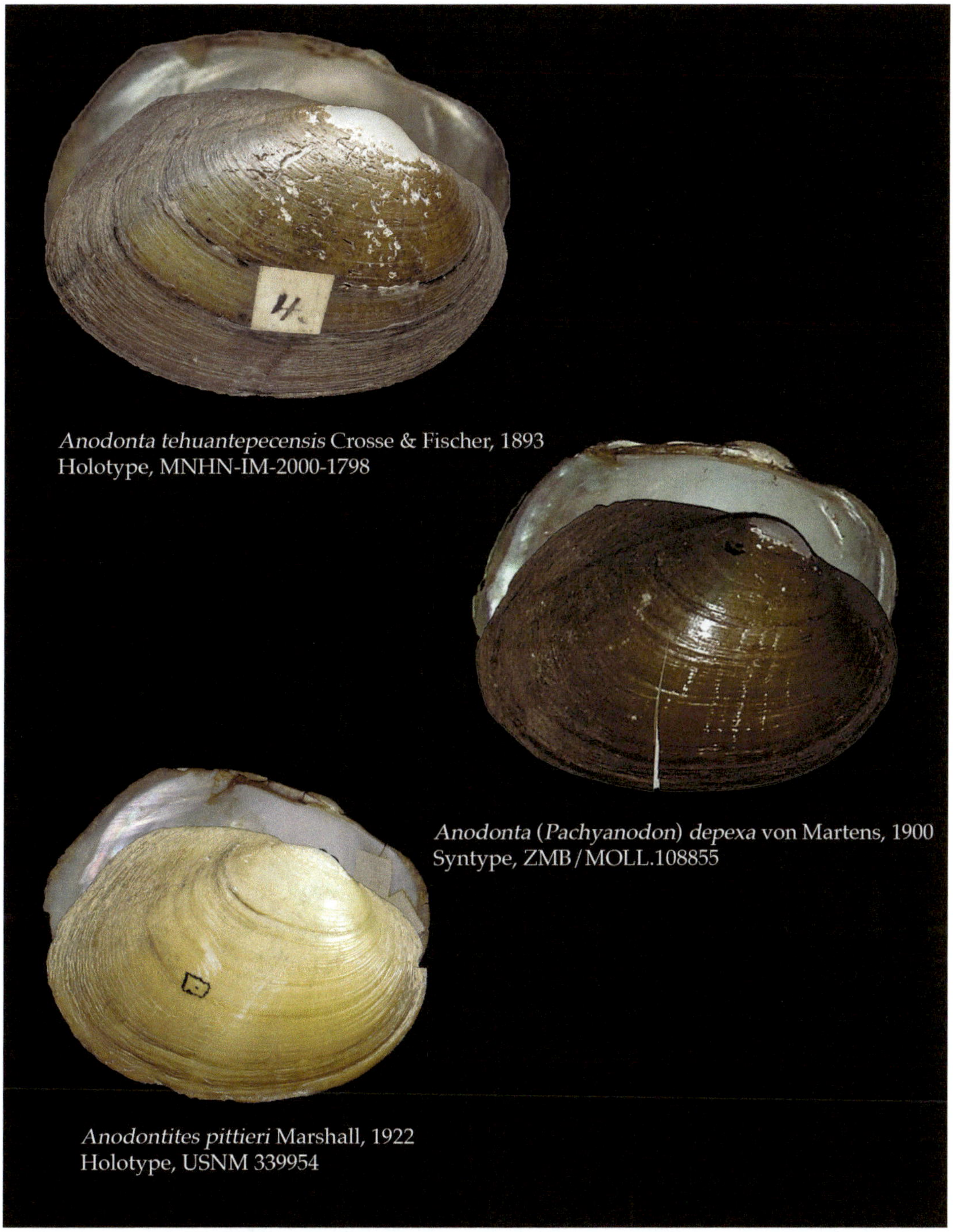

PLATE 77 ***Anodontites tehuantepecensis* (Crosse & Fischer, 1893). Top Left:** *Anodonta tehuantepecensis* Crosse & Fischer, 1893. Holotype, MNHN-IM-2000-1798 (56 mm L, 37 mm H). Río Novillero (Pacific Dr.), Caco Prieto, 6.6 kilometers ENE San Pedro Tapanatepec, Isthmus of Tehuantepec, Oaxaca, México [L = 16.3910, -94.1348]. **Middle Right:** *Anodonta* (*Pachyanodon*) *depexa* von Martens, 1900. Figured Syntype, ZMB/MOLL.108855 (67 mm L, 45 mm H). Río Michatoya (Pacific Dr.), Paso San Antonio, Escuintla Department, Guatemala [L = 14.0617, -90.6809]. **Bottom Left:** *Anodontites pittieri* Marshall, 1922. Holotype, USNM 339954 (50 mm L, 35 mm H). Guaremales, (Caribbean–Urama Dr.), Carabobo, Venezuela [L = 10.4444, -68.3908].

Anodontites tehuantepecensis (Crosse & Fischer, 1893)

Plate 77, Map 93

Province(s): San Juan; Oaxaca, México to Colombia.

Distribution and Status: Found in scattered Caribbean and Pacific drainages from the Isthmus of Tehuantepec in Oaxaca, México south to the Lago Maracaibo basin of northern Venezuela. A few records exist from the Orinoco Basin in Guarico, Venezuela (Martínez-Escarbassiere et al. 2004:295; Lasso et al. 2009:72, as *Anodontites pittieri* Marshall, 1922). Not assessed (IUCN 2024). Found alive in the San Juan (Caribbean Dr.) in Costa Rica in 2024.

Anatomy: Shell oval, slightly inflated, relatively solid, and edentate. Anterior and posterior ends rounded. Dorsal margin nearly straight. Ventral margin curved and convex. Beaks, small and often eroded. Periostracum tan, greenish brown, or brown, typically rayless, with fine, concentric striae. Nacre bluish white and iridescent.

Biology: Inhabits small to medium-sized caños in sand and mud in Venezuela.

Adult Size: Length to 8 cm.

Similar Species: *Anodontites montezuma* (Lea, 1841) and *Anodontites trapesialis* (Lamarck, 1819) co-occur with *Anodontites tehuantepecensis* (Crosse & Fischer, 1893) in the Lake Nicaragua/San Juan basin. *Anodontites tortilis* (Lea, 1852) co-occurs with *A. tehuantepecensis* in the Río Michatoya, but it is thinner and smaller in size as an adult.

MAP 93 Distribution of *Anodontites tehuantepecensis* (Crosse & Fischer, 1893).

Anodonta tehuantepecensis Crosse & Fischer, 1893. Description d'un *Bulimulus* et d'un *Anodonta* nouveaux, provenant du Mexique. *Journal de Conchyliologie* 41(1):31–32. Page 32. Figured in Fischer and Crosse (1894a: 526, pl. 70, figs. 1–1a).

Verbatim Type Locality: "prope Cacoprieto, in Isthmo de Tehuantepecensi, in provincia Oajaca dicta, Reipublicæ Mexicanæ."

Type Locality: Río Novillero (Pacific Dr.), Caco Prieto, 6.6 kilometers ENE San Pedro Tapanatepec, Isthmus of Tehuantepec, Oaxaca, México [L = 16.3910, -94.1348].

Type specimen: Holotype by monotypy, MNHN-IM-2000-1798.

Taxonomic History:

Anodonta (*Mesanodon*) *tehuantepecensis* Crosse & Fischer, 1893 — Fischer & Crosse (1894a)
Anodonta (*Pachyanodon*) *tehuantepecensis* Crosse & Fischer, 1893 — von Martens (1900d)
Glabaris tehuantepecensis (Crosse & Fischer, 1893) — Simpson (1900)
Anodontites (s.s.) *tehuantepecensis* (Crosse & Fischer, 1893) — Simpson (1914)
Anodontites tehuantepecensis (Crosse & Fischer, 1893) — Frierson (1927), Graf & Cummings (2007; 2021), Czaja et al. (2023)
Anodonta tehuantepecensis Crosse & Fischer, 1893 — Cruz & Quesada (2017)

Anodonta (*Pachyanodon*) *depexa* von Martens, 1900 Land and freshwater Mollusca. *In*: F.D. Godman & O. Salvin (eds.). *Biologia Centrali-Americana* Part 158, Fascile 68–71:537–568, plates 38–39. Page 538, and Part 159, Fascile 72–75:569–600, plates 40–41. Pl. 41, figs. 3, 3a–b.

Verbatim Type Locality: "W. Guatemala: Paso San Antonio, in the lower part of the Rio Michatoya."

Type Locality: Río Michatoya (Pacific Dr.), Paso San Antonio, Escuintla Department, Guatemala [L = 14.0617, -90.6809].

Type Specimens: Lectotype, ZMB/MOLL.108855, here designated (the figured specimen). Paralectotypes, ZMB/MOLL.305850 (*n*=2).

Taxonomic History:

Anodontites (s.s.) *depexus* (von Martens, 1900) — Simpson (1914), Haas (1969)
Anodontites depexa (von Martens, 1900) — Frierson (1927), Graf & Cummings (2021)
Anodontites depexus (von Martens, 1900) — Graf & Cummings (2007)
Anodonta depexa von Martens, 1900 — Cruz & Quesada (2017)
Anodontites inaequivalvis [sic] (Lea, 1868) — Mansur et al. (2020)

Anodontites pittieri Marshall, 1922. New pearly freshwater mussels from South America. *Proceedings of the United States National Museum* 61(2437):1–9 + 3 plates. Page 6, pl. 1, figs. 9, 11; pl. 2, figs. 1, 12; pl. 3, fig. 6. Also figured in Pointier (2015:174, pl. 89, fig. 1), Martínez-Escarbassiere et al. (2004:295, fig. 1j), and Martínez-Escarbassiere et al. (2015:171, pl. 89, figs. 1a, 2a–c).

Verbatim Type Locality: "Guaremales, Venezuela."

Type Locality: Guaremales [= Guaremal], (Caribbean–Urama Dr.), Carabobo, Venezuela [L = 10.4444, -68.3908].

Type Specimens: Holotype, USNM 339954, by original designation (Johnson 1974b:12).

Taxonomic History:

Anodontites pittieri Marshall, 1922 — Marshall (1930), H.B. Baker (1930), Martínez-Escarbassiere et al. (2004), Graf & Cummings (2007; 2021), Lasso et al. (2009), Mansur et al. (2020)
Anodontites (s.s.) *pittieri* Marshall, 1922 — Pereira et al. (2014)

PLATE 78 ***Anodontites tortilis*** **(Lea, 1852). Top:** *Anodonta tortilis* Lea, 1852. Holotype, USNM 86405 (36 mm L, 23 mm H). **Middle Left:** *Anodonta trautwiniana* Lea, 1852. Lectotype, USNM 86684 (63 mm L, 38 mm H). Both, Cartagena (Caribbean Dr.), Bolivar Department, Colombia [L=10.3997, -75.5144]. **Middle Right:** *Anodonta luteola* Lea, 1858. Holotype, USNM 86766 (43 mm L, 28 mm H). Isthmus of Darien, Darién Province, Panamá [D=9.0300, -79.6344]. **Bottom Left:** *Anodonta carinata* Dunker, 1858. Syntype, ZMB/MOLL.108725 (68 mm L, 41 mm H). Río Choco (Caribbean–Atrato Dr.), Chocó Department, Colombia [W=7.2113, -76.6802]. **Bottom Right:** *Anodontites palmeri* Marshall, 1930. Holotype, USNM 380695 (75 mm L, 42 mm H). Small stream (Río Magdalena–Lebrija Dr.), 13 km S Puerto Santos, Province of Santander del Norte, Colombia [L=7.3944, -73.3975].

Anodontites tortilis (Lea, 1852)

Plate 78, Map 94

Province(s): Gulf of Honduras & Northern Caribbean; Oaxaca, San Juan; México to Colombia.

Distribution and Status: A widespread and common species in Caribbean and Pacific drainages Central and northern South America stretching from Guatemala south to the Magdalena and Orinoco basins of Colombia and Venezuela. Not assessed (IUCN 2024). Found live in the Rio Sabanas (Pacific–Lara Dr.) near Arimae, Darién Province, Panamá and the Rio Conaca (Pacific–Santa Maria Dr.), near Conaca, Veraguas Province, Panamá in 2023.

Anatomy: Shell small, thin, oblong, and edentate. Beaks moderately elevated. Anterior and posterior ends rounded. Ventral margin curved and convex. Surface with irregular growth lines and often rough in texture. Periostracum greenish-yellow or olive. Nacre white, bluish white, or pink.

Biology: Inhabits small rivers in mud and sand. Often abundant.

Adult Size: Length to 6 cm.

Similar Species: *Anodontites tehuantepecensis* (Crosse & Fischer, 1893) and *Anodontites trapesialis* (Lamarck, 1819).

MAP 94 Distribution of *Anodontites tortilis* (Lea, 1852).

Anodonta tortilis Lea, 1852. Descriptions of new species of the family Unionidae. *Transactions of the American Philosophical Society* 10(N.S.):253–294 + plates 12–29. Page 291, pl. 28, fig. 54.

Verbatim Type Locality: "Carthagena, South America."

Type Locality: Cartagena (Caribbean Dr.), Bolivar Department, Colombia [L = 10.3997, -75.5144].

Type Specimens: Holotype by monotypy, USNM 86405 (Johnson 1974:146; Breure et al. 2022:227, fig. 307).

Remarks: Simone (2006:280, pl. 1005) considered *A. tortilis* as valid and stated that USNM 86404 was the holotype. However, that specimen is *Anodontites schomburgiana* (Sowerby, 1870) from the Río Yuruari in the Orinoco drainage in Venezuela, not *Anodonta tortilis* Lea, 1852 from Cartagena, Colombia.

Taxonomic History:

Margaron (Anodonta) tortilis (Lea, 1852) — Lea (1852; 1870)
Anodonta tortilis Lea, 1852 — Hanley (1856), Hupé (1857)
Anodonta (Symphynota) tortilis Lea, 1852 — Adams & Adams (1857)
Anodon tortilis (Lea, 1852) — Sowerby (1870)
Glabaris (s.s.) *tortilis* (Lea, 1852) — Simpson (1900)
Anodontites (s.s.) *tortilis* (Lea, 1852) — Simpson (1914), Haas (1969), Pereira et al. (2014)
Anodontites tortilis (Lea, 1852) — Frierson (1927), Marshall (1930), H.B. Baker (1930), Mansur & Valer (1992), Taylor (1993), Simone (2006), Graf & Cummings (2007; 2021), Lasso et al. (2009), Mansur et al. (2020), Breure et al. (2022)

Anodonta trautwiniana Lea, 1852. Descriptions of new species of the family Unionidae. *Transactions of the American Philosophical Society* 10(N.S.):253–294+plates 12–29. Page 287, pl. 26, fig. 48.

Verbatim Type Locality: "Carthagena, South America."

Type Locality: Cartagena (Caribbean Dr.), Bolivar Department, Colombia [L=10.3997, -75.5144].

Type Specimens: Lectotype, USNM 86684 designated by Johnson and Baker (1973:173; Johnson 1974a:147, both as the holotype). Paralectotypes, ANSP 41841 (n2½); USNM 86684a (n=4) (Johnson and Baker 1973:173; Johnson 1974a:147, as paratypes).

Taxonomic History:

Margaron (Anodonta) troutwiniana [sic] (Lea, 1852) — Lea (1852)
Anodonta trantwiniana [sic] Lea, 1852 — Hupé (1857)
Anodonta troutwiniana [sic] Lea, 1852 — Adams & Adams (1857)
Anodon trautwinianus (Lea, 1852) — Sowerby (1870)
Margaron (Anodonta) trautwiniana (Lea, 1852) — Lea (1870)
Glabaris (s.s.) *trautwinianus* (Lea, 1852) — Simpson (1900)
Anodontites (s.s.) *trautwinianus* (Lea, 1852) — Simpson (1914)
Anodontites (s.s.) *trigonus trautwinianus* (Lea, 1852) — Haas (1931; 1969)
Anodontites trigonus trigonus (Spix in Wagner, 1827) — Ramírez et al. (2003)
Anodontites trigonus (Spix in Wagner, 1827) — Simone (2006)

Anodonta carinata Dunker, 1858. Einige neue species der naiaden. *Malakozoologische Blätter* 5:225–229. Page 225. Figured in Breure et al. (2022:225, fig. 304) and Mansur et al. (2020:428, fig. 11.127–D).

Verbatim Type Locality: "in flumine Rio Choco Novae Granádae."

Type Locality: Río Choco (Caribbean–Atrato Dr.), Chocó Department, Colombia [W=7.2113, -76.6802].

Type Specimens: Lectotype, ZMB/MOLL.108725 (n=2) (Haas 1931:92, fig. 36, as "Der Typ. Presumably, the larger of two specimens identified as the type and paratype in Berlin; Breure et al. 2022:225 as a syntype); Paralectotype, SMF 3871 (n=1) (Haas 1931:92, Zilch 1967:130)."

Remarks: Dunker (1858) was explicit that he had multiple specimens and only provided measurements for the largest one. Breure et al. (2022:225, figs. 303–304) figured the holotype of *Anodonta* *napoensis* Lea, 1868 (USNM 25429) and the lectotype of *Anodonta carinata* Dunker, 1858 (ZMB/MOLL.108725), but reversed the figure captions.

Taxonomic History:

Glabaris carinata (Dunker, 1858) — Simpson (1900)
Anodontites carinata (Dunker, 1858) — Simpson (1914), Marshall (1930) Graf & Cummings (2021)
Anodontites (s.s.) *carinatus carinatus* (Dunker, 1858) — Haas (1931; 1969)
Anodontites crispatus Bruguière, 1792 — Simone (2006)
Anodontites (s.s.) *carinatus* (Dunker, 1858) — Pereira et al. (2014)
Anodontites carinatus (Dunker, 1858) — Mansur et al. (2020)
Anodontites carinata (Dunker, 1858) — Graf & Cummings (2021), Breure et al. (2022)

Anodonta luteola Lea, 1858. Description of four new fresh-water molluscs, from the Isthmus of Darien and Honduras. *Proceedings of the Academy of Natural Sciences of Philadelphia* 10(1858):118. Figured in Lea (1860a JANSP Vol. 4, Part 3(7):267; 1860c Obs. 7:85, pl. 43, fig. 147).

Verbatim Type Locality: "Isthmus of Darien."

Type Locality: Isthmus of Darien, Darién Province, Panamá [D=9.0300, -79.6344].

Type Specimens: Holotype by monotypy, USNM 86766 (Johnson 1974:87).

Remarks: Pilsbry (1920b:10) stated that "Lea's type of this species [*A. luteola*] was said to be from the Isthmus of Darien, but his figures were drawn from a larger specimen from Lake Nicaragua (Gabb), No. 41833, A.N.S.P." An examination of the holotype in the USNM does not support that assumption.

Taxonomic History:

Anodonta luteola Lea, 1858 — Lea (1860a), Tate (1870), Clessin (1874a)
Anodon luteolus (Lea, 1858) — Sowerby (1870)
Margaron (*Anodonta*) *luteola* (Lea, 1858) — Lea (1870)
Anodonta (*Pachyanodon*) *luteola* Lea, 1858 — von Martens (1900d)
Glabaris (s.s.) *luteolus* (Lea, 1858) — Simpson (1900)
Anodontites (s.s.) *tortilis* (Lea, 1852) — Simpson (1914)
Anodontites luteolus (Lea, 1858) — Pilsbry (1920b)
Anodontites tortilis (Lea, 1852) — Frierson (1927), H.B. Baker (1930)

Anodontites palmeri Marshall, 1930. New land and fresh-water mollusks from South America. *Proceedings of the United States National Museum* 77(2825):1–7+2 plates. Page. 6, pl. 2, figs. 1, 4, 7.

Verbatim Type Locality: "It comes from a small stream, in a dense jungle, 13 kilometers south of Puerto Santos, Province of Santander del Norte, Republic of Colombia."

Type Locality: Small stream (Río Magdalena–Lebrija Dr.), 13 km S Puerto Santos, Province of Santander del Norte, Colombia [L=7.3944, -73.3975].

Type Specimens: Holotype, USNM 380695. Paratype, USNM 380702; both by original designation (Johnson 1974b:11).

Taxonomic History:

Anodontites (s.s.) *carinatus carinatus* (Dunker, 1858) — Haas (1931; 1969)
Anodontites crispatus Bruguière, 1792 — Simone (2006)

PLATE 79 ***Anodontites trapesialis* (Lamarck, 1919). Top Left:** *Anodonta trapesialis* Lamarck, 1819. Lectotype, MHNG-MOLL-50581. Locality Unknown. **Middle Right:** *Anodonta glauca* Lamarck, 1819. Paralectotype, MNHN-IM-2000-1706. Acapulco (Pacific Dr.) Guerrero, México [L=16.8500, -99.9166]. **Middle Left:** *Anodonta nicaraguae* Philippi, 1849. Paralectotype, MCZ 184530. Lake Nicaragua (Caribbean–San Juan Dr.), Río San Juan Department, Nicaragua [W=11.1350, -84.8367]. **Bottom Right:** *Anodon ciconia* Gould, 1851. Lectotype, MCZ 169084 (109 mm L, 62 mm H). México?

PLATE 80 ***Anodontites trapesialis*** **(Lamarck, 1919). Top Left:** *Anodonta bridgesii* Lea, 1868. Lectotype, USNM 86691 (118 mm L, 64 mm H). **Middle Right:** *Anodonta jewettiana* Lea, 1868. Holotype, USNM 86681 (109 mm L, 63 mm H). Both, Lake Nicaragua (Caribbean–San Juan Dr.), Río San Juan Department, Nicaragua [W = 11.1350, -84.8367]. **Middle Left:** *Anodonta glauca* var. *sinaloensis* Crosse & Fischer, 1884. Syntype, MCZ 87549. Río Presidio (Pacific Dr.), near Mazatlan, Sinaloa, México [L = 23.1016, -106.2864]. **Bottom Right:** *Anodontites flucki* Walker, 1924. Holotype, UMMZ 112641 (87 mm L, 47 mm H). Río Coco (Caribbean Dr.), near Sangsang, Región Autónoma del Atlántico Norte, Nicaragua [L = 14.6685, -84.6303].

Anodontites trapesialis (Lamarck, 1819)

Plates 79–80, Map 95

Province(s): Mesa Central and Balsas; Gulf of Honduras & Northern Caribbean; San Juan; Oaxaca, México to Colombia.

Distribution and Status: The most widely distributed of all freshwater mussels ranging from México south to Argentina and often common where found. *Anodontites trapesialis* is known in Central America primarily along the Pacific coast from Sinaloa, Jalisco, and Guerrero, México south to Nicaragua and Panamá. Also found in a few river basins (San Juan–Lake Nicaragua, Coco, Sixaola, and Patuca) that drain into the Caribbean in Costa Rica, Nicaragua, and Honduras. There is a considerable gap between the populations in México and the rest of Central America, and it is conceivable that there is more than one species represented in the region. Not assessed (IUCN 2024).

Anatomy: Shell highly variable but typically large, elliptical, edentate, thin, inflated, with full prominent beaks. A slight wing is present on the post-dorsal part of the shell and occasionally a small one in front of the beaks. Gaping decidedly in front and slightly behind. A triangular ligamental sinus is present on the hinge line. Hinge line straight or slightly sinuous. Periostracum smooth, typically yellow or olive with green rays. Nacre bluish-silver and iridescent. Animal with the marsupium filling the inner gills. Branchial opening small, plicate, but not crenulate or papillose; anal opening large, smooth, separated from the branchial.

Biology: Often abundant in small to large rivers in mud and sand. Reported fish hosts include: *Oreochromis niloticus* (Linnaeus, 1758); *Geophagus brasiliensis* (Quoy & Gaimard, 1824); *Coptodon rendalli* (Boulenger, 1897) (Cichlidae); *Piaractus brachypomus* (Cuvier, 1818) (Serrasalmidae); and *Cyprinus carpio* (Linnaeus, 1758) (Cyprinidae); *Psalidodon fasciatus* (Cuvier, 1819) (*Characidae*); *Prochilodus lineatus* (Valenciennes, 1837) (Prochilodontidae); *Hoplias malabaricus* (Bloch, 1794) (Erythrindae); *Plecostomus* sp. (Loricariidae); *Pimelodella gracilis* (Valenciennes, 1835) (Heptapteridae); *Corydoras paleatus* (Jenyns, 1842) (Callichthyidae); *Synodontis clarias* (Linnaeus, 1758) (Mochokidae); *Jenynsia lineata* (Jenyns, 1842) (Anablepidae) (Bonetto and Ezcurra 1962:198; Parodiz and Bonetto. 1963:184; Silva-Souza and Eiras. 2002:432; Castro-Rojas and Rodríguez-Pulido 2003:24).

Adult Size: Length to 17.5 cm.

Similar Species: *Brachyanodon impura* (Say, 1829) and *Anodontites trapesialis* (Lamarck, 1819) co-occur in the Ameca and in Lago de Chapala in Jalisco, México. They can be separated by the presence of a triangular ligamental sinus at the posterior end of the hinge in *A. trapesialis* (Figure 2). *Anodontites tortilis* (Lea, 1852) has a rough periostracum and is much smaller as an adult.

Remarks: This species has an extensive synonymy. The taxa below are the nominal taxon and those that have been described from Central America, should some of those populations warrant specific status. Simone (1994) described the anatomy and discussed the systematics of *Anodontites trapesialis* (Lamarck, 1819) in South America.

Anodonta trapesialis Lamarck, 1819. Anodonte (*Anodonta*). Pp. 83–88 *In* Les Nayades. Histoire Naturelle des Animaux sans Vertèbres. Paris. 6(1):vi–343 pp. Page 87, species 11. Figured in Küster (1838:185, pl. 86, fig. 762).

Verbatim Type Locality: "Habite..... des caux douces étrangères à celles de l'Europe?"

Type Locality: Unknown.

MAP 95 Distribution of *Anodontites trapesialis* (Lamarck, 1919) in Central America and northwestern South America.

Type Specimens: Lectotype and Paralectotype, MHNG-MOLL-50581 (n=2) (Johnson 1953:94, as the holotype and paratype; Emmanuel Tardy, Geneva Museum, pers. comm.; Breure et al. 2022:227, fig. 308, as syntypes).

Remarks: Although USNM 341372 were called "Probable syntypes" by Simone (2006:281, pl. 1006b, fig. F), these specimens are from the Río Yi at Durazno, Uruguay and were collected by F.F. Felippone well after Lamarck described the species and are not types.

Taxonomic History:

Anodonta trapezialis [sic] Lamarck, 1819 — Blainville (1825), — Bory de St. Vincent (1827), Menke (1828; 1830), Lea (1834), Hanley (1843), Hupé (1857)
Iridina trapezialis d'Orbigny (*non* Lea) — d'Orbigny (1835)
Margarita (*Anodonta*) *trapezialis* [sic] (Lamarck, 1819) — Lea (1836)
Anodonta exotica Lamarck, 1819 — d'Orbigny (1846b)
Anodon trapezialis [sic] (Lamarck, 1819) — Swainson (1840), Sowerby (1867)
Margaron (*Anodonta*) *trapezialis* [sic] (Lamarck, 1819) — Lea (1852; 1870)
Anodonta (*Patularia*) *trapezialis* [sic] Lamarck, 1819 — Adams & Adams (1857)
Anodonta trazesialis [sic] Lamarck, 1819 — von Ihering (1890)
Glabaris (s.s.) *trapesialis* (Lamarck, 1819) — Simpson (1900)
Glabaris trapezialis [sic] (Lamarck, 1819) — von Ihering (1910)
Anodontites (s.s.) *trapesialis* (Lamarck, 1819) — Simpson (1914), Haas (1916), Rumi et al. (2008), Pereira et al. (2014), Torres et al. (2018)
Anodontites trapezialis [sic] (Lamarck, 1819) — Marshall (1915), H.B. Baker (1930)
Anodontites (s.s.) *trapesialis trapesialis* (Lamarck, 1819) — Haas (1931; 1969), Figueiras (1965)
Glabaris trapesialis (Lamarck, 1819) — Modell (1942)
Anodontites (*Pachyanodon*) *trapezialis trapezialis* [sic] (Lamarck, 1819) — de Morretes (1949)

Anodontites trapesialis (Lamarck, 1819) — de Lucena (1951), Graf (2000), Martínez-Escarbassiere et al. (2004), Mansur & Pereira (2006), Simone (2006), Graf & Cummings (2007; 2021), Avelar & Vianna (2008), Lasso et al. (2009), Guerra et al. (2017), Huang et al. (2018; 2019), Mansur et al. (2020), Bieler (2021), Agudo-Padrón (2022), Breure et al. (2022), Czaja et al. (2023)
Anodontites trapesialis trapesialis (Lamarck, 1819) — Mansur & Valer (1992), Ramírez et al. (2003)
Anodontites patagonicus (Lamarck, 1819) — Taylor (1993)

Anodonta glauca Lamarck, 1819. Anodonte (*Anodonta*). Pp. 83–88 *In* Les Nayades. Histoire Naturelle des Animaux sans Vertèbres. Paris. 6(1):vi–343 pp. Page 87, species 13. Figured in Fischer and Crosse (1894a:533, pl. 69, figs. 1–1a).

Verbatim Type Locality: "Habit en Amérique, dans des eaux douces voisines d'Acapulco."

Type Locality: Acapulco (Pacific drainage) Guerrero, México [L = 16.8500, -99.9166].

Type Specimens: Lectotype, MHNG-MOLL-50591 designated by Johnson (1953:90, as the figured holotype). Paralectotypes, MNHN-IM-2000-1706 ($n = 2$) (Johnson 1953:90, without catalogue number, presumably from the type lot).

Remarks: Lamarck (1819) described *Anodonta glauca* and cited *A. glauca* crediting Valenciennes with the name. He stated that the specimens were from the Humboldt and Bonpland expedition and that his specimens were in his cabinet. "Collection de MM. le baron de Humbolt et Bonpland. Belle espèce, très - distincte, à coquille mince, très-fragile. Largeur, 98 millimètres. Mon cabinet." Valenciennes redescribed it in 1827, but Lamarck's name has priority and *Anodonta glauca* Valenciennes in Humbolt & Bonpland, 1827 is a junior subjective synonym.

A note with the specimens at MHNG: forme typique: 1 spécimen à l'origine (variété b: voir n° 50576); "1 échantillon (douteux) de 80 mm, probablement figuré par Chenu, Manuel de Conchyliologie, fig. 733." Translated: typical form: originally 1 specimen (variety b: see no. 50576); "1 (doubtful) sample of 80 mm, probably figured by Chenu, Manuel de Conchyliologie, fig. 733."

Taxonomic History:

Anodonta glauca Lamarck, 1819 — Valenciennes in Humboldt & Bonpland (1827), Lea (1834), Ferussac (1835), Delessert (1841), Hanley (1843), Conrad (1853), Adams & Adams (1857), Cruz & Quesada (2017)
Margarita (*Anodonta*) *glauca* (Lamarck, 1819) — Lea (1836)
Margaron (*Anodonta*) *glauca* (Lamarck, 1819) — Lea (1852; 1870)
Anodon glaucus (Lamarck, 1819) — Sowerby (1870)
Anodonta (sec. *Patularia*) *glauca* Lamarck, 1819 — Fischer (1886)
Anodonta (*Pyganodon*) *glauca* Lamarck, 1819 — Fischer & Crosse (1894a)
Anodonta (*Patularia*) *glauca* Lamarck, 1819 — von Martens (1900d)
Glabaris (s.s.) *glaucus* (Lamarck, 1819) — Simpson (1900)
Anodontites (s.s.) *glauca* (Lamarck, 1819) — Simpson (1914)
Leila (*Pseudoleila*) *trapezialis-glauca* [sic] (Lamarck, 1819) — Frierson (1927)
Anodontites trapezialis [sic] *glaucus* (Lamarck, 1819) — Haas (1929)
Anodontites (s.s.) *trapesialis glaucus* (Lamarck, 1819) — Haas (1931; 1969)
Anodontites (*Patularia*) *glauca* (Lamarck 1819) — Morrison (1968)
Anodontites trapesialis glaucus (Lamarck, 1819) — Taylor (1993)

Anodonta Nicaraguae Philippi, 1849. Centuria altera Testaceorum novorum. *Zeitschrift für Malakozoologie* 5(9):129–144. Page 130. Not figured.

Verbatim Type Locality: "Nicaragua."

Type Locality: Lake Nicaragua (Caribbean–San Juan Dr.), Río San Juan Department, Nicaragua [W=11.1350, -84.8367].

Type Specimens: Lectotype designated by Haas (1930:322, fig. 1), Muséum d'Histoire Naturelle de Rouen, France MALA.2015.0.619.0.MHN (Johnson 1951b:32; Coan and Kabat 2017:71; Nagel 2025:221). Paralectotypes, MCZ 184530 (*n*=1) (Johnson 1951a:81, plate 5, fig. 1, as the lectotype; Johnson 1951b:32, as a paratype), MCZ 17071 (*n*=1) (Johnson 1951a:81; Johnson 1951b:32; 1956:125, both as a paratype), SMF 5106 (*n*=1) (Zilch 1967:131, as a syntype; Nagel 2025:221); Muséum of Rouen (*n*=2) (Coan and Kabat 2017:71).

Remarks: Johnson's selection of a lectotype (1951a:81) was invalid as one had already been designated by Haas in 1930. Johnson later (1951b:32) corrected his error crediting Haas (1930).

Taxonomic History:

Anodonta nicaraguae Philippi, 1849 — Adams & Adams (1857), Haas (1930), Cruz & Quesada (2017)
Anodonta nicaragua [sic] Philippi, 1849 —Tate (1870)
Anodonta (*Glabaris*) *nicaragua* [sic] Philippi, 1849 — von Martens (1900d)
Glabaris nicaragua [sic] (Philippi, 1849) — Simpson (1900)
Anodontites nicaragua [sic] (Philippi, 1849) — Simpson (1914)
Leila (*Pseudoleila*) *nicaragua* [sic] (Philippi, 1849) — Frierson (1927)
Anodontites trapezialis [sic] *bridgesii* (Philippi, 1849) — Haas (1929)
Anodontites (s.s.) *trapesialis nicaraguae* (Philippi, 1849) — Haas (1930)
Anodontites (s.s.) *trapesialis glaucus* (Valenciennes in Humbolt & Bonpland, 1827) — Haas (1931; 1969), Zilch (1967)
Anodontites nicaraguae (Philippi, 1849) — Johnson (1951a)
Anodontites trapesialis (Lamarck, 1819) — Coan & Kabat (2017), Nagel (2025)
Leila nicaragua [sic] (Philippi, 1849) — Bieler (2021)

Anodon ciconia Gould, 1851. [Descriptions of new species of Mollusca from California and México]. *Proceedings of the Boston Society of Natural History* 4(6):87–93. Page 92. Figured in Johnson (1964:54, pl. 34, fig. 1) and von Martens (1900f, pl. 41, fig. 4, ZMB/MOLL.305851).

Verbatim Type Locality: "Mexico?"

Type Locality: México?

Type Specimens: Lectotype, MCZ 169084 designated by Johnson (1964:54, as the holotype and figured as MCZ 169048; Johnson 1971:81, as the holotype and as MCZ 189084). Paralectotypes, MCZ 169085 (*n*=2), USNM 611266 (*n*=1) (Johnson 1964:54; Johnson 1971:81, all as paratypes).

Remarks: NHMUK 1964397 (*n*=1) was figured in Sowerby 1870 (pl. 29, species 115a) as *Anodon ciconia* Gould, 1851. Johnson (1964:54; Johnson 1971:81, as a paratype) noted that "The specimen figured by Sowerby, 1867, in Reeve, Conchologia Iconica, vol. 17, *Anodon*, pl. 29, fig. 115, has been examined by the present author and is not a type." According to the NHMUK catalogue, the specimen was collected in the USA. This specimen clearly has a ligamental sinus and is *A. trapesialis* and thus the published locality data are incorrect. Johnson also stated that the second specimen figured by Sowerby as *Anodon ciconia* was NHMUK 1964389 [sic] (actually NHMUK 1964398) was not the same species but did not say what he thought it was.

Taxonomic History:

Anodon ciconia Gould, 1851 — Sowerby (1870)
Margaron (*Anodonta*) *trapezialis* [sic] (Lamarck, 1819) — Lea (1870)
Anodonta (*Pseudoleila*) *ciconia* (Gould, 1851) — Fischer & Crosse (1894a), von Martens (1900d)
Glabaris (s.s.) *trapesialis* (Lamarck, 1819) — Simpson (1900)

Glabaris (s.s.) *glaucus* (Valenciennes, 1833) — Simpson (1900)
Anodontites (s.s.) *trapesialis* (Lamarck, 1819) — Simpson (1914)
Leila (*Pseudoleila*) *ciconia* (Gould, 1851) — Frierson (1927)
Anodontites (s.s.) *trapesialis nicaraguae* (Philippi, 1849) — Haas (1930)
Anodontites (s.s.) *trapesialis glaucus* (Valenciennes in Humbolt & Bonpland, 1827) — Haas (1931; 1969)
Anodonta ciconia (Gould, 1851) — Cruz & Quesada (2017)

Anodonta bridgesii Lea, 1868. Description of nine species of Unionidae from Lake Nicaragua, Central America. *Proceedings of the Academy of Natural Sciences of Philadelphia* 20:94–96. Page 95. Figured in Lea (1868a JANSP Vol. 6:291, 1869a Obs. 12:51, pl. 42, fig. 104).

Verbatim Type Locality: "Lake Nicaragua, Cent. Amer."

Type Locality: Lake Nicaragua (Caribbean–San Juan Dr.), Río San Juan Department, Nicaragua [W = 11.1350, -84.8367].

Type Specimens: Lectotype, USNM 86691 designated by Johnson (1951a:81; Johnson and Baker 1973:149; Johnson 1974a:23, all as the holotype). Paralectotypes, ANSP 41850 ($n = 2$) (Johnson and Baker 1973:149, as paratypes).

Taxonomic History:

Anodonta bridgesii Lea, 1868 — Tate (1870), Clessin (1874b)
Margaron (*Anodonta*) *bridgesii* (Lea, 1868) — Lea (1870)
Anodonta (*Glabaris*) *nicaragua* [sic] Philippi, 1849 — von Martens (1900d)
Glabaris (s.s.) *bridgesii* (Lea, 1868) — Simpson (1900)
Anodontites (s.s.) *bridgesii* (Lea, 1868) — Simpson (1914)
Leila (*Pseudoleila*) *nicaragua* [sic] (Philippi, 1849) — Frierson (1927)
Anodontites trapezialis [sic] *bridgesii* (Lea, 1868) — Haas (1929)
Anodontites (s.s.) *trapesialis nicaraguae* (Philippi, 1849) — Haas (1930)
Anodontites (s.s.) *trapesialis glaucus* (Valenciennes in Humbolt & Bonpland, 1827) — Haas (1931; 1969)
Anodontites nicaraguae (Philippi, 1849) — Johnson (1951a)

Anodonta jewettiana Lea, 1868. Description of nine species of Unionidae from Lake Nicaragua, Central America. *Proceedings of the Academy of Natural Sciences of Philadelphia* 20:94–96. Page 95. Figured in Lea (1868a JANSP Vol. 6:289, 1869a Obs. 12:49, pl. 41, fig. 101, as *Anodonta jewettii*).

Verbatim Type Locality: "Lake Nicaragua, Cent. Amer."

Type Locality: Lake Nicaragua (Caribbean–San Juan Dr.), Río San Juan Department, Nicaragua [W = 11.1350, -84.8367].

Type Specimen: Holotype by monotypy, USNM 86681 (Johnson 1951a:81; Johnson 1974a:78).

Remarks: Bieler (2021:21) noted that Lea (1868c:95) described a species of freshwater mussel from Lake Nicaragua as *Anodonta jewettiana*, as a patronym for "Col. E. Jewett." Lea (1868a: 289, pl. 41, fig. 101; copied in 1869a: 49, pl. 41, fig. 101) changed the spelling to *Anodonta jewettii*. While perhaps an intentional change of the original spelling, Lea's (1868a) action does not fulfill the requirements of a demonstrably intentional emendation under Art. 33.2 (Lea referenced the original publication but did not cite the original spelling).

Taxonomic History:

Anodonta jewettii [sic] Lea, 1868 — Lea (1868a), Clessin (1874b)
Margaron (*Anodonta*) *jewettii* [sic] (Lea, 1868) — Lea (1870)
Anodonta jewettii [sic] Lea, 1868 — Sowerby (1870), Wright (1888)

Anodonta jewettii [sic] Lea, 1868 — Wright (1888)
Anodonta (Glabaris) jewettii [sic] Lea, 1868 — von Martens (1900d)
Glabaris (s.s.) *jewittianus* [sic] (Lea, 1868) — Simpson (1900)
Anodontites (s.s.) *jewettianus* [sic] (Lea, 1868) — Simpson (1914)
Leila (*Pseudoleila*) *nicaragua* [sic] (Philippi, 1848) — Frierson (1927, as *jewittiana* and *jewitii*)
Anodontites (s.s.) *trapesialis nicaraguae* (Philippi, 1849) — Haas (1930)
Anodontites (s.s.) *trapesialis glaucus* (Valenciennes in Humbolt & Bonpland, 1827) — Haas (1931)
Anodontites nicaraguae (Philippi, 1849) — Johnson (1951a)
Anodontites (s.s.) *trapesialis glaucus* (Valenciennes, 1833) — Haas (1969)
Anodonta jewettiana Lea, 1868 — Cruz & Quesada (2017)
Anodontites trapesialis (Lamarck, 1819) — Bieler (2021)

Anodonta glauca var. *β*, *sinaloensis* Crosse & Fischer, 1884. Description d'une variété nouvelle d'*Anodonta*, recueillie par M.A. Forrer, dans l'Etat de Sinaloa (Mexique). *Journal de Conchyliologie* 31(3):219–221. Page 219. Not figured.

Verbatim Type Locality: "environs de Mazatlan, dans l'etat de Sinaloa."

Type Locality: Río Presidio (Pacific Dr.), near Mazatlan, Sinaloa, México [L=23.1016, -106.2864].

Type Specimens: Syntypes, MCZ 87549 (*n*=1) (Johnson 1956:135); MNHN-IM-2000-1707 (*n*=1).

Remarks: River data from the label with the syntype at MNHN.

Taxonomic History:

Anodonta (*Pseudoleila*) *ciconia* (Gould, 1851) — Fischer & Crosse (1894a), von Martens (1900d)
Glabaris (s.s.) *glaucus* var. *sinaloensis* (Crosse & Fischer, 1884) — Simpson (1900)
Glabaris glaucus var. *sinaloensis* (Crosse & Fischer, 1884) — de Leon (1912)
Anodontites (s.s.) *glauca* var. *sinaloensis* (Crosse & Fischer, 1884) — Simpson (1914)
Leila (*Pseudoleila*) *ciconia* (Gould, 1851) — Frierson (1927)
Anodontites trapezialis [sic] *sinaloensis* (Crosse & Fischer, 1884) — Haas (1929)
Anodontites (s.s.) *trapesialis nicaraguae* (Philippi, 1849) — Haas (1930)
Anodontites (s.s.) *trapesialis glaucus* (Valenciennes in Humbolt & Bonpland, 1827) — Haas (1931; 1969)
Anodontites sinaloensis (Crosse & Fischer, 1884) — Contreras-Arquieta (2000)

Anodontites flucki Walker, 1924. Two new naiades from Nicaragua. *The Nautilus* 38(2):51–54 + plate 1, Page 53, pl. 1, figs. 3–4.

Verbatim Type Locality: "the Wanks River or Rio Coco, Nicaragua."

Type Locality: Río Coco [= Wanks River] (Caribbean Dr.), near Sangsang, Región Autónoma del Atlántico Norte, Nicaragua [L=14.6685, -84.6303].

Type Specimens: Holotype, UMMZ 112641, by original designation (ex. Walker 74610). Paratypes, MCZ 167989 (*n*=3) (Johnson 1956:115).

Remarks: The holotype was considered lost by Johnson (1979:32) who gave the Walker number (74610) as the UMMZ catalogue number. Specific locality data from the label with the holotype.

Taxonomic History:

Anodontites (s.s.) *trapesialis glaucus* (Valenciennes in Humbolt & Bonpland, 1827) — Haas (1969)

PLATE 81 ***Nomina dubia.*** **Top Left:** *Unio callosus* Lea, 1841. Holotype, AMNH 30561 (56 mm L, 36 mm H). Ohio Canal, 12 miles below Columbus. **Top Right:** *Unio melleus* Lea, 1859, Holotype, NHMUK 1965175 (51 mm L, 28 mm H). Locality unknown. **Middle Left:** *Unio oregonensis* Lea, 1852. Lectotype, USNM 85073 (81 mm L, 46 mm H). Columbia River, Oregon. **Middle Right:** *Unio sphenorhynchus* Crosse & Fischer, 1894. Holotype, MNHN-IM-2000-1792 (71 mm L, 37 mm H). republica Mexicana. **Bottom Left:** *Unio* (*Simonaias*) *cuprinus* var. *subventralis* von Martens, 1900. Holotype, ZMB/MOLL.108852 (55 mm L, 29 mm H). S.E. México: Tabasco.

Nomina dubia

The following are taxa that were previously considered as valid by Graf and Cummings (2021) and others, but are here regarded as *Nomina dubia* or doubtful species of uncertain affinities. Taxa are listed alphabetically by specific epithet, and those with type specimens are shown on Plate 81.

Unio callosus Lea, 1841. On fresh water and land shells (cont.). *Proceedings of the American Philosophical Society* 2(17):30–34. Page 31. Figured in Lea (1842a TAPS Vol. 8:329; 1842b Obs. 3:77, pl. 23, fig. 54) and Boyko and Sage (1996:24, fig. 8).

Verbatim Type Locality: "Ohio Canal, 12 miles below Columbus."

Type specimen: Holotype by monotypy, AMNH 30561 (ex. Jay Collection 1524) (Boyko and Sage 1996:24).

Remarks: Johnson (1974a:26) stated that the type specimen was lost, but it was subsequently found in the AMNH in 1994 (Boyko and Sage 1996:24). Without examining a type specimen, Simpson (1900:701) concluded that this was a México species stating "I am sure that the locality is wrong, and it appears to be a Mexican form" saying it was almost identical to *Unio coloratus* Küster, 1856 (another taxon of dubious identity). Frierson (1927:37) equated it with *Unio tehuantepecensis* Crosse & Fischer 1894. Boyko and Sage (1996:24) suggested that "This species is a possible senior synonym of *Nephronaias* (*Sphenonaias*) *mexicana* (Philippi, 1847)" citing Haas (1969:206), but provided no new evidence that it was in fact collected in México. Given that the published type locality was in Ohio, USA, the supposition that the holotype was collected in México is pure conjecture. We follow Inoue et al. (2020a:476) in synonymizing this under *Unio ligamentina* Lamarck, 1819.

Taxonomic History:

Unio callosus Lea, 1841 — Conrad (1853), von Martens (1901c)
Margaron (*Unio*) *callosus* (Lea, 1841) — Lea (1852; 1870)
Unio (*Niäa*) *callosus* Lea, 1841 — Adams & Adams (1857)
Unio (sec. *Elliptio*) *callosus* Lea, 1841 — Simpson (1900; 1914)
Unio callorus [sic] Lea, 1841 — de Leon (1912)
Elliptio (*Sphenonaias*) *callosus* (Lea, 1841) — Frierson (1927)
Nephronaias (*Sphenonaias*) *mexicana* (Philippi, 1847) — Haas (1969)
Sphenonaias callosa (Lea, 1841) — Graf & Cummings (2007; 2021)
Unio ligamentina Lamarck, 1819 — Inoue et al. (2020a)

Unio coloratus Küster, 1856. Die flusperlmuscheln (*Unio* et *Hyria*). *In* Abbildungen nach der Natur mit Beschreibungen. Systematisches Conchylien-Cabinet von Martini und Chemnitz. Küster edition. 9(Part 2):137–160, pls. 45–54. Page 155, pl. 44, fig. 6.

Verbatim Type Locality: "Mexiko."

Type Specimens: Unknown. Presumed lost.

Remarks: Early workers considered this taxon as valid, and most have placed it in *Nephronaias* (Fischer & Crosse (1894b) or *Simonaias* (von Martens 1900b:506). Everyone from Baker (1922) on have placed it in *Sphenonaias* Crosse & Fischer, 1894. von Martens (1865:59) gave the locality as "Fluss Medelin, bei Veracruz" citing Uhde. Given the lack of a specific type locality and a type specimen, this taxon is unidentifiable based upon Küster's description and figure. Attributed by Küster to "Charpentier mss."

Taxonomic History:

Unio coloratus Küster, 1856 — von Martens (1865), de Leon (1912)
Unio (*Nephronaias*) *coloratus* Küster, 1856 — Fischer & Crosse (1894b)
Unio (*Simonaias*) *cuprinus* var. *coloratus* Küster, 1856 — von Martens (1900b)
Unio (sec. *Elliptio*) *coloratus* Küster, 1856 — Simpson (1900; 1914)
Elliptio (*Sphenonaias*) *plexus* (Conrad, 1838) — H.B. Baker (1922)
Elliptio (*Sphenonaias*) *coloratus* (Küster, 1856) — Frierson (1927)
Nephronaias (*Sphenonaias*) *colorata* (Küster, 1856) — Haas (1969)
Sphenonaias colorata (Küster, 1856) — Graf & Cummings (2007; 2021), Czaja et al. (2023)

Unio mexicanus Philippi 1847. Testaceorum novorum centuria. *Zeitschrift für Malakozoologie* 4(6):84–96. Page 95. Figured in Philippi (1849:28, pl. 6, fig. 3) and Küster (1862:285, pl. 95, fig. 7).

Verbatim Type Locality: "México."

Type Specimens: Unknown. Presumed lost.

Remarks: All previous workers have considered this taxon as valid and almost everyone from Frierson (1927) to the present have placed it in *Sphenonaias* Crosse & Fischer, 1894. von Martens (1900b:517) remarked that *U. opacatus* was "perhaps only a shorter variety of *U. mexicanus*." Simpson (1914) apparently considered *U. mexicanus* as practically unidentifiable and stated that "Philippi gives a miserable colored figure of this species, which is so blurred that it is almost useless for purposes of identification, while his brief Latin description helps out but little." He said there was a single, rather young shell in the Lea collection that had the shell characteristics of a young *U. opacatus*. Given the brief description, the lack of a specific type locality or any type specimens *U. mexicanus* could be any number of things and this taxon is unidentifiable.

Taxonomic History:

Unio mexicanus Philippi, 1847 — Philippi (1849), Küster (1862), de Leon (1912), H.B. Baker (1922)
Margaron (*Unio*) *mexicanus* (Philippi, 1847) — Lea (1870)
Unio (*Uniomerus*) *mexicanus* Philippi, 1847 — Fischer & Crosse (1894b)
Unio (*Elliptio*) *mexicanus* Philippi, 1847 — von Martens (1900b)
Unio (sec. *Elliptio*) *mexicanus* Philippi, 1847 — Simpson (1900; 1914)
Actinonaias mexicanus (Philippi, 1847) — H.B. Baker (1922)
Elliptio (*Sphenonaias*) *mexicanus* (Philippi, 1847) — Frierson (1927)
Elliptio (*Nephronaias*) *mexicanus* (Philippi, 1847) — Haas (1929)
Nephronaias (*Sphenonaias*) *mexicana* (Philippi, 1847) — Haas (1969)
Sphenonaias mexicana (Philippi, 1847) — Graf & Cummings (2007; 2021), Coan & Kabat (2017), Czaja et al. (2023)

Unio melleus Lea, 1859. Descriptions of twenty-one new species of exotic Unionidae. *Proceedings of the Academy of Natural Sciences of Philadelphia* 11(1859):151–154. Page 152. Figured in Lea (1860a JANSP Vol. 4, Part 3(7):250; 1860c Obs. 7:68, pl. 38, fig. 129).

Verbatim Type Locality: "Hab.—."

Type Specimens: Holotype by monotypy, NHMUK 1965175 (Johnson 1971:88; Johnson 1974:90; ex. Hugh Cuming Collection).

Remarks: Lea (1860c:68) mentioned having a single specimen sent to him by Hugh Cuming without a locality, and the holotype is in the NHMUK. For some reason, all previous workers have considered this taxon as valid and as coming from Central America without data to support that contention. Although stated by Simpson (1900:597) to be from "Mexico or Central America, no doubt," it looks like nothing we have seen from Central America.

Taxonomic History:

Unio melleus Lea, 1859 — Lea (1860a), Reeve (1865)
Margaron (*Unio*) *melleus* (Lea, 1859) — Lea (1870)
Nephronaias melleus (Lea, 1859) — Simpson (1900), de Leon (1912), Frierson (1917)
Nephronaias mellea (Lea, 1859) — Simpson (1914), Graf & Cummings (2007; 2021), Czaja et al. (2023)
Elliptio (*Nephronaias*) *melleus* (Lea, 1859) — Frierson (1927)
Nephronaias (s.s.) *mellea* (Lea, 1859) — Haas (1969)

Unio oregonensis Lea, 1852. Descriptions of new species of the family Unionidae. *Transactions of the American Philosophical Society* 10(N.S.):253–294 + plates 12–29. Page 275, pl. 22, fig. 33.

Verbatim Type Locality: "Columbia River, Oregon."

Type Specimens: Lectotype, USNM 85073 designated by Johnson and Baker (1973:164; Johnson 1974:104, both as the holotype).

Remarks: The locality "Columbia R., Oregon" is penciled in the left valve. Despite this and the published type locality of the Columbia River, Oregon, Simpson (1900:594) stated that this taxon was from "Nicaragua" and "Nicaragua; Guatemala" (Simpson 1914:261), without any supporting data. An old pencil label (presumably in Simpson's hand) with the lectotype states that it is "undoubtedly an old *Unio rowellii* Lea from Cent. Am." The catalogue entry for *Unio oregonensis* NHMUK 1907.12.30.52 (figured in Sowerby 1868, pl. 74, species 383) is "Columbia River, Oregon." The locality "Hayden Island" was also added to the online catalogue entry. It is most certainly not from Oregon, but there are no data to support the contention that the specimens were collected in either Nicaragua or Guatemala either. Given the published type locality, we regard this as a *nomen dubium* for Central America.

Taxonomic History:

Unio oregonensis Lea, 1852 — Sowerby (1868)
Margaron (*Unio*) *oregonensis* (Lea, 1852) — Lea (1852; 1870)
Unio (*Hyridella*) *oregonensis* Lea, 1852 — Adams & Adams (1857)
Naphronaias [sic] *oregonensis* (Lea, 1852) — Simpson (1900)
Nephronaias oregonensis (Lea, 1852) — Simpson (1914), Graf & Cummings (2007; 2021)
Elliptio (*Nephronaias*) *oregonensis* (Lea, 1852) — Frierson (1927)
Nephronaias (s.s.) *oregonensis* (Lea, 1852) — Haas (1969)

Psoronaias plexoideus Frierson, 1927. *A Classification and Annotated Check List of the North American Naiades*. Baylor University Press, Waco, Texas. 111 pp. Page 63, sp. 273. Not figured.

Verbatim Type Locality: "Siam."

Type specimen(s): NHMUK, lost (Johnson 1972a:147, but see below under *rusticus*).

Remarks: This is Frierson's replacement name for *Unio rusticus* Sowerby, 1868 (pl. 64, species 324). Johnson (1972:170, plate 31, fig. 2) reproduced Sowerby's figure and stated that the holotype of *Psoronaias plexoides* [sic] was from the Río Usumacinta, Guatemala despite the published type locality of "Siam." Although Simpson (1900:702; 1914:591, as *Unio crocodilarum* [sic] Morelet, 1849), Frierson (1927:63), Haas [1969:247, as *Barynaias plexus* (Conrad, 1838)], and Johnson

(1972a:170, pl. 31, fig. 2) have opined that Sowerby's figured specimen is a Central American species, there are no data to support that assumption.

Taxonomic History:

Elliptio (Barynaias) plexoideus (Frierson, 1927) — Haas (1930)
Barynaias plexus (Conrad, 1838) — Haas (1969)
Psoronaias plexoides [sic] Frierson, 1927 — Johnson (1972)
Psoronaias plexoideus Frierson, 1927 — Johnson (1972)

Unio purpuriatus Say, 1831. New terrestrial and fluviatile shells of North America. (cont.). The Disseminator. [2nd Series]. New Harmony, Indiana 1(29): no pagination, 15 January 1831. Not figured.

Verbatim Type Locality: "México....in a stream a few leagues from Vera-Cruz."

Type Locality: Stream (Río Jamapa?), about 15 km from Veracruz (Gulf of México–Jamapa Dr.), Veracruz, México [W = 19.0378, -96.1502].

Type Specimens: Unknown. Presumed lost.

Remarks: Most previous workers have synonymized Say's *Unio purpuriatus* Say, 1831 with *Actinonaias medellina* (Lea, 1838). For some reason, Frierson (1927:83) decided that *Unio purpuriatus* was valid despite the fact that it was never figured, the type specimens lost, the type locality vague, and that it is unrecognizable based solely upon the original description. Fischer and Crosse (1894b:604) remarked that "Conrad a cru devoir identifier l'*U. Medellinus*, Lea, avec l'*U. purpuriatus*, Say. Il nous paraît bien difficile d'être édifiés à ce sujet, l'espèce de Say n'ayant jamais été figurée et portant un nom qui s'applique vraisemblablement à la coloration de la face interne de ses valves, bien différente de celle qui est assignée par I. Lea au type de l'*Unio Medellinus*" [Conrad thought he had to identify the *U. Medellinus*, Lea, with the *U. purpuriatus*, Say. It seems to us very difficult to be edified on this subject, the species of Say having never been figured and bearing a name which probably applies to the coloring of the internal surface of its valves, very different from that which is assigned by I. Lea to the type of the *Unio Medellinus*]. Simpson (1900:592) also recognized this over 100 years ago stating that "Believed by Conrad to be the same as Lea's *medellinus*. Fischer and Crosse in Mission Scientifique are doubtful about this, as Say's species was never figured, and his description does not just agree with Lea's species. As the type is not known to be in existence I think under the circumstances, since we cannot be sure of what Say described, it is best to use Lea's name." von Martens (1900b:503) thought it might be synonymous with *Nephronaias aztecorum* (Philippi, 1847). Given the uncertainty surrounding the description, type locality, and lack of any type specimens, we consider this taxon as a *nomen dubium*.

Taxonomic History:

Unio purpuriatus Say, 1831 — Ferussac (1835), Say (1840), Conrad (1853), Singley (1893), Simpson (1914), Strecker (1931)
Margaron (Unio) medellinus (Lea, 1838) — Lea (1852; 1870)
Unio medellinus Lea, 1838 — Küster (1861)
Unio (Graphonaias) medellinus Lea, 1838 — Fischer & Crosse (1894b)
Unio (Elliptio) medellinus Lea, 1838 — von Martens (1900b)
? Nephronaias medellinus (Lea, 1838) — Simpson (1900)
Nephronaias medellina (Lea, 1838) — Simpson (1914)
Actinonaias medellina (Lea, 1838) — H.B. Baker (1922)
Lampsilis (Disconaias) purpuriata (Say, 1831) — Frierson (1927), Haas (1929; 1969)
Disconaias purpuriata (Say 1831) — Graf & Cummings (2007; 2021), Czaja et al. (2023)

Unio rugulosus Küster, 1856. Die flusperlmuscheln (*Unio* et *Hyria*.). *In* Abbildungen nach der Natur mit Beschreibungen. Systematisches Conchylien-Cabinet von Martini und Chemnitz. Küster edition. 9(Part 2):137–160, pls. 45–54. Page 154, pl. 44, fig. 5. Plate 4 published in 1855.

Verbatim Type Locality: "Neuholland."

Type Specimens: Unknown. Presumed lost.

Remarks: Primary homonym of *Unio rugulosus* Wood, 1828. This species is not recognizable based upon the short description, the published figure, and the fact that the type locality is "Neuholland" [= Australia]. Attributed to "Charpentier" by Küster (1856).

Taxonomic History:

Unio rugulosus Charpentier in Küster, 1856 — Smith (1882)
Unio (*Simonaias*) *dysoni* Lea, 1859 — von Martens (1900b)
Nephronaias rugulosus (Charpentier in Küster, 1856) — Simpson (1900), de Leon (1912)
Nephronaias tabascoensis (Küster, 1856) — von Ihering (1901)
Nephronaias rugulosa (Charpentier in Küster, 1856) — Simpson (1914)
Elliptio (*Nephronaias*) *rugolosus* [sic] (Charpentier in Küster, 1856) — Frierson (1927)
Elliptio (*Nephronaias*) *rugulosa* (Charpentier in Küster) — Haas (1929)
Nephronaias (s.s.) *rugulosa* (Charpentier in Küster, 1856) — Haas (1969)
Nephronaias rugulosa (Charpentier in Küster, 1856) — Graf & Cummings (2007)
Nephronaias rugulosa (Küster, 1856) — Graf & Cummings (2021)

Unio rusticus Sowerby, 1868. Monograph of the genus *Unio*. *In* L. Reeve and G.B. Sowerby (eds.). *Conchologia Iconica* 16: [unpaginated]+plates 1, 61–96. Plate 64, species 324.

Verbatim Type Locality: "Siam."

Type Specimens: NHMUK, lost (Johnson 1971:91; Johnson 1972a:147, as *plexoides* [sic]). Possibly NHMUK 1874.12.11.15 from Mrs. T. Lombe Taylor. Cited as coming from Siam via "Mus. Taylor" in Sowerby.

Remarks: Primary homonym of *Unio rusticus* Lea, 1856. Renamed *Psoronaias plexoideus* by Frierson, 1927. Previous workers have considered this be any number of things: *Psoronaias plexoideus* Frierson, 1927, *Psoronaias crocodilorum* (Morelet, 1849), or *Sphenonaias plexus* (Conrad, 1838), despite the published type locality of Siam. Without a type and the published type locality, it is pure speculation that it is a Central American species.

Taxonomic History:

? Unio (*sec. Elliptio*) *crocodilarum* [sic] Morelet, 1849 — Simpson (1900; 1914)
Psoronaias plexoideus Frierson, 1927 — Frierson (1927)
Barynaias plexus (Conrad, 1838) — Haas (1969)

Unio sphenorhynchus Crosse & Fischer, 1894. Diagnoses molluscorum novorum, reipublicæ Mexicanæ incolarum. *Journal de Conchyliologie* 41(4):293–297. Page 296. Figured in Fischer and Crosse (1894b:617, pl. 66, figs. 2–2a).

Verbatim Type Locality: "republica Mexicana."

Type Specimens: Holotype by monotypy, MNHN-IM-2000-1792.

Remarks: Described from "republica Mexicana." "San Gabriel" was added as a locality on the (MNHN on-line catalogue). However, San Gabriel was not mentioned in the original description or subsequent monograph. Most previous workers have considered this taxon as valid and considered *Unio ortmanni* Frierson, 1913 from the Río Conchins [= Motagua] (Caribbean–Gulf of Honduras Dr.), Izabal Department, Guatemala to be a junior synonym. von Martens (1900b:501; 1901c:650) even thought it might be *Unio acutirostris* Lea, 1866 from South America. He stated that "It is to be regretted that no definite indication of locality is given in either case." Fischer and Crosse (1894b:618) noted that the thickness and outline of the shell was different than anything they had seen from México, but other than its overall shape, there is no compelling reason to unite it with *Nephronaias ortmanni* (Frierson, 1913) from Guatemala. *Nephronaias ortmanni* (Frierson, 1913) is heavily ribbed, whereas *Unio sphenorhynchus* is smooth. Without any specific locality data, it is impossible to tell what this is. The date of publication is often cited as 1893 but is 1894 according to Winckworth (1936:156).

Taxonomic History:

Unio (Sphenonaias) sphenorhynchus Crosse & Fischer, 1893 — Fischer & Crosse (1894b)
Unio (Sphenonaias) acutirostris Lea, 1866 — von Martens (1900b; 1901c)
Unio (sec. *Elliptio*) *sphenorhynchus* Crosse & Fischer, 1893 — Simpson (1900; 1914)
Unio sphenorhynchus Crosse & Fischer, 1893 — de Leon (1912)
Elliptio sphenorhynchus (Crosse & Fischer, 1893) — H.B. Baker (1922)
Elliptio (Nephronaias) sphenorhynchus (Crosse & Fischer, 1893) — Frierson (1927), Haas (1929)
Nephronaias (s.s.) *sphenorhynchus* Crosse & Fischer, 1893) — Haas (1969)
Nephronaias sphenorhyncha (Crosse & Fischer, 1893) — Graf & Cummings (2007; 2021), Czaja et al. (2023)

Unio (Simonaias) cuprinus var. *subventralis* von Martens, 1900. Land and freshwater Mollusca. *In* F.D. Godman and O. Salvin (eds.). *Biologia Centrali-Americana* Part 155, Fascile 63–65:497–520, plates 33–35. Page 506, and Part 157, Fascile 67:529–536, plates 36–37. Pl. 36, figs. 1, 1a–b.

Verbatim Type Locality: "S.E. Mexico: Tabasco."

Type Specimens: Holotype by monotypy, ZMB/MOLL.108852.

Remarks: Given the imprecise locality data, this taxon is unidentifiable. von Martens (1900b:506) described *Unio (Simonaias) cuprinus* var. *subventralis* from a single specimen collected in Tabasco, México. He included it as one of six subspecies of *Unio cuprinus* Lea, 1831 (type locality "Mexico"). The subgenus *Simonaias* (considered here as a synonym of *Nephronaias*) has *Unio tabascoensis* Küster 1856 as its type, another taxon we listed above as unidentifiable. There is only one suspect record of any *Nephronaias* in Tabasco (UF 270358, see above under *N. ravistella*). *Unio (Simonaias) cuprinus* var. *subventralis* von Martens, 1900 looks like nothing we have seen from Tabasco, but that state needs much additional work.

Taxonomic History:

Nephronaias calamitarum var. *subventralis* (von Martens, 1900) — Simpson (1914)
Elliptio (Nephronaias) subventralis (von Martens, 1900) — Frierson (1927)
Nephronaias (Popenaias) metallica (Say, 1831) — Haas (1969)

Unio tabascoënsis Küster, 1856. Die flusperlmuscheln (*Unio* et *Hyria*.). *In* Abbildungen nach der Natur mit Beschreibungen. Systematisches Conchylien-Cabinet von Martini und Chemnitz. Küster edition. 9(Part 2):137–160, pls. 45–54. Page 153, pl. 44, fig. 3.

Verbatim Type Locality: "Tabasco in Mexiko."

Type Specimens: Unknown. Presumed lost.

Remarks: Previous workers have considered this taxon valid as *Nephronaias tabascoensis* (Küster, 1856) or *Nephronaias calamitarum* (Morelet, 1849). A.A. Hinkley distributed specimens from Lago de Izabal (Caribbean–Gulf of Honduras Dr.), Jocolo, Izabal Department, Guatemala under this name to various museums, but we have seen no specimens from any collection that were collected in Tabasco, México or anywhere in México that look like the figured specimen. Given that the type is lost, and it is known only from the published figure and a short Latin description, we consider it unidentifiable. If it is indeed from the Usumacinta basin, it is possibly either *Nephronaias aeruginosa* (Morelet, 1849) or *Nephronaias calamitarum* (Morelet, 1849), but without a type, it is impossible to tell. Said by Fischer and Crosse (1894b:613) to belong to the same group as *Nephronaias calamitarum.*

Taxonomic History:

Margaron (Unio) tabascoensis (Küster, 1856) — Lea (1870)
Unio (Simonaias) tabascoensis Küster, 1856 — Fischer & Crosse (1894b)
Unio (Simonaias) cuprinus var. *tabascoensis* Küster, 1856 — von Martens (1900b)
Nephronaias tabascoensis (Küster, 1856) — Simpson (1900; 1914), von Ihering (1901)
Nephronaias calamitarum (Morelet, 1849) — Simpson (1900), de Leon (1912)
Unio cuprinus var. *tabascoensis* Lea [sic], 1856 — de Leon (1912)
Nephronaias tabascoensis (Küster, 1856) — Hinkley (1920), Graf & Cummings (2007; 2021), Czaja et al. (2023)
Elliptio (Nephronaias) tabascoensis (Küster, 1856) — Frierson (1927), Haas (1929)
Nephronaias (s.s.) *tabascoensis* (Küster, 1856) — Haas (1969)

Unio taumilapanus Conrad, 1855. Descriptions of three new species of *Unio. Proceedings of the Academy of Natural Sciences of Philadelphia* 7(7):256. Not figured.

Verbatim Type Locality: "San Juan River, Taumilapas [sic]."

Type Specimens: Unknown. Presumed lost. (Johnson 1999a:19).

Remarks: Previous workers have considered *Sphenonaias taumilapana* (Conrad, 1855) as valid. Given the fact that *Unio taumilapanus* Conrad, 1855 was never figured, the type is lost, and that it is unrecognizable from the description, there is no way to discern what *Unio taumilapanus* actually is. Fischer and Crosse (1894b:622) recognized this and stated that "Espèce non figurée et décrite de la manière la plus insuffisante." [Species not figured and described in the most insufficient way]. This was later reiterated by Coney and Taylor (1986:13) "*Unio taumilapana* Conrad, 1855, from the Rio San Juan of the Rio Grande drainage, Mexico, was not sufficiently described to permit its application to any known species with certainty, has never been illustrated, and the type is apparently lost." We also have been unable to locate any specimens identified as the type of *Unio taumilapanus* Conrad, 1855 and regard it as *nomen dubium* (Graf and Cummings 2021:9).

Taxonomic History:

Unio tamaulipasensis [sic] Conrad, 1855 — Fischer & Crosse (1894b)
Unio tamaulipanus [sic] Conrad, 1855 — von Martens (1900c, *nomen dubium*, 1901c)
Unio taumilapanus Conrad, 1855 — Simpson (1914)
Elliptio (Barynaias) tamaulipasensis [sic] (Conrad, 1855) — Frierson (1927)
Nephronaias (Sphenonaias) taumilapana (Conrad, 1855) — Haas (1969)
Quincuncina mitchelli (Simpson in Dall, 1896) — Howells et al. (1996), Johnson (1999a)
Sphenonaias taumilapana (Conrad, 1855) — Graf & Cummings (2007), Cummings & Graf (2010), Pfeiffer et al. (2016), Smith et al. (2021), Czaja et al. (2023)
Sphenonaias taumilapana (Conrad, 1855) — Graf & Cummings (2021, *nomen dubium*)

Species Accounts for the Rio Grande del Norte and Colorado River

TABLE 16

Freshwater Mussels (Unionidae) of the Río Grande and Colorado Drainages, México and North America.

Total: 16 species (13 native, 5 endemic [E], 3 introduced [I]). Extirpated [EX]

Order Unionoida

Superfamily Unionoidea

Family Unionidae

Subfamily Unioninae

Tribe Anodontini

Anodonta californiensis Lea, 1852

Pyganodon grandis (Say, 1829) [I]

Utterbackia imbecillis (Say, 1829)

Subfamily Ambleminae

Tribe Lampsilini

Cyrtonaias berlandierii (Lea, 1857)

Disconaias conchos Taylor, 1997 [E, EX]

Lampsilis teres (Rafinesque, 1820)

Potamilus metnecktayi Johnson, 1998 [E]

Potamilus purpuratus (Lamarck, 1819) [I]

Toxolasma parvum (Barnes, 1823)

Toxolasma texasiense (Lea, 1857)

Truncilla cognata (Lea, 1860) [E]

Tribe Popenaiadini

Popenaias popeii (Lea, 1857) [E]

Tribe Quadrulini

Megalonaias nervosa (Rafinesque, 1820)

Quadrula couchiana (Lea, 1860) [E EX]

Quadrula quadrula (Rafinesque, 1820) [I]

Uniomerus tetralasmus (Say, 1831)

DOI: 10.1201/9781003314448-24

Order Unionoida
Superfamily Unionoidea Rafinesque, 1820
Family Unionidae Rafinesque, 1820
Subfamily Unioninae Rafinesque, 1820
Tribe Anodontini Rafinesque, 1820
Genus *Anodonta* Lamarck, 1799

Anodonta Lamarck, 1799. Prodrome d'une nouvelle classification des coquilles. *Mémoires de la Société d'Histoire Naturelle de Paris* 1799:63–91. Page 87.

Diagnosis: "Coq. transverse, ayant trois impressions musculaires; charnière simple, sans aucune dent." [Shell transverse, having three muscular impressions; simple hinge, without any teeth.]

Type Species: *Mytilus cygneus* Linnaeus, 1758, by monotypy.

Taxonomic History:
Anodonta Lamarck, 1799 — Menke (1830), Ferussac (1835), Küster (1838), d'Orbigny (1846b), Gray (1847), Hannibal (1912a), Simpson (1914), Ortmann & Walker (1922), Thiele (1934), Modell (1945; 1964), Heard & Guckert (1970), Graf & Cummings (2006; 2007; 2021), Williams et al. (2008; 2017), Cummings & Graf (2010), Lopes-Lima et al. (2019; 2020), Nagel (2025)
Margarita (*Anodonta*) Lamarck, 1799 — Lea (1836; 1838a)
Anodonta (s.s.) — von Martens (1900c), Simpson (1900), Frierson (1927), Haas (1930; 1969)

Arnoldina Hannibal, 1912. A synopsis of the Recent and Tertiary freshwater Mollusca of the Californian Province, based upon an ontogenetic classification. *Proceedings of the Malacological Society of London* 10(2):112–165. Page 128.

Diagnosis: "Shell of rather large size, averaging 80 mm. in length, anodontiform, sub-solid, elongate-elliptical, distinctly broader posteriorly, moderately inflated, beaks which are not elevated above general curvature of shell, marked by prominent calycules and a few coarse doubly looped ridges alternating with pits which lie along the postero-ventral ridge, early growth elongate-ovate, hinge edentate and gently curved; habitat lacustro-fluviatile."

Type Species: *Anodonta dejecta* Lewis, 1875, by original designation.

Remarks: "*Arnoldina*, named after Dr. Ralph Arnold, includes, so far as known, the sole species, *Anodonta dejecta* Lewis, a mollusc which has proved an anomaly to everyone who has studied it. The development, peculiar beak-sculpture, and an indescribable velvety texture of the shell preclude its reference to *Anodonta* or any allied genus" (Hannibal, 1912a).

Taxonomic History:
Anodonta Lamarck, 1799 — Simpson (1914), Thiele (1934), Heard & Guckert (1970), Williams et al. (2008)
Anodonta (*Arnoldina*) Hannibal, 1912 — Frierson (1927), Haas (1969)
Arnoldina Hannibal, 1912 — Modell (1945; 1964)

Distribution: There are currently five recognized species in the genus *Anodonta*, in the Western Hemisphere that range from Alaska south to México. However, these species are widely disjunct from Eurasian *Anodonta* (Chong et al., 2008; Lopes-Lima et al., 2017). *Anodonta* is represented in this work by one species, *Anodonta californiensis* Lea, 1852, found in Colorado Province.

Remarks: There are numerous generic synonyms under *Anodonta*. We list only the following name that pertains to Central America.

PLATE 82 ***Anodonta californiensis*** **Lea, 1852. Top Left**: *Anodonta californiensis* Lea, 1852. Lectotype. USNM 86393 (56mm L, 33mm H). Colorado River (Gulf of California Dr.), California, USA [W=33.6083, -114.5290]. **Middle Right**: *Anodon micans* Anthony, 1865. Lectotype. MCZ 187290 (70mm L, 41mm H). Mouth of Carrizo Creek (Endorheic–Salton Sea–New Dr.), Imperial County, California [L=33.0982, -115.9277]. **Middle Left**: *Anodonta dejecta* Lewis, 1875. Lectotype, USNM 30721. Gila River (Gulf of California–Colorado Dr.), Camp Goodwin, Graham County, Arizona, USA [L=33.0949, -110.0566]. **Bottom Right**: *Anodonta mearnsiana* Simpson, 1893. Lectotype. USNM 610313 (85mm L, 45mm H). Rancho San Bernadino (Gulf of California–Yaqui Dr.), Cochise County, Arizona [L=31.3369, -109.2788].

Anodonta californiensis (Lea, 1852)

Plate 82, Map 96

Province(s): Colorado.

Distribution and Status: Known from northern California and Utah south to the Yaqui River basin in Arizona, USA and Chihuahua, México and the Colorado River drainage (Gulf of California Dr.) in Arizona and California, south to northern Baja California. The status of *A. californiensis* has been given as follows: U—Undetermined (Williams et al. 1993:10); VU—Vulnerable (Blevins et al. 2016; IUCN 2024; as a synonym of *Anodonta nuttalliana* Lea, 1838); G3—Vulnerable (NatureServe 2024). Taylor (1981:142) remarked that most of the natural populations in California have been eradicated. In Arizona, *A. californiensis* is only known from a few kilometers of perennially flowing waters in the Upper Black River of the Colorado River system (Chong et al. 2008:1306) and Chevelon Creek in the Little Colorado River system (Jepsen et al. 2010:15). Jepsen et al. (2010:15, citing Terry Myers) suggested that *A. californiensis* has likely been extirpated from the following places in Arizona: the lower mainstem of the Colorado River, the lower Gila River watershed, the Arizona portion of the Santa Cruz River watershed, the Tonto Basin, Phoenix Basin, New River (Salt River Dr.), Verde River watershed, the West Fork Black River (Black River Dr.), and the San Pedro River system.

Anatomy: Shell thin, elliptical or ovate in shape and edentate. A slight posterior dorsal wing is often present. Beaks low barely rising above the hinge line with double-looped sculpture. Periostracum smooth, yellow, olive, or brown occasionally with two or three green rays on the posterior slope. Nacre typically white but sometimes salmon, bluish white, or purple. O'Brien et al. (2019:22–23) compared the glochidia of *Anodonta californiensis* and *A. nuttalliana* and noted that the shell size and fine structure differed between the two, which supported the distinctiveness of the two species.

Biology: Found in lakes, reservoirs, perennial streams (Taylor 1981:142). The reported host fish from a population from Pajaro River, near Hollister, California was the Western Mosquitofish, *Gambusia affinis* (Baird & Girard, 1853) (d'Eliscu 1972:57). If the population in the Columbia River basin of Oregon is the same species, reported fish hosts include: *Oncorhynchus mykiss* (Walbaum, 1792) (Salmonidae); *Ictalurus punctatus* (Rafinesque, 1819) (Ictaluridae); *Rhinichthys osculus* (Girard, 1856) and *Richardsonius balteatus* (Richardson, 1836) (Leuciscidae); *Gambusia holbrooki* Girard, 1859 (Poecillidae); *Lepomis macrochirus* Rafinesque, 1819, *Micropterus salmoides* (Lacepède, 1802) and *Pomoxis nigromaculatus* (Lesueur, 1829) (Centrarchidae); and *Cottus* spp. (Cottidae) Maine and O'Brien (2022:206).

Adult Size: Length to 12 cm.

Similar Species: *Pyganodon grandis* (Say, 1829) and *Utterbackia imbecillis* (Say, 1829).

Remarks: *Anodonta californiensis* was found to be sister to *Anodonta nuttalliana* Lea, 1838 by Chong et al. (2008:1307). They found that *Anodonta californiensis/A. nuttalliana* and *A. oregonensis* Lea, 1838/*A. kennerlyi* Lea, 1860 were distinct, highly divergent clades, and that *Beringiana beringiana* (Middendorff, 1851) was more closely allied with *Sinanodonta woodiana* (Lea, 1834), an Asian species, than either of the other two western North American clades. They noted that the divergence between the *A. californiensis/A. nuttalliana* clade (13.2%) was comparable to or greater than the divergence between different genera of other anodontines (e.g., *Alasmidonta marginata* Say, 1818 vs. *Pseudodontoideus subvexus* (Conrad, 1834), 10.8% divergence; *Lasmigona compressa* (Lea, 1829) vs. *Pseudanodonta complanata* (Rossmässler, 1835), 12.5% divergence). They were unable to resolve the placement of these three clades with respect to other anodontines and suggested the need for a broader phylogenetic framework.

MAP 96 Distribution of *Anodonta californiensis* Lea, 1852.

Anodonta californiensis Lea, 1852. Descriptions of new species of the family Unionidae. *Transactions of the American Philosophical Society* 10(N.S.):253–294+plates 12–29. Page 286, pl. 25, fig. 47.

Verbatim Type Locality: "Rio Colorado, California."

Type Locality: Colorado River (Gulf of California Dr.), California [W=33.6083, -114.5290].

Type Specimens: Lectotype, USNM 86393 designated by Johnson (1974a:26, as the holotype). Paralectotypes, MCZ 161905 ($n=3$) (Johnson 1956:109, as paratypes).

Remarks: Taylor (1981:142) remarked "Type locality 'Rio Colorado' actually a former distributary of the river, approximately New River, Imperial County, California." The New River drains to the Salton Sea, an endorheic basin adjacent to the Colorado River.

Taxonomic History:

Anodonta californiensis Lea, 1852 — Conrad (1853), Hanley (1856), Hemphill (1891), Simpson (1900; 1914), Burch (1975), Taylor (1981), Turgeon et al. (1988; 1998), Williams et al. (1993; 2017), Contreras-Arquieta (2000), Graf & Cummings (2007; 2021), Chong et al. (2008), Cummings & Graf (2010), Haag (2012), Mock et al. (2013), Czaja et al. (2023)

Margaron (Anodonta) californiensis (Lea, 1852) — Lea (1852; 1870)

Anodonta (Symphynota) californiensis Lea, 1852 — Adams & Adams (1857)

Anodon californiensis (Lea, 1852) — Sowerby (1870)

Anodonta cygnea impura Say, 1829 — Hannibal (1912a; 1912b, fig. 3) [in part]

Anodonta (s.s.) *californiensis* Lea, 1852 — Frierson (1927)

Anodonta marginata californiensis Lea, 1852 — Modell (1945)

Anodonta (Anodonta) californiensis (Lea, 1852) — Bequaert & Miller (1973)

Anodonta nuttalliana Lea, 1838 — Czaja et al. (2023)

Anodon micans Anthony, 1865. Descriptions of new species of North American Unionidae. *American Journal of Conchology* 1(2):155–164+plates 12–16. Page 162, pl. 16, fig. 1.

Verbatim Type Locality: "Texas."

Type Locality: Mouth of Carrizo Creek (Endorheic–Salton Sea–New Dr.), Imperial County, California, USA [L=33.0982, -115.9277].

Type Specimens: Lectotype, MCZ 187290 designated by Johnson (1956:124; 1971:88, both as the holotype). Paralectotype, NHMUK 1968652 (Johnson 1971:88, as a paratype, figured by Sowerby 1867, pl. 15, species 53).

Remarks: Taylor (1981:142) stated that "type locality 'Texas' in error; probably from the former course of New River, near mouth of Carrizo Creek, Imperial County, California."

Taxonomic History:

Anodon micans J.G. Anthony, 1865 — Sowerby (1867)
Margaron (*Anodonta*) *micans* (J.G. Anthony, 1865) — Lea (1870)
Anodonta micans (J.G. Anthony, 1865) — Singley (1893), Simpson (1900; 1914)
Anodonta (*Pyganodon*) *grandis-lugubris* Say, 1832 — Frierson (1927)
Anodonta (*Pyganodon*) *grandis lugubris* Say, 1832 — Haas (1969)
Anodonta grandis Say, 1829 — Strecker (1931), Howells et al. (1996)
Pyganodon grandis (Say, 1829) — Williams et al. (2008)
Pyganodon grandis grandis (Say, 1829) — Watters et al. (2009)

Anodonta dejecta Lewis, 1875. Description of a new species of *Anodonta. Field and Forest* 1(3–4):26–27. Page 26. Figured in Dall (1897:372, pl. 32, figs. 4–5).

Verbatim Type Locality: "Arkansas River or its tributaries west of the 100th meridian."

Type Locality: Gila River (Gulf of California–Colorado Dr.), Camp Goodwin, Graham County, Arizona [L=33.0949, -110.0566].

Type Specimens: Lectotype, USNM 30721 designated by Johnson (1975:26).

Remarks: When Simpson was examining specimens for his 1894 paper, he remarked that "I am almost certain that the locality given by Dr. Yarrow is wrong....it is quite probable that the types of *A. dejecta* came from the Colorado drainage basin" (Simpson 1894:53). Simpson was actually looking at mussels collected by E.A. Mearns during the second Boundary Survey and collected from the Río San Bernardino at Rancho San Bernardino that is located at the Arizona–Sonora border. Simpson erroneously believed the Río San Bernardino was a part of the Colorado River basin, which led to the San Bernardino specimens being treated as the types for *A. dejecta* by Simpson, Hannibal, and others. In addition, when he figured *A. dejecta* in his contribution to Dall's 1896 paper, Simpson figured the Mearns material from Río San Bernardino, not the material collected during the Wheeler Expedition in the Gila River that formed the true *A. dejecta* type material for the species described by Lewis.

Information provided to and by Lewis regarding the source (collector and locality) of the *A. dejecta* type material he received from Yarrow was erroneous. The material was not collected from Arkansas River basin west of the 100th Meridian, nor was it collected by H.C. Yarrow. The year of collection, 1874, however, does appear to have been correct. An extensive review of field and collection reports and museum records indicates the *dejecta* material was collected from the Gila River (or perhaps one of its tributaries) in Arizona during 1874. Based on field numbers, the location at which the *dejecta* type material was obtained appears to have been along the Gila River in the vicinity of Camp Goodwin during early October 1874 (Terry Myers pers. comm. 2023).

Taxonomic History:

Anodonta dejecta Lewis, 1875 — Simpson (1900; 1914), de Leon (1912), Thiele (1934), Henderson (1935), Burch (1975), Turgeon et al. (1998), Contreras-Arquieta (2000), Graf & Cummings (2007), Chong et al. (2008), Cummings & Graf (2010), Mock et al. (2013)
Arnoldina dejecta (Lewis, 1875) — Hannibal (1912a)
"*Anodonta*" *dejecta* (Lewis, 1875) — Hannibal (1912b)
Anodonta (*Arnoldina*) *dejecta* Lewis, 1875 — Frierson (1927), Haas (1969)
Anodonta (*Anodonta*) *californiensis* (Lea, 1852) — Bequaert & Miller (1973)
Anodonta californiensis Lea, 1852 — Graf & Cummings (2021)

Anodonta mearnsiana Simpson, 1893. A new *Anodonta*. *The Nautilus* 6(12):134–135. Page 134. Not figured.

Verbatim Type Locality: "San Bernadino Ranch, Mexican Boundary Line, Arizona."

Type Locality: Rancho San Bernadino (Gulf of California–Yaqui Dr.), Cochise County, Arizona, USA [L=31.3369, -109.2788].

Type Specimens: Lectotype, USNM 610313 designated by Johnson (1975:14, as the holotype). Paralectotypes, MCZ 247226 (*n*=10), USNM 130173 (*n*=13), USNM 129973 (*n*=1).

Taxonomic History:

Anodonta dejecta Lewis, 1875 — Simpson (1900; 1914), de Leon (1912), Burch (1975)
Arnoldina dejecta (Lewis, 1875) — Hannibal (1912a)
Anodonta (*Arnoldina*) *dejecta* Lewis, 1875 — Frierson (1927), Haas (1969)
Anodonta (*Anodonta*) *californiensis* (Lea, 1852) — Bequaert & Miller (1973)

PLATE 83 ***Pyganodon grandis* (Say, 1829) Top Left:** *Anodonta grandis* Say, 1829. "Neotype" SMF 4300 (168 mm L, 101 mm H). Ohio, USA. **Middle Right:** *Anodonta texasensis* Lea, 1859. Holotype, USNM 86680 (51 mm L, 31 mm H). Texas, USA. **Middle Left:** *Anodonta leonensis* Lea, 1862. Lectotype, USNM 87806 (89 mm L, 51 mm H). **Bottom Right:** *Anodonta bealei* Lea, 1863. Lectotype, USNM 86690 (43 mm L, 25 mm H). Both, Leon County, Texas [D=31.3360, -95.6571].

Pyganodon grandis (Say, 1829)

Plate 83, Map 97

Province(s): Río Grande del Norte; Colorado (introduced).

Distribution and Status: Widespread in central and eastern North America. Reaches its southern limit in the Río Grande basin. Though widely distributed and often common in Texas and elsewhere in the United States (Howells et al. 1996), there have been few records of *P. grandis* from the Río Grande including the El Toro Cement Agency Lake, El Paso County, Texas, 1969 (USNM 681635 and ANSP 471629); fossil material from Billingslea Draw near Toyah off the Pecos River, Reeves County, Texas (Metcalf 1982:48); and Granjeno Lake (Strecker 1931) and canals at Mercedes (Ellis et al. 1931:512), both in Hidalgo County, Texas. None were taken in any surveys conducted since 1992. However, in 2000, Lang (2001) found living specimens in the Pecos River upstream of Avalon Reservoir and recently dead shells downstream of that impoundment, in Eddy County, New Mexico (Howells 2003:63). *Pyganodon grandis* (Say, 1829) has also been introduced and is known from various locations in Colorado Province around Flagstaff, Lake Mary and Kinnikinik Lake, Arizona. *Pyganodon grandis* Say, 1829, has inhabited reservoirs in the Little Colorado River system near Flagstaff since at least the early 1950s. Valves of this species have also been found at Lake Theodore Roosevelt (Tonto and Salt Rivers) in 1995, where its status is unknown (Myers 2009:22). Thought to have been introduced into the Colorado during the stocking of non-native fishes (Myers 2009:30). Graf and Cummings (2021:15) listed *Pyganodon grandis* as native to Central America (Region CA1) by including *Pyganodon globosa* under *P. grandis*. CS—Currently Stable (Williams et al. 1993:14); LC—Least Concern (IUCN 2024); G5—Secure (NatureServe 2024).

Anatomy: Shell extremely variable but usually large, smooth, somewhat elliptical or elongate, inflated, and edentate. Shell extremely thin in young mussels, somewhat thicker in older individuals. Hinge line slightly thickened and curved. Anterior end broadly rounded; posterior end bluntly pointed. Ventral margin straight or slightly curved. Beaks full, usually located in the anterior third of the shell but located more toward the center in the large river form (*P. grandis* var. *corpulenta*) and elevated above the hinge line. Beak sculpture consists of 3–5 double-looped ridges. Beak cavity broad and moderately deep. Periostracum light yellow or yellowish green with rays visible in younger individuals; dark green to brown (almost black) in adults. Nacre white, bluish white, pink, salmon, or copper-colored and iridescent.

Anal and supra-anal openings separated by a very long mantle connection. The branchial opening has distinct papillae, the inner edge in front smooth. Anal opening crenulated on the inner edge; palpi connected posteriorly one-third to one half their length. Gills long and rather narrow, the inner extending beyond the outer gill, its anterior edge being between the palpi and the inner gill; inner lamina of inner gills attached to abdominal mass only at anterior end; outer gills only marsupial. Lateral water tubes developed; the ova are contained in the central ovisacs that are not divided; placentae not present, the glochidia discharged in loose, irregular masses. Color of animal generally white or cream-colored, orange or yellowish on foot; palpi and gills brown; mantle edged with brown and black. When filled with eggs, the marsupium is white, which usually changes to brown when the glochidia have developed (Baker 1928:154).

Biology: Found in a wide variety of habitats in mud and sand in slow-flowing or standing water. The species is bradytictic, the breeding season extending from August to April or May. Glochidia discharged very early. Gravid females were observed in Mirror Lake, Wisconsin on July 20 (Baker 1928:154). A host generalist with numerous species reported including various members of the Atherinopsidae, Catostomidae, Centrarchidae, Cichlidae, Clupeidae, Cyprinidae, Ictaluridae, Lepisosteidae, Moronidae, Percidae, and Poeciliidae (Freshwater Mussel Host Database 2023: https://mollusk.inhs.illinois.edu/57-2/).

Adult Size: Length to 20 cm.

Similar Species: Co-occurs with the edentate *Anodonta californiensis* (Lea, 1852) in the Colorado River basin and with *Utterbackia imbecillis* (Say, 1829) in the Colorado in Arizona and the Rio Grande in Texas. *Utterbackia imbecillis* can be separated from *A. californiensis* in that the umbo does not project above the hinge line. *Pyganodon grandis* is also typically darker and more globose than *Anodonta californiensis.*

Remarks: This species has an extensive synonymy. The names included below are the nominal taxon and those described from Texas.

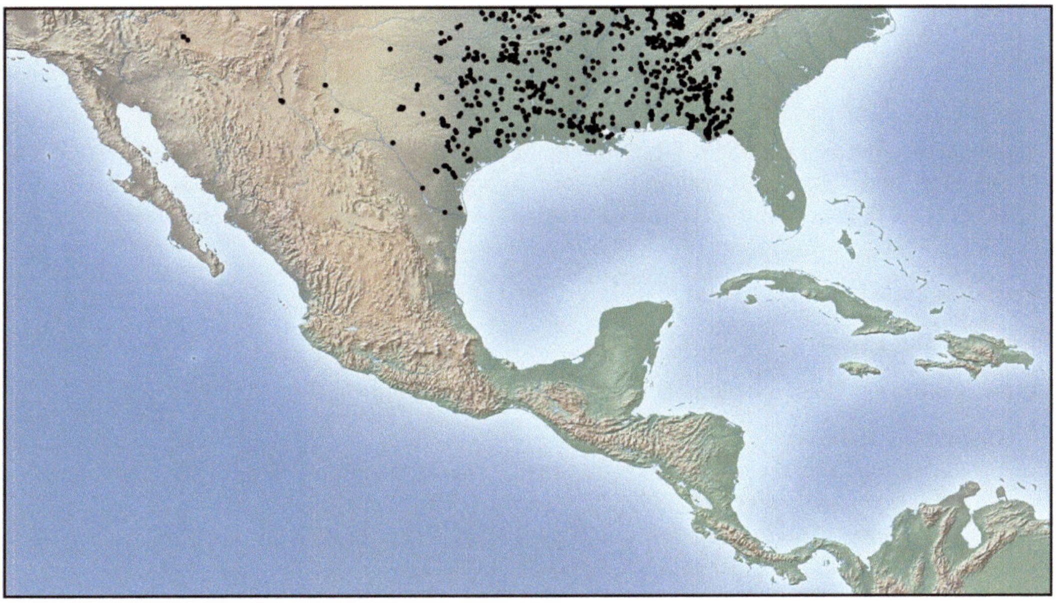

MAP 97 Distribution of *Pyganodon grandis* (Say, 1829).

Anodonta grandis Say, 1829. Descriptions of some new terrestrial and fluviatile shells of North America. (cont.). The Disseminator of Useful Knowledge; containing hints to the youth of the U. States, from the School of Industry 2(22):339–341. 4 November 1829. Page 341, sp. 3.

Verbatim Type Locality: "Fox River of the Wabash."

Type Locality: Fox River (Wabash River Dr.), opposite New Harmony, Indiana, White County, Illinois, USA [L=38.1574, -87.9362].

Type Specimens: Unknown. Presumed lost.

Remarks: Haas (1930:326) designated a neotype (SMF 4300), which is invalid due to the selection of a specimen from Ohio, USA and not from the Wabash River. Evidence that the neotype came as nearly as practicable from the original type locality is a qualifying condition under Article 75.3.6 of the International Code of Zoological Nomenclature (ICZN).

Taxonomic History:

Anodonta grandis Say, 1829 — Ferussac (1835), Hanley (1843), Conrad (1853), Adams & Adams (1857), Singley (1893), Simpson (1900; 1914), Ortmann (1910), Turgeon et al. (1988), Howells et al. (1996)

Margarita (Anodonta) grandis (Say, 1829) — Lea (1836)
Margaron (Anodonta) grandis (Say, 1829) — Lea (1852; 1870)
Anodon grandis (Say, 1829) — Sowerby (1870)
Anodonta (Pyganodon) corpulenta Say, 1829 — Bequaert & Miller (1973)
Anodonta (Pyganodon) grandis Say, 1829 — Frierson (1927), Zilch (1967), Johnson (1999a)
Anodonta (s.s.) *grandis* Say, 1829 — Haas (1930)
Pyganodon grandis (Say, 1829) — Modell (1945), Cummings & Mayer (1992), Williams et al. (1993; 2017), Turgeon et al. (1998), Graf & Cummings (2007; 2021), Cummings & Graf (2010), Inoue et al. (2013; 2018), Pfeiffer & Graf (2013; 2015), Lopes-Lima et al. (2017; 2019; 2020), Pfeiffer et al. (2019a)
Anodonta (Pyganodon) grandis grandis Say, 1829 — Haas (1969)
Anodonta grandis grandis Say, 1829 — Clarke (1973), Burch (1975), Johnson (1980), Neck & Metcalf (1988), Liu et al. (1996)

Anodonta texasensis Lea, 1859. Descriptions of eight new species of Unionidae from Georgia, Mississippi, and Texas. *Proceedings of the Academy of Natural Sciences of Philadelphia* 11(1859):112–113.Page 194. Figured in Lea (1860b JANSP Vol. 4, Part 4(13):366; 1860d Obs. 8:48, pl. 63, fig. 191).

Verbatim Type Locality: "Texas."

Type Locality: Texas.

Type Specimens: Holotype by monotypy, USNM 86680 (Johnson 1974:145).

Taxonomic History:

Anodonta texasensis Lea, 1859 — Singley (1893), Simpson (1900; 1914)
Anodon texasensis (Lea, 1859) — Sowerby (1870)
Margaron (Anodonta) texasensis (Lea, 1859) — Lea (1870)
Anodonta danielsii Lea, 1860 — Utterback (1915)
Anodonta (Pyganodon) grandis-lugubris Say, 1832 — Frierson (1927)
Anodonta (Pyganodon) grandis lugubris Say, 1832 — Haas (1969)
Anodonta grandis grandis Say, 1829 — Johnson (1980)
Anodonta grandis Say, 1829 — Strecker (1931), Howells et al. (1996)
Pyganodon grandis (Say, 1829) — Williams et al. (2008)
Pyganodon grandis grandis (Say, 1829) — Watters et al. (2009)

Anodonta leonensis Lea, 1862. Descriptions of ten new species of Unionidae of the United States. *Proceedings of the Academy of Natural Sciences of Philadelphia* 14:168–169. Page 160. Figured in Lea (1866 JANSP Vol. 6:25; 1867 Obs. 11:29, pl. 9, fig. 24).

Verbatim Type Locality: "Leon County, Texas."

Type Locality: Leon County, Texas [D=31.3360, -95.6571].

Type Specimens: Lectotype, USNM 87806 designated by Johnson (1974a:83, as the holotype).

Taxonomic History:

Anodonta leonensis Lea, 1862 — Lea (1866), Wright (1888), Singley (1893)
Margaron (Anodonta) leonensis (Lea, 1862) — Lea (1870)
Anodonta grandis var. *leonensis* Lea, 1862 — Simpson (1900; 1914)
Anodonta (Pyganodon) grandis-lugubris Say, 1832 — Frierson (1927)
Anodonta (Pyganodon) grandis lugubris Say, 1832 — Haas (1969)
Anodonta grandis grandis Say, 1829 — Johnson (1980)

Anodonta grandis Say, 1829 — Strecker (1931), Howells et al. (1996)
Pyganodon grandis (Say, 1829) — Williams et al. (2008)
Pyganodon grandis grandis (Say, 1829) — Watters et al. (2009)

Anodonta bealei Lea, 1863. Descriptions of twenty-four new species of Unionidae of the United States. *Proceedings of the Academy of Natural Sciences of Philadelphia* 15:191–194. Page 194. Figured in Lea (1866 JANSP Vol. 6:26; 1867 Obs. 11:30, pl. 9, fig. 25).

Verbatim Type Locality: "Leon County, Texas."

Type Locality: Leon County, Texas [D=31.3360, -95.6571].

Type Specimens: Lectotype, USNM 86690 designated by Johnson (1974a:19, as the holotype).

Taxonomic History:

Anodonta bealei Lea, 1863 — Lea (1866), Simpson (1900; 1914), Utterback (1915)
Margaron (*Anodonta*) *bealei* (Lea, 1863) — Lea (1870)
Anodonta bealii [sic] Lea, 1863 — Singley (1893)
Anodonta (*Pyganodon*) *grandis-bealei* Lea, 1863 — Frierson (1927)
Anodonta (*Pyganodon*) *grandis bealei* Lea, 1863 — Haas (1969)
Anodonta grandis grandis Say, 1829 — Johnson (1980)
Anodonta grandis Say, 1829 — Howells et al. (1996)
Anodonta (*Pyganodon*) *grandis* (Say, 1829) — Johnson (1999a)
Pyganodon grandis (Say, 1829) — Williams et al. (2008)
Pyganodon grandis grandis (Say, 1829) — Watters et al. (2009)

Order Unionoida
Superfamily Unionoidea Rafinesque, 1820
Family Unionidae Rafinesque, 1820
Subfamily Ambleminae Rafinesque, 1820
Tribe Lampsilini von Ihering, 1901
Genus *Cyrtonaias* Crosse & Fischer, 1894

Cyrtonaias Crosse & Fischer, 1894. *In* P. Fischer and H. Crosse. 1894. Études sur les mollusques terrestres et fluviatiles du Mexique et du Guatemala. *In* M.H.–Milne Edwards (ed.). *Mission Scientifique au Mexique et dans l'Amérique centrale. Recherches Zoologiques pour servir á l'Histoire de la Faune de l'Amérique Centrale et du Mexique.* 15th Livraison. Feuilles 62–72, pp. 489–576, plates 59–62, Imprimerie Nationale, Paris Volume 2. 731 pp. Page 556. Introduced as subgenus of *Unio* Philipsson, 1788.

Diagnosis: "Coquille ovale ou subtétragone, épaisse; sommets généralement renflés; surface lisse ou striée concentriquement (*U. Berlandieri*, Lea)." Shell oval or sub tetragonal, thick; usually bulging beaks; smooth or concentrically ribbed surface (*U. Berlandieri*, Lea).

Type Species: *Unio berlandierii* Lea, 1857, by original designation.

Taxonomic History:

Unio (*Lampsilis*) Rafinesque, 1820 — von Martens (1900b) [in part]
Lampsilis (*Proptera*) Rafinesque, 1819 — Simpson (1900; 1914)
Lampsilis (*Cyrtonais*) [sic] Crosse & Fischer, 1894 — Frierson (1927)
Lampsilis (*Cyrtonaias*) Crosse & Fischer, 1894 — Haas (1929; 1969)
Lampsilis (sec. *Cyrtonaias*) Crosse & Fischer, 1894 — Thiele (1934)
Cyrtonaias Crosse & Fischer, 1894 — Modell (1942; 1964), Heard & Guckert (1970), Burch (1975) [in part], Williams et al. (1993), Cummings & Graf (2010), Pfeiffer et al. (2019a–b), Tiemann et al. (2020), Graf & Cummings (2007; 2021) [in part]

Distribution: The genus *Cyrtonaias* formerly contained four species (Graf and Cummings 2021:17), and it was thought that the members of the genus were found from east-central Texas south to the Usumacinta basin in Guatemala. However, Pfeiffer et al. (2019a:112) noted that the genus *Cyrtonaias* was not recovered as monophyletic. *Cyrtonaias explicata* (from the Usumacinta basin) was nested deeply within the Lampsilini and very closely related to *Sphenonaias microdon* (von Martens, 1887), whereas *Cyrtonaias "tampicoensis"* (from the Colorado River in Texas) was recovered in its customary position as one of the earlier diverging lineages of the Lampsilini and sister to *Glebula rotundata* (Lamarck, 1819). Haas (1929:313) placed *Unio microdon* von Martens, 1887 in *Cyrtonaias* as *Lampsilis* (*Cyrtonaias*) *microdon* and the same pairing was obtained in molecular studies by Bogan and Hoeh (2000:164), Hoeh et al. (2001:266; 2002:130), Roe and Hoeh (2003:101), Campbell et al. (2005:142), Chapman et al. (2008:306), Zanatta and Murphy (2008:201), Burlakova et al. (2019:353), and Inoue et al. (2020a:463). As a result, we recognize the genus *Cyrtonaias* as monotypic with *Cyrtonaias berlandierii* (Lea, 1857) the only member. The three remaining former members all occur south of the Río Grande and are now placed in the genus *Mesonaias* and treated above.

PLATE 84 ***Cyrtonaias berlandierii* (Lea, 1857). Top:** *Unio berlandierii* Lea, 1857. Lectotype, USNM 84427 (86 mm L, 63 mm H). Río Grande del Norte, Matamoros, Tamaulipas, México [L=25.8829, -97.4769]. **Bottom:** *Unio heermannii* Lea, 1862. Lectotype, USNM 83932 (63 mm L, 41 mm H). Medina River (San Antonio Dr.), Bexar/Medina County, Texas, USA [W=29.2341, -98.4075].

Cyrtonaias berlandierii (Lea, 1857)

Plate 84, Map 98

Province(s): Río Grande del Norte.

Distribution and Status: Widely distributed from the Brazos River in Texas south to the Río Grande in Texas and Tamaulipas, Nuevo León, and Coahuila, México (Johnson 1999a:27; Howells 2003:60). Suspected to have been introduced into the Trinity River of Texas (Howells et al. 1996:48; Johnson 1999a:27) and Lake Texoma (Red River basin) (Mather 1989:564). Subfossil and fossil shells have been reported in the Pecos River in Chavez and Eddy counties, New Mexico (Metcalf 1982:50). Howells (2003:61) reported it to be the most abundant mussel remaining in the Río Grande. It is now thought to be absent from areas upstream of Amistad Reservoir and Texas tributaries downstream to Starr County. CS—Currently Stable (Williams et al. 1993:11); Not assessed (IUCN 2024); G5—Secure (NatureServe 2024, as *C. tampicoensis*).

Anatomy: Shell large, round to oblong with full, high beaks. Beak cavity deep. Left valve with two pseudocardinals and two short laterals. Right valve with two pseudocardinals and one lateral. Surface of the shell smooth with irregular growth lines and rayless. Periostracum ranges from a dull yellowish-brown to dark brown, and occasionally black. Nacre white, salmon, pink, or purple.

Biology: Inhabits slow-flowing stream or reservoirs in mud, sand, and gravel. Females with glochidia have been found in central Texas in May and June (Howells et al. 1996:49). Fish host reported to be the longnose gar, *Lepisosteus osseus* (Linnaeus, 1758). Glochidia were attached to the gills and transformed in 14 days at 18° C. Other fish species to which glochidia attached but failed to transform included bluegill, redear sunfish, warmouth, redbreast sunfish, longear sunfish, green sunfish, largemouth bass, channel catfish, and gray redhorse (Howells 1997a:258).

Adult Size: Length to 13 cm.

Similar Species: *Potamilus metnecktayi* Johnson, 1998 and *Potamilus purpuratus* (Lamarck, 1819).

MAP 98 Distribution of *Cyrtonaias berlandierii* (Lea, 1857).

Unio berlandierii Lea, 1857. Description of six new species of fresh water and land shells of Texas and Tamaulipas, from the collection of the Smithsonian Institution. *Proceedings of the Academy of Natural Sciences of Philadelphia* 9:101–102. Page 101. Figured in Lea (1860b JANSP Vol. 4, Part 4(13):369; 1860d Obs. 8:51, pl. 65, fig. 194) and Reeve (1865, pl. 23, species 108).

Verbatim Type Locality: "Matamoras, Tamaulipas, Mexico."

Type Locality: Río Grande del Norte (Gulf of Mexico Dr.), Matamoros, Tamaulipas, México [L=25.8829, -97.4769].

Type Specimens: Lectotype, USNM 84427 designated by Johnson (1974a:20, as the holotype; Johnson 1998:448, pl. 24, fig. 1; Johnson 1999a:27). Paralectotypes, USNM 122379 (*n*=1); ANSP A14547 (*n*=3, soft parts in alcohol).

Taxonomic History:

Unio berlandierii Lea, 1857 — Reeve (1865), Singley (1893)
Margaron (*Unio*) *berlandierii* (Lea, 1857) — Lea (1870)
Unio (sec. *Naidea*) *berlandieri* [sic] Lea, 1857 — Paetel (1890)
Unio tampicoensis Lea, 1838 — Singley (1893)
Unio (*Cyrtonaias*) *berlandieri* [sic] Lea, 1857 — Fischer & Crosse (1894b)
Unio (*Lampsilis*) *berlandieri* [sic] Lea, 1857 — von Martens (1900b)
Unio (*Lampsilis*) *tampicoensis* Lea, 1838 — von Martens (1901c)
Lampsilis (*Proptera*) *berlandierii* (Lea, 1857) — Simpson (1900; 1914)
Lampsilis berlandieri [sic] (Lea, 1857) — de Leon (1912)
Actinonaias berlandierii (Lea, 1857) — H.B. Baker (1922)
Lampsilis (*Cyrtonaias*) *tampicoensis-berlandieri* [sic] (Lea, 1857) — Frierson (1927)
Lampsilis tampicoensis berlandierii (Lea, 1857) — Strecker (1931)
Lampsilis (*Cyrtonaias*) *tampicoensis berlandierii* (Lea, 1857) — Haas (1929; 1969)
Lampsilis (sec. *Cyrtonaias*) *berlandierii* (Lea, 1857) — Thiele (1934)
Cyrtonaias berlandieri [sic] (Lea, 1857) — Morrison (1968)
Cyrtonaias tampicoensis berlandieri [sic] (Lea, 1857) — Neck & Metcalf (1988)

Cyrtonaias tampicoensis (Lea, 1838) — Burch (1975), Fuller (1975), Turgeon et al. (1988), Williams et al. (1993; 2017) [in part], Howells et al. (1996), Johnson (1998; 1999a), Graf & Cummings (2007; 2021) [in part], Burlakova et al. (2012; 2019), Pfeiffer et al. (2016; 2019a) [in part], Smith et al. (2019), Inoue et al. (2020a), Tiemann et al. (2020), Czaja et al. (2023) [in part]

Unio saladoensis Lea, 1860. Descriptions of three new species of Uniones from Mexico. *Proceedings of the Academy of Natural Sciences of Philadelphia* 12:305–306. Page 305. Figured in Lea (1860b JANSP Vol.4, Part 4(13):370; 1860d Obs. 8:52, pl. 65, fig. 195).

Verbatim Type Locality: "Rio Saldo, New Leon, Mexico."

Type Locality: Río Salado (Gulf of México–Río Grande del Norte Dr.), Nuevo León, México [L=27.1944, -100.1091].

Type Specimens: Unknown. Presumed lost (Johnson 1974a:128; Johnson 1998:446, pl. 23, fig. 2; Johnson 1999a:27).

Taxonomic History:

Margaron (*Unio*) *saladoensis* (Lea, 1860) — Lea (1870)
Unio (*Cyrtonaias*) *saladoensis* Lea, 1860 — Fischer & Crosse (1894b)
Lampsilis (*Proptera*) *saladoensis* (Lea, 1860) — Simpson (1900; 1914)
Unio saladoensis Lea, 1860 — von Martens (1900c, *nomen dubium*, 1901c)
Unio (*Metaptera*) *saladoensis* Lea, 1860 — von Martens (1901c)

Lampsilis saladoensis (Lea, 1860) — de Leon (1912)
Lampsilis (*Cyrtonaias*) *saladoensis* (Lea, 1860) — Frierson (1927), Haas (1969)
Cyrtonaias saladoensis (Lea, 1860) — Graf & Cummings (2007) Cummings & Graf (2010)
Cyrtonaias tampicoensis (Lea, 1838) — Metcalf (1982), Johnson (1998; 1999a), Graf & Cummings (2021)

Unio heermannii Lea, 1862. Descriptions of eleven new species of the genus *Unio* from the United States. *Proceedings of the Academy of Natural Sciences of Philadelphia* 13:391–393. Page 392. Figured in Lea (1862 JANSP Vol. 5:194; 1863 Obs. 9:16, pl. 26, fig. 263) and Johnson (1998:48, pl. 24, fig. 2).

Verbatim Type Locality: "Medina River, Texas."

Type Locality: Medina River (Gulf of México–San Antonio Dr.), Bexar/Medina County, Texas, USA [W=29.2341, -98.4075].

Type Specimens: Lectotype, USNM 83932 designated by Johnson and Baker (1973:157; Johnson 1974a:69, both as the holotype; Johnson 1998:448, pl. 24, fig. 2; Johnson 1999a:27). Paralectotype, ANSP lost (Johnson and Baker 1973:157).

Taxonomic History:
Unio heermanni Lea, 1862 — Sowerby (1868)
Margaron (*Unio*) *heermannii* (Lea, 1862) — Lea (1870)
Unio heermanii [sic] Lea, 1862 — Singley (1893)
Lampsilis (*Proptera*) *tampecoensis* [sic] (Lea, 1838) — Simpson (1900)
Lampsilis tampicoensis (Lea, 1838) — de Leon (1912, as *hermanni*)
Lampsilis (*Cyrtonaias*) *tampicoensis-heermanni* (Lea, 1862) — Frierson (1927)
Lampsilis tampicoensis heermani [sic] (Lea, 1862) — Strecker (1931)
Lampsilis (*Cyrtonaias*) *tampicoensis heermannii* (Lea, 1862) — Haas (1969)
Cyrtonaias tampicoensis (Lea, 1838) — Howells et al. (1996), Johnson (1998; 1999a, as both as *heermanii*)

Unio grandensis Conrad, 1855. Descriptions of three new species of *Unio*. *Proceedings of the Academy of Natural Sciences of Philadelphia* 7(7):256. Not figured.

Verbatim Type Locality: "Rio Grande, Texas."

Type Locality: Río Grande del Norte, Texas.

Type Specimens: Unknown. Presumed lost (Clench and Turner 1956:192; Johnson and Baker 1973:157; Johnson 1999a:26).

Remarks: Various authors have considered *Unio grandensis* Conrad, 1855 to be a synonym of *Unio rotundata* Lamarck, 1819 (= *Glebula rotundata*). That is not tenable as *Glebula* does not occur in the Río Grande. Morrison (1968:51) noted it was a synonym of *Cyrtonaias berlandieri* [sic] (Lea, 1857). Howells et al. (1996:48) and Johnson (1999a:27) thought it likely to be *Cyrtonaias*. Given that it was never figured and there is no known type, it will never be known for certain what it is.

Glebula rotundata (Lamarck, 1819) — Simpson (1900; 1914, both as *granadensis*), Clench & Turner (1956), Haas (1969), Burch (1975), Brim Box & Williams (2000)
Cyrtonaias berlandieri [sic] (Lea, 1857) — Morrison (1968)
Cyrtonaias grandensis (Conrad, 1855) — Metcalf (1974)
Cyrtonaias tampicoensis (Lea, 1838) — Metcalf (1982), Howells et al. (1996), Johnson (1999a)

PLATE 85 ***Disconaias conchos* Taylor, 1997**. **Top:** Holotype, LACM 2257 (124 mm L, 70 mm H). Río Conchos (Río Grande del Norte Dr.), about 0.5 km W Julimes, Chihuahua, México [L=28.4240, -105.4373]. **Middle:** USNM 218306. Río Conchos, Chihuahua, México [W=29.5833, -104.4166]. **Bottom:** Río Conchos, México (114 mm L, 119 mm L).

Disconaias conchos (Taylor, 1997)

Plate 85, Map 99

Province(s): Río Grande del Norte.

Distribution and Status: Known from the ríos Conchos, Sabinas, and Salado de Nadadores in the Río Grande del Norte drainage of México. Shell fragments (UTEP 3588) referable to *D. conchos* have also been found in Pleistocene deposits along the Pecos River in Eddy County, New Mexico. *Disconaias conchos* has not been found alive in decades. Ten unmatched valves were collected in the Río Conchos near Julimes in 1994, but except for a single juvenile, all were relict shells (Howells 2003:61). A 2018 survey of the Río Conchos found *D. conchos* at three sites, as relict shells only: two in the Río Conchos and one in a tributary to the river at Coyame. The lack of recent live or fresh-dead individuals suggests that it has been extirpated or possibly extinct (Tiemann et al. 2020:180). Not assessed (IUCN 2024).

Anatomy: Shell smooth, elongate-oval, compressed, thick and solid. Anterior margin rounded, broadly convex below. Dorsal margin broadly curved. Beaks and slightly projecting above the hinge line. Beak sculpture removed by erosion. Beak cavity shallow. Right valve with two pseudocardinal teeth, and a weakly curved, short, strong, lateral tooth. Left valve with two strong pseudocardinals and two short, strong, slightly curved lateral teeth striate on their opposing faces. Pallial line strong. Periostracum dark brown to black. Nacre white, with a faint salmon tinge in some individuals.

Biology: Unknown.

Adult Size: Length to 14 cm.

Similar Species: Both *Potamilus metnecktayi* and *Potamilus purpuratus* tend to be deeper bodied and have finer, more compressed pseudocardinal teeth. The lateral teeth in *Potamilus purpuratus* are more elongate and the dorsal muscle scars positioned more inside the beak cavity. *Cyrtonaias berlandierii* is much less elongate and has dorsal muscle scars positioned well inside the beak cavity.

MAP 99 Distribution of *Disconaias conchos* Taylor, 1997.

Disconaias conchos Taylor, 1997. A new mussel, *Disconaias conchos* (Bivalvia: Unionidae) from the Rio Conchos of the Rio Grande system, México. Occasional Papers on Mollusks, Museum of Comparative Zoology, Harvard University 5(75):419–425. Page 420, pl. 21, fig. 1.

Verbatim Type Locality: "Rio Conchos, about 0.5 km W of Julimes, Chihuahua [State], México."

Type Locality: Río Conchos (Río Grande del Norte Dr.), about 0.5 km W Julimes, Chihuahua, México [L=28.4240, -105.4373].

Type Specimens: Holotype, LACM 2257, by original designation (Johnson 1999a:31). Various specimens have been cited or identified as paratypes, LACM 2849 (*n*=2) (Groves 2012:144); LACM 1969.239.1 (*n*=2); MCZ 316166 (*n*=1) (Kabat and Johnson 2008:203).

Remarks: Taylor only specifically designated a holotype by catalogue number but stated that there were "Paratypes -- From the same locality" and that "Six adult specimens were collected at the type locality." This enigmatic species was described based upon conchological characters and placed in the genus *Disconaias* because Taylor thought that the "species is much like that figured by Martens (1900: 502) as *Unio aztecorum*, forma *major* (plate 21, figure 2) [= *Disconaias fimbriata* (Frierson, 1907)], from Playa de Misantla, Verz [sic] Cruz." Taylor went on to state that "Classification of species and details of synonymy in *Disconaias* cannot be resolved at this time. While it is unclear where *Disconaias conchos* belongs, it is almost certainly not in *Disconaias*. Soft tissue characteristics, glochidial morphology, and genetic comparisons to other unionids are needed to clearly define its taxonomic affinities and ascertain its placement in the family. Unfortunately, it is now possibly extinct, and we may never know where it belongs.

Taxonomic History:

Disconaias conchos Taylor, 1997 — Johnson (1999a), Contreras-Arquieta (2000), Graf & Cummings (2007; 2021), Cummings & Graf (2010), Czaja et al. (2023)

Order Unionoida
Superfamily Unionoidea Rafinesque, 1820
Family Unionidae Rafinesque, 1820
Subfamily Ambleminae Rafinesque, 1820
Tribe Lampsilini von Ihering, 1901
Genus *Lampsilis* Rafinesque, 1820

Lampsilis Rafinesque, 1820. Monographie des coquilles bivalves fluviatiles de la Riviere Ohio, contenant douze genres et soixante-huit especes. *Annales Générales des Sciences Physiques, Bruxelles* 5(13):287–322+plates 80–82. Page 298.

Diagnosis: "Shell oval; ligament curved, cardinal tooth furrowed; lamellar tooth curved, flexuous; axis extramedial; marginal contour thickened; three muscular impressions; animal similar to that of the *Unio*, but with syphons apparent, short. This name is modified from *Lasmacampsilis*, which signifies flexuolated lamellar tooth, the essential character of this genus." Translation of Rafinesque from Poulson (1832:26–27).

Type Species: *Unio ovatus* Say, 1817, by subsequent designation (Herrmannsen 1847a:575).

Taxonomic History:

Unio (*Lampsilis*) Rafinesque, 1820 — Conrad (1853), Adams & Adams (1857), von Martens (1900a; 1900b)

Lampsilis (s.s.) Rafinesque, 1820 — Simpson (1900), Simpson (1914), Frierson (1927), Haas (1930), Morrison (1968), Haas (1969), Haas (1969), R.I. Johnson (1970), R.I. Johnson (1972)
Lampsilis (sec. *Eurynia*) Rafinesque, 1820 — Simpson (1900)
Lampsilis Rafinesque, 1820 — Simpson (1900), Ihering (1901), Pilsbry (1904; 1910), Ortmann (1912), Thiele (1934), Modell (1942), Graf & Cummings (2006; 2007; 2021), Cummings & Graf (2010), Pfeiffer & Graf (2013), Inoue et al. (2014; 2020), Lopes-Lima et al. (2017), Williams et al. (2017), Pfeiffer et al. (2019b), Nagel (2025)
Lampsilis (*Eurynia*) Rafinesque, 1820 — Simpson (1914)
Elliptio (s.s.) — Frierson (1927), Haas (1969)
Lampsilis (*Ligumia*) — Frierson (1927), Haas (1930)
Lampsilis (*Ortmanniana*) — Frierson (1927), Haas (1969)

Aeglia Swainson, 1840. *A Treatise on Malacology: or the Natural Classification of Shells and Shell-fish*. London. Printed for Longman, Orme, Brown, Green, & Longman, Peternoster Row 419 pp. Page 265 (as Æglia).

Type Species: *Unio ovatus* Say, 1817 by subsequent designation (Herrmannsen 1846a:20).

Aglia 'Swains.' Gray, 1847. A list of the genera of recent Mollusca, their synonyma and types. *Proceedings of the Zoological Society of London* 1847(15):129–219. Page 196.
Remarks: Error for *Aeglia* Swainson, 1840.

Lampsida Landacre, 1902. On a visual area in *Lampsida ventricosus*. *Ohio Naturalist* 3(1): 320–321. Page 320.
Remarks: Error for *Lampsilis* Rafinesque, 1820.

Lampsilus Mansfield in Cook & Mansfield, 1933. A new mollusk from the Chadron Formation (Oligocene) of Nebraska. *Journal of the Washington Academy of Sciences* 23:263–267. Page 267.
Remarks: Error for *Lampsilis* Rafinesque, 1820.

Lamsilis van der Schalie, 1934. *Lampsilis jonesi*, a new naiad from southeastern Alabama. *The Nautilus* 47(4):125–127. Plate. 15.
Remarks: Error for *Lampsilis* Rafinesque, 1820.

Simpsonunio Starobogatov, 1970. *Fauna Mollyuskov i Zoogeograficheskoe Raionirovanie Kontinental'nykh Vodoemov Zemnogo Shara*. Akademiya Nauk SSSR, Leningrad. 371 pp. Page 288. Introduced as subgenus of *Lampsilis* Rafinesque, 1820.

Type Species: *Unio anodontoides* Lea, 1831. By original designation.

Taxonomic History:
Lampsilis Rafinesque, 1820 — Williams et al. (2008)

Distribution & Remarks: The genus *Lampsilis* is widely distributed in North America east of the Rocky Mountains from Canada south to Florida and west to Texas.

PLATE 86 ***Lampsilis teres* (Rafinesque, 1820)**. **Top:** *Unio* (*Eurynia*) *teres* Rafinesque, 1820. Neotype, INHS 89410. Wabash River (Mississippi–Ohio Dr.), 3.4 km SSE Calvin, I-64 bridge, White County, Illinois, USA [38.2286, -87.9851]. **Bottom:** *Lampsilis fallaciosus* Simpson, 1900. Holotype, USNM 30023a (90 mm L, 40 mm H). Ohio River (Mississippi Dr.), Sedamsville, Hamilton County, Ohio, USA [L=39.0871, -84.5657].

Lampsilis teres (Rafinesque, 1820)

Plate 86, Map 100

Province(s): Río Grande del Norte.

Distribution and Status: Widespread in rivers of the Interior Basin and the Western Gulf Coastal Plain, USA, southwest to the Rió Grande del Norte south of the Pecos River in Texas and México. Reported from the Río Sabinas in Coahuila, the Río Salado de Nadadores in Nuevo León and Tamaulipas, México (Metcalf 1982:54; Johnson 1999a:33). CS—Currently Stable (Williams et al. 1993:12); LC—Least Concern (IUCN 2012); G5—Secure (NatureServe 2024). Although common in other drainages in Texas and historically widespread in the Río Grande basin, none have been collected alive there in decades (Howells 2003:62).

Anatomy: Shell smooth, relatively large, elongate, moderately thick, and slightly to moderately inflated. Anterior end rounded, posterior end pointed in males, saber-shaped in females. Dorsal margin straight, ventral margin straight or occasionally concave in the middle. Beaks only slightly elevated above the hinge line. Beak sculpture of 4–6 double-looped ridges. Beak cavity moderately deep. Pseudocardinal teeth elongate and compressed; two in the left valve, one in the right (occasionally with a thin ridgelike tooth in front). Lateral teeth long and straight to slightly curved; two in the left valve, one in the right. Shell smooth, very shiny when young becoming dull with age. Periostracum yellow usually with green rays in young shells, rayless in older individuals. Nacre silvery white, occasionally tinged with cream or salmon near the beak cavity, iridescent posteriorly.

Anal and supra-anal openings of about equal length, the mantle connection very much shorter than either. The branchial opening rather large and papillose. The anal opening is finely crenulated. The inner edge of the mantle in the female, anterior to the branchial opening, forms a long, ribbon-like expansion of a blackish or brownish color, covered with crenulations, but without papillae. This flap extends less than a third the length of the ventral margin; inner laminae of inner gills connected their entire length to abdominal sac; palpi wide, connected for about one-fourth their length at the base; marsupium kidney-shaped, occupying more than half of the posterior part of the outer gill, when gravid forming a thick, pad-like structure of upwards of 30 ovisacs. The margin of marsupium is black. The color of the soft parts white or cream-colored (Baker 1928:267). Glochidia semi-elliptical, ventral margin rounded, hinge line short and slightly undulate. Length, 0.185, height 0.210 mm (Surber 1912:9).

Biology: Found in medium to large rivers in sand or fine gravel. Ortmann (1912:346) examined two gravid females from the Colorado and Río Grande Rivers in Texas collected in May 1907, five specimens from the Kansas River, Lawrence, Douglas County, Kansas, and one gravid female from the Ouachita River, Arkadelphia, Clark County, Arkansas, collected on 6 February 1911. The gravid females all had glochidia and showed that the species is bradytictic, carrying the larvae over the winter. Reported hosts include various species of gars: *Lepisosteus osseus* (Linnaeus, 1758), *Lepisosteus platostomus* Rafinesque, 1820 (Coker et al. 1921:121; Barney 1922:54; Arey 1932:212), and *Atractosteus spatula* (Lacepède, 1803) (Wilson 1916:344). Mature glochidia were found in every month of the year by Coker et al. (1921:141).

Adult Size: Length to 15 cm.

Similar Species: None.

MAP 100 Distribution of *Lampsilis teres* (Rafinesque, 1820).

Unio (*Eurynia*) *teres* Rafinesque, 1820. Monographie des coquilles bivalves fluviatiles de la Riviere Ohio, contenant douze genres et soixante-huit especes. *Annales Générales des Sciences Physiques, Bruxelles* 5(13):287–322+plates 80–82. Page 321. Figured in Conrad (1836:52, pl. 28) and Küster (1853:68, pl. 17, figs. 1, 4).

Verbatim Type Locality: "Dans la rivière Wabash."

Type Locality: Wabash River (Mississippi–Ohio Dr.), 3.4 km SSE of Calvin, White County, Illinois, USA [W=38.2286, -87.9851].

Type Specimens: Neotype, INHS 89410 designated and type locality fixed by Keogh and Simons (2019:188, fig. 6).

Remarks: The original "type" at the ANSP was considered lost (Johnson 1972b:242; Johnson and Baker 1973:172). Johnson (1972b:242) selected the figure on plate 28 (not 38 as indicated by Johnson) of Conrad (1836:52–53) as the type figure. The specimen came from the Poulson Collection and was labeled by Rafinesque. There was no mention as to the locality of the figured specimen. That specimen has not been located.

Taxonomic History:

Unio teres Rafinesque, 1820 — Ferussac (1835), Conrad (1836; 1853), Küster (1853), Adams & Adams (1857)
Margarita (*Unio*) *anodontoides* (Lea, 1831) — Lea (1836)
Margaron (*Unio*) *anodontoides* (Lea, 1831) — Lea (1852; 1870)
Lampsilis (*Ligumia*) *teres* (Rafinesque, 1820) — Frierson (1927), Haas (1930)
Ligumia teres teres (Rafinesque, 1820) — Haas (1969)
Lampsilis teres (Rafinesque, 1820) — Strecker (1931), Burch (1975), Neck & Metcalf (1988), Turgeon et al. (1988; 1998), Williams et al. (1993; 2017), Howells et al. (1996), Johnson (1999a), Contreras-Arquieta (2000), Graf & Cummings (2007;

2021), Cummings & Graf (2010), Pfeiffer et al. (2016), Burlakova et al. (2019), Keogh & Simons (2019), Inoue et al. (2020b), Czaja et al. (2023)
Lampsilis (s.s.) *teres* (Rafinesque, 1820) — Johnson (1972b)

Unio anodontoides Lea, 1831. Observations on the naïades, and descriptions of new species of that and other families. *Transactions of the American Philosophical Society* 4(N.S.):63–121 + plates iii–xviii. Page 81, pl. 8, fig. 11. Also figured in Lea (1834, Obs. 1:91, pl. 8, fig. 11).

Verbatim Type Locality: "Mississippi river,... Alabama river, ... Ohio river."

Type Locality: Mississippi, Ohio and Alabama rivers (Gulf of Mexico Dr.), USA.

Type Specimens: Unknown. Presumed lost (Johnson 1972b:242; Johnson and Baker 1973:148).

Taxonomic History:

Unio anodontoides Lea, 1831 — Jay (1835), Hanley (1843), Reeve (1865), Singley (1893), de Gregorio (1914), Henderson (1935)
Margarita (*Unio*) *anodontoides* (Lea, 1831) — Lea (1836)
Margaron (*Unio*) *anodontoides* (Lea, 1831) — Lea (1852; 1870)
Lampsilis (sec. *Eurynia*) *anodontoides* (Lea, 1831) — Simpson (1900)
Lampsilis anodontoides (Lea, 1831) — de Leon (1912), Ortmann (1912)
Lampsilis (*Eurynia*) *anodontoides* (Lea, 1831) — Simpson (1914)
Lampsilis teres (Rafinesque, 1820) — Keogh & Simons (2019)

Lampsilis fallaciosus Simpson, 1900. New and unfigured Unionidae. *Proceedings of the Academy of Natural Sciences of Philadelphia* 52:74–86 + 5 plates. Page 74, pl. 2. fig. 5.

Verbatim Type Locality: "Upper Mississippi drainage, south to the Cumberland and Arkansas rivers; Red River of the North."

Type Locality: Ohio River, Sedamsville, Hamilton County, Ohio, USA [L = 39.0871, -84.5657].

Type Specimens: Holotype by monotypy, USNM 30023a (Johnson 1975:27, Keough and Simons 2019:188).

Remarks: *Lampsilis fallaciosus* Smith, 1899 is a *nomen nudem*. It was named and figured, but not described. Smith (1899:291) did mention that was taken at Muscatine, Iowa as reiterated by Johnson (1972b:243). Simpson (1900:74) formally described and figured *Lampsilis fallaciosus*, but no type specimen was designated or type locality given. Johnson (1975:27) identified the type and provided the specific locality data.

Taxonomic History:

Lampsilis fallaciosus Smith, 1899 — Smith (1899) *nomen nudem*, Simpson (1900), Utterback (1916)
Lampsilis (sec. *Eurynia*) *fallaciosus* Smith, 1899 — Simpson (1900)
Lampsilis anodontoides (Lea, 1831) — Ortmann (1912)
Lampsilis (*Eurynia*) *fallaciosus* Smith, 1899 — Simpson (1914)
Lampsilis anodontoides fallaciosa Smith, 1899 — Ortmann (1925)
Lampsilis (*Ligumia*) *teres-fallaciosa* Smith, 1899 — Frierson (1927)
Lampsilis fallaciosa Smith, 1899 — F.C. Baker (1928)
Ligumia teres teres (Rafinesque, 1820) — Haas (1969)
Lampsilis (s.s.) *teres* (Rafinesque, 1820) — Johnson (1972)
Lampsilis teres (Rafinesque, 1820) — Burch (1975), Johnson (1980), Howells et al. (1996), Williams et al. (2008), Keogh & Simons (2019)
Lampsilis teres fallaciosa Smith, 1899 — Johnson (1999a)

Order Unionoida
Superfamily Unionoidea Rafinesque, 1820
Family Unionidae Rafinesque, 1820
Subfamily Ambleminae Rafinesque, 1820
Tribe Lampsilini von Ihering, 1901
Genus *Potamilus* Rafinesque, 1820

Potamilus Rafinesque, 1818. Discoveries in natural history, made during a journey through the Western region of the United States. *American Monthly Magazine and Critical Review* 3:354–356. Page 355.

Diagnosis: "Bivalve. Shell equivalve unequilateral, commonly transverse, rugose transversely, sloping posteriorly, shape variable, margin thickened, two muscular impressions, an epidermis surrounding the margin by a membranaceous brim, connective oblong convex membranaceous. Ligament with two teeth on one side, and a deep furrow on the other, between two carina in the left shell, while the right shell has two unequal teeth, and two unequal carinas. Animal with a mantle open and bilobe, branchias as a second interior mantle, body compressed tough, two openings or siphons anterior on each side, not tubular, one foot on each side commonly bilamellose, next to the openings."

Type Species: *Unio alatus* Say, 1817, by subsequent designation (Morrison 1969:24).

Taxonomic History:

Lampsilis (*Proptera*) Rafinesque, 1819 — Simpson (1900: 1914)
Lastena Rafinesque, 1820 — Simpson (1914)
Proptera Rafinesque, 1819 — Ortmann (1910; 1912), Haas (1969), Heard & Guckert (1970), Johnson (1980), Contreras-Arquieta (2000)
Anodonta (*Lastena*) Rafinesque, 1820 — Frierson (1927), Haas (1969)
Lampsilis (*Leptodea*) Rafinesque, 1820 — Frierson (1927)
Potamilus Rafinesque, 1818 — Morrison (1969), Johnson (1980; 1998; 1999a), Turgeon et al. (1988), Cummings & Mayer (1992), Roe & Hoeh (2003), Graf & Cummings (2007; 2021), Williams et al. (2008), Pfeiffer et al. (2019a–b), Inoue et al. (2020b), Tiemann et al. (2020), Nagel (2025)
Potamilis [sic] Rafinesque, 1818 — Contreras-Arquieta (2000)

Potamila Rafinesque, 1818. General account of the discoveries made in the zoology of the Western States. *American Monthly Magazine and Critical Review* 4:106–107. Page 107.
Remarks: Error for *Potamilus* Rafinesque, 1818.

Proptera Rafinesque, 1819. Prodrome de 70 nouveaux genres d'Animaux découverts dans l'intérieur des États-Unis d'Amérique, durant l'année 1818. Journal de Physique, de Chimie, d'Histoire Naturelle 88:417–428. Page 426. Introduced as subgenus of *Unio* Philipsson, 1788.

Type Species: *Unio alatus* Say, 1817

Remarks: Renamed *Metaptera* Rafinesque, 1820.

Lastena Rafinesque, 1820. Monographie des coquilles bivalves fluviatiles de la Riviere Ohio, contenant douze genres et soixante-huit especes. *Annales Générales des Sciences Physiques, Bruxelles* 5(13):287–322+plates 80–82. Page 316. Introduced as subgenus of *Anodonta* Lamarck, 1799.

Type Species: *Anodonta* (*Lastena*) *ohiensis* Rafinesque, 1820, by subsequent designation (Herrmannsen 1847a:577, 1846a:58, as a synonym of *Anodonta*).

Remarks: Renamed *Stenelasma* by Herrmannsen (1849a:502). Herrmannsen replaced Rafinesque's name without explanation and is an unnecessary replacement name.

Leptodea Rafinesque, 1820 Monographie des coquilles bivalves fluviatiles de la Riviere Ohio, contenant douze genres et soixante-huit especes. *Annales Générales des Sciences Physiques, Bruxelles* 5(13):287–322+plates 80–82. Page 295. Introduced as subgenus of *Unio* Philipsson, 1788.

Type Species: *Unio* (*Leptodea*) *fragilis* Rafinesque, 1820 by subsequent designation (Herrmannsen 1847a:584).

Metaptera Rafinesque, 1820. Monographie des coquilles bivalves fluviatiles de la Riviere Ohio, contenant douze genres et soixante-huit especes. *Annales Générales des Sciences Physiques, Bruxelles* 5(13):287–322+plates 80–82. Page 299.

Type Species: *Unio alatus* Say, 1817, by subsequent designation (Herrmannsen 1847b:41)

Remarks: Herrmannsen (1849b:668) listed *Metaptera* as a synonym of *Unio* Retzius, 1788. New name for *Proptera* Rafinesque, 1819. None of the species included in 1819 in this generic nomen was available at the time *Unio alatus* Say was chosen as the type for *Metaptera* by Herrmannsen, which is an objective synonym of *Proptera*. *Metaptera* is an unnecessary replacement.

Symphynota Lea, 1829. Description of a new genus of the family of naïades, including eight new species, four of which are new; also the description of eleven new species of the genus *Unio* from the rivers of the United States: with observations on some of the characters of the naïades. *Transactions of the American Philosophical Society* 3(N.S.):403–457+plates vii–xiv. Page 424, 442.

Type Species: *Unio alatus* Say, 1817, by original designation.

Lasmonos Rafinesque, 1831. Continuation of a monograph of the bivalve shells of the river Ohio and other rivers of the western states. By Prof. C.S. Rafinesque. Published at Brussels, September 1820. Containing 46 species, from No. 76 to no. 121. Including an appendix on some bivalve shells of the rivers of Hindostan, with a supplement on the fossil bivalves of the Western states, and the *Tulosites*, a new genus of fossils. Philadelphia, Pennsylvania. 8 pp. Page 5.

Type Species: *Lasmonos fragilis* Rafinesque, 1831, by original designation.

Remarks: Renamed *Monelasmus* Agassiz, 1846.

Naidea Swainson, 1840. *A Treatise on Malacology: or the Natural Classification of Shells and Shell-fish*. London. Printed for Longman, Orme, Brown, Green, & Longman, Peternoster Row 419 pp. Pages 276, 378. Introduced as subgenus of *Aeglia* Swainson, 1840.

Type Species: *Unio ater* Lea, 1829, by monotypy.

Paraptera Ortmann, 1911. A monograph of the najades of Pennsylvania. Parts I and II. *Memoirs of the Carnegie Museum* 4(6):279–347+4 plates. Page 301.

Type Species: *Unio gracilis* Barnes, 1823, by original designation.

Distribution & Remarks: This genus contains eight species, one of which *Potamilus metnecktayi* Johnson, 1998, is endemic to the Río Grande del Norte in Texas, USA and Coahuila and Tamaulipas, México. *Potamilus purpuratus* (Lamarck, 1819) is thought to have been introduced in the Río Grande (Howells 2003:63). Czaja et al. (2023:88) reported *Potamilus alatus* (Say, 1817) from Tabasco, México, which is way out of the range and certainly an error.

PLATE 87 ***Potamilus metnecktayi* Johnson, 1998**. **Top Left:** Holotype, UMMZ 255018 (109 mm L, 65 mm H). Río Salado (Río Grande del Norte Dr.), 72 km S Nuevo Laredo, Tamaulipas, México [L=26.7866, -99.4221]. **Middle Right & Left:** NRI 8015.2 and NRI 8388.14 Río Grande del Norte, Black Gap Wildlife Management Area, Brewster County, Texas, USA [L=29.6670, -102.7085]. **Bottom Right: *Potamilus purpuratus* (Lamarck, 1819).** *Unio coloradoensis* Lea, 1856. Lectotype, USNM 84845 (124 mm L, 78 mm H). Colorado River (TX) (Gulf of Mexico Dr.), Texas, USA [W=30.2622, -97.7450].

Potamilus metnecktayi (Johnson, 1998)

Plate 87, Map 101

Province(s): Río Grande del Norte.

Distribution and Status: This species is endemic to the Río Grande del Norte basin of Texas and México (Johnson 1998:430). *Potamilus metnecktayi* historically occurred from the confluence of the Rio Conchos with the Rio Grande (Presidio County, Texas) downstream to just below the current location of Falcon Dam (Starr County, Texas), in the lower Pecos near its confluence with the Rio Grande, and in tributaries of the Rio Grande in the northern part of México (Metcalf 1982; Neck and Metcalf 1988; Johnson 1999; Howells 1999, 2001). This species has experienced a significant decline within its historical range. Presumed extirpated from the Río Salado de Nadadores in Tamaulipas, México where it was last documented in the late 1970s. In Texas, it has been extirpated from the Pecos River and the middle portion of the Río Grande. Its current range is thought to be limited to the Lower Canyons portion of the Río Grande from Big Bend National Park to reaches located just upstream of Lake Amistad, a length of river that is no more than 200 km, and only about a third of which is protected by the Big Bend National Park and the adjacent Area Natural Protegida Maderas del Carmen, Coahuila, México (Randklev et al. 2018:738). Burlakova et al. (2011:161) suggested that the IUCN list it as Critically Endangered. On 25 July 2023, the U.S. Fish and Wildlife Service (USFWS) proposed to list *Potamilus metnecktayi* as endangered species under the Endangered Species Act of 1973, as amended (Federal Register 2023:47952). The USFWS noted that *P. metnecktayi* is currently restricted to approximately 16 percent of its historical range in the United States and México and that only one extant population is known in the Lower Canyons area of the Río Grande between Big Bend National Park and Amistad Reservoir (Federal Register 2023:47964). Not assessed (IUCN 2024); G1—Critically Imperiled (NatureServe 2024).

Anatomy: Shell, solid, moderately large and somewhat obovate. Valves moderately inflated. Beaks slightly elevated above the hinge line. Beak cavity rather shallow. Anterior end regularly rounded, posterior end more broadly rounded, somewhat pointed in males; and more broadly rounded in females. Ventral margin straight or slightly curved. Dorsal margin broadly curved, somewhat convex with small wing. Posterior slope with double, sometimes triple, radiating ridges. Left valve with two pseudocardinal teeth, and two short, slightly curved lateral teeth. Right valve with two pseudocardinal teeth, the anterior one tiny, with a large, trigonal posterior tooth, and a single lateral one. Surface of the shell with fine concentric raised threads and irregular ridges. Periostracum brown, tan, or black with hints of green, and with faint rays, especially in young specimens. Nacre white to bluish white, and iridescent.

Johnson (1998:430) noted that Ralph Taylor studied a single male *P. metnecktayi* and observed that the mantle margin and gill structure had a similar "preforate [sic] mantle between the anal and supra-anal openings, a structure previously unknown," and suggested that these two species should be placed in a new sub-genus.

Biology: *Potamilus metnecktayi* occurs in medium to large rivers, generally in nearshore habitats and crevices, undercut riverbanks, travertine shelves, and under large boulders adjacent to runs (Howells et al. 1996:130–134; Karatayev et al. 2012:210; Randklev et al. 2017:157, 159). Found along the shores, in soft sediments (a mixture of silt and clay) at 0.5–1.2 m depth (at low flows ~30m^3/s) (Karatayev et al. 2012:210). Freshwater Drum (*Aplodinotus grunniens* Rafinesque, 1819) has been identified as a suitable host fish. However, this was the only fish species tested in laboratory experiments and other species could serve as hosts in the wild. The glochidia remained encysted for 13–28 days during transformation to the juvenile stage (Bosman et al. 2015:15).

Adult Size: Length to 12 cm.

Similar Species: *Potamilus purpuratus* (Lamarck, 1819) and *Cyrtonaias berlandierii* (Lea, 1857).

MAP 101 Distribution of *Potamilus metnecktayi* Johnson, 1998.

Potamilus metnecktayi Johnson, 1998. A new mussel, *Potamilus metnecktayi* (Bivalvia: Unionidae) from the Río Grande system, México and Texas with notes on Mexican *Disconaias*. Occasional Papers on Mollusks, Museum of Comparative Zoology, Harvard University 5(76):427–455. Page 429, pl. 22, fig. 1 (holotype), fig. 2 (allotype).

Verbatim Type Locality: "Rio Salado, 45 mi. S Nuevo Laredo, State of Tamaulipas, México."

Type Locality: Río Salado (Gulf of México–Río Grande del Norte Dr.), 72 km S Nuevo Laredo, Tamaulipas, México [L=26.7866, -99.4221].

Type Specimens: Holotype (male), UMMZ 255018 by original designation. Paratypes, UMMZ 255019 ($n=1$) (female allotype), UMMZ 66993 ($n=1$), from the type locality; UTEP 2519 (female) from Río Grande, 9.6 km W Del Rio, Val Verde County, Texas. All by original designation.

Remarks: Metcalf (1982:48–49) identified fossil material from the Pecos River drainage as *L. salinasensis*, and Neck and Metcalf (1988:265) referred to weathered dead shells from the lower Rio Grande downstream of Falcon Dam as *Potamilus salinasensis*. Both of these records likely represent *P. metnecktayi* specimens (Johnson 1998:428, 433).

Taxonomic History:

Potamilus metnecktayi Johnson, 1998 — Johnson (1999a), Contreras-Arquieta (2000), Graf & Cummings (2007; 2021), Cummings & Graf (2010), Williams et al. (2017), Burlakova et al. (2019), Inoue et al. (2020b), Smith et al. (2019), Smith et al. (2020), Tiemann et al. (2020), Czaja et al. (2023)

Lampsilis salinasensis Simpson, 1908 — Metcalf (1982)

Potamilus salinasensis (Simpson, 1908) — Howells et al. (1996)

Potamilus purpuratus (Lamarck, 1819)

Plate 87, Map 102

Province(s): Río Grande del Norte. Introduced.

Distribution & & Status: The native range of this species is the lower Mississippi basin and Gulf Coastal Region, not extending beyond the Guadalupe River drainage of Texas (Howells et al. 1996). Now found in the Río Grande System, where it is thought to have been introduced. Shells and living specimens were found in Amistad Reservoir, Val Verde County, Texas, in 1994 and 1995 and again in 1998 (Howells 2003:63). Collected upstream of Amistad Reservoir and Devils River in 2013. Early reports of this species in the Río Grande represent misidentified *Cyrtonaias berlandierii* (Howells 1997b:80). CS—Currently Stable (Williams et al. 1993:14); Not assessed (IUCN 2024); G5—Secure (NatureServe 2024).

Anatomy: Shell smooth, rectangular to oval, moderately to greatly inflated, thin in young shells to moderately thick in older individuals. Anterior end rounded, posterior end slightly pointed in males and bluntly squared or truncated in females. Beaks elevated above the hinge line. Beak sculpture of 3–4 double-looped ridges, visible only in young shells. Beak cavity shallow. A low wing may be present behind the umbo. Pseudocardinal teeth small, roughed and divergent; two in the left valve, two in the right. Lateral teeth long, thin, and curved; two in the left valve, one in the right. Periostracum brown, becoming black in older individuals. Young mussels typically marked with dark green rays that become fainter with age. Nacre purple and highly iridescent.

Biology: Found in streams or impoundments in quiet pools in mud or gravel. Specimens with glochidia were found in June, October, November, and January in Texas (Howells et al. 1996:101). Considered bradytictic. Glochidia are axe-head shaped. Fish hosts include Freshwater Drum, *Aplodinotus grunniens* Rafinesque, 1819 (Wilson 1916:338); Golden shiner, *Notemigonus crysoleucas* (Mitchill, 1814); Warmouth, *Lepomis gulosus* (Cuvier, 1829); and Red Shiners, *Cyprinella lutrensis* (Baird & Girard, 1853) (Howells 1995:1; Marshall et al. 2018:46).

Adult Size: Length to 20 cm.

Similar Species: *Potamilus metnecktayi* Johnson, 1998 and *Cyrtonaias berlandierii* (Lea, 1857).

MAP 102 Distribution of *Potamilus purpuratus* (Lamarck, 1819).

Unio purpurata Lamarck, 1819. Pp. 67–81 *In* Les Nayades. Mulette (*Unio*). Histoire Naturelle des Animaux sans Vertèbres 6(1):vi–343 pp. Page 71, sp. 6.

Verbatim Type Locality: "Habite... Je la crois des grandes rivières de l'Afrique."

Type Locality: It has been commonly accepted that the published type locality of Africa is erroneous, but it has never been corrected or fixed.

Type Specimens: Muséum d'histoire naturelle, Genève (MHNG) without a catalogue number (Johnson 1953:92, as the holotype).

Remarks: Lamarck, by pointing to specimens in the Paris Museum as well as his own cabinet, suggests that he has multiple specimens, although he only gave a single set of measurements. The synonymy below includes the nominal taxon and those described from Texas.

Taxonomic History:

Unio purpurata Lamarck, 1819 — Jay (1835)
Unio purpuratus Lamarck, 1819 — Ferussac (1835), Hanley (1843; 1856), Conrad (1853), Küster (1861), Reeve (1865), Singley (1893)
Margarita (*Unio*) *purpuratus* (Lamarck, 1819) — Lea (1836)
Margaron (*Unio*) *purpuratus* (Lamarck, 1819) — Lea (1852; 1870)
Unio (sec. *Naïa*) *purpuratus* Lamarck, 1819 — Paetel (1890)
Lampsilis (*Proptera*) *purpuratus* (Lamarck, 1819) — Simpson (1900)
Proptera purpurata (Lamarck, 1819) — Ortmann (1912), Frierson (1927), Strecker (1931), Haas (1969), Johnson (1980).
Lampsilis (*Proptera*) *purpurata* (Lamarck, 1819) — Simpson (1914)
Lampsilis purpurata (Lamarck, 1819) — Johnson (1953)
? Potamilus salinasensis (Simpson, 1908) — Neck & Metcalf (1988), Howells et al. (1996), Haag (2012)
Potamilus purpuratus (Lamarck, 1819) — Turgeon et al. (1988; 1998), Cummings & Mayer (1992), Williams et al. (1993; 2017), Howells et al. (1996), Johnson (1999a), Graf & Cummings (2007; 2021), Cummings & Graf (2010), Inoue et al. (2020b), Smith et al. (2020)
Potamilus purpurata (Lamarck, 1819) — Johnson (1998), Smith et al. (2019)

Unio coloradoensis Lea, 1856. Description of four new species of exotic Uniones. *Proceedings of the Academy of Natural Sciences of Philadelphia* 8(3):103. Figured in Lea (1858a JANSP Vol. 3:314; 1858b Obs. 6:34, pl. 31, fig. 29).

Verbatim Type Locality: "Rio Colorado, Texas."

Type Locality: Colorado River, Texas [W=30.2622, -97.7450].

Type Specimens: Lectotype, USNM 84845 designated by Johnson (1974a:32, as the holotype; Johnson 1998:446, pl. 24, fig. 1).

Taxonomic History:

Unio coloradoensis Lea, 1856 — Singley (1893)
Margaron (*Unio*) *coloradoensis* (Lea, 1856) — Lea (1870)
Lampsilis (*Proptera*) *coloradoensis* (Lea, 1856) — Simpson (1900; 1914)
Cyrtonaias coloradoensis (Lea, 1856) — Morrison (1968)
Proptera purpurata (Lamarck, 1819) — Frierson (1927), Haas (1969)
Potamilus purpuratus (Lamarck, 1819) — Howells et al. (1996), Williams et al. (2008)
Potamilus purpurata (Lamarck, 1819) — Johnson (1998)
Potamilus purpuratus coloradoensis (Lea, 1856) — Roe & Lydeard (1998)

Order Unionoida
Superfamily Unionoidea Rafinesque, 1820
Family Unionidae Rafinesque, 1820
Subfamily Ambleminae Rafinesque, 1820
Tribe Lampsilini von Ihering, 1901
Genus *Toxolasma* Rafinesque, 1831

Toxolasma Rafinesque, 1831. Continuation of a monograph of the bivalve shells of the river Ohio and other rivers of the western states. By Prof. C.S. Rafinesque. (Published at Brussels, September 1820) Containing 46 species, from No. 76 to no. 121. Including an appendix on some bivalve shells of the rivers of Hindostan, with a supplement on the fossil bivalves of the Western states, and the *Tulosites*, a new genus of fossils. Philadelphia, Pennsylvania. 8 pp. Page 2.

Diagnosis: "*Toxolasma*, differs from *Amblema*, *Plagiola* and *Sintoxia*, by lamellar tooth not obliqual but arched parallel with the back, axis nearly terminal, general form rounded, back curved."

Type Species: *Toxolasma lividus* Rafinesque, 1831, by subsequent designation (Frierson 1914:7).

Taxonomic History:

Lampsilis (sec. *Eurynia*) Rafinesque, 1819 — Simpson (1900)
Lampsilis (*Carunculina*) Simpson, 1898 — Simpson (1900; 1914)
Lampsilis Rafinesque, 1820 — Simpson (1900) [in part]
Carunculina Simpson, 1898 — Ortmann (1910; 1915), Haas (1969), Johnson (1980)
Toxolasma Rafinesque, 1831 — Herrmannsen (1849a), Morrison (1968; 1969), Williams et al. (1993; 2008), Johnson (1999a), Roe & Hoeh (2003), Graf & Cummings (2007; 2021), Pfeiffer et al. (2019b)
Toxelasma [sic] Rafinesque, 1831 — Thiele (1934)

Distribution & Remarks: This genus contains seven species, two of which, *Toxolasma parvum* (Barnes, 1823) and *Toxolasma texasiense* (Lea, 1857), are known from the Río Grande del Norte. Only *Toxolasma texasiense* has been found in México.

PLATE 88 ***Toxolasma parvum* (Barnes, 1823). Top:** *Carunculina parva cahni* F.C. Baker, 1927. Holotype, INHS 82281 (45 mm L, 25 mm H). Neosho Mill Pond (Mississippi–Rock–Rubicon Dr.), Dodge County, Wisconsin, USA [L=43.3124, -88.5127]. ***Truncilla cognata* (Lea, 1860). Middle:** *Unio cognatus* Lea, 1860. Holotype, USNM 85004 (43 mm L, 28 mm H). Río Salado (Río Grande del Norte Dr.), Nuevo León, México [W=26.8449, -99.5837]. **Bottom:** *Truncilla cognata* (Lea, 1860). NRI 8383.9. Río Grande del Norte, upstream of La Bota, Laredo, Webb County, Texas [L -27.6022, -99.5805].

Toxolasma parvum (Barnes, 1823)

Plate 88, Map 103

Province(s): Río Grande del Norte.

Distribution and Status: A widespread and common species in the eastern United States. Occurs in Gulf of Mexico tributaries from the Ochlockonee River southwest to the San Antonio basin with a disjunct population in the lower Río Grande del Norte drainage in Texas. There are no verified records of *Toxolasma parvum* in México. CS—Currently Stable (Williams et al. 1993:14); LC—Least Concern (IUCN 2024); G5—Secure (NatureServe 2024).

Anatomy: Shell small, elliptical or cylindrical, relatively solid, and inflated. The anterior and posterior ends are evenly rounded. Beaks inflated and slightly elevated above the hinge line. Beak sculpture consists of 5–6 distinct angled concentric ridges. Beak cavity moderately deep. Pseudocardinal teeth thin, elevated, compressed, and serrated; two in the left valve, single or double in the right. Lateral teeth long, thin and straight; two in the left valve, one in the right. Surface of the shell smooth, with a clothlike texture. Periostracum with a cloth-like texture, dark green, brown, to black and rayless. Nacre silvery or bluish white and highly iridescent. The mantle margins of females possess a circular caruncle just posterior to the branchial opening. This caruncle is reddish in color and is used a lure to attract host fishes. The glochidia were described by Lea (1874) as pouch shaped and without hooks.

Biology: This species is commonly found in both lentic and lotic environments from small creeks to large rivers in mud, sand, or fine gravel. This species is probably a long-term brooder (bradytictic). Glochidia have been observed May–July (Baker 1928:252). Host fishes include the Johnny Darter, *Etheostoma nigrum* Rafinesque, 1820 (Watters et al. 2005:12) and Warmouth, *Lepomis gulosus* (Cuvier, 1829) (Wilson 1916:338).

Adult Size: Length to 4 cm.

Similar Species: Differs from *Toxolasma texasiense* (Lea, 1857) by being smaller in size and lacking pronounced sexual dimorphism.

MAP 103 Distribution of *Toxolasma parvum* (Barnes, 1823).

Unio parvus Barnes, 1823. On the genera *Unio* and *Alasmodonta*; with introductory remarks. *American Journal of Science and Arts* 6(1):107–127; 6(2):258–280+13 plates. Page 274 [174], Plate 13, fig. 9.

Verbatim Type Locality: "Inhabits the Fox River."

Type Locality: Fox River, Wisconsin, USA [W=44.4488, -88.0602].

Type Specimens: Unknown. Presumed lost (Johnson 1967:128; Johnson 1972b:230).

Taxonomic History:

Unio parvus Barnes, 1823 — Lea (1829), Ferussac (1835), Hanley (1843; 1856), Küster (1848), Sowerby (1866), Singley (1893)
Margarita (*Unio*) *parvus* (Barnes, 1823) — Lea (1836)
Margaron (*Unio*) *parvus* (Barnes, 1823) — Lea (1852; 1870)
Unio (*Hyridella*) *parvus* Barnes, 1823 — Adams & Adams (1857)
Lampsilis (*Carunculina*) *parvus* (Barnes, 1823) — Simpson (1900)
Carunculina parva (Barnes, 1823) — Frierson (1927)
Lampsilis (*Carunculina*) *parva* (Barnes, 1823) — Simpson (1914)
Carunculina parva parva (Barnes, 1823) — Haas (1969)
Toxolasma parvus (Barnes, 1823) — Neck & Metcalf (1988), Turgeon et al. (1988), Cummings & Mayer (1992), Williams et al. (1993), Howells et al. (1996), Johnson (1999a), Graf & Cummings (2007), Cummings & Graf (2010)
Toxolasma parvum (Barnes, 1823) — Williams et al. (2008), Graf & Cummings (2021)

Carunculina parva cahni F.C. Baker, 1927. On the division of the Sphaeriidae into two subfamilies: and the descriptions of a new genus of Unionidae, with descriptions of new varieties. *American Midland Naturalist* 10(7):220–223. Page 222. Figured in Baker (1928:253, pl. 105, fig. 14, holotype, figs. 15–18, paratypes).

Verbatim Type Locality: "Neosha Mill Pond, Dodge Co., Wis."

Type Locality: Neosho Mill Pond (Mississippi–Rock–Rubicon Dr.), Dodge County, Wisconsin [L=43.3124, -88.5127].

Type Specimens: Holotype, INHS 82281 (ex. UIMNH 17341). Paratypes, INHS 82282 (*n*=4) (ex. UIMNH 17342) (Baker 1927:222, both by original designation; 1928:253; Franzen 1956:27; Johnson 1972b:231).

Taxonomic History:

Carunculina parva cahni F.C. Baker, 1927 — F.C. Baker (1928)
Carunculina parva parva (Barnes, 1823) — Haas (1969)
Carunculina parva (Barnes, 1823) — Johnson (1972b), Burch (1975), Johnson (1980)
Toxolasma parvum (Barnes, 1823) — Williams et al. (2008)

Unio pertennis de Gregorio, 1914. Su taluni molluschi di acqua dolce di America (American fresh water shells of America). Il Naturalista Siciliano Nuova Serie. 2:31–72+plates 3–12. Page 61, pl. 6, figs. 4a–d.

Remarks: de Gregorio regarded this as a new name for *U. phaselus* Lea, 1856 which he spelled as "*Phaseolus*" and treated as a homonym of *U. phaseolus* Hildreth. However, *phaselus* is the proper spelling. Moreover, although one would expect *U. pertennis* to be an objective synonym of *U. phaselus*, it is not. Lea's species is from southeast Asia, while the specimen figured in de Gregorio is perhaps a *Toxolasma* from North America (although no locality was reported).

Taxonomic History:

Unio pertenuis [sic] de Gregorio, 1914 — Watters et al. (2009)

PLATE 89 ***Toxolasma texasiense*** **(Lea, 1857). Top Left:** *Unio texasiensis* Lea, 1857. Lectotype, USNM 85302 (male) (35 mm L, 20 mm H). Guadalupe River, DeWitt County, Texas, USA [W=29.0904, -97.3296]. **Middle Right:** *Unio bairdianus* Lea, 1857. Lectotype, USNM 85303 (female) (30 mm L, 18 mm H). Devils River (Río Grande del Norte–Pecos Dr.), Texas [W=29.4588, -101.0591]. **Middle Left:** *Unio bealei* Lea, 1862. Lectotype, USNM 85305 (42 mm L, 25 mm H). Trinity River, Leon County, Texas [W=31.3360, -95.6571]. **Bottom Right:** *Lampsilis mearnsi* Simpson, 1900. Lectotype, USNM 151549 (53 mm L, 30 mm H). Elm Creek (Río Grande del Norte–Las Moras Dr.), 43.4 km from Ft. Clark, Kinney County, Texas [L=29.2079, -100.5135]. **Bottom Left:** *Lampsilis* (*Carunculina*) *texasensis* var. *compressus* Simpson, 1900. Lectotype, USNM 152059. Las Moras Creek (Río Grande del Norte Dr.), Fort Clark Springs, Brackettville, Kinney County, Texas [L=29.3075, -100.4181].

Toxolasma texasiense (Lea, 1857)

Plate 89, Map 104

Province(s): Río Grande del Norte.

Distribution and Status: A widespread and common species on the Coastal Plain of the southern United States from southern Illinois south to Alabama and southwest to the Río Grande basin of Texas and México. CS—Currently Stable (Williams et al. 1993:14); Not assessed (IUCN 2024); G4—Apparently Secure (NatureServe 2024).

Anatomy: Shell elongate, relatively thin and moderately inflated. Anterior end rounded. In males, the dorsal edge of the shell behind the umbo slopes down and results in the posterior end being pointed. In females, the posterior end is broadly truncated and slanted at an angle from the ventral margin to the posterior dorsal edge of the shell. Beaks even with or only slightly elevated above the hinge line. Beak sculpture with 5–6 strong angular ridges. Beak cavity shallow. Pseudocardinal teeth relatively thin and compressed, elevated and serrated; two in the left valve, one in the right. Lateral teeth long, straight to curved; two in the left valve, one in the right. Periostracum dark brown to black with cloth-like texture. Nacre white, occasionally tinged with salmon in the beak cavity and center of the shell.

Biology: Found in small rivers or creeks in mud and sand. Gravid females found in July in Texas (Howells et al. 1996:134). Naturally occurring infections were found on Bluegill, *Lepomis macrochirus* Rafinesque, 1819 and Warmouth, *Lepomis gulosus* (Cuvier, 1829). The glochidia are hookless and were attached to the gill filaments. The total number of glochidia per fish ranged from 2 to 36. The encystment was primarily confined to the distal region of the gill filament (Stern and Felder 1978:235).

Adult Size: Length to 6 cm.

Similar Species: *Toxolasma parvum* (Barnes, 1823) is typically smaller and oval in outline without pronounced sexual dimorphism.

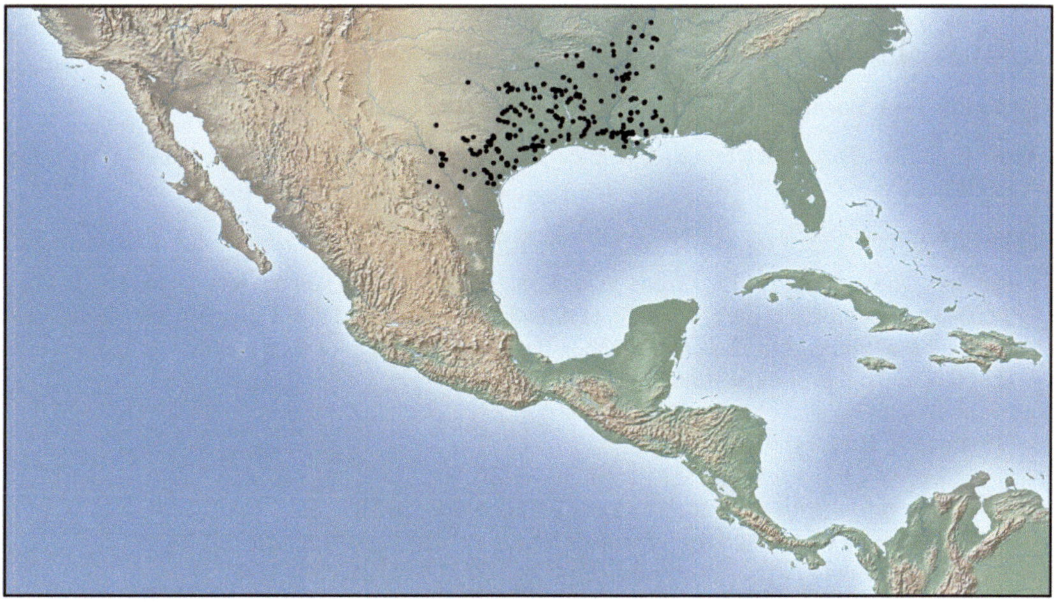

MAP 104 Distribution of *Toxolasma texasiense* (Lea, 1857).

Unio texasiensis Lea, 1857. Description of eight new species of naiades from various parts of the United States. *Proceedings of the Academy of Natural Sciences of Philadelphia* 9:84. Figured in Lea (1860b JANSP Vol. 4, Part 4(13):359; 1860d Obs. 8:41, pl. 61, fig. 184, as *Unio texasensis*).

Verbatim Type Locality: "DeWitt County, Texas."

Type Locality: Guadalupe River (Gulf of Mexico–Guadalupe Dr.), DeWitt County, Texas, USA [W=29.0904, -97.3296].

Type Specimens: Lectotype, USNM 85302 designated by Johnson (1974a:145, as the holotype; Johnson 1999a:34, 58, pl. 4, fig. 1, as *texasensis*). Paralectotypes, MCZ 178921 (*n*=4) (Johnson 1956:138, as *texasensis* and as paratypes).

Remarks: Johnson (1974a:145) added Guadalupe River to the type locality.

Taxonomic History:

Unio texasianus [sic] Lea, 1857 — Sowerby (1866)
Margaron (*Unio*) *texasensis* [sic] (Lea, 1857) — Lea (1870)
Unio texasensis [sic] Lea, 1857 — Singley (1893)
Eurynia (*Carunculina*) *texasensis* [sic] (Lea, 1857) — Ortmann (1912)
Lampsilis (*Carunculina*) *texasensis* [sic] (Lea, 1857) — Simpson (1914)
Carunculina texasensis [sic] (Lea, 1857) — Ortmann (1915)
Carunculina parva-texasensis [sic] (Lea, 1857) — Frierson (1927)
Carunculina parva texasensis [sic] (Lea, 1857) — Strecker (1931), Haas (1969)
Toxolasma texasensis [sic] (Lea, 1857) — Neck & Metcalf (1988), Cummings & Mayer (1992), Williams et al. (1993)
Toxolasma texasiensis (Lea, 1857) — Turgeon et al. (1998), Graf & Cummings (2007), Cummings & Graf (2010)
Toxolasma texasense [sic] (Lea, 1857) — McMurray et al. (2012)
Toxolasma texasiense (Lea, 1857) — Williams et al. (2017), Graf & Cummings (2021)

Unio bairdianus Lea, 1857. Description of six new species of fresh water and land shells of Texas and Tamaulipas, from the collection of the Smithsonian Institution. *Proceedings of the Academy of Natural Sciences of Philadelphia* 9:101–102. Page 102. Figured in Lea (1860b JANSP Vol. 4, part 4(13):361; 1860d Obs. 8:43, pl. 61, fig. 186).

Verbatim Type Locality: "Devil's River, Texas."

Type Locality: Devils River (Río Grande del Norte–Pecos Dr.), Texas, USA [W=29.4588, -101.0591].

Type Specimens: Lectotype, USNM 85303, here designated.

Remarks: The types of *Unio bairdianus* Lea, 1857 were presumed lost (Johnson 1974a:18; 1999a:34), but later discovered. Johnson (1999a:34) listed MCZ 316413 (*n*=1) from "Mississippi" as a probable syntype citing Lea (1860b:362), but it is not from a locality mentioned in the original description and it has no type status.

Taxonomic History:

Unio bairdianus Lea, 1857 — Singley (1893)
Margaron (*Unio*) *bairdianus* (Lea, 1857) — Lea (1870)
Lampsilis (*Carunculina*) *texasensis* [sic] (Lea, 1857) — Simpson (1914)
Carunculina parva-texasensis [sic] (Lea, 1857) — Frierson (1927)
Carunculina parva texasensis [sic] (Lea, 1857) — Strecker (1931), Haas (1969)
Carunculina parva (Barnes, 1823) — Burch (1975)
Toxolasma parvus (Barnes, 1823) — Johnson (1999a)

Unio bealei Lea, 1862. Descriptions of ten new species of Unionidae of the United States. *Proceedings of the Academy of Natural Sciences of Philadelphia* 14:168–169. Page 169. Figured in Lea (1862a JANSP Vol. 5:204; 1863 Obs. 9:26, pl. 30, fig. 273).

Verbatim Type Locality: "Leon County, Texas."

Type Locality: Trinity River (Gulf of Mexico Dr.), Leon County, Texas, USA [W=31.3360, -95.6571].

Type Specimens: Lectotype, USNM 85305 designated by Johnson (1974a:19, as the holotype; Johnson 1999a:35, 58, pl. 4, fig. 3, who added Trinity River to the type locality).

Taxonomic History:

Margaron (*Unio*) *bealei* (Lea, 1862) — Lea (1870)
Unio bealii [sic] Lea, 1862 — Singley (1893)
Lampsilis (*Carunculina*) *texasensis* [sic] (Lea, 1857) — Simpson (1914)
Carunculina parva-texasensis [sic] (Lea, 1857) — Frierson (1927)
Carunculina parva texasensis [sic] (Lea, 1857) — Strecker (1931), Haas (1969)
Carunculina parva (Barnes, 1823) — Burch (1975)
Toxolasma parvus (Barnes, 1823) — Johnson (1999a)

Lampsilis mearnsi Simpson, 1900. New and unfigured Unionidae. *Proceedings of the Academy of Natural Sciences of Philadelphia* 52:74–86+5 plates. Page 75, pl. 1, fig. 4.

Verbatim Type Locality: "Near Fort Clark, southwest Texas."

Type Locality: Elm Creek (Río Grande del Norte–Las Moras Dr.), 43.4 km from Ft. Clark, Kinney County, Texas, USA [L=29.2079, -100.5135].

Type Specimens: Lectotype, USNM 151549 designated by Johnson (1975:14, as the holotype; Johnson 1999a:35, 60, pl. 5, fig. 2). Paralectotypes, MCZ 87962 (*n*=4) (Johnson 1956:124, 1975:14, as paratypes).

Remarks: Johnson (1975:14; 1999a:35) stated that the type locality was Las Moras Creek, pool, 20 miles S Fort Clark, Kinney County, Texas, which is just adjacent to Elm Creek (a tributary of Las Moras Creek). The stream is also listed on the label with the lectotype. Johnson initially called MCZ 87962 paratypes, but later (1999a:35) changed his designation to topotypes.

Taxonomic History:

Lampsilis (*Carunculina*) *mearnsi* Simpson, 1900 — Simpson (1900; 1914)
Carunculina parva-mearnsi (Simpson, 1900) — Frierson (1927)
Carunculina parva mearnsi (Simpson, 1900) — Strecker (1931)
Carunculina parva parva (Barnes, 1823) — Haas (1969)
? Carunculina parva (Barnes, 1823) — Burch (1975)
Toxolasma mearnsi (Simpson, 1900) — Turgeon et al. (1988; 1998), Williams et al. (1993), Graf & Cummings (2007), Cummings & Graf (2010)
Toxolasma texasensis [sic] (Lea, 1857) — Howells et al. (1996)
Toxolasma parvus (Barnes, 1823) — Johnson (1999a)
Toxolasma texasiense (Lea, 1857) — Williams et al. (2017), Graf & Cummings (2021)

Lampsilis (*Carunculina*) *texasensis* var. *compressus* Simpson, 1900. Synopsis of the naiades, or pearly fresh-water mussels. *Proceedings of the United States National Museum* 22(1205):501–1044. Page 564. Not figured.

Verbatim Type Locality: "... the variety in southwest Texas."

Type Locality: Las Moras Creek (Río Grande del Norte Dr.), Fort Clark Springs, Brackettville, Kinney County, Texas, USA [L=29.3075, -100.4181].

Type Specimens: Lectotype, USNM 152059 designated and type locality fixed by Johnson (1975:11, pl. 1, fig. 5; Johnson 1999a:35, 60, pl. 5, fig. 1).

Remarks: Simpson (1914:150) noted that "All the specimens I have seen of the variety seem to be a little diseased, yet the form differs so widely from the type that I cannot believe it to be the result of pathologic influences."

Taxonomic History:

Lampsilis (Carunculina) texasensis var. *compressa* Simpson, 1900 — Simpson (1914)
Carunculina parva-compressa (Simpson, 1900) — Frierson (1927)
Carunculina parva compressa (Simpson, 1900) — Strecker (1931)
Carunculina parva parva (Barnes, 1823) — Haas (1969)
Toxolasma parvus (Barnes, 1823) — Johnson (1999a)

Order Unionoida
Superfamily Unionoidea Rafinesque, 1820
Family Unionidae Rafinesque, 1820
Subfamily Ambleminae Rafinesque, 1820
Tribe Lampsilini von Ihering, 1901
Genus *Truncilla* Rafinesque, 1819

Truncilla Rafinesque, 1819. Prodrome de 70 nouveaux genres d'Animaux découverts dans l'intérieur des États–Unis d'Amérique, durant l'année 1818. *Journal de Physique, de Chimie, d'Histoire Naturelle* 88:417–428. Page 427. Introduced as a subgenus of *Unio* Philipsson, 1788.

Diagnosis: "Valves bombées, tronquées, antérieurement. Dent postérieure semi-lamellaire dentée, dent lamellaire, oblique, courte, axe presque médial." [Valves bulging, truncated, anteriorly. Serrated semi-lamellar posterior tooth, lamellar tooth, oblique, short, axis almost medial].

Type Species: *Truncilla truncata* Rafinesque, 1820, by subsequent designation (Herrmannsen 1849b:628).

Taxonomic History:

Unio Philipsson, 1788 — Menke (1830)
Truncilla Rafinesque, 1819 — Herrmannsen (1849b)
Unio (*Lampsilis*) Rafinesque, 1820 — Adams & Adams (1857), von Martens (1900b) [in part]
Truncilla (s.s.) Rafinesque, 1819 — Simpson (1900; 1914), Frierson (1927), Thiele (1934), Modell (1942; 1964), Morrison (1968), Haas (1969), Cummings & Mayer (1992), Williams et al. (1993), Graf & Cummings (2007; 2021), Pfeiffer et al. (2019b)

Amygdalonaias Crosse & Fischer, 1894. *In* P. Fischer and H. Crosse. 1894. Études sur les mollusques terrestres et fluviatiles du Mexique et du Guatemala. *In* M.H.–Milne Edwards (ed.). *Mission Scientifique au Mexique et dans l'Amérique centrale. Recherches Zoologiques pour servir á l'Histoire de la Faune de l'Amérique Centrale et du Mexique.* 15th Livraison. Feuilles 62–72, pp. 489–576, plates 59–62, Imprimerie Nationale, Paris Volume 2. 731 pp. Page 557. Introduced as a subgenus of *Unio* Philipsson, 1788.

Diagnosis: "Coquille ovale, épaisse, lisse; côté postérieur atténué; épiderme rayonné; dents latérales antérieures épaisses (*U. cognatus*, Lea)." [Shell oval, thick, smooth; posterior side attenuated; radiated epidermis; thick anterior lateral teeth (*U. cognatus*, Lea)].

Type Species: *Unio cognatus* Lea, 1860, by original designation.

Taxonomic History:

Unio (*Lampsilis*) Rafinesque, 1820 —von Martens (1900a) [in part]
Plagiola (*Amygdalonaias*) Crosse & Fischer, 1894 — Simpson (1900; 1914)
Truncilla Rafinesque, 1819 — Modell (1964)

Distribution & Remarks: The genus *Truncilla* contains four species and is widely distributed in eastern North America. Of the four species, *Truncilla cognata* (Lea, 1860) is found in the Río Grande in Texas, USA, and Nuevo León and Tamaulipas, México.

Truncilla cognata (Lea, 1860)

Plate 88, Map 105

Province(s): Río Grande del Norte.

Distribution and Status: *Truncilla cognata* historically occurred in the Rio Grande basin from the confluence of the Pecos River and the Rio Grande in Val Verde County, Texas, to just downstream of Falcon Dam in Starr County, Texas, and in the Rio Salado of México. The current status in the Rio Salado is unknown and presumed extirpated (Burlakova et al. 2019:346). Many places within its historical range in the Rio Grande are now inundated by Lake Amistad. However, the species is currently restricted to the Rio Grande from Eagle Pass, Texas to near San Ygnacio, Texas. Burlakova et al. (2011:161) recommended that *T. cognata* be listed by the IUCN as Critically Endangered. Considered a Threatened Species in Texas. On 25 July 2023 the U.S. Fish and Wildlife Service, proposed to list *Truncilla cognata* as endangered species under the Endangered Species Act of 1973, as amended (Federal Register 2023:47952). The USFWS noted that *T. cognata* is currently restricted to approximately 48 percent of its historical range in the United States and México and is comprised of only one extant population in the Lower Rio Grande between Amistad Reservoir and Laredo, Texas. (Karatayev et al. 2012:211; Federal Register 2023:47965). E–Endangered (Williams et al. 1993:14); Not assessed (IUCN 2024); G1–Critically Imperiled (NatureServe 2024).

Anatomy: Shell is elongate, elliptical, and laterally inflated, with a well-developed posterior ridge and high beaks sculptured with a few double-looped bars. Beak cavity shallow. Dorsal margin slightly curved. Ventral margin convex. Left valve with two triangular pseudocardinals, the anterior one smaller, and two laterals. Right valve with one pseudocardinal and one lateral. Periostracum smooth, yellowish green with dark green chevron markings. Nacre bluish white.

Biology: Unknown. The primary host fish for the Mexican Fawnsfoot is unknown; however, based on other *Truncilla* spp., it is likely Freshwater Drum, *Aplodinotus grunniens* Rafinesque, 1819 (Haag 2012:178–179).

Adult Size: Length to 4.5 cm.

Similar Species: None.

Remarks: Johnson (1999a:38) synonymized *Truncilla cognata* and *T. macrodon* (Lea, 1859) with *T. donaciformis* (Lea, 1828), and Howells et al. (1996:142) hypothesized that all three may be conspecific. Smith et al. (2020:512) recovered *Truncilla cognata* as sister to *T. macrodon* and that duo as sister to *T. donaciformis*. Two recent genetic analyses found them to be three distinct species with *T. cognata* sister to *T. donaciformis* (Burlakova et al. 2019:353; Saxon et al. 2023:7, fig. 4). Of the four species of *Truncilla* they examined, Saxon et al. (2023:6) found that *Truncilla truncata* Rafinesque, 1820 was the most genetically diverse species (Hd = 0.929; π = 0.00765) and had a haplotype diversity equal to the remaining members of *Truncilla* (Hd = 0.933). *Truncilla cognata* was the least genetically diverse (Hd = 0.105; π = 0.00018), and *T. donaciformis* (Hd = 0.708) and *T. macrodon* (Hd = 0.676; π = 0.00564) had an intermediate level of diversity.

MAP 105 Distribution of *Truncilla cognata* (Lea, 1860).

Unio cognatus Lea, 1860. Descriptions of three new species of Uniones from Mexico. *Proceedings of the Academy of Natural Sciences of Philadelphia* 12:305–306. Page 306. Figured in Lea (1860b JANSP Vol. 4, Part 4(13):368; 1860d Obs. 8:50, pl. 65, fig. 193).

Verbatim Type Locality: "Rio Salado, New Leon, Mexico."

Type Locality: Río Salado (Gulf of México–Río Grande del Norte Dr.), Nuevo León, México [W=26.8449, -99.5837].

Type specimen: Holotype by monotypy, USNM 85004 (*n*=½) (Johnson 1974a:32; Johnson 1999a:38, 64, pl. 7, fig. 5).

Taxonomic History:

Margaron (*Unio*) *cognatus* (Lea, 1860) — Lea (1870)
Unio (*Amygdalonaias*) *cognatus* Lea, 1860 — Fischer & Crosse (1894b)
Unio (*Lampsilis*) *cognatus* Lea, 1860 — von Martens (1900b)
Plagiola (*Amygdalonaias*) *cognata* (Lea, 1860) — Simpson (1900; 1914)
Plagiola cognata (Lea, 1860) — de Leon (1912)
Truncilla cognata (Lea, 1860) — Frierson (1927), Haas (1969), Neck & Metcalf (1988), Turgeon et al. (1988; 1998), Williams et al. (1993; 2017), Howells et al. (1996), Graf & Cummings (2007; 2021), Burlakova et al. (2019), Inoue et al. (2020b), Smith et al. (2019; 2020), Czaja et al. (2023), Saxon et al. (2023)
Truncilla donaciformis (Lea, 1828) — Johnson (1999a)
Truncilla cognatus (Lea, 1860) — Haag (2012)

PLATE 90 ***Popenaias popeii* (Lea, 1857). Top:** *Unio popeii* Lea, 1857. Lectotype, USNM 85895 (58 mm L, 28 mm H). Río Salado (Río Grande del Norte Dr.), Nuevo León, México [W=27.1944, -100.1091]. **Middle:** USNM 118394. Devils River, Texas, USA [W=29.4588, -101.0591]. **Bottom:** ANSP 341891. Devils River (Río Grande del Norte–Pecos Dr.), 3.2 km below Texas Route 163, nine miles north of Comstock, Val Verde Co., Texas [L=29.9366, -101.0932].

Order Unionoida
Superfamily Unionoidea Rafinesque, 1820
Family Unionidae Rafinesque, 1820
Subfamily Ambleminae Rafinesque, 1820
Tribe Popenaiadini Heard & Guckert, 1970
Genus *Popenaias* Frierson, 1927

Popenaias popeii (Lea, 1857)

Plate 90, Map 106

Province(s): Río Grande del Norte.

Distribution and Status: Endemic to the Río Grande del Norte basin in New Mexico, Texas, and México. Historically, this species occurred in the lower Pecos River of New Mexico, downstream throughout the Lower Río Grande River and major tributaries in Texas. T—Threatened (Williams et al., 1993:13); CR—Critically Endangered (IUCN 2024); G1—Critically Imperiled (NatureServe 2024). Listed as Federally Endangered in 2018 (Federal Register 2018:5720). *Popenaias popeii* has declined significantly in overall distribution, currently occupying approximately 15% of its historical range in the USA. At the time of its listing, there were five known populations within the historical range of the species in the United States (Black River, Lower Pecos River, Río Grande–Lower Canyons, Río Grande–Laredo, and Devils River) (Federal Register 2018). Subsequently, a small population was discovered in the confluence of Río San Diego, México (Hein et al. 2018), for a total of six populations known within the historical range of the species. All extant populations show some evidence of recent reproduction, except the Pecos River population (Federal Register 2018:5728). It was historically present in the ríos Sabinas and Salado de Nadadores in Coahuila and Nuevo León and the Río San Juan in Nuevo León, México where it is presumably rare if still extant.

Anatomy: Shell elongated and subtrapezoidal, rather compressed. Posterior slope obliquely subtruncate above. Ventral margin straight or slightly concave in the middle. Beaks rather low, their sculpture consisting of a few double-looped ridges. Beak cavity shallow. Left valve with two small pseudocardinals and two long, curved or nearly straight laterals. Right valve with one pseudocardinal, sometimes with a faint second one above. Periostracum olive-green or brown, sometimes with slight rays. Nacre dull purple or occasionally white.

Biology: Typically inhabits medium to large rivers, in crevices, undercut riverbanks, travertine shelves, and under large boulders that contain suitable amounts of small-grained substrate, such as clay, silt, and sand. Smith et al. (2003) described the gametogenetic cycle, reproductive anatomy, and glochidia of *Popenaias popeii* from July 1997 to June 1998 in the Black River, New Mexico, USA. They reported that *P. popeii* was an asynchronous short-term brooder. Smith et al. noted gravid females from 29 April to 30 July with all stages of embryos and larvae present on every sampling date during this period. Although spawning can extend over 8 months, an individual brood (from oviposition to release of glochidia) can be produced in 4–6 weeks, and Smith et al. (2003:337) suggested that it might be possible for a single female to reproduce more than once during the breeding season. They noted that although *P. popeii* had been identified as ectobranchous (Ortmann 1912; Heard and Guckert 1970), 15% of their samples contained individuals that brooded larvae in portions of the inner demibranchs despite the lack of marsupial structure. They speculated that the inner demibranchs probably function in a facultative capacity as an overflow reservoir for glochidia unable to fit into full marsupia of the specialized outer demibranchs due to occasional elevated fecundity

MAP 106 Distribution of *Popenaias popeii* (Lea, 1857)

of a particular female. Glochidia of *P. popeii* are semi-elliptical with pitted valves and measure about 220mm in length and 200mm in height. The inner part of the occlusal surface of each valve possesses rows of simple, conical denticles on a platform just inside the lip (Smith et al. 2003:337, fig. 4). Presumptive host fish for *P. popeii* were stated to be: *Cyprinella lutrensis* (Baird & Girard, 1853) (Leuciscidae), *Carpiodes carpio* (Rafinesque, 1820) (Catostomidae), and *Moxostoma congestum* (Baird & Girard, 1854) (Catostomidae) (Rangaswami et al. 2023:1). Teiga-Teixeira et al. (2020) described the complete mitochondrial genome of *Popenaias popeii* collected from the Río Grande at Laredo, Texas.

Adult Size: Length to 11 cm.

Similar Species: *Lampsilis teres* (Rafinesque, 1820) or *Uniomerus tetralasmus* (Say, 1831).

Unio popeii Lea, 1857. Description of six new species of fresh water and land shells of Texas and Tamaulipas, from the collection of the Smithsonian Institution. *Proceedings of the Academy of Natural Sciences of Philadelphia* 9:101–102. Figured in Lea (1860a JANSP Vol. 4, Part 4(13):372; 1860d Obs. 8:54, pl. 66, fig. 197) and Sowerby (1868, pl. 81, sp. 430).

Verbatim Type Locality: "Devils River and Río Salado, Texas."

Type Locality: Río Salado (Gulf of México–Río Grande del Norte Dr.), Nuevo León, México [W=27.1944, -100.1091].

Type Specimens: Lectotype, USNM 85895, designated and the type locality fixed by Johnson (1999a:21, 62, pl. 6, fig. 1, as *popei* and incorrectly as USNM 85859; Johnson 1974a:115, as the figured holotype). Paralectotypes, USNM 85895a ($n=2$).

Taxonomic History:

Unio popei Lea, 1857 — Lea (1860a), Sowerby (1868), Singley (1893), de Leon (1912)
Margaron (*Unio*) *popeii* (Lea, 1857) — Lea (1870)
Unio (*Graphonaias*) *popei* Lea, 1857 — Fischer & Crosse (1894b)
Unio (*Nephronaias*) *popei* Lea, 1857 — von Martens (1900b; 1901c)
Unio (sec. *Elliptio*) *popeii* Lea, 1857 — Simpson (1900; 1914)
Lampsilis popei (Lea, 1857) — Pilsbry (1910) [in part]
Elliptio (*Popenaias*) *popei* (Lea, 1857) — Frierson (1927)
Elliptio popei (Lea, 1857) — Strecker (1931)
Elliptio (sec. *Popenaias*) *popei* (Lea, 1857) — Thiele (1934)
Nephronaias (*Popenaias*) *popeii* (Lea, 1857) — Haas (1969)
Popenaias popei (Lea, 1857) — Heard & Guckert (1970), Burch (1975), Neck & Metcalf (1988), Turgeon et al. (1988), Williams et al. (1993), Howells et al. (1996), Johnson (1999a), Contreras-Arquieta (2000), Haag (2012)
Popenaias popeii (Lea, 1857) — Turgeon et al. (1998), Strenth et al. (2004), Graf & Cummings (2007 [in part]; 2021), Cummings & Graf (2010), Burlakova et al. (2012; 2019), Inoue et al. (2013; 2015; 2018), Williams et al. (2017), Pfeiffer et al. (2019b [in part]), Smith et al. (2020), Teiga-Teixeira et al. (2020), Czaja et al. (2023)

PLATE 91 ***Megalonaias nervosa*** **(Rafinesque, 1820). Top:** *Unio heros* Say, 1829. "Neotype," SMF 4307 (159 mm L, 127 mm H). North America. **Bottom:** *Unio eightsii* Lea, 1860. Holotype, USNM 83991 (114 mm L, 82 mm H). Río Sabinas (Río Grande del Norte–Salado de Nadadores Dr.), Nuevo León, México [W = 26.8590, -99.5907].

Megalonaias nervosa (Rafinesque, 1820)

Plate 91, Map 107

Province(s): Río Grande del Norte.

Distribution and Status: Widely distributed from eastern North America southwest to the Río Grande basin in Texas. In the Río Grande, fossil shells have been found as far upstream as Eddy County, New Mexico (Metcalf 1982:53). It historically ranged from a fragment at Elm Creek, Maverick County (Metcalf 1982:46) and Las Moras Creek, Kinney County (Strecker 1931) to fresh shells in the Río Grande at Chapeno downstream of Falcon Dam in the mid-1970s (Neck and Metcalf 1988). Mexican reports include the Río Sabinas (Johnson 1999a) and fossil material from the ríos Sabinas and Salado de Nadadores, Nuevo León, and Coahuila (Metcalf 1982:54) (Howells 2001:5). Neck and Metcalf (1988:261) reported that it was rare in the Río Grande and possibly extinct. It appears to be rare in the Río Grande basin today. CS—Currently Stable (Williams et al. 1993:13); LC—Least Concern (IUCN 2024); G5—Secure (NatureServe 2024).

Anatomy: Shell large, thick, rectangular, and compressed to moderately inflated. Anterior end rounded. Posterior end truncated. Dorsal margin straight. Ventral margin straight to slightly curved. A small wing with several ridges is present behind the umbo (most evident in small shells, becoming indistinct in older individuals). Beaks may be slightly elevated above the hinge line. Beak cavity deep. Surface of the shell covered with both large and small corrugations or V-shaped ridges (covering the shell in juveniles). A series of large folds present on the posterior half of the shell. Pseudocardinal teeth well developed and serrated; two in the left valve, one in the right. Lateral teeth straight to slightly curved; two in the left, one in the right. Periostracum rayless, light brown in juveniles, dark brown or black in adults. Nacre white, often with purple or copper-colored blotches, iridescent posteriorly.

Biology: Reported fish hosts include: Longnose Gar, *Lepisosteus osseus* (Linnaeus, 1758); Green Sunfish, *Lepomis cyanellus* Rafinesque, 1819; Bluegill, *Lepomis macrochirus* Rafinesque, 1819; Longear Sunfish, *Lepomis megalotis* (Rafinesque, 1820); Largemouth Bass, *Micropterus salmoides* (Lacepède, 1802); Black Crappie, *Pomoxis nigromaculatus* (Lesueur, 1829); White Crappie, *Pomoxis annularis* Rafinesque, 1819; Yellow Perch, *Perca flavescens* (Mitchell, 1814); Slenderhead Darter, *Percina phoxocephala* (Nelson, 1876); Logperch, *Percina caprodes* (Rafinesque, 1818); Black Bullhead, *Ameiurus melas* (Rafinesque, 1820); Brown Bullhead, *Ameiurus nebulosus* (Lesueur, 1829); Channel Catfish, *Ictalurus punctatus* (Rafinesque, 1818); Flathead Catfish, *Pylodictis olivaris* (Rafinesque, 1818); and Freshwater Drum, *Aplodinotus grunniens* Rafinesque, 1819 (Coker, et al. 1921:152; O'Dee and Watters.2000:79).

Adult Size: Length to 28 cm.

Similar Species: None.

MAP 107 Distribution of *Megalonaias nervosa* Rafinesque, 1820 in the southern USA and México.

Unio (Leptodea) nervosa Rafinesque, 1820. Rafinesque, C.S. 1820. Monographie des coquilles bivalves fluviatiles de la Riviere Ohio, contenant douze genres et soixante-huit especes. *Annales Générales des Sciences Physiques, Bruxelles* 5(13):287–322+plates 80–82. Page 296, pl. 80, figs. 8–10.

Verbatim Type Locality: "aux rapides de l'Ohio."

Type Locality: Falls of the Ohio River, Louisville, Jefferson County, Kentucky, USA [L=38.2724, -85.7630].

Type Specimens: Unknown. Presumed lost (Johnson 1999a:15).

Remarks: This species has an extensive synonymy, and only the type species and those taxa described from Central America are treated below.

Taxonomic History:

Potamilus nervosus — Rafinesque (1818) *nomen nudem*
Unio nervosus Rafinesque, 1820 — Ferussac (1835), Conrad (1853), Küster (1861)
Lampsilis (Ligumia) nervosa (Rafinesque, 1820) — Frierson (1927)
Ligumia nervosa (Rafinesque, 1820) — Haas (1969)
Megalonaias gigantea (Barnes, 1823) — Morrison (1968), Neck & Metcalf (1988)
Megalonaias nervosa (Rafinesque, 1820) — Turgeon et al. (1988; 1998), Cummings & Mayer (1992), Williams et al. (1993; 2017), Howells et al. (1996), Johnson (1999a), Contreras-Arquieta (2000), Graf & Cummings (2007; 2021), Cummings & Graf (2010), Johnson et al. (2018), Pfeiffer et al. (2018; 2019a), Czaja et al. (2023)

Unio heros Say, 1829 Descriptions of some new terrestrial and fluviatile shells of North America. (cont.). The Disseminator of Useful Knowledge; containing hints to the youth of the U. States, from the School of Industry 2(19):291–293. 23 September 1829. Page 291. Figured in Nagel (2025:165, fig. 160, SMF 4307, as the neotype).

Verbatim Type Locality: "Fox River of the Wabash."

Type Locality: Fox River (Wabash River Dr.), opposite New Harmony, Indiana, USA [L=38.0928, -87.9811].

Type Specimens: Unknown. Presumed lost.

Remarks: Haas (1930:328) (see also Zilch 1967:108; Johnson and Baker 1973:157) designated a neotype (SMF 4307), which is invalid due to the selection of a specimen from the "N. Amerika" and not from the Fox River. Evidence that the neotype came as nearly as practicable from the original type locality is a qualifying condition under Article 75.3.6 of the International Code of Zoological Nomenclature (ICZN).

Taxonomic History:

Unio heros Say, 1829— Ferussac (1835), Conrad (1840), Hanley (1843; 1856), Reeve (1864), Wright (1888)
Margarita (*Unio*) *multiplicatus* (Lea, 1831) — Lea (1836; 1838a)
Complanaria heros (Say, 1829) — Agassiz (1852)
Margaron (*Unio*) *multiplicatus* (Lea, 1831) — Lea (1852; 1870)
Quadrula (sec. *Crenodonta*) *heros* (Say, 1829) — Simpson (1900; 1914)
Quadrula heros (Say, 1829) — de Leon (1912)
Crenodonta heros (Say, 1829) — Ortmann (1912)
Megalonaias heros (Say, 1829) — Utterback (1915)
Megalonaias gigantea (Barnes, 1823) — Ortmann & Walker (1922), F.C. Baker (1928), Zilch (1967), Haas (1969), Burch (1975), Johnson (1980)
Amblema (*Megalonaias*) *gigantea* (Barnes, 1823) — Frierson (1927), Haas (1930)
Amblema (sec. *Megalonaias*) *heros* (Say, 1829) — Thiele (1934)
Megalonaias gigantea gigantea (Barnes, 1823) — Haas (1969)
Megalonaias nervosa (Rafinesque, 1820) — Pfeiffer et al. (2018; 2019a), Graf & Cummings (2007; 2021), Nagel (2025)

Unio eightsii Lea, 1860. Descriptions of three new species of Uniones from Mexico. *Proceedings of the Academy of Natural Sciences of Philadelphia* 12:305–306. Page 306. Figured in Lea (1860a JANSP Vol. 4, Part 4(13):367; 1860d Obs. 8:49, pl. 64, fig. 192).

Verbatim Type Locality: "Texas and Sabinas River, New Leon, Mexico."

Type Locality: Río Sabinas (Río Grande del Norte–Salado de Nadadores Dr.), Nuevo León, México [W=26.8590, -99.5907].

Type Specimens: Lectotype, USNM 83991 designated by Johnson (1974a:51, as the holotype; Johnson 1999a:15).

Taxonomic History:

Unio eightsii Lea, 1860 — Singley (1893)
Margaron (*Unio*) *eightsii* (Lea, 1860) — Lea (1870)
Unio (*Plectomerus*) *eightsii* Lea, 1860 — Fischer & Crosse (1894b)
Unio (*Crenodonta*) *digitatus* Morelet, 1851 — von Martens (1900a) [in part]
Quadrula (sec. *Crenodonta*) *heros* (Say, 1829) — Simpson (1900; 1914)
Quadrula heros (Say, 1829) — de Leon (1912)
Amblema (*Megalonaias*) *gigantea* (Barnes, 1823) — Frierson (1927)
Amblema gigantea (Barnes, 1823) — Strecker (1931)
Megalonaias gigantea gigantea (Barnes, 1823) — Haas (1969)
Megalonaias nervosa (Rafinesque, 1820) — Howells et al. (1996), Johnson (1999a), Pfeiffer et al. (2018)

Order Unionoida
Superfamily Unionoidea Rafinesque, 1820
Family Unionidae Rafinesque, 1820
Subfamily Ambleminae Rafinesque, 1820
Tribe Quadrulini von Ihering, 1901
Genus *Quadrula* Rafinesque, 1820

Quadrula Rafinesque, 1820. Monographie des coquilles bivalves fluviatiles de la Riviere Ohio, contenant douze genres et soixante-huit especes. *Annales Générales des Sciences Physiques, Bruxelles* 5(13):287–322+plates 80–82. Page 305. Introduced as a subgenus of *Obliquaria* Rafinesque, 1820.

Diagnosis: "Forme écarrie, mais arrondie antérieurement, á peine transversale." [Squared shape, but rounded anteriorly, barely transverse].

Type Species: *Obliquaria* (*Quadrula*) *quadrula* Rafinesque, 1820, by subsequent designation (Herrmannsen 1848:384). Herrmannsen (1849b:668) listed this as a synonym of *Unio* Retzius, 1788.

Taxonomic History:

Quadrula Rafinesque, 1820 — Herrmannsen (1848), Simpson (1900), Pilsbry (1910), Utterback (1915), Modell (1942; 1964), Cummings & Mayer (1992), Williams et al. (1993; 2017), Johnson (1999a), Graf & Cummings (2007; 2021), Bouchet & Rocroi (2010), Cummings & Graf (2010), Lopes-Lima et al. (2017), Pfeiffer et al. (2019b)
Unio (*Quadrula*) Rafinesque, 1820 — Adams & Adams (1857)
Quadrula (s.s.) Rafinesque, 1820 — Simpson (1900; 1914), Frierson (1927), Haas (1969)

Amphinaias Crosse & Fischer, 1894. *In* P. Fischer and H. Crosse. 1894. Études sur les mollusques terrestres et fluviatiles du Mexique et du Guatemala. M.H.–Milne Edwards (ed.). *Mission Scientifique au Mexique et dans l'Amérique centrale. Recherches Zoologiques pour servir á l'Histoire de la Faune de l'Amérique Centrale et du Mexique.* 15th Livraison. Feuilles 62–72, pp. 489–576, plates 59–62, Imprimerie Nationale, Paris Volume 2. 731 pp. Page 556. Introduced as a subgenus of *Unio* Philipsson, 1788.

Diagnosis: "Coquille épaisse, subquadrangulaire, renflée, épaisse, lisse en avant, plissée en arrière; dents fortes; bord ventral sinueux (*U. Couchianus*, Lea)." [Shell thick, subquadrangular, swollen, thick, smooth in front, pleated behind; strong teeth; sinuous ventral edge (*U. Couchianus*, Lea)].

Type Species: *Unio couchianus* Lea, 1860, by original designation.

Taxonomic History:

Unio (*Pleurobema*) Rafinesque, 1819 — von Martens (1900b)
Quadrula (sec. *Amphinaias*) Crosse & Fischer, 1894 — Thiele (1934)
Amphinaias Crosse & Fischer, 1894 — Graf & Cummings (2007), Cummings & Graf (2010)
Quadrula Rafinesque, 1820 — Lopes-Lima et al. (2019)

Distribution & Remarks: The genus *Quadrula* ranges from the eastern United States and Canada southwest to the Río Grande basin of México. There are three species in the genus (Graf and Cummings 2021:20), two of which, *Quadrula couchiana* (Lea, 1860) and *Quadrula quadrula* (Rafinesque, 1820), occur in the Rio Grande.

PLATE 92 ***Quadrula quadrula* (Rafinesque, 1820). Top Left:** *Obliquaria* (*Quadrula*) *quadrula* Rafinesque, 1820. Neotype, ANSP 20224. Salt River (Mississippi–Ohio Dr.), Kentucky, USA [W=37.9851, -85.7169]. **Middle Right:** *Unio forsheyi* Lea, 1859. Lectotype, USNM 84169 (51 mm L, 43 mm H). Fayette County (Colorado (TX) Dr.), Texas, USA [L=29.9005, -96.8871]. **Middle Left:** *Unio speciosus* Lea, 1862. Lectotype, USNM 84172 (38 mm L, 31 mm H). Leon County (Trinity Dr.), Texas [L=31.3360, -95.6571]. **Bottom Right: *Quadrula couchiana* (Lea, 1860).** *Unio couchianus* Lea, 1860. Lectotype, USNM 85728 (46 mm L, 33 mm H). Río Salado de Nadadores (Río Grande del Norte Dr.), Nuevo León, México [W=27.1944, -100.1091].

Quadrula quadrula (Rafinesque, 1820)

Plate 92, Map 108

Province(s): Río Grande del Norte. Introduced.

Distribution and Status: Widespread in the USA in the Interior Basin and Gulf of Mexico drainages from Minnesota south to Mississippi and Louisiana southwest through Texas. Howells et al. (1996:105) reported that it "occurs in all major drainage systems" of Texas. Karatayev et al. (2012:213) considered *Quadrula apiculata* (Say, 1829), as introduced in the Río Grande and noted a slow, upriver range extension of *Q. apiculata* in Casa Blanca and Falcon reservoirs. CS—Currently Stable (Williams et al. 1993:14); LC—Least Concern (IUCN 2024); G5—Secure (NatureServe 2024, as *Quadrula quadrula*).

Anatomy: Shell quadrate, thick and somewhat inflated. Anterior end rounded. Posterior end truncated. Dorsal margin straight. Ventral margin curved in the anterior half, straight or arched posteriorly. Beaks small and only slightly elevated above the hinge line. Beak sculpture of two rows of raised bumps or nodules that continue downward on the surface of the shell, separated by a furrow or sulcus. Beak cavity deep. Posterior slope with radiating rows of ridges or pustules. Pseudocardinal teeth well developed, serrated; two in the left valve, one in the right. Lateral teeth striated, long, and straight; two in the left valve, one in the right. Periostracum variable from yellowish green to light brown, becoming greenish brown, chestnut, or dark brown in larger individuals. Faint green rays occasionally present. Nacre pearly white, iridescent posteriorly.

Biology: Inhabits small to large rivers in mud, sand, and gravel. Reported fish hosts are the Ictalurid catfish *Ameiurus natalis* (Lesueur, 1819), *Ictalurus punctatus*, and *Pylodictis olivaris* (Howard and Anson. 1922).

Adult Size: Length to 10 cm.

Similar Species: *Quadrula couchiana* (Lea, 1860).

MAP 108 Distribution of *Quadrula quadrula* (Rafinesque, 1820).

Remarks: The pustulose form, *Quadrula apiculata* (Say, 1829), has been recently synonymized under *Quadrula quadrula* following Lopes-Lima et al. (2019). *Quadrula apiculata* is very common in southwest Texas (Burlakova et al. 2011:158). *Quadrula quadrula* has an extensive synonymy, and only the type species and those described from Texas or México are included below.

Obliquaria (*Quadrula*) *quadrula* Rafinesque, 1820. Monographie des coquilles bivalves fluviatiles de la Riviere Ohio, contenant douze genres et soixante-huit especes. *Annales Générales des Sciences Physiques, Bruxelles* 5(13):287–322+plates 80–82. Page 307. Also, as *Unio quadrula*.

Verbatim Type Locality: "commune dans l'Ohio."

Type Locality: Salt River, Kentucky [W=37.9851, -85.7169].

Type Specimens: Neotype, ANSP 20224 selected and type locality fixed by Johnson and Baker (1973:167, pl. 2, fig. 2).

Taxonomic History:

Unio quadrulus Rafinesque, 1820 — Ferussac (1835), Conrad (1840), Küster (1861)
Obliquaria (*Quadrula*) *quadrula* Rafinesque, 1820 — Lea (1836), Simpson (1900; 1914)
Unio (*Theliderma*) *quadrulus* Rafinesque, 1820 — Conrad (1853)
Unio quadratus [sic] Rafinesque, 1820 — Reeve (1864)
Quadrula quadrula (Rafinesque, 1820) — Utterback (1915), Ortmann & Walker (1922), Ortmann (1925), Thiele (1934), Modell (1942), Burch (1975), Turgeon et al. (1988; 1998), Williams et al. (1993; 2008; 2017), Howells et al. (1996), Graf & Cummings (2006; 2007; 2021), Cummings & Graf (2010), Inoue et al. (2014), Lopes-Lima et al. (2017; 2019; 2020), Smith et al. (2019; 2020), Teiga-Teixeira et al. (2020)
Quadrula (s.s.) *quadrula* (Rafinesque, 1820) — Frierson (1927), Haas (1969)
Quadrula (s.s.) *gigantea* [sic] (Rafinesque, 1820) — Haas (1930)
Quadrula (s.s.) *quadrula quadrula* (Rafinesque, 1820) — Haas (1969)
Quadrula quedrula [sic] (Rafinesque, 1820) — Contreras-Arquieta (2000)

Unio apiculatus Say, 1829. Descriptions of some new terrestrial and fluviatile shells of North America. (cont.). The Disseminator of Useful Knowledge; containing hints to the youth of the U. States, from the School of Industry 2(20):308–310. 7 October 1829. Page 309. Figured in Conrad (1837:78, pl. 44, fig. 1) and Reeve (1864, pl. 3, species 11).

Verbatim Type Locality: "Lake Ponchartrain, near N. Orleans."

Type Locality: Lake Ponchartrain (Gulf of Mexico Dr.), near New Orleans, Orleans Parrish, USA [W=30.0312, -90.0839].

Type Specimens: Unknown. Presumed lost.

Taxonomic History:

Unio apiculatus Say, 1829 — Ferussac (1835), Conrad (1837), Hanley (1843; 1856), Reeve (1864), Wright (1888), Singley (1893)
Margarita (*Unio*) *apiculatus* (Say, 1829) — Lea (1836; 1838a)
Margaron (*Unio*) *apiculatus* (Say, 1829) — Lea (1852; 1870)
Unio (*Theliderma*) *apiculatus* Say, 1829 — Conrad (1853)
Unio (*Quadrula*) *apiculatus* Say, 1829 — Adams & Adams (1857)
Unio (sec. *Quadrula*) *apiculatus* Say, 1829 — Paetel (1890)
Quadrula (sec. *Theliderma*) *apiculata* (Say, 1829) — Simpson (1900; 1914)
Quadrula apiculata (Say, 1829) — Ortmann (1910), Johnson (1980), Turgeon et al. (1988), Graf & Cummings (2007), Pfeiffer et al. (2018; 2019a–b), Inoue et al. (2020b)

Quadrula (s.s.) *quadrula-apiculata* (Say, 1829) — Frierson (1927)
Quadrula quadrula apiculata (Say, 1829) — Strecker (1931)
Quadrula (s.s.) *quadrula apiculata* (Say, 1829) — Haas (1969)
Quadrula quadrula (Rafinesque, 1820) — Burch (1975), Lopes-Lima et al. (2019), Graf & Cummings (2021)

Unio forsheyi Lea, 1859. Descriptions of seven new species of Uniones from South Carolina, Florida, Alabama and Texas. *Proceedings of the Academy of Natural Sciences of Philadelphia* 11(1859):154–155. Page 155. Figured in Lea (1860a JANSP Vol. 4, Part 4(13):357; 1860d Obs. 8:39, pl. 60, fig. 182).

Verbatim Type Locality: "Fayette Co., Texas....and Alabama."

Type Locality: Fayette County (Gulf of Mexico–Colorado (TX) Dr.), Texas, USA [L=29.9005, -96.8871].

Type Specimens: Lectotype, USNM 84169 designated and the type locality fixed by Johnson (1974a:59, as the figured holotype). Paralectotypes, MCZ 178785 (*n*=8), from Alabama (Johnson 1957:115, as paratypes); UMMZ 167845 (*n*=1) (Johnson 1979:32, from Anthony, ex. MCZ, without locality. Alabama from UMMZ catalogue).

Taxonomic History:

Unio forsheyi Lea, 1859 — Lea (1860a), Reeve (1864), Wright (1888), Singley (1893)
Margaron (*Unio*) *forsheyi* (Lea, 1859) — Lea (1870)
Quadrula (sec. *Theliderma*) *forsheyi* (Lea, 1859) — Simpson (1900; 1914)
Quadrula forsheyi (Lea, 1859) — Ortmann (1910)
Quadrula (s.s.) *quadrula-forsheyi* (Lea, 1859) — Frierson (1927)
Quadrula quadrula forsheyi (Lea, 1859) — Strecker (1931)
Quadrula (s.s.) *quadrula forsheyi* (Lea, 1859) — Haas (1969)
Quadrula quadrula (Rafinesque, 1820) — Burch (1975), Howells et al. (1996), Lopes-Lima et al. (2019)
Quadrula apiculata (Say, 1829) — Williams et al. (2008)

Unio speciosus Lea, 1862. Descriptions of ten new species of Unionidae of the United States. *Proceedings of the Academy of Natural Sciences of Philadelphia* 14:168–169. Page 168. Figured in Lea (1862 JANSP Vol. 5:207; 1863 Obs. 9:29, pl. 31, fig. 276).

Verbatim Type Locality: "Colorado River, near Lagrange, Texas......and Leon Co., Texas."

Type Locality: Leon County (Trinity Dr.), Texas, USA [L=31.3360, -95.6571].

Type Specimens: Lectotype, USNM 84172 designated and the type locality fixed by Johnson (1974a:134, as the figured holotype, without locality). Paralectotype USNM 84172a (*n*=1) from Colorado River, near La Grange, Texas.

Remarks: Locality data of the lectotype and paralectotype from the label and catalogue.

Taxonomic History:

Unio speciosus Lea, 1862 — Sowerby (1868), Wright (1888), Singley (1893), Call (1895)
Margaron (*Unio*) *speciosus* (Lea, 1862) — Lea (1870)
Quadrula (sec. *Theliderma*) *speciosa* (Lea, 1862) — Simpson (1900; 1914)
Quadrula speciosa (Lea, 1862) — Ortmann (1910)
Quadrula (s.s.) *quadrula-speciosa* (Lea, 1862) — Frierson (1927)
Quadrula (s.s.) *quadrula speciosa* (Lea, 1862) — Haas (1969)
Quadrula quadrula (Rafinesque, 1820) — Burch (1975)
Quadrula apiculata (Say, 1829) — Johnson (1999a)

Quadrula couchiana (Lea, 1860)

Plate 92, Map 109

Province(s): Río Grande del Norte.

Distribution and Status: Historically known from the Pecos, Sabinas and Salado de Nadadores in the Río Grande Basin. Neck (1984:11) and Howells et al. (1996:110) also listed the Río Conchos, Chihuahua, México on the authority of Metcalf, but we have no verified records of it from that basin. Metcalf (1982:47) noted that there is a lot (USNM 524035) from Chacon Creek (Nueces Dr.), 1 mi below La Pryor-Eagle Pass Road Bridge, Zavala County, Texas collected on 18 June 1919 by A.G. Maddren. Recent surveys in the Rio Grande, including tributaries where this species historically occurred, failed to find any evidence of extant populations in Texas (Karatayev et al. 2012:212; Randklev et al. 2018). However, recent surveys in the Rio Salado drainage of México reported shell, of unknown condition, for this species (David Berg, pers. comm.), which may indicate that a population of *Q. couchiana* could still persist within this system. A limited resurvey of eleven sites in the Conchos was conducted in May 2018, and no *Quadrula couchiana* were found alive or dead (Tiemann et al. 2020:182). E—Endangered (Williams et al. 1993:14); CR—Critically Endangered (IUCN 2024); GH—Possibly Extinct (NatureServe 2024). Williams et al. (2017:47) speculated that *Quadrula couchiana* is now extinct.

Anatomy: Shell of medium size, subrhomboid, and rather solid. Beaks full and elevated, their sculpture consisting of strong double-looped or zigzag ridges. Beak cavity moderately deep. Posterior ridge moderately developed, generally double, and indistinctly angled. Pseudocardinals elevated, ragged and uneven. Lateral teeth in the right valve single or slightly double. A shallow sulcus is present in the middle of the shell. Shell surface with fine, concentric growth lines and randomly scattered pustules. Two rows of pronounced pustules mid-disk are present adjacent to the sulcus and run parallel from the umbo to the ventral margin. Periostracum tan or slightly greenish in the young individuals, brown in adults. Nacre silvery white and iridescent.

MAP 109 Distribution of *Quadrula couchiana* (Lea, 1860).

Biology: Unknown.

Adult Size: Length to 6 cm.

Similar Species: *Quadrula quadrula* (Rafinesque, 1820).

Unio couchianus Lea, 1860. Descriptions of three new species of Uniones from Mexico. *Proceedings of the Academy of Natural Sciences of Philadelphia* 12:305–306. Page 305. Figured in Lea (1860a JANSP Vol. 4, Part 4(13):371; 1860d Obs. 8:53, pl. 66, fig. 196) and Sowerby (1868, pl. 81, sp. 429).

Verbatim Type Locality: "Rio Salado, New Leon, Mexico."

Type Locality: Río Salado de Nadadores (Gulf of México–Río Grande del Norte Dr.), Nuevo León, México [W=27.1944, -100.1091].

Type Specimens: Lectotype, USNM 85728 designated by Johnson (1974a:38, as the holotype; Johnson 1999a:15, 54, pl. 2, fig. 1).

Taxonomic History:

Unio couchianus Lea, 1860 — Sowerby (1868)
Margaron (*Unio*) *couchianus* (Lea, 1860) — Lea (1870)
Unio (*Amphinaias*) *couchianus* Lea, 1860 — Fischer & Crosse (1894b)
Unio (*Pleurobema*) *couchianus* Lea, 1860 — von Martens (1900b)
Quadrula (sec. *Theliderma*) *couchiana* (Lea, 1860) — Simpson (1900; 1914)
Unio conchiana [sic] Lea, 1860 — de Leon (1912)
Quadrula (*Bullata*) *couchiana* (Lea, 1860) — Frierson (1927)
Quadrula (sec. *Amphinaias*) *couchiana* (Lea, 1860) — Thiele (1934)
Quadrula couchiana (Lea, 1860) — Strecker (1931), Morrison (1968), Neck (1984) Neck & Metcalf (1988), Turgeon et al. (1988; 1998), Williams et al. (1993; 2017), Howells et al. (1996), Johnson (1999a), Haag (2012), Lopes-Lima et al. (2019), Tiemann et al. (2020), Graf & Cummings (2021), Czaja et al. (2023)
Quadrula (*Pustulosa*) *couchiana* (Lea, 1860) — Haas (1969)
Quadrula counchiana [sic] (Lea, 1860) — Contreras-Arquieta (2000)
Amphinaias couchiana (Lea, 1860) — Graf & Cummings (2007), Cummings & Graf (2010)

Order Unionoida
Superfamily Unionoidea Rafinesque, 1820
Family Unionidae Rafinesque, 1820
Subfamily Ambleminae Rafinesque, 1820
Tribe Quadrulini von Ihering, 1901
Genus *Uniomerus* Conrad, 1853

Uniomerus Conrad, 1853. A synopsis of the family of Naïades of North America, with notes, and a table of some of the genera and sub-genera of the family, according to their geographical distribution, and descriptions of genera and sub-genera. *Proceedings of the Academy of Natural Sciences of Philadelphia* 6(7):243–269. Page 268.

Diagnosis: "Shell elongated, not thick, with parallel dorsal and basal margins; beaks with obliquely concentric plicae. *U. declivis*, *camptodon*, *subcroceus*, *Sayii*, *rivularis*, *porrectus*, *symmetricus*, and *excultus*."

Type Species: *Unio declivis* Say, 1831, by subsequent designation (Frierson 1927:10).

Taxonomic History:

Unio (*Uniomerus*) Conrad, 1853 — Fischer & Crosse (1894a), von Martens (1900a)
Unio (sec. *Uniomerus*) Conrad, 1853 — Simpson (1900; 1914)
Uniomerus Conrad, 1853 — Utterback (1915), Modell (1942; 1964), Haas (1969), Johnson (1970; 1999a), Cummings & Mayer (1992), Williams et al. (1993; 2017), Graf & Cummings (2007; 2021), Cummings & Graf (2010), Pfeiffer et al. (2018; 2019b)
Elliptio (*Uniomerus*) Conrad, 1853 — Frierson (1927), Haas (1929)
Elliptio (sec. *Uniomerus*) Conrad, 1853 — Thiele (1934)

Distribution & Remarks: The genus *Uniomerus* is widely distributed in eastern North America in Atlantic, Mississippi, and Gulf of Mexico drainages. *Unio tetralasmus* Say, 1831 had been considered by some authors to be the type species of the genus. However, as pointed out by Clench and Turner (1956:176), that species was not among those originally included by Conrad. Clench and Turner (1956) also incorrectly selected *Unio excultus* Conrad, 1838 as the type species, which was also not mentioned in the original erection of the genus.

PLATE 93 ***Uniomerus tetralasmus*** **(Say, 1831). Top:** *Unio manubius* Gould, 1855. Lectotype, MCZ 169447 (94 mm L, 46 mm H). **Middle:** Paralectotype, MCZ 169448. **Bottom:** Paralectotype, USNM 85709. All three, Río Aqualeguas (Río Grande del Norte Dr.), ca. 4.8 km NE of General Trevino, Nuevo León, México [L=26.2635, -99.4369].

Uniomerus tetralasmus (Say, 1831)

Plate 93, Map 110

Province(s): Río Grande del Norte.

Distribution and Status: Widely distributed throughout much of the central and lower Mississippi basin, Great Lakes, Southern Atlantic Slope, Florida, and Gulf Coast, southwest to the Río Grande Basin, Texas and México (Johnson 1999a:25). Apparently extirpated from the Rio Grande basin. Neck (1987:151) noted that, "No form of *Uniomerus* is known today from the lowland, heavy clay areas of the Lower Rio Grande Valley." CS—Currently Stable (Williams et al. 1993:14); LC—Least Concern (IUCN 2024); G5—Secure (NatureServe 2024).

Anatomy: Shell relatively thin, elongate, and compressed. Anterior end rounded. Posterior end sharply pointed. Dorsal margin straight. Ventral margin straight, rarely curved. Beaks low, even, or slightly elevated above the hinge line. Beak sculpture with 4–5 concentric ridges. Beak cavity shallow. Two shallow grooves present on the posterior slope giving rise to a short ridge. Shell smooth and shiny in small shells, becoming rougher and dull in older individuals. Pseudocardinal teeth small and thin; two in the left valve, one in the right. Lateral teeth relatively thin, short, and straight to slightly curved. Periostracum yellowish brown in small shells, dark brown to black in adults and rayless. Nacre white, occasionally with a tinge of blue or salmon.

The supra-anal is separated from the anal by a long mantle-connection. Inner edge of anal with fine crenulations. Branchial opening with papilla. Inner lamina of inner gills free, except anteriorly. Posterior margins of palpi connected for about one-fourth of their length. Marsupium formed by the outer gills. Color of soft parts whitish.

Biology: A common inhabitant of habitats with low oxygen and highly turbid water such as ponds and slow-flowing streams. Although *U. tetralasmus* is often found with other typical lowland species, its ability to withstand desiccation allows it to live in ephemeral streams that support few other mussel species. Strecker (1931:15) noted that "I once found many living specimens of *Elliptio tetralasmus* imbedded in the banks of an artificial lake several months after the lake had been drained." Morrison (1976) stated that "*tetralasmus* individuals all possess the female type of gill structure, and hence are evidently either parthenogenetic or hermaphroditic." Ten gravid females were found in March, April, and June in Mississippi (Jones et al. 2019:312). The only reported fish host is the Golden Shiner, *Notemigonus crysoleucas* (Mitchill, 1814). Encysted glochidia were observed on the gill filaments and gill rakers, with an average infection of 12 glochidia per host fish (Stern and Felder 1978:233–234).

Adult Size: Length to 12 cm.

Similar Species: *Lampsilis teres* (Rafinesque, 1820).

Remarks: As with *Quadrula quadrula*, this species has an extensive synonymy. The names included below are the nominal taxon and those described from Texas or México.

MAP 110 Distribution of *Uniomerus tetralasmus* (Say, 1831) in the southern USA and México.

Unio tetralasmus Say, 1831. American Conchology, or descriptions of the shells of North America. Illustrated by coloured figures from original drawings executed from nature. New Harmony, Indiana 1(3): unpaginated+plates 21–30. Plate 23.

Verbatim Type Locality: "Bayou St. John, near New Orleans."

Type Locality: Bayou St. John, New Orleans, Orleans Parrish, Louisiana [L=29.9847, -90.0892].

Type Specimens: Unknown. Presumed lost (Johnson 1970:339; Johnson 1972b:220).

Taxonomic History:

Unio tetralasmus Say, 1831 — Ferussac (1835), Conrad (1840), Hanley (1843; 1856), Conrad (1853), Singley (1893), de Leon (1912)
Margarita (*Unio*) *tetralasmus* (Say, 1831) — Lea (1836)
Margaron (*Unio*) *tetralasmus* (Say, 1831) — Lea (1852; 1870)
Unio (*Niäa*) *tetralasmus* Say, 1831 Say, 1831 — Adams & Adams (1857)
Unio (sec. *Uniomerus*) *tetralasmus* Say, 1831 — Simpson (1900; 1914)
Uniomerus tetralasmus (Say, 1831) — Ortmann (1912), Johnson (1980; 1999), Neck (1987), Neck & Metcalf (1988), Turgeon et al. (1988; 1998), Cummings & Mayer (1992), Williams et al. (1993; 2017), Howells et al. (1996), Johnson (1999a), Graf & Cummings (2007; 2021), Cummings & Graf (2010), Johnson et al. (2018), Inoue et al. (2018), Lopes-Lima et al. (2019), Pfeiffer et al. (2019a), Teiga-Teixeira et al. (2020), Czaja et al. (2023)
Elliptio (*Uniomerus*) *tetralasmus* (Say, 1831) — Frierson (1927), Haas (1929)
Elliptio tetralasmus (Say, 1831) — Strecker (1931)
Elliptio (sec. *Uniomerus*) *tetralasmus* (Say, 1831) — Thiele (1934)
Uniomerus declivus (Say, 1932 [sic]) — Morrison (1976)
Uniomerus tretalasmus [sic] (Say, 1831) — Contreras-Arquieta (2000)

Unio manubius Gould, 1855. New species of land and fresh-water shells from western (N.) America. (Cont.). *Proceedings of the Boston Society of Natural History* 5(15):228–229. Page 229. Figured in Johnson (1964:108, pl. 32, fig. 5).

Verbatim Type Locality: "Chihuahua, 60 mi from Camp Ringgold."

Type Locality: Río Aqualeguas (Gulf of México–Río Grande del Norte Dr.), ca. 4.8 km NE of General Trevino, Nuevo León, México [L=26.2635, -99.4369].

Type Specimens: Lectotype, MCZ 169447 designated by Johnson (1964:108, 1970:340, 1972b:222, all as the holotype; Johnson 1999a:24, 62, pl. 6, fig. 2). Paralectotypes, MCZ 169448 (*n*=1); USNM 85709 (*n*=½) (Johnson 1964:108, as paratypes).

Remarks: Johnson (1999a:24) gave the locality as Puntiagudo. According to Taylor (1967:154), Puntiagudo is "30 mi SW Mier." Puntiagudo is now known as Villa de General Treviño, about 40 km west of Mier.

Taxonomic History:

Unio manubrius [sic] Gould, 1855 — Paetel (1890)
Unio manubius Gould, 1855 — Wright (1888), Singley (1893)
Margaron (*Unio*) *manubius* (Gould, 1855) — Lea (1870)
Unio (*Uniomerus*) *manubius* Gould, 1855 — Fischer & Crosse (1894b), von Martens (1900b)
Unio (s.s.) *tetralasmus* Say, 1831 — Call (1895, as *manubrius*)
Unio (sec. *Uniomerus*) *tetralasmus* var. *manubius* Gould, 1855 — Simpson (1900; 1914)
Unio (*Elliptio*) *manubius* Gould, 1855 — von Martens (1901c)
Unio tetralasmus Say, 1831 — de Leon (1912)
Elliptio (*Uniomerus*) *tetralasmus-manubius* (Gould, 1855) — Frierson (1927)
Elliptio tetralasmus manubius (Gould, 1855) — Strecker (1931)
Uniomerus tetralasmus (Say, 1831) — Haas (1969), Johnson (1970; 1972), Burch (1975), Neck (1987), Johnson (1999a), Williams et al. (2008)
Uniomerus tetralasmus manubius (Gould, 1855) — Neck & Metcalf (1988)

Species Erroneously Attributed to the Río Grande del Norté

Fusconaia mitchelli (Simpson in Dall, 1895)

Distribution and Status: The distribution of *F. mitchelli* was thought to occur from "Southern Texas to New Leon, Mexico" (Simpson 1900:711), which included the Brazos, Colorado, Guadalupe, and Rio Grande drainages. Various authorities have considered the range of *F. mitchelli* to extend west to the Río Grande drainage based on the hypothesis that it was synonymous with *Sphenonaias taumilapana* (Conrad, 1855) (Howells et al. 1996:127; Johnson 1999a:20; Randklev et al. 2013:248). With the elevation of *F. iheringi* for individuals in the Brazos and Colorado drainages, the distribution of *F. mitchelli* is restricted to the Guadalupe drainage in Texas, USA (Smith et al. 2021:69). T—Threatened (Williams et al. 1993:14); Proposed for Federal Endangered Status (Federal Register 2018); CR—Critically Endangered (IUCN 2024); GNR–Unranked—Global Range. Not assessed (NatureServe 2024). Considered extinct until its recent rediscovery in the Guadalupe River near Gonzales, Texas (Randklev et al. 2012:327; 2013:247).

Unio mitchelli Simpson in Dall, 1895. Diagnoses of new mollusks from the survey of the Mexican boundary. *Proceedings of the United States National Museum* 18(1033):1–6. Page 5. Figured in Dall (1897:371, pl. 32, figs. 1–3).

Verbatim Type Locality: "Guadalupe River, Victoria County, Texas,...... Rio Salado, near New Leon, Mexico."

Type Locality: Guadalupe River (Gulf of México–Guadalupe Dr.), Victoria, Victoria County, Texas, USA [D=28.7996, -97.0187].

Type Specimens: Lectotype, USNM 128364 designated and the type locality fixed by Johnson (1975:15, as the holotype; Johnson 1999a:19). Paralectotypes, MCZ 165695 (n=2) (Johnson 1956:124, as paratypes).

Remarks: Although *Unio iheringi* Wright, 1888 and *Unio elongatus* Simpson, 1914 were placed in synonymy with *Fusconaia mitchelli* by Howells et al. (1996:127), phylogenetic and morphometric analyses supported the elevation of *Fusconaia iheringi* from synonymy with *F. mitchelli* (Smith et al., 2021:69). The analyses showed genetic divergence between populations in the Guadalupe drainage and those in the Brazos and Colorado drainages, which was further supported by available biogeographic information. Morphometric analyses and coalescent-based species delimitation models integrating both DNA sequence and morphological data also provided strong support for the divergence observed between the two geographically isolated clades of *F. mitchelli* (Smith et al., 2021).

Taxonomic History:

Unio (sec. *Elliptio*) *mitchelli* Simpson in Dall, 1895 — Simpson (1900; 1914)
Unio (*Metaptera*) *mitchelli* Simpson in Dall, 1895 — von Martens (1900c, *nomen dubium*, 1901c)
Unio mitchelli Simpson in Dall, 1895 — de Leon (1912)
Elliptio (*Barynaias*) *tamaulipasensis* [sic] (Conrad, 1855) — Frierson (1927)

DOI: 10.1201/9781003314448-25

Elliptio tamaulipasensis [sic] (Conrad, 1855) — Strecker (1931)
Nephronaias (*Sphenonaias*) *taumilapana* (Conrad, 1855) — Haas (1969)
Quincuncina mitchelli (Simpson in Dall, 1895)—Turgeon et al. (1988; 1998), Williams et al. (1993), Howells et al. (1996), Johnson (1999a), Graf & Cummings (2007), Cummings & Graf (2010), Karatayev et al. (2012)
Sphenonaias tanaulipana [sic] Conrad, 1855 — Contreras-Arquieta (2000)
Quadrula mitchelli (Simpson in Dall, 1895) — Haag (2012)
Fusconaia mitchelli (Simpson in Dall, 1895) — Pfeiffer et al. (2016), Williams et al. (2017), Inoue et al. (2018), Pieri et al. (2018), Keogh & Simons (2019), Smith et al. (2021), Graf & Cummings (2021), Czaja et al. (2023)

Unio (sec. *Elliptio*) *mitchelli* var. *elongatus* Simpson, 1914. A descriptive catalogue of the naiades, or pearly fresh-water mussels. Parts I–III. Bryant Walker, Detroit, Michigan xii + 1540 pp. Page 623. Not figured.

Verbatim Type Locality: "Guadalupe River, Kerr County, Texas."

Type Locality: Guadalupe River, near Kerrville, Kerr County, Texas [L = 30.0472, -99.1400].

Type Specimens: Lectotype, USNM 251917 designated by Johnson (1975:12).

Remarks: The catalogue number of the lectotype was incorrectly given as USNM 128364 by Johnson (1999a:54, pl. 2, fig. 3) in the figure legend. Primary homonym of *Unio elongata* Lamarck, 1819. Introduced as intra-sp. of *Unio mitchelli* Simpson in Dall, 1895.

Taxonomic History:

Elliptio (*Barynaias*) *tamaulipasensis-elongatus* (Simpson, 1914) — Frierson (1927)
Elliptio tamaulipasensis [sic] *elongatus* (Simpson, 1914) — Strecker (1931)
Nephronaias (*Sphenonaias*) *taumilapana* (Conrad, 1855) — Haas (1969)
Quincuncina mitchelli (Simpson in Dall, 1895) — Howells et al. (1996), Johnson (1999a)
Fusconaia mitchelli (Simpson in Dall, 1895) — Smith et al. (2021)

Quincuncina (*Quincuncina*) *guadalupensis* Wurtz, 1950. *Quadrula* (*Quincuncina*) *guadalupensis* sp. nov. (Unionidae: Pelecypoda). *Notulae Naturae* 224:1–3. Page 2, figs. 1–5.

Verbatim Type Locality: "Guadalupe River above Seguin between Routes 123 and 90, Guadalupe County, Texas."

Type Locality: Guadalupe River (Gulf of México–Guadalupe Dr.), above Seguin between Routes 123 and 90, Guadalupe County, Texas, USA [L = 29.5591, -97.9602].

Type Specimens: Holotype, ANSP 185974 by monotypy and original designation.

Taxonomic History:

Quincuncina guadalupensis Wurtz, 1950 — Burch (1975)
Quincuncina mitchelli (Simpson in Dall, 1895) — Howells et al. (1996), Johnson (1999a)
Fusconaia mitchelli (Simpson in Dall, 1895) — Smith et al. (2021)

Literature Cited

Abbott, R.T., and C.B. Wurtz. 1971. Horace Burrington Baker (1889–1971)–an obituary. *The Nautilus* 85(1):1–4.

Abbott, R.T., and M.E. Young (eds.). 1973. *American Malacologists: A National Register of Professional and Amateur Malacologists and Private Shell Collectors and Biographies of Early American Mollusk Workers Born Between 1618 and 1900*. American Malacologists, Falls Church, Virginia. Consolidated/Drake Press, Pennsylvania, Philadelphia. 494 pp.

Abell, R., D. Olson, E. Dinerstein, P. Hurley, J. Diggs, W. Eichbaum, S. Walters, W. Wettengel, T. Allnutt, C. Loucks, P. Hedao, and C. Taylor. 2000. *Freshwater Ecoregions of North America: A Conservation Assessment*. Island Press, Washington, DC. World Wildlife Fund Ecoregion Assessments Vol. 2. 368 pp.

Abell, R., M.L. Thieme, C. Revenga, M. Bryer, M. Kottelat, N. Bogutskaya, B. Coad, N. Mandrak, S. Contreras-Balderas, W. Bussing, M.L.J. Stiassny, P. Skelton, G.R. Allen, P. Unmack, A. Naseka, R. Ng, N. Sindorf, J. Robertson, E. Armijo, J.V. Higgins, T.J. Heibel, E. Wikramanayake, D. Olson, H.L. López, R.E. Reis, J.G. Lundberg, M.H. Sabaj Pérez, and P. Petry. 2008. Freshwater ecoregions of the world: A new map of biogeographic units for freshwater biodiversity conservation. *Bioscience* 58(5):403–414.

Adams, H., and A. Adams. 1857. *The Genera of Recent Mollusca; Arranged According to Their Organization*. John van Vorst, Paternoster Row, London Vol II. 661 pp.

Agassiz, L. 1852. Ueber die Gattungen unter den Nord Amerikanischen Najaden. *Archiv für Naturgeschichte* 18:41–52.

Agudo-Padrón, A.I. 2022. Update of available knowledge about the non-marine molluscs occurring in the "Republic of El Salvador" Central America. *Brazilian Journal of Animal and Environmental Research* 5(4):3876–3886.

Agudo-Padrón, I. 2019. Los moluscos no marinos ocurrentes en la República de El Salvador, América Central: una breve revisión panorámica introductoria de su actual conocimiento. *BIOMA. La naturaleza en tus manos* 5(50):48–53.

Allen, W.R. 1914. The food and feeding habits of freshwater mussels. *Biological Bulletin* 27:127–146.

Andrews, E.W., IV. 1969. *The Archaeological Use and Distribution of Mollusca in the Maya Lowlands*. Middle American Research Institute, Tulane University Publication New Orleans, Louisiana Vol. 34. 115 pp.

Anon. 1887. Sketch of Isaac Lea. *Popular Science Monthly* 31:404–411.

Arango, R. 1865. Catalogo de los moluscos terrestres y fluviales de la Isla de Cuba. Continuacion. Pp. 123–149 *In* F. Poey (ed.). *Repertorio Fisico-Natural de la Isla de Cuba*. Havana, Cuba, Vol. 1.

Arey, L.B. 1932. The formation and structure of the glochidial cyst. *Biological Bulletin (Woods Hole)* 62(2):212–221.

Avelar, W.E.P., and M.P. Vianna. 2008. Hyriidae and Mycetopodidae. Pp. 193–199, 207–208, 205–216, 218–221, 223–224 *In* A.B.M. Machado, G.M. Drummond, and A.P. Paglia (eds.). *Livro Vermelho da Fauna Brasileira Ameaçada de Extinção*. Ministério do Meio Ambiente e Fundação Biodiversitas, Brasília. 293 pp. + 8 plates.

Baker, F.C. 1927. On the division of the Sphaeriidae into two subfamilies: And the description of a new genus of Unionidae, with descriptions of new varieties. *American Midland Naturalist* 10(7):220–223.

Baker, F.C. 1928. The fresh water Mollusca of Wisconsin. Part II. Pelecypoda. *Bulletin of the Wisconsin Geological and Natural History Survey*. University of Wisconsin 70(2):vi–495 + plates 29–55.

Baker, H.B. 1922. The Mollusca collected by the University of Michigan-Walker Expedition in Southern Vera Cruz, Mexico. I. Occasional Papers of the Museum of Zoology, University of Michigan No. 106. 95 pp.

Baker, H.B. 1930. The Mollusca collected by the University of Michigan-Williamson Expedition in Venezuela. VI. Occasional Papers of the Museum of Zoology, University of Michigan No. 210. 95 pp.

Baker, H.B. 1957. Henry Augustus Pilsbry 1862–1957. American Malacological Union, Inc. Annual Reports 1957(23):1 + 1 plate.

Barney, R.L. 1922. An indication of the value of artificial propagation of pearly mussels. *The Nautilus* 36(2):53–58.

Barrios, J.E., J.A. Rodríguez-Pineda, and M. De La Maza Benigos. 2009. Integrated river basin management in the Conchos River basin, Mexico: A case study of freshwater climate change adaptation. *Climate and Development* 1:249–260.

Bauer, W.G., D.T. Stewart, R. Quesada Céspedes, S. Arias Valverde, and R.H. Easy. 2021. DNA barcoding evidence of the tropical invasive lineage of *Sinanodonta woodiana* in Costa Rica. *Neotropical Naturalist* 5:1–8.

Beltran-López, R.G., A.B. García-Andrade, and C.P. Ornelas-García. 2023. Chapter 7. Mexican freshwater fishes in the Anthropocene. Pp. 129–152 *In* R.W. Jones, C.P. Ornelas-García, R. Pineda-López, and F. Álvarez (eds.). *Mexican Fauna in the Anthropocene*. Springer Nature, Cham, Switzerland. xvii+597 pp.

Bequaert, J.C., and W.B. Miller. 1973. *The Mollusks of the Arid Southwest: With an Arizona Check List*. University of Arizona Press, Tucson, Arizona. xvi+271 pp.

Bieler, R. 2021. Isaac Lea's 1792–1886 substitutions and other modifications of his own names of molluscan species. *Malacologia* 64(1):1–56.

Bieler, R., J.G. Carter, and E.V. Coan. 2010. Classification of bivalve families. Pp. 113–133 *In* P. Bouchet, and J.-P. Rocroi (eds.). *Nomenclator of Bivalve Families. Malacologia* 52(2):1–184. Also published by ConchBooks, Hackenheim, Germany.

Blainville, H M.D. de. 1825. *Manuel de Malacologie et de Conchyliologie*. F.G. Levrault, Paris. 2 vols. 647 pp.+87 plates.

Blevins, E., S. Jepsen, J. Brim Box, and D. Nez. 2016. *Anodonta nuttalliana* (errata version published in 2017). The IUCN Red List of Threatened Species 2016: e.T91149898A114130670.

Blinn, D.W., and N.L. Poff. 2005. Chapter 11. Colorado River Basin. Pp. 483–538 *In* A.C. Benke, and C.E. Cushing (eds.). *Rivers of North America*. Academic Press, Elsevier Inc., Amsterdam. 1168 pp.

Bogan, A.E., and W.R. Hoeh. 2000. On becoming cemented: Evolutionary relationships among the genera in the freshwater bivalve family Etheriidae (Bivalvia: Unionoida). Pp. 159–168 *In* E.M. Harper, J.D. Taylor, and J.A. Crame (eds.). *The Evolutionary Biology of the Bivalvia*. Geological Society, Special Publication No. 177, London. 494 pp.

Bogan, A.E., and P.D. Johnson. 2011. Obituary – Herbert David Athearn (1923–2011). *Ellipsaria* 13(1):31–33.

Böhm, M., N.I. Dewhurst-Richman, M. Seddon, S.E.H. Ledger, C. Albrecht, D. Allen, A.E. Bogan, J. Cordeiro, K.S. Cummings, A. Cuttelod, G. Darrigran, W. Darwall, Z. Fehér, C. Gibson, D.L. Graf, F. Köhler, M. Lopes-Lima, G. Pastorino, K.E. Perez, K. Smith, D. van Damme, M.V. Vinarski, T. von Proschwitz, T. von Rintelen, D.C. Aldridge, N.A. Aravind, P.B. Budha, C. Clavijo, D. Van Tu, O. Gargominy, M. Ghamizi, M. Haase, C. Hilton-Taylor, P.D. Johnson, Ü. Kebapçı, J. Lajtner, C.N. Lange, D.A.W. Lepitzki, A. Martínez-Ortí, E.A. Moorkens, E. Neubert, C.M. Pollock, V. Prié, C. Radea, R. Ramírez, M.A. Ramos, S.B. Santos, R. Slapnik, M.O. Son, A.-S. Stensgaard, and B. Collen. 2020. The conservation status of the world's freshwater molluscs. *Hydrobiologia* 848(12–13):3231–3245.

Bonetto, A.A. 1962. Especies del genéro *Mycetopoda* en el sistema hidrográfico del Rio de la Plata. Revista del Museo Argentino de Ciencias Naturales "Bernardino Rivadavia" e Instituto Nacional de Investigacion de las Ciencias Naturales. *Ciencias Zoológicas* 8(14):173–182.

Bonetto, A.A., and I.D. Ezcurra. 1962. El desarrollo del lasidium de *Anodontites trapesialis forbesianus* (Lea) (Moll. Lamell.). *Physis (Buenos Aires)* 23(65):195–203.

Bory de Saint-Vincent. 1827. Tableau Encyclopédique et Methodique des trois règnes de la nature. Vers, coquilles, mollusques et polypiers. Paris, Agasse, Imprimeur-Libraire. *Livraison* 98 part 29:83–84, 133–180.

Bosman, B., B. Christie, C.M. Hart, J. Morton, and C. Randklev. 2015. Host confirmed for *Potamilus amphichaenus* and *Potamilus metnecktayi*. *Ellipsaria: Newsletter of the Freshwater Mollusk Conservation Society* 17(4):15–16.

Bouchet, P., and J.-P. Rocroi. 2010. Nomenclator of bivalve families; with a classification of bivalve families. *In* Bieler, R., J.G. Carter, and E.V. Coan (eds.). *Malacologia* 52(2):1–184.

Boyko, C.B., and W.E. Sage III. 1996. Catalog of recent type specimens in the Department of Invertebrates, American Museum of Natural History. II. Mollusca Part 1 (Classes Aplacophora, Polyplacophora, Gastropoda [Subclass Opsthobranchia], Bivalvia, and Scaphopoda. *American Museum Novitates* 3170:1–50.

Breure, A.S.H., C. Audibert, and J.D. Ablett. 2018. *Pierre Marie Arthur Morelet (1809–1892) and His Contributions to Malacology*. Nederlandse Malacologische Vereniging, Leiden. 544 pp.

Breure, A.S.H., M.T. Roosen, and J.D. Ablett. 2022. Land and freshwater molluscs of mainland Ecuador: An illustrated checklist. Moluscos terrestres y dulceacuícolas de Ecuador continental: un listado ilustrado. *Iberus* 40(1):1–290.

Brim Box, J., and J.D. Williams. 2000. Unionid mollusks of the Apalachicola Basin in Alabama, Florida, and Georgia. *Bulletin of the Alabama Museum of Natural History* 21:1–143.

Brooks, S.T., and B.W. Brooks. 1931. List of types of Pelecypoda in the Carnegie Museum on January 1, 1931. *Annals of the Carnegie Museum* 20:171–177.

Brown, M.K. 2001. Investigations of Middle Preclassic public architecture at the site of Blackman Eddy, Belize. FAMSI. Foundation for the Advancement of Mesoamerican Studies. 25 pp.

Burch, J.B. 1975. *Freshwater Unionacean clams (Mollusca: Pelecypoda) of North America*. Revised Edition. Malacological Publications, Hamburg, Michigan. 204 pp.

Burkhead, N.M. 2012. Extinction rates in North American freshwater fishes, 1900–2010. *BioScience* 62(9):798–808.

Burlakova, L.E., A.Y. Karatayev, V.A. Karatayev, M.E. May, D.L. Bennett, and M.J. Cook. 2011. Endemic species: Contribution to community uniqueness, effect of habitat alteration, and conservation priorities. *Biological Conservation* 144(1):155–165.

Burlakova, L.E., D. Campbell, and A.Y. Karatayev. 2019. Status of rare endemic species: Molecular phylogeny, distribution and conservation of freshwater molluscs *Truncilla macrodon* and *Truncilla cognata* in *Texas. Malacologia* 62(2):345–363.

Burlakova, L.E., D. Campbell, A.Y Karatayev, and D. Barclay. 2012. Distribution, genetic analysis and conservation priorities for rare Texas freshwater molluscs in the genera *Fusconaia* and *Pleurobema* (Bivalvia:Unionidae). *Aquatic Biosystems* 8(12):1–15.

Burr, B.M., and R.L. Mayden. 1992. Phylogenetics and North American freshwater fishes. Pp. 18–75 *In* R.L. Mayden (ed.). *Systematics, Historical Ecology, and North American Freshwater Fishes.* Stanford University Press, Stanford, California. 969 pp.

Bussing, W., and J. Hales. 2019. Mosquitia. ©World Wildlife Fund/The Nature Conservancy (feow.org/ecoregions/details/203).

Bussing, W., and S. Smith. 2019. San Juan (Nicaragua/Costa Rica). ©World Wildlife Fund/The Nature Conservancy (feow.org/ecoregions/details/205).

Bussing, W.A. 1976. Geographic distribution of the San Juan ichthyofauna of Central America with remarks on its origin and ecology. Pp. 157–175 *In* T.B. Thorson (ed.). *Investigations of the Ichthyofauna of Nicaraguan Lakes.* University of Nebraska, Lincoln.

Calamusso, B., J.N. Rinne, and R.J. Edwards. 2005. Historic changes in the Rio Grande fish fauna: Status, threats, and management of native species. Pp. 205–223 *In* J.N. Rinne, R.M. Hughes, and B. Calamusso (eds.). *Historical Changes in Large River Fish Assemblages of the Americas.* American Fisheries Society, Symposium 45, Bethesda, Maryland. 612 pp.

Call, R.E. 1895. A study of the Unionidae of Arkansas, with incidental reference to their distribution in the Mississippi valley. *Transactions of the Academy of Science of St. Louis* 7(1):1–64+21 plates.

Campbell, D.C., J.M. Serb, J.E. Buhay, K.J. Roe, R.L. Minton, and C. Lydeard. 2005. Phylogeny of North American amblemines (Bivalvia, Unionoida): Prodigious polyphyly proves pervasive across genera. *Invertebrate Biology* 124(2):131–164.

Castro-Rojas, G.V., and J.A. Rodríguez-Pulido. 2003. Estudio bioecológico de la almeja de aqua dulce (*Anodontites* sp). *Orinoquia* 7(1–2):1–27.

Chakrabarty, P. 2004. Cichlid biogeography: Comment and review. *Fish and Fisheries* 5(2):97–184.

Chamberlain, T.K. 1933. Ages and shell measurements of two large specimens of *Megalonaias gigantea* (Barnes). *The Nautilus* 47:29–30.

Chapman, E.G., M.E. Gordon, J.M. Walker, B.K. Lang, D.C. Campbell, G.T. Watters, J.P. Curole, H. Piontkivska, and W.R. Hoeh. 2008. Evolutionary relationships of *Popenaias popeii* and the early evolution of Lampsiline bivalves (Unionidae): Phylogenetic analyses of DNA and amino acid sequences from F and M mitochondrial genomes. *Malacologia* 50(1–2):303–318.

Chernoff, B., R.R. Miller, and C.R. Gilbert. 1982. *Notropis orca* and *Notropis simus*, cyprinid fishes from the American Southwest, with description of a new subspecies. Occasional Papers of the Museum of Zoology, University of Michigan No.698. 49 pp.

Chong, J.P., J.C. Brim Box, J.K. Howard, D. Wolf, T.L. Myers, and K.E. Mock. 2008. Three deeply divided lineages of the freshwater mussel genus *Anodonta* in Western North America. *Conservation Genetics* 9(5):1303–1309.

Clarke, A.H. 1973. The freshwater molluscs of the Canadian Interior Basin. *Malacologia* 13(1–2):1–509.

Clench, W.J., and R.D. Turner. 1956. Freshwater mollusks of Alabama, Georgia, and Florida from the Escambia to the Suwannee River. *Bulletin of the Florida State Museum, Biological Sciences* 1(3):97–239.

Clessin, S. 1874a. Die Gattung *Anodonta* nebst den übrigen najaden mit unvollkommenem Schloss. *In Abbildungen nach der Natur mit Beschreibungen. Systematisches Conchylien-Cabinet von Martini und Chemnitz.* Küster Edition. 9(Part 1):113–128, plates 34–39.

Clessin, S. 1874b. Die Gattung *Anodonta* nebst den übrigen najaden mit unvollkommenem Schloss. *In Abbildungen nach der Natur mit Beschreibungen. Systematisches Conchylien-Cabinet von Martini und Chemnitz.* Küster Edition. 9(Part 1):129–144, plates 40–45.

Clessin, S. 1874c. Die Gattung *Anodonta* nebst den übrigen najaden mit unvollkommenem Schloss. *In Abbildungen nach der Natur mit Beschreibungen. Systematisches Conchylien-Cabinet von Martini und Chemnitz.* Küster Edition. 9(Part 1):145–168, plates 46–51.

Clessin, S. 1875. Die Gattung *Anodonta* nebst den übrigen najaden mit unvollkommenem Schloss. *In Abbildungen nach der Natur mit Beschreibungen. Systematisches Conchylien-Cabinet von Martini und Chemnitz*. Küster Edition. 9(Part 1):193–216, plates 64–69.
Clessin, S. 1876. Die Gattung *Anodonta* nebst den übrigen najaden mit unvollkommenem Schloss. *In Abbildungen nach der Natur mit Beschreibungen. Systematisches Conchylien-Cabinet von Martini und Chemnitz*. Küster Edition. 9(Part 1):217–240, plates 70–75.
Coan, E., and M.G. Harasewych. 1993. Publication dates of the Nautilus. *The Nautilus* 106(4):174–180.
Coan, E.V., and A.R. Kabat. 2017. The malacological contributions of Rudolph Amandus Philippi (1808–1904). *Malacologia* 60(1–2):31–322.
Coan, E.V., and A.R. Kabat. 2024a. *Annex 1: Collations of Malacological Significance*. On-line Edition. February 2, 2024. 148 pp.
Coan, E.V., and A.R. Kabat. 2024b. *2,400 Years of Malacology*, Twenty-first Edition, February 2, 2024, 2,037 pp. + 148 pp. [Annex 1 – Book Collations] + 65 pp. [Annex 2 – Küster Collation], 73 pp. [Annex 3 – Journal Collations]. American Malacological Society.
Coe, W.R. 1959. *Piedras Negras Archaeology: Artifacts, Caches, and Burials*. University of Pennsylvania Museum, Monographs No. 18, Philadelphia, Pennsylvania. 176 pp + 69 plates.
Coker, R.E., A.F. Shira, H.W. Clark, and A.D. Howard. 1921. Natural history and propagation of fresh-water mussels. *Bulletin of the Bureau of Fisheries. [Issued separately as U.S. Bureau of Fisheries Document* 893] 37(1919–20):75–181 + 17 plates.
Coney, C.C., and D.W. Taylor. 1986. Systematic position of *Quincuncina mitchelli* (Simpson, 1896) (Unionidae). (Abstract). *Western Society of Malacologists. Annual Report* 18:12–13.
Conrad, T.A. 1836. *Monography of the Family Unionidae, or Naiades of Lamarck, (Fresh Water Bivalve Shells) or North America, Illustrated by Figures Drawn on Stone from Nature*. J. Dobson, Philadelphia, Pennsylvania. 6:49–56, plates 26–31.
Conrad, T.A. 1837. *Monography of the Family Unionidae, or Naiades of Lamarck, (Fresh Water Bivalve Shells) or North America, Illustrated by Figures Drawn on Stone from Nature*. J. Dobson, Philadelphia, Pennsylvania. 9:73–80, plates 41–45.
Conrad, T.A. 1840. *Monography of the Family Unionidae, or Naiades of Lamarck, (Fresh Water Bivalve Shells) or North America, Illustrated by Figures Drawn on Stone from Nature*. J. Dobson, Philadelphia, Pennsylvania. 12:103–110, plates 57–60.
Conrad, T.A. 1853. A synopsis of the family of Naïades of North America, with notes, and a table of some of the genera and sub-genera of the family, according to their geographical distribution, and descriptions of genera and sub-genera. *Proceedings of the Academy of Natural Sciences of Philadelphia* 6(7):243–269.
Contreras, F. 1930. Moluscos del Lago de Xochimilco, D.F. *Anales del Instituto de Biologia Universidad Nacional Autónoma de México Serie Zoologia* 1(1):39–46.
Contreras-Arquieta, A. 2000. Bibliografía y lista taxonómica de las especies de moluscos dulceacuícolas en México. *Mexicoa* 2(1):40–53.
Contreras-Balderas, S., and J. Hales. 2019a. Lerma-Chapala. ©World Wildlife Fund/The Nature Conservancy (feow.org/ecoregions/details/165).
Contreras-Balderas, S., and J. Hales. 2019b. Rio Balsas. ©World Wildlife Fund/The Nature Conservancy (feow.org/ecoregions/details/169).
Contreras-Balderas, S., H. Obregón, and M. Lourdes Lozano. 1996. Punta de Morro, an interesting barrier for distributional patterns of continental fishes in north and central Veracruz, Mexico. *Acta Biologica Venezuelica* 16(4):37–42.
Contreras-Balderas, S., and W. Bussing. 2019. Quintana Roo-Motagua. ©World Wildlife Fund/The Nature Conservancy (feow.org/ecoregions/details/202).
Contreras-MacBeath, T., M.B. Rodríguez, V. Sorani, C. Goldspink, and G. McGregor Reid. 2014. Richness and endemism of the freshwater fishes of Mexico. *Journal of Threatened Taxa* 6(2):5421–5433.
Cooke, A.H. 1899. In Memorium: M.H. Crosse. *Journal of Malacology* 7(1):4–7.
Cosgrove, P.J., M.R. Young, L.C. Hastie, M. Gaywood, and P.J. Boon. 2000. The status of the freshwater pearl mussel *Margaritifera margaritifera* Linn, in Scotland. *Aquatic Conservation in Marine and Freshwater Ecosystems* 10:197–208.
Crosnier, A., and P.F. Clark. 1998. Publication dates of Recherches zoologiques pour servir á l'histoire de la faune de l'Amérique centrale et du Mexique. *Archives of Natural History* 25(1):87–101.
Crosse, H. 1893. Nécrologie. *Journal de Conchyliologie* 41(1):75–77.
Cruz, R.A., and R. Quesada. 2017. Taxonomic position and glochidia from the freshwater mussel, *Anodonta luteola* (Unionoida: Unionidae), in Costa Rica. *UNED Research Journal* 9(1):39–43.

Cummings, K.S., and C.A. Mayer. 1992. *Field Guide to Freshwater Mussels of the Midwest.* Illinois Natural History Survey, Manual 5, Champaign, Illinois. 194 pp.

Cummings, K.S., and D.L. Graf. 2010. Mollusca: Bivalvia. Pp. 309–384 *In* J.H. Thorp, and A.P. Covich (eds.). *Ecology and Classification of North American Freshwater Invertebrates.* Third Edition. Academic Press, Elsevier Inc., Amsterdam. xiv + 1021 pp.

Cummings, K.S., H.A. Jones, and M. Lopes-Lima. 2016. Rapid bioassessment methods for freshwater molluscs. Pp. 185–207 *In* T.H. Larsen (ed.). *Core Standardized Methods for Rapid Biological Field Assessment.* Conservation International, Arlington, Virginia. 209 pp.

Czaja, A., A.P. Covich, J.L. Becerra-Lopez, D.G. Cordero-Torres, and J.L. Estrada-Rodríguez. 2023. Chapter 5. The freshwater mollusks of Mexico: Can we still prevent their silent extinction? Pp. 81–103 *In* R.W. Jones, C.P. Ornelas-García, R. Pineda-López, and F. Álvarez (eds.). *Mexican Fauna in the Anthropocene.* Springer Nature, Cham, Switzerland. xvii + 597 pp.

d'Eliscu, P.N. 1972. Observation of the glochidium, metamorphosis, and juvenile of *Anodonta californiensis* Lea, 1857. *The Veliger* 15(1):57–58.

d'Orbigny, A. 1835. Synopsis terrestrium et fluviatilium molluscorum, in suo per Americam meridionalem itinere, ab. A. d'Orbigny, collectorum. *Revue et Magasin de Zoologie Pure et Appliquée (Paris)* 5(61–62):1–44.

d'Orbigny, A. 1846a. Voyage dans l'Amérique Méridionale (le Brésil, la République orientale de l'Uruguay, la République Argentine, la Patagonie, la République du Chili, la République de Bolivia, la République du Pérou), exécuté pendant les années 1826, 1827, 1828, 1829, 1830, 1831, 1832 et 1833. Pp. 529–600 *In* Chez P. Bertrand (ed.). *Tome V, part 3, Mollusques.* Libraire de la Société Géologique de France, Paris. Strasbourg, Chez V. Levrault.

d'Orbigny, A. 1846b. Voyage dans l'Amérique Méridionale (le Brésil, la République orientale de l'Uruguay, la République Argentine, la Patagonie, la République du Chili, la République de Bolivia, la République du Pérou), exécuté pendant les années 1826, 1827, 1828, 1829, 1830, 1831, 1832 et 1833. Pp. 601–656 *In* Chez P. Bertrand (ed.). *Tome V, part 3, Mollusques.* Libraire de la Société Géologique de France, Paris. Strasbourg, Chez V. Levrault.

Dahm, C.N., R.J. Edwards, and F.P. Gelwick. 2005. Chapter 5. Gulf Coast rivers of the Southwestern United States. Pp. 181–228 *In* A.C. Benke, and C.E. Cushing (eds.). Rivers of North America. Academic Press, Elsevier Inc., Amsterdam. 1168 pp.

Dall, W.H. 1897. Report on the mollusks collected by the International Boundary Commission of the United States and Mexico, 1892–1894. *Proceedings of the United States National Museum* 19(1111):333–379 + plates 31–33.

Dall, W.H. 1908. Descriptions and figures of some land and fresh-water shells from Mexico, believed to be new. *Proceedings of the United States National Museum* 35(1642):177–182 + plates 29–30.

Darwin, C. 1878. Transplantation of shells. *Nature. A Weekly Illustrated Journal of Science* 18(448):120–121.

Degerman, E., S. Alexanderson, J. Bergengren, L. Henrikson, B.E. Johansson, B.M. Larsen, and H. Söderberg. 2009. *Restoration of Freshwater Pearl Mussel Streams.* WWF Sweden, Solna. 62 pp.

Delessert, B. 1841. *Recueil de coquilles décrites par Lamarck dans son Histoire naturelle des animaux sans vertebres et non encore figurées.* Chez, Fortin, Masson, et Cie. Libraires, Paris. 40 plates and captions.

Downing, W.L., J. Shostell, and J.A. Downing. 1992. Non-annual external annuli in the freshwater mussels *Anodonta grandis grandis* and *Lampsilis radiata siliquoidea. Freshwater Biology* 28(3):309–317.

Drouët, H. 1892. Nécrologie–Arthur Morelet. *Nachrichtsblatt der Deutschen Malakozoologischen Gesellschaft* 24(11–12):204–206.

Drouët, H. 1893. Notice sur Arthur Morelet. *Mémoires de L'Académie des Sciences, Arts et Belles-Lettres de Dijon Quatrième Série Tome* 4:11–36 + portrait.

Duncan, F.M. 1937. On the dates of publication of the Society's 'Proceedings', 1859–1926. With an Appendix containing the dates of publication of 'Proceedings', 1830–1858, compiled by the late F.H. Waterhouse, and of the 'Transactions', 1833–1868, by the late Henry Peavot, originally published in P.Z.S. 1893, 1913. *Proceedings of the Zoological Society of London, Series A* 107(1):71–84.

Elías, D.J., C.D. McMahan, and K.R. Piller. 2022. Molecular data elucidate cryptic diversity within the widespread Threadfin Shad (*Dorosoma petenense*: Clupeidae) across the Nearctic and Northern Neotropics. *Hydrobiologia* 849:89–111.

Ellis, M.M., A.D. Merrick, and M.D. Ellis. 1931. The blood of North American fresh-water mussels under normal and adverse conditions. *Bulletin of the U.S. Bureau of Fisheries* 46:509–542.

Emery, K.F. 2008. A zooarchaeological test for dietary resource depression at the end of the classical period in the Petexbatun, Guatemala. *Human Ecology* 36:617–634.

Fechter, R. 1983. Das Typenmaterial der von J.B. v. Spix in Brasilien gesammelten. Unionacea. *Spixiana Supplement* 9:225–255.

Federal Register. 2018. CFR Part 17. Endangered and Threatened Wildlife and Plants; Endangered species status for Texas Hornshell. *Federal Register* 83(28):5720–5735.

Federal Register. 2023. CFR Part 17. Endangered and Threatened Wildlife and Plants; Endangered Species Status for Salina Mucket and Mexican Fawnsfoot and Designation of Critical Habitat. *Federal Register* 88(141):47952–47988.

Ferussac, A.E. de. 1835. Observations. Adressées en forme de lettre a M.M. Th. Say, C.-S. Rafinesque, Is. Lea, S.-P. Hildreth, T.-A. Conrad, et C.-A. Poulson, sur la synonymie de coquilles bivalves de l'Amerique Septentrionale, et Essai d'une table de concordance a ce sujet. *Revue et Magasin de Zoologie Pure et Appliquée (Paris)* 5(59–60):1–36.

Figueiras, A. 1965. La malacofauna dulceacuícola del Uruguay (Parte II). *Comunicaciones de la Sociedad Malacológica del Uruguay* 1(8):223–270.

Fischer, P. 1887. *Manuel de Conchyliologie et de Paléontologie Conchyliologique ou histoire naturelle des mollusques vivants et fossiles*. F. Savy, Paris. 1369 pp. + 23 plates.

Fischer, P., and H. Crosse. 1870–1878. Études sur les mollusques terrestres et fluviatiles du Mexique et du Guatemala. *In* M.H. Milne Edwards (ed.). *Mission Scientifique au Mexique et dans l'Amérique Centrale. Recherches Zoologiques pour servir á l'Histoire de la Faune de l'Amérique Centrale et du Mexique*. Septiéme Partie, Imprimerie Nationale, Paris Vol. 1. 702 pp.

Fischer, P., and H. Crosse. 1878–1902. Études sur les mollusques terrestres et fluviatiles du Mexique et du Guatemala. *In* M.H. Milne Edwards (ed.). *Mission Scientifique au Mexique et dans l'Amérique Centrale. Recherches Zoologiques pour servir á l'Histoire de la Faune de l'Amérique Centrale et du Mexique*. Septiéme Partie, Imprimerie Nationale, Paris Vol. 2. 731 pp. + 72 plates.

Fischer, P., and H. Crosse. 1894a. Études sur les mollusques terrestres et fluviatiles du Mexique et du Guatemala. *In* M.H. Milne Edwards (ed.). *Mission Scientifique au Mexique et dans l'Amérique Centrale. Recherches Zoologiques pour servir á l'Histoire de la Faune de l'Amérique Centrale et du Mexique*. Septiéme Partie. 15th Livraison. Feuilles 62–72, pp. 489–576, plates 59–62, Imprimerie Nationale, Paris Vol. 2. 731 pp.

Fischer, P., and H. Crosse. 1894b. Études sur les mollusques terrestres et fluviatiles du Mexique et du Guatemala. *In* M.H. Milne Edwards (ed.). *Mission Scientifique au Mexique et dans l'Amérique Centrale. Recherches Zoologiques pour servir á l'Histoire de la Faune de l'Amérique Centrale et du Mexique*. Septiéme Partie. 16th Livraison. Feuilles 73–82, pp. 577–656, plates 63–66, Imprimerie Nationale, Paris Vol. 2. 731 pp.

Fischer, P., and H. Crosse. 1902. Études sur les mollusques terrestres et fluviatiles du Mexique et du Guatemala. *In* M.H. Milne Edwards (ed.). *Mission Scientifique au Mexique et dans l'Amérique Centrale. Recherches Zoologiques pour servir á l'Histoire de la Faune de l'Amérique Centrale et du Mexique*. Septiéme Partie. 17th Livraison. Feuilles 83–92, pp. 657–731, plates 67–72, Imprimerie Nationale, Paris Vol. 2. 731 pp.

Fish, J.E. 1977. Karst hydrogeology and geomorphology of the Sierra de el Abra and the Valles-San Luis Potosi region, Mexico. Ph.D. Dissertation. McMaster University, Hamilton, Ontario. 469 pp.

Fox, W.J. 1913. Dates of publication. Pp. vii–xiv *In* E.L. Nolan (ed.). *An Index to the Scientific Contents of the Journal and Proceedings of the Academy of Natural Sciences of Philadelphia*. The Academy of Natural Sciences, Philadelphia, Pennsylvania, xiv + 1419 pp.

Franzen, D.S. 1956. Types of mollusks described by F.C. Baker Part I, University of Illinois. *The Nautilus* 70(1):21–27.

Frierson, L.S. 1907. A new Mexican mussel, *Lampsilis fimbriata*. *The Nautilus* 21(8):86–87 + 1 plate.

Frierson, L.S. 1912. *Unio* (*Lampsilis*?) *purpuriatus* Say. *The Nautilus* 26(2):22–23 + 1 plate.

Frierson, L.S. 1913. *Unio* (*Nephronaias*) *ortmanni*, N. sp. *The Nautilus* 27(2):14–15. Page 14, pl. 2. [Plate 2 appears in Vol. 27(3)].

Frierson, L.S. 1914. Remarks on classification of the Unionidae. *The Nautilus* 28(1):6–8.

Frierson, L.S. 1917. New genera and species of Central American naiades. *The Nautilus* 31(2):47–49 + plate 7.

Frierson, L.S. 1922. Observations on the genera *Leila* and *Anodontites*. *The Nautilus* 36(1):7–10.

Frierson, L.S. 1927. *A Classification and Annotated Check List of the North American Naiades*. Baylor University Press, Waco, Texas. 111 pp.

Frierson, L.S. 1928. Illustrations of Unionidae. *The Nautilus* 41(4):138–139 + 3 plates.

Fuller, S.L.H. 1975. The systematic position of *Cyrtonaias* (Bivalvia: Unionidae). *Malacological Review* 8(1–2):81–89.

Gill, T., and J.F. Bransford. 1877. Synopsis of the fishes of Lake Nicaragua. *Proceedings of the Academy of Natural Sciences of Philadelphia* 29(1877):175–191.

Goodrich, C., and H. van der Schalie. 1937. Mollusca of Petén and North Alta Vera Paz, Guatemala. Miscellaneous Publications, Museum of Zoology, University of Michigan, Ann Arbor, Michigan. No. 34. 50 pp. + 1 plate and 1 map.

Graf, D.L. 2000. The Etherioidea revisited: A phylogenetic analysis of hyriid relationships (Mollusca: Bivalvia: Paleoheterodonta: Unionoida). *Occasional Papers of the Museum of Zoology, University of Michigan* No. 729. 21 pp.

Graf, D.L., and D. Ó Foighil. 2000. The evolution of brooding characters among the freshwater pearly mussels (Mollusca: Bivalvia: Unionoidea) of North America. *Journal of Molluscan Studies* 66(2):157–170.

Graf, D.L., and K.S. Cummings. 2006. Palaeoheterodont diversity (Mollusca: Trigonioida + Unionoida): What we know and what we wish we knew about freshwater mussel evolution. *Zoological Journal of the Linnean Society* 148(3):343–394.

Graf, D.L., and K.S. Cummings. 2007. Review of the systematics and global diversity of freshwater mussel species (Bivalvia: Unionoida). *Journal of Molluscan Studies* 73(3):291–314.

Graf, D.L., and K.S. Cummings. 2019a. Chapter 38. Unionidae Rafinesque, 1820 and the general Unionida. Pp. 202–209 *In* C. Lydeard, and K.S. Cummings (eds.). *Freshwater Mollusk Families of the World*. Johns Hopkins University Press, Baltimore, Maryland. 264 pp.

Graf, D.L., and K.S. Cummings. 2019b. Chapter 42. Mycetopodidae Gray, 1840. Pp. 222–224 *In* C. Lydeard, and K.S. Cummings (eds.). *Freshwater Mollusk Families of the World*. Johns Hopkins University Press, Baltimore, Maryland. 264 pp.

Graf, D.L., and K.S. Cummings. 2021. A 'big data' approach to global freshwater mussel diversity (Bivalvia: Unionoida), with an updated checklist of genera and species. *Journal of Molluscan Studies* 87(1):1–36.

Gray, J.E. 1840. Shells of molluscous animals. Pp. 105–152 *In Synopsis of the Contents of the British Museum*. Forty-Second Edition. Woodfall and Son, London. 370 pp.

Gray, J.E. 1847. A list of the genera of recent Mollusca, their synonyma and types. *Proceedings of the Zoological Society of London* 1847(15):129–219.

de Gregorio, A. 1914. Su taluni molluschi di acqua dolce di America (American fresh water shells of America). *Il Naturalista Siciliano Nuova Serie* 2:31–72 + plates 3–12.

Groves, L.T. 2012. Catalog of Recent molluscan types in the Natural History Museum of Los Angeles County. Natural History Museum of Los Angeles County, Malacology Section 160 pp.

Guerra, D., F. Plazzi, D.T. Stewart, A.E. Bogan, W.R. Hoeh, and S. Breton. 2017. Evolution of sex-dependent mtDNA transmission in freshwater mussels (Bivalvia: Unionida). *Scientific Reports* 7(1151):1–13.

Guerra, D., M. Lopes-Lima, E. Froufe, H.M. Gan, P. Ondina, R. Amaro, M.W. Klunzinger, C. Callil, V. Prié, A.E. Bogan, D.T. Stewart, and S. Breton. 2019. Variability of mitochondrial ORFans hints at possible differences in the system of doubly uniparental inheritance of mitochondria among families of freshwater mussels (Bivalvia: Unionida). *BMC Evolutionary Biology* 19(229):1–22.

Günther, A. 1905. Carl Eduard von Martens. *Proceedings of the Linnean Society of London* 117(1):43–45.

Gutíerrez, M., and P. Borrego 1999. Water quality assessment of the Rio Conchos, Chihuahua, Mexico. *Environment International* 25:573–583.

Haag, W.R. 2012. *North American Freshwater Mussels. Natural History, Ecology and Conservation*. Cambridge University Press, Cambridge, United Kingdom. xvi + 505 pp.

Haag, W.R., and A.M. Commons-Carson. 2008. Testing the assumption of annual shell ring deposition in freshwater mussels. *Canadian Journal of Fisheries and Aquatic Science* 65(3):493–508.

Haas, F. 1916. *Náyades del viaje al Pacífico, verificado de 1862 á 1865 por una comisión de naturalistas enviada por el Gobierno Español*. Trabajos del Museo Nacional Ciencias Naturales, Serie Zoológica, Madrid 25:1–63 + 2 plates.

Haas, F. 1929. Bemerkungen über mittelamerikanische Najaden. *Senckenbergiana* 11(5–6):310–344 + 21 text figures.

Haas, F. 1930. Über nord–und mittelamerikanische Najaden. *Senckenbergiana* 12(6):317–330.

Haas, F. 1931. Versuch einer kritischen Sichtung der südamerikanischen Najaden, hauptsächlich an Hand der Sammlung des Senckenberg–Museums III. *Senckenbergiana* 13(2):87–110.

Haas, F. 1939. On the life habits of some tropical fresh-water mussels. *The Nautilus* 53(2):53–56.

Haas, F. 1940. A tentative classification of the Palearctic unionids. *Zoological Series of Field Museum of Natural History* 24(11):115–141.

Haas, F. 1969. Superfamilia Unionacea. *Das Tierreich (Berlin)* 88:x + 663 pp.

Hanley, S. 1843. *An Illustrated and Descriptive Catalogue of Recent Bivalve Shells, with 960 Figures by Wood and Sowerby Forming and Appendix to the Index Testaceologicus*. Williams and Norgate, London. pp. i–xviii, 33–272.

Hanley, S. 1856. *An Illustrated and Descriptive Catalogue of Recent Bivalve Shells, with 960 Figures by Wood and Sowerby Forming and Appendix to the Index Testaceologicus*. Williams and Norgate, London. pp. i–xviii, 273–392; plates 20–24.

Hannibal, H. 1912a. A synopsis of the recent and tertiary freshwater Mollusca of the Californian Province based upon an ontogenetic classification. *Proceedings of the Malacological Society,* London 10(2):112–166.

Hannibal, H. 1912b. The aquatic molluscs of Southern California and adjacent regions, a transition fauna. *Bulletin of the Southern California Academy of Sciences* 15(1):18–46.

Hanson, J.M., W.C. MacKay, and E.E. Prepas. 1989. Effect of size-selective predation by muskrats (*Ondatra zibethicus*) on a population of unionid clams (*Anodonta grandis simpsoniana*). *Journal of Animal Ecology* 58:15–28.

Harmon, J.L., and J.E. Joy. 1990. Growth rates of the freshwater mussel, *Anodonta imbecillis* Say 1829, in five West Virginia Wildlife Station ponds. *American Midland Naturalist* 124(2):372–378.

Heard, W.H. 1975. Sexuality and other aspects of reproduction in *Anodonta* (Pelecypoda: Unionidae). *Malacologia* 15(1):81–103.

Heard, W.H., and R.H. Guckert. 1970. A re-evaluation of the recent Unionacea (Pelecypoda) of North America. *Malacologia* 10(2):333–355.

Hebert, P.D., A. Cywinska, and S.L. Ball. 2003. Biological identifications through DNA barcodes. *Proceedings of the Royal Society of London. Series B: Biological Sciences* 270:313–321.

Hein, S.R., M.P. Jones, D.A. Trujillo, N.E. Strenth, and D.J. Berg. 2018. Mussel surveys of Rio Grande tributaries in Coahuila, Mexico. Report to U.S. Fish and Wildlife Service, Albuquerque, New Mexico. 12 pp.

Hemphill, H. 1891. A collector's notes on variation in shells, with some new varieties. *Zoe* 1(11):321–337 + 1 plate.

Henderson, J. 1935. Fossil non-marine Mollusca of North America. *Geological Society of America Special Paper* 3:1–313.

Hendrickson, D.A., W.L. Minckley, R.R. Miller, D.J. Siebert, and P. Haddock Minckley. 1980. Fishes of the Río Yaqui Basin, México and United States. *Journal of the Arizona-Nevada Academy of Science* 15(3):65–106.

Herrmannsen, A.N. 1846a. Indicis Generum Malacozoorum Primordia. Nomina subgenerum, generum, familiarum, tribuum, ordinum, classium; adjectis auctoribus, temporibus, locis systematicis atque literariis, etymis, synonymis. *Cassellis. Sumptibus & Typis Theodori Fischeri* Vol. 1:i-xxviii + 1–104.

Herrmannsen, A.N. 1846b. Indicis Generum Malacozoorum Primordia. Nomina subgenerum, generum, familiarum, tribuum, ordinum, classium; adjectis auctoribus, temporibus, locis systematicis atque literariis, etymis, synonymis. *Cassellis. Sumptibus & Typis Theodori Fischeri* Vol. I:105–232.

Herrmannsen, A.N. 1847a. Indicis Generum Malacozoorum Primordia. Nomina subgenerum, generum, familiarum, tribuum, ordinum, classium; adjectis auctoribus, temporibus, locis systematicis atque literariis, etymis, synonymis. *Cassellis. Sumptibus & Typis Theodori Fischeri* Vol. 1:489–616.

Herrmannsen, A.N. 1847b. Indicis Generum Malacozoorum Primordia. Nomina subgenerum, generum, familiarum, tribuum, ordinum, classium; adjectis auctoribus, temporibus, locis systematicis atque literariis, etymis, synonymis. *Cassellis. Sumptibus & Typis Theodori Fischeri* Vol. 2:1–104.

Herrmannsen, A.N. 1847c. Indicis Generum Malacozoorum Primordia. Nomina subgenerum, generum, familiarum, tribuum, ordinum, classium; adjectis auctoribus, temporibus, locis systematicis atque literariis, etymis, synonymis. *Cassellis. Sumptibus & Typis Theodori Fischeri* Vol. 2:105–232.

Herrmannsen, A.N. 1848. Indicis Generum Malacozoorum Primordia. Nomina subgenerum, generum, familiarum, tribuum, ordinum, classium; adjectis auctoribus, temporibus, locis systematicis atque literariis, etymis, synonymis. *Cassellis. Sumptibus & Typis Theodori Fischeri* Vol. 2:353–492.

Herrmannsen, A.N. 1849a. Indicis Generum Malacozoorum Primordia. Nomina subgenerum, generum, familiarum, tribuum, ordinum, classium; adjectis auctoribus, temporibus, locis systematicis atque literariis, etymis, synonymis. *Cassellis. Sumptibus & Typis Theodori Fischeri* Vol. 2:493–612.

Herrmannsen, A.N. 1849b. Indicis Generum Malacozoorum Primordia. Nomina subgenerum, generum, familiarum, tribuum, ordinum, classium; adjectis auctoribus, temporibus, locis systematicis atque literariis, etymis, synonymis. *Cassellis. Sumptibus & Typis Theodori Fischeri* Vol. 2:613–717 + xxix–xlii.

Herrmannsen, A.N. 1852. *Indicis Generum Malacozoorum. Supplementa et Corrigenda. Cassellis.* Sumptibus Theodori Fischer. i–v + 1–140.

Hinkley, A.A. 1907a. Shells collected in northeastern Mexico. *The Nautilus* 21(6):68–72.

Hinkley, A.A. 1907b. Shells collected in northeastern Mexico. (cont.). *The Nautilus* 21(7):76–80.

Hinkley, A.A. 1920. Guatemala mollusca. *The Nautilus* 34(2):37–55.

Hoeh, W.R. 1990. Phylogenetic relationships among Eastern North American *Anodonta* (Bivalvia: Unionidae). *Malacological Review* 23(1–2):63–82.

Hoeh, W.R., A.E. Bogan, K.S. Cummings, and S.I. Guttman. 2002. Evolutionary relationships among the higher taxa of freshwater mussels (Bivalvia: Unionoida): Inferences on phylogeny and character evolution from analyses of DNA sequence data. *Malacological Review* 31/32(2):123–141.

Hoeh, W.R., A.E. Bogan, and W.H. Heard. 2001. A phylogenetic perspective on the evolution of morphological and reproductive characteristics in the Unionoida. Pp. 257–280 *In* G. Bauer, and K. Wächtler (eds.). *Ecology and Evolution of the Freshwater Mussels Unionoida.* Springer-Verlag, Berlin, Ecological Studies Vol. 145. 394 pp.

Hoeh, W.R., K.S. Frazer, E. Naranjo-García, M.B. Black, D.J. Berg, and S.I. Guttman. 1998. Correlation between mating system and distribution of genetic variation in *Utterbackia* (Bivalvia: Unionidae). *Journal of Shellfish Research* 17(5): 1383–1393.

Hoeh, W.R., K.S. Frazer, E. Naranjo-Garcia, and R.J. Trdan. 1995. A phylogenetic perspective on the evolution of simultaneous hermaphroditism in a freshwater mussel clade (Bivalvia: Unionidae: *Utterbackia*). *Malacological Review* 28(1/2): 25–42.

Howard, A.D., and B.J. Anson. 1922. Phases in the parasitism of the Unionidae. *The Journal of Parasitology* 9(2):68–82.

Howells, R.G. 1995. Rio Grande bleuffer. *Info-Mussel Newsletter* 3(1):1.

Howells, R.G. 1997a. New fish hosts for nine freshwater mussels (Bivalvia: Unionidae) in Texas. *Texas Journal of Science* 49(3):255–258.

Howells, R.G. 1997b. Range extension of the freshwater mussels *Potamilus purpuratus* (Bivalvia: Unionidae) in Texas. *Texas Journal of Science* 49(1):79–82.

Howells, R.G. 2001. *Status of Freshwater Mussels of the Rio Grande, with Comments on Other Bivalves.* Texas Parks and Wildlife Department, Inland Fisheries Division, Austin, Texas. 81 pp.

Howells, R.G. 2003. Declining status of freshwater mussels of the Rio Grande, with comments on other bivalves. Pp. 59–73 *In* G.P. Garrett, and N.L. Allen (eds.). *Aquatic Fauna of the Northern Chihuahuan Desert.* Museum of Texas Tech University, Special Publication No. 46, Lubbock, Texas. 160 pp.

Howells, R.G., R.W. Neck, and H.D. Murray. 1996. *Freshwater Mussels of Texas.* Texas Parks and Wildlife Press, Austin, Texas. iv+218 pp.

Huang, X.-C., J.-H. Su, J.-X. Ouyang, S. Ouyang, C.-H. Zhou, and X.-P. Wu. 2019. Towards a global phylogeny of freshwater mussels (Bivalvia: Unionida): Species delimitation of Chinese taxa, mitochondrial phylogenomics, and diversification patterns. *Molecular Phylogenetics and Evolution* 130:45–59.

Huang, X.-C, J. Rong, Y. Liu, M.-H. Zhang, Y. Wan, S. Ouyang, C.-H. Zhou, and X.-P. Wu. 2013. The complete maternally and paternally inherited mitochondrial genomes of the endangered freshwater mussel *Solenaia carinatus* (Bivalvia: Unionidae) and implications for Unionidae taxonomy. *PLoS ONE* 8(12):e84352. 13 pp.

Huang, X.-C., R.-W. Wu, C.-T. An, G.-L. Xie, J.-H. Su, S. Ouyang, C.-H. Zhou, and X.-P. Wu. 2018. Reclassification of *Lamprotula rochechouartii* as *Margaritifera rochechouartii* comb. nov. (Bivalvia: Margaritiferidae) revealed by time-calibrated multi-locus phylogenetic analyses and mitochondrial phylogenomics of Unionoida. *Molecular Phylogenetics and Evolution* 120:297–306.

Hudson, P.F., D.A. Hendrickson, A.C. Benke, A. Varela-Romero, R. Rodiles-Hernández, and W. Minckley. 2005. Chapter 23. Rivers of México. Pp. 1030–1084 *In* A.C. Benke, and C.E. Cushing (eds.). Rivers of North America. Academic Press, Elsevier Inc., Amsterdam. 1168 pp.

Hulsey, C.D., and H. López-Fernández. 2011. Chapter Seventeen. Nuclear Central America Pp. 279–291 *In* James Albert and Roberto E. Reis (eds.) Historical Biogeography of Neotropical Freshwater Fishes. University of California Press. Berkeley, Los Angeles, London. xv+388 pp.

Hupé, H. 1857. Mollusques de L'Amerique du Sud. Pp. 1–103 *In* F. De Castelnau, P. Gervais, H. Lucas, and H. Hupé (eds.). *Animaux nouveaux ou rares recueillis pendant l'expédition dans les parties centrales de l'Amérique du Sud, de Rio de Janeiro a Lima, et de Lima au Para.* Paris, Chez P. Bertrand, Libraire - Éditeur. 103 pp+20 plates.

International Commission on Zoological Nomenclature (Online). 2025. W.D.L. Ride, H.G. Cogger, C. Dupuis, O. Kraus, A. Minelli, F. C. Thompson, and P.K. Tubbs (eds.). Published by The International Trust for Zoological Nomenclature. ISBN 0 85301 006 4. https://www.iczn.org/the-code/the-code-online/

von Ihering, H. 1890. Revision der von Spix in Brasilien gesammelten Najaden. *Archiv für Naturgeschichte* 56(1):117–170+1 plate.

von Ihering, H. 1901. The Unionidae of North America. (cont.). *The Nautilus* 15(5):50–53.

von Ihering, H. 1910. Über Brasilianische Najaden. *Abhandlungen der Senckenbergischen Naturforschenden Gesellschaft* 32:111–140.

Inomata, T., J.C. Fernandez-Diaz, D. Triadan, M. García Mollinedo, F. Pinzón, M. García Hernández, A. Flores, A. Sharpe, T. Beach, G.W.L. Hodgins, J.J. Durón Díaz, A. Guerra Luna, L. Guerrero Chávez, M. de Lourdes Hernández Jiménez, and M. Moreno Díaz. 2021. Origins and spread of formal ceremonial complexes in the Olmec and Maya regions revealed by airborne lidar. *Nature Human Behaviour* 5(11):1487–1501.

Inoue, K., A.L. McQueen, J.L. Harris, and D.J. Berg. 2014. Molecular phylogenetics and morphological variation reveal recent speciation in freshwater mussels of the genera *Arcidens* and *Arkansia* (Bivalvia: Unionidae). *Biological Journal of the Linnean Society* 112(3):535–545.

Inoue, K., B.K. Lang, and D.J. Berg. 2013. Development and characterization of 20 polymorphic microsatellite markers for the Texas Hornshell, *Popenaias popeii* (Bivalvia: Unionidae), through next-generation sequencing. *Conservation Genetics Resources* 5(1):195–198.

Inoue, K., B.K. Lang, and D.J. Berg. 2015. Past climate change drives current genetic structure of an endangered freshwater mussel species. *Molecular Ecology* 24(8):1910–1926.

Inoue, K., D.M. Hayes, J.L. Harris, N.A. Johnson, C.L. Morrison, M.S. Eackles, T.L. King, J.W. Jones, E.M. Hallerman, A.D. Christian, and C.R. Randklev. 2018. The Pleurobemini (Bivalvia: Unionida) revisited: Molecular species delineation using a mitochondrial DNA gene reveals multiple conspecifics and undescribed species. *Invertebrate Systematics* 32(3):689–702.

Inoue, K., J.L. Harris, C.R. Robertson, N.A. Johnson, and C.R. Randklev. 2020b. A comprehensive approach uncovers hidden diversity in freshwater mussels (Bivalvia: Unionidae) with the description of a novel species. *Cladistics* 36(1):88–113.

Inoue, K., K.S. Cummings, J.S. Tiemann, T.D. Miller, N.A. Johnson, C.H. Smith, and C.R. Randklev. 2020a. A new species of freshwater mussel in the genus *Popenaias* Frierson, 1927, from Gulf coastal rivers of central Mexico (Bivalvia: Unionida: Unionidae) with comments on the genus. *Zootaxa* 4816(4):457–490.

International Code of Zoological Nomenclature (ICZN). 1999. *International Code of Zoological Nomenclature*. Fourth Edition. International Trust for Zoological Nomenclature, London. xxix + 306 pp.

International Union for Conservation of Nature (IUCN). 2024. The IUCN Red List of Threatened Species. Version 2021–3. Available https://www.iucnredlist.org.

Ituarte, C., and M.C.D. Mansur. 2020. Chapter 11. Class Bivalvia. Pp. 378–390 *In* C. Damborenea, D.C. Rogers, and J. Thorp (eds.). *Thorp and Covich's Freshwater Invertebrates (Fourth Edition) Volume 5: Keys to Neotropical and Antarctic Fauna*. Elsevier Inc., Amsterdam. 1017 pp.

Jay, J.C. 1835. *Catalogue of Recent Shells in the Cabinet of John C.* Jay, M.D. D. Fanshaw, New York, New York. 55 pp.

Johnson, N.A., C.H. Smith, J.M. Pfeiffer, C.R. Randklev, J.D. Williams, and J.D. Austin. 2018. Integrative taxonomy resolves taxonomic uncertainty for freshwater mussels being considered for protection under the U.S. Endangered Species Act. *Scientific Reports* 8(15892):1–16.

Johnson, R.I. 1951a. A clarification of some Central American Unionidae described by R.A. Philippi. *The Nautilus* 64(3):77–83 + plate 5.

Johnson, R.I. 1951b. A rectification of some Central American Unionidae. *The Nautilus* 65(1):32–33.

Johnson, R.I. 1952. A study of Lamarck's types of Unionidae and Mutelidae. *The Nautilus* 66(2):63–67.

Johnson, R.I. 1953. A study of Lamarck's types of Unionidae and Mutelidae (cont.). *The Nautilus* 66(3):90–95.

Johnson, R.I. 1956. The types of naiades (Mollusca: Unionidae) in the Museum of Comparative Zoology. *Bulletin of the Museum of Comparative Zoology* 115(4):101–142 + 2 plates.

Johnson, R.I. 1964. The recent Mollusca of Augustus Addison Gould. *Bulletin of the United States National Museum* 239:v–182.

Johnson, R.I. 1967. *Carunculina pulla* (Conrad), an overlooked Atlantic drainage unionid. *The Nautilus* 80(4):127–131.

Johnson, R.I. 1968. Martini and Chemnitz (Kuester's Edition) Systematisches Conchylien-Cabinet, 1837–1920, a complete collation. *Journal of the Society for the Bibliography of Natural History* 4(7):363–367.

Johnson, R.I. 1969. Illustrations of Lamarck's types of North American Unionidae mostly in the Paris Museum. *The Nautilus* 83(2):52–61.

Johnson, R.I. 1970. The systematics and zoogeography of the Unionidae (Mollusca: Bivalvia) of the southern Atlantic Slope Region. *Bulletin of the Museum of Comparative Zoology* 140(6):263–449.

Johnson, R.I. 1971. The types and figured specimens of Unionacea (Mollusca: Bivalvia) in the British Museum (Natural History). *Bulletin of the British Museum (Natural History)* 20(3):75–108 + 2 plates.

Johnson, R.I. 1972a. Illustrations of all of the mollusks described by Lorraine Screven Frierson. *Occasional Papers on Mollusks, Museum of Comparative Zoology, Harvard University* 3(41):137–173.

Johnson, R.I. 1972b. The Unionidae (Mollusca: Bivalvia) of peninsular Florida. *Bulletin of the Florida State Museum, Biological Sciences* 16(4):181–249 + addendum.

Johnson, R.I. 1973. The types of Unionidae (Mollusca: Bivalvia) described by C. S. Rafinesque in the Museum National d'Histoire Naturelle, Paris. *Journal de Conchyliologie* 110(2):35–37+2 plates.

Johnson, R.I. 1974a. Lea's unionid types or recent and fossil taxa of Unionacea and Mutelacea introduced by Isaac Lea, including the location of all the extant types. Special Occasional Publication, Museum of Comparative Zoology, Harvard University No. 2. 159 pp.

Johnson, R.I. 1974b. Marshall's unionid types or types of recent and fossil Unionacea and Mutelacea introduced by William B. Marshall, including a bibliography of all his writings on mollusks. Special Occasional Publication, Museum of Comparative Zoology, Harvard University No. 3. 14 pp.

Johnson, R.I. 1975. Simpson's unionid types and miscellaneous unionid types in the National Museum of Natural History. Special Occasional Papers, Museum of Comparative Zoology, Harvard University No. 4. 56 pp.+3 plates.

Johnson, R.I. 1979. The types of Unionacea (Mollusca: Unionidae) in the Museum of Zoology, The University of Michigan. *Malacological Review* 12(1–2):29–36.

Johnson, R.I. 1980. The types of Unionacea (Mollusca: Bivalvia) in the Academy of Natural Sciences of Philadelphia: Additions and corrections. *Proceedings of the Academy of Natural Sciences of Philadelphia* 132:277–278.

Johnson, R.I. 1981. Recent and fossil Unionacea and Mutelacea (freshwater bivalves) of the Caribbean Islands. *Occasional Papers on Mollusks, Museum of Comparative Zoology, Harvard University* 4(60):269–288.

Johnson, R.I. 1998. A new mussel, *Potamilus metnecktayi* (Bivalvia: Unionidae) from the Rio Grande system, Mexico and Texas with notes on Mexican *Disconaias*. *Occasional Papers on Mollusks, Museum of Comparative Zoology, Harvard University* 5(76):427–455.

Johnson, R.I. 1999a. Unionidae of the Rio Grande (Rio Bravo del Norte) system of Texas and Mexico. *Occasional Papers on Mollusks, Museum of Comparative Zoology, Harvard University* 6(77):1–65.

Johnson, R.I. 1999b. Taxa omitted from "Unionacea" by Fritz Haas, 1969. *Occasional Papers on Mollusks, Museum of Comparative Zoology, Harvard University* 6(77):66.

Johnson, R.I., and H.B. Baker. 1973. The types of Unionacea (Mollusca: Bivalvia) in the Academy of Natural Sciences of Philadelphia. *Proceedings of the Academy of Natural Sciences of Philadelphia* 125:145–186.

Kabat, A.R., and E.V. Coan. 2016. Zeitschrift für Malakozoologie (1844–1853): A history and collation. *The Nautilus* 130(1):17–22.

Kabat, A.R., and E.V. Coan. 2017. The life and work of Rudolph Amandus Philippi (1808–1904). *Malacologia* 60(1–2):1–30.

Kabat, A.R., and K.J. Boss. 1997. *Karl Eduard von Martens (*1831–1904*): His Life and Works*. Department of Mollusks, Museum of Comparative Zoology, Cambridge, Massachusetts. vii+417 pp.

Kabat, A.R., and R.I. Johnson. 2008. Dwight Willard Taylor (1932–2006): His life and malacological research. *Malacologia* 50(1–2):175–218.

Karatayev, A.Y., T.D. Miller, and L.E. Burlakova. 2012. Long-term changes in unionid assemblages in the Rio Grande, one of the world's top 10 rivers at risk. *Aquatic Conservation: Marine and Freshwater Ecosystems* 22(2):206–219.

Keogh, S.M., and A.M. Simons. 2019. Molecules and morphology reveal 'new' widespread North American freshwater mussel species (Bivalvia: Unionidae). *Molecular Phylogenetics and Evolution* 138:182–192.

Kessler, D.H, and J.A. Downing. 1997. Internal shell annuli yield inaccurate growth estimates in the freshwater mussels *Elliptio complanata* and *Lampsilis radiata*. *Freshwater Biology* 37:325–332.

Kiser, A.H., K.S. Cummings, J.S. Tiemann, C.H. Smith, N.A. Johnson, R.R. Lopez, and C.R. Randklev. 2022. Using a multi-model ensemble approach to determine biodiversity hotspots with limited occurrence data in understudied areas: An example using freshwater mussels in México. *Ecology and Evolution* 12:e8909. 14 pp.

Kobelt, W. 1893. *Iconographie der Land- und Süsswasser-Mollusken, mit vorzüglicher Berücksichtigung der Europäischen noch nicht abgebildeten Arten*. Wiesbaden. C.W. Kreidel's Verlag Neue Folge Vol. VI. 102 pp.+plates 151–180.

Küster, H.C. 1838. Die Gattung *Anodonta* nebst den übrigen najaden mit unvollkommenem Schloss. *In Abbildungen nach der Natur mit Beschreibungen. Systematisches Conchylien-Cabinet von Martini und Chemnitz*. Küster Edition. 9(Part 1). 287 pp.+plates 20–87, A, 11.

Küster, H.C. 1853. Die flusperlmuscheln (*Unio* et *Hyria*.). *In Abbildungen nach der Natur mit Beschreibungen. Systematisches Conchylien-Cabinet von Martini und Chemnitz* Fortgesetzt von Hofrath Dr. G. v. Schubert und Professor Dr. J.A. Wagner 9(Part 2):65–88, plates 18–22.

Küster, H.C. 1856. Die flusperlmuscheln (*Unio* et *Hyria*.). *In Abbildungen nach der Natur mit Beschreibungen. Systematisches Conchylien-Cabinet von Martini und Chemnitz* Fortgesetzt von Hofrath Dr. G. v. Schubert und Professor Dr. J.A. Wagner 9(Part 2):137–160, plates 45–54.

Küster, H.C. 1861. Die flusperlmuscheln (*Unio* et *Hyria.*). *In Abbildungen nach der Natur mit Beschreibungen. Systematisches Conchylien-Cabinet von Martini und Chemnitz* Fortgesetzt von Hofrath Dr. G. v. Schubert und Professor Dr. J.A. Wagner 9(Part 2):161–208, plates 72–76, 78–83, 85.

Küster, H.C. 1862. Die flusperlmuscheln (*Unio* et *Hyria.*). *In Abbildungen nach der Natur mit Beschreibungen. Systematisches Conchylien-Cabinet von Martini und Chemnitz* Fortgesetzt von Hofrath Dr. G. v. Schubert und Professor Dr. J.A. Wagner 9(Part 2):257–318, plates 68, 87–100.

Lang, B.K. 2001. Status of the Texas Hornshell and Native Freshwater Mussels (Unionoidea) in the Rio Grande and Pecos River of New Mexico and Texas. Completion Report, New Mexico Department of Game and Fish, Conservation Services Division, Santa Fe, New Mexico.

Lasso, C.A., R. Martínez-Escarbassiere, J.C. Capelo, M.A. Morales-Betancourt, and A. Sánchez-Maya. 2009. Lista de los moluscos (Gastropoda-Bivalvia) dulceacuícolas y estuarinos de la cuenca del Orinoco (Venezuela). *Biota Colombiana* 10(1–2):63–74.

Lea, I. 1829. Description of a new genus of the family of naïades, including eight new species, four of which are new; also the description of eleven new species of the genus *Unio* from the rivers of the United States: With observations on some of the characters of the naïades. *Transactions of the American Philosophical Society* 3(N.S.):403–457 + plates 7–14.

Lea, I. 1831. Observations on the naïades, and descriptions of new species of that and other families. *Transactions of the American Philosophical Society* 4(N.S.):63–121 + plates 3–18.

Lea, I. 1834. *Observations on the Genus Unio, Together with Descriptions of New Genera and Species in the Families Naïades, Conchæ, Colimacea, Lymnæana, Melaniana and Peristomiana: Consisting of Four Memoirs Read before the American Philosophical Society from 1827 to 1834, and Originally Published in Their Transactions. With Coloured Plates.* Printed for the author, James Kay, Jun. and Co., Philadelphia, Pennsylvania. Vol. I. 231 pp. + 46 plates. Title, dedication (to Philip Houlbrooke Nicklin), introduction, TOC, index, errata.

Lea, I. 1836. *A Synopsis of the Family of Naiades.* Cary, Lea and Blanchard. London, John Miller, Philadelphia, Pennsylvania. viii 1–59 + 1 plate.

Lea, I. 1838a. *A Synopsis of the Family of Naiades.* Second Edition enlarged and improved. Carey, Lea & Blanchard, Philadelphia, Pennsylvania. 44 pp.

Lea, I. 1838b. Description of new freshwater and land shells. *Transactions of the American Philosophical Society* 6(N.S.):1–154 + plates 1–24.

Lea, I. 1838c. *Observations on the Genus Unio, Together with Descriptions of New Genera and Species in the Families Naïades, Colimacea, Lymnæana, Melaniana and Peristomiana.* Printed for the Author, Philadelphia, Pennsylvania. Vol. 2. Dedication (to Thomas G. Lea). 154 pp. + plates 1–24.

Lea, I. 1842a. Description of new fresh water and land shells. *Transactions of the American Philosophical Society* 8(N.S.):163–250 + 23 plates.

Lea, I. 1842b. *Observations on the Genus Unio, Together with Descriptions of New Species in the Families Naiades, Conchae, Colimacea, Lymnæana, Melaniana and Peristomiana.* Printed for the Author, Philadelphia, Pennsylvania. Vol. 3. viii + 88 pp. + plates 5–27.

Lea, I. 1852. *A Synopsis of the Family of Naiades.* Third Edition. Blanchard and Lea, Philadelphia, Pennsylvania. xx;17–88 pp.

Lea, I. 1858a. Descriptions of exotic genera and species of the family Unionidae. *Journal of the Academy of Natural Sciences of Philadelphia Second Series* 3, Part 4(19):289–321 + plates 21–33.

Lea, I. 1858b. *Observations of the Genus Unio, Together with Descriptions of New Species in the Family Unionidae.* Printed for the Author. Marrihew & Thompson, Printers, Philadelphia, Pennsylvania. Vol. 6, Part 1: errata, pp. 9–41 + plates 21–33.

Lea, I. 1860a. Descriptions of exotic Unionidae. *Journal of the Academy of Natural Sciences of Philadelphia Second Series* 4, Part 3(7):235–273 + plates 33–45.

Lea, I. 1860b. New Unionidae of the United States and Northern Mexico. *Journal of the Academy of Natural Sciences of Philadelphia Second Series* 4, Part 4(13):327–374 + plates 51–66.

Lea, I. 1860c. *Observations on the Genus Unio, Together with Descriptions of New Species, Their Soft Parts, and Embryonic Forms, in the Family Unionidae.* Printed for the Author, Philadelphia, Pennsylvania. Vol. 7. Part 2. TOC, errata, Introduction [2 pp.] pp. 53–91 + plates 33–45.

Lea, I. 1860d. *Observations on the Genus Unio, Together with Descriptions of New Species, Their Soft Parts, and Embryonic Forms, in the Family Unionidae.* Printed for the Author, Philadelphia, Pennsylvania. Vol. 8. Part 1. pp. 9–56 + plates 51–66.

Lea, I. 1862. New Unionidae of the United States and Arctic America. *Journal of the Academy of Natural Sciences of Philadelphia Second Series* 5, Part 2(5):187–216 + plates 24–33.

Lea, I. 1863. *Observations on the Genus Unio, Together with Descriptions of New Species, Their Soft Parts, and Embryonic Forms, in the Family Unionidae and Descriptions of New Genera and Species of the Melanidæ*. Printed for the Author, Philadelphia, Pennsylvania. Vol. 9. 180 pp. Introduction [1 p.] pp. 9–178 + plates 24–39.

Lea, I. 1866. New Unionidae, Melanidae, etc., chiefly of the United States. *Journal of the Academy of Natural Sciences of Philadelphia Second Series* 6, Part 1(1):5–65 + plates 1–21.

Lea, I. 1867. *Observations on the Genus Unio, Together with Descriptions of New Species in the Family Unionidae, and Descriptions of New Species of the Melanidæ, Limneidæ, Paludinæ, and Helicidæ*. Printed for the Author, Philadelphia, Pennsylvania. Vol. 11. 146 pp. + plates 1–21.

Lea, I. 1868a. New Unionidae, Melanidae, etc., chiefly of the United States. *Journal of the Academy of Natural Sciences of Philadelphia Second Series* 6, Part 3(6):249–302 + plates 29–45.

Lea, I. 1868b. New Unionidae, Melanidae, etc., chiefly of the United States. *Journal of the Academy of Natural Sciences of Philadelphia Second Series* 6, Part 4(7):303–343 + plates 46–54.

Lea, I. 1868c. Description of nine species of Unionidae from Lake Nicaragua, Central America. *Proceedings of the Academy of Natural Sciences of Philadelphia* 20:94–96.

Lea, I. 1869a. *Observations on the Genus Unio, Together with Descriptions of New Species in the Family Unionidæ, and Descriptions of New Species of the Melanidæ, Paludinæ, and Helicidæ*. Printed for the Author, Philadelphia, Pennsylvania. Vol. 12. Title, TOC, dedication (to E.R. Schowalter, M.D.) 105 pp. + plates 29–54.

Lea, I. 1869b. *Index to Vol.* XII. *and Supplementary Index to Vols. I to XI. of Observations on the Genus Unio, Together with Description of New Species of the Family Unionidæ, and Description of New Species of the Melanidæ, Paludinæ, Helicidæ, etc. Read before the American Philosophical Society and the Academy of Natural Sciences of Philadelphia, from 1827 to 1868*. Printed for the author, Philadelphia, Pennsylvania. Vol. II. 23 pp.

Lea, I. 1870. *A Synopsis of the Family Unionidae*. Fourth Edition, very greatly enlarged and improved. Henry C. Lea, Philadelphia, Pennsylvania. xxx + 184 pp.

Lea, I. 1874a. Descriptions of fifty-two species of Unionidae. *Journal of the Academy of Natural Sciences of Philadelphia Second Series* 8, Part 1(1):5–54 + plates 1–18.

Lea, I. 1874b. *Observations on the Genus Unio, Together with Descriptions of New Species in the Family Unionidæ, and Descriptions of Embryonic Forms and Soft Parts, Also, New Species of Steptomatidæ and Limnæidæ*. Printed for the Author. Collins, Printer, Philadelphia, Pennsylvania. Vol. 13. 75 pp. + plates 1–22.

de Leon, J.D. 1912. Mollusca. Catalogus Molluscarum Maxicanæ Reipublicæ hucusque descripta. *La Naturaleza, Tercera Serie* 1(4):13–139.

Liu, H.-P., J.B. Mitton, and S.J. Herrmann. 1996. Genetic differentiation in and management recommendations for the freshwater mussel, *Pyganodon grandis* (Say, 1829). *American Malacological Bulletin* 13(1–2):117–124.

Lopes-Lima, M., A. Hattori, T. Kondo, J.H. Lee, S.K. Kim, A. Shirai, H. Hayashi, T. Usui, K. Sakuma, T. Toriya, Y. Sunamura, H. Ishikawa, N. Hoshino, Y. Kusano, H. Kumaki, Y. Utsugi, S. Yabe, Y. Yoshinari, H. Hiruma, A. Tanaka, K. Sao, T. Ueda, I. Sano, J.-I. Miyazaki, D.V. Gonçalves, O.K. Klishko, E.S. Konopleva, I.V. Vikhrev, A.V. Kondakov, M.Y. Gofarov, I.N. Bolotov, E.M. Sayenko, M. Soroka, A. Zieritz, A.E. Bogan, and E. Froufe. 2020. Freshwater mussels (Bivalvia: Unionidae) from the rising sun (Far East Asia): Phylogeny, systematics, and distribution. *Molecular Phylogenetics and Evolution* 146:1–27.

Lopes-Lima, M., E. Froufe, V. Tu Do, M. Ghamizi, K.E. Mock, Ü. Kebapçı, O. Klishko, S. Kovitvadhi, U. Kovitvadhi, O.S. Paulo, J.M. Pfeiffer III, M. Raley, N. Riccardi, H. Şereflişan, R. Sousa, A. Teixeira, S. Varandas, X. Wu, D.T. Zanatta, A. Zieritz, and A.E. Bogan. 2017. Phylogeny of the most species-rich freshwater bivalve family (Bivalvia: Unionida: Unionidae): Defining modern subfamilies and tribes. *Molecular Phylogenetics and Evolution* 106:174–191.

Lopes-Lima, M., L. Burlakova, A. Karatayev, A. Gomes-dos-Santos, A. Zieritz, E. Froufe, and A.E. Bogan. 2019. Revisiting the North American freshwater mussel genus *Quadrula* sensu lato (Bivalvia Unionidae): Phylogeny, taxonomy and species delineation. *Zoologica Scripta* 48(3):313–336.

López-López, E., and J. Paulo-Maya. 2001. Changes in the fish assemblages in the Upper Río Ameca, Mexico. *Journal of Freshwater Ecology* 16(2):179–187.

Lozada-Flores, O. 2011a. Inventario de especies y catálogo ilustrado de los bivalvos de la "Colección Malacológica Dr. Antonio García-Cubas." *Apéndice* 1. 46 pp.

Lozada-Flores, O. 2011b. Catálogo Ilustrado Colección Malacológica "Dr. Antonio García-Cubas." 42 pp.

de Lucena, D.T. 1951. Lista de moluscos do Nordeste, com um apendice sobre algumas espécies de outras regiões. *Papéis Avulsos do Departamento de Zoologia* 10(3):93–104.

Luebke, C.M., and D.L. Graf. 2013. Curation of the Daniel J. Bereza Collection of freshwater mollusks at the Smithsonian National Museum of Natural History: Phase I. Poster presented at the University of Wisconsin Stevens-Point College of Letters and Sciences Undergraduate Research Symposium.

Lydeard, C., R.H. Cowie, W.F. Ponder, A.E. Bogan, P. Bouchet, S.A. Clark, K.S. Cummings, T.J. Frest, O. Gargominy, D.G. Herbert, R. Hershler, K.E. Perez, B. Roth, M. Seddon, E.E. Strong, and F.G. Thompson. 2004. The global decline of nonmarine mollusks. *BioScience* 54(4):321–330.

Lyons, T.J., L. Máiz-Tomé, M. Tognelli, A. Daniels, C. Meredith, R. Bullock, and I. Harrison. 2020. *The Status and Distribution of Freshwater Fishes in Mexico*. IUCN, Cambridge, United Kingdom and Albuquerque, New Mexico and ABQ BioPark. vii:1–80.

Maine, A.N., and C. O'Brien. 2022. Use of native and nonnative fish hosts by the freshwater mussel *Anodonta californiensis* (California Floater) in the Columbia River Basin. *Malacologia* 64(2):203–213.

Mansur, M.C.D., and C.P. dos Santos. 2008. *Mycetopoda siliquosa* (Spix, 1827). Pp. 231–232 *In* A.B.M. Machado, G.M. Drummond, and A.P. Paglia (eds.). *Livro Vermelho da Fauna Brasileira Ameaçada de Extinção*. Ministério do Meio Ambiente e Fundação Biodiversitas, Brasília. 293 pp. + 8 plates.

Mansur, M.C.D., and D. Pereira. 2006. Limnic bivalves of the Sinos river basin, Rio Grande do Sul, Brazil (Bivalvia, Unionoida, Veneroida and Mytiloida). *Revista Brasileira de Zoologia* 23(4):1123–1147.

Mansur, M.C.D., I.C. Miyahira, J.O. Arruda, T.N. Antoniazzi, and D.M. Pimpão. 2020. Chapter 11. Phylum Mollusca. Key to Unionida. Pp. 409–430 *In* C. Damborenea, D.C. Rogers, and J. Thorp (eds.). *Thorp and Covich's Freshwater Invertebrates (Fourth Edition) Volume 5: Keys to Neotropical and Antarctic Fauna*. Elsevier Inc. 1017 pp.

Mansur, M.C.D., and R.M. Valer. 1992. Moluscos bivalves do Rio Uraricoera e Rio Branco, Roraima, Brazil. Amazoniana. *Limnologia et Oecologia Regionalis Systemae Fluminis Amazonas* 12(1):85–100.

Marshall, N.T., J.A. Banta, L.R. Williams, M.G. Williams, and J.S. Placyk, Jr. 2018. DNA barcoding permits identification of potential fish hosts of unionid freshwater mussels. *American Malacological Bulletin* 36(1):42–56.

Marshall, W.B. 1915. Three new species of *Anodontites* from Brazil. *Proceedings of the United States National Museum* 49(2122):527–529 + 3 plates.

Marshall, W.B. 1930. New land and fresh-water mollusks from South America. *Proceedings of the United States National Museum* 77(2825):1–7 + 2 plates.

von Martens, E. 1855. Ueber die Verbreitung der europäischen Land- und Süsswasser-Gastropoden. Inaugural-Dissertation zur Erlangung der Doktorwürde in der Medicin und Chirurgie. Tübingen. iv + 144 pages. [Also published in the Jahreshefte des Vereins für vaterländische Naturkunde in Württemberg. 11(2):129–272].

von Martens, E. 1865. Ueber die mexikanischen Binnen-Conchylien aus den Sammlungen von Deppe und Uhde im Berliner Museum. *Malakozoologische Blätter* 12:1–78 + 1 plate.

von Martens, E. 1890–1901. Land and freshwater Mollusca. *In* F.D. Godman, and O. Salvin (eds.). *Biologia Centrali-Americana*. Vol. 9, xxviii + 706 pp. + 44 plates. R. H. Porter, London. https://www.biodiversitylibrary.org/bibliography/730

von Martens, E. 1900a. Land and freshwater Mollusca. *In* F.D. Godman, and O. Salvin (eds.). *Biologia Centrali-Americana* Part 154, Fascile 61–62:481–496, plates 29–32. R. H. Porter, London. https://www.biodiversitylibrary.org/bibliography/730

von Martens, E. 1900b. Land and freshwater Mollusca. *In* F.D. Godman, and O. Salvin (eds.). *Biologia Centrali-Americana* Part 155, Fascile 63–65:497–520, plates 33–35. R. H. Porter, London. https://www.biodiversitylibrary.org/bibliography/730

von Martens, E. 1900c. Land and freshwater Mollusca. *In* F.D. Godman, and O. Salvin (eds.). *Biologia Centrali-Americana* Part 156, Fascile 66:521–528. R. H. Porter, London. https://www.biodiversitylibrary.org/bibliography/730

von Martens, E. 1900d. Land and freshwater Mollusca. *In* F.D. Godman, and O. Salvin (eds.). *Biologia Centrali-Americana* Part 157, Fascile 67:529–536, plates 36–37. R. H. Porter, London. https://www.biodiversitylibrary.org/bibliography/730

von Martens, E. 1900e. Land and freshwater Mollusca. *In* F.D. Godman, and O. Salvin (eds.). *Biologia Centrali-Americana* Part 158, Fascile 68–71:537–568, plates 38–39. R. H. Porter, London. https://www.biodiversitylibrary.org/bibliography/730

von Martens, E. 1900f. Land and freshwater Mollusca. *In* F.D. Godman, and O. Salvin (eds.). *Biologia Centrali-Americana* Part 159, Fascile 72–75:569–600, plates 40–41. R. H. Porter, London. https://www.biodiversitylibrary.org/bibliography/730

von Martens, E. 1901a. Land and freshwater Mollusca. *In* F.D. Godman, and O. Salvin (eds.). *Biologia Centrali-Americana* Part 161, Fascile 77–79:609–632, plates 42–43. R. H. Porter, London. https://www.biodiversitylibrary.org/bibliography/730

von Martens, E. 1901b. Land and freshwater Mollusca. *In* F.D. Godman, and O. Salvin (eds.). *Biologia Centrali-Americana* Part 161, Fascile 80–81:633–648, plate 44. R. H. Porter, London. https://www.biodiversitylibrary.org/bibliography/730

von Martens, E. 1901c. Land and freshwater Mollusca. *In* F.D. Godman, and O. Salvin (eds.). *Biologia Centrali-Americana* Part 161, Fascile 82–85:649–674. R. H. Porter, London. https://www.biodiversitylibrary.org/bibliography/730

Martínez-Escarbassiere, R., A. Cabrera, and C.A. Lasso. 2004. Moluscos bivalvos (Unionacea y Mutelacea) de la cuenca del río Orinoco, Venezuela. *Memoria de la Fundación La Salle de Sciencias Naturales* 159–160:283–303.

Martínez-Escarbassiere, R., H. Piñago, and J.-P. Pointier. 2015. Chapter 10.2. Bivalvia. Pp. 170–205 *In* J.-P. Pointier (ed.). *Freshwater Molluscs of Venezuela and Their Medical and Veterinary Importance.* ConchBooks, Harxheim, Germany. 228 pp.

Mather, C.M. 1989. New bivalve records from Lake Texoma, Oklahoma-Texas. *Southwestern Naturalist* 34(4):563–564.

McMurray, S.E., J.S. Faiman, A. Roberts, B. Simmons, and M.C. Barnhart. 2012. *A Guide to Missouri's Freshwater Mussels.* Missouri Department of Conservation, Jefferson City, Missouri. 94 pp.

Meek, S.E. 1902. A contribution to the ichthyology of Mexico. *Field Columbian Museum Zoological Series* 3(6):63–128.

Meek, S.E. 1903. Distribution of the fresh-water fishes of Mexico. *American Naturalist* 37(443):771–784.

Menke, C.T. 1828. *Synopsis methodica Molluscorum generum omnium et specierum earum, quae in Museo Menkeano adservantur; cum synonymia critica et novarum specierum diagosibus.* Henrici Gelpke, Pyrmonti. 91 pp.

Menke, C.T. 1830. *Synopsis methodica Molluscorum generum omnium et specierum earum, quae in Museo Menkeano adservantur; cum synonymia critica et novarum specierum diagnosibus.* Georgi Uslar, Pyrmonti. 68 pp.

Metcalf, A.L. 1974. Fossil and living freshwater mussels (Unionacea) from the Pecos River, New Mexico and Texas. (Abstract). *Bulletin of the American Malacological Union, Inc.* 1973:47–48.

Metcalf, A.L. 1982. Fossil unionacean bivalves from three tributaries of the Rio Grande. Pp. 43–58 *In* J.R. Davis (ed.). *Proceedings of the Symposium on Recent Benthological Investigations in Texas and Adjacent States.* Aquatic Science Section, Texas Academy of Science, Austin, Texas. 278 pp.

Metcalf, A.L., and R. Smartt. 1972. Records of introduced mollusks: New Mexico and western Texas. *The Nautilus* 85(4):144–145.

Miller, R.R. 1966, Geographical distribution of Central American freshwater fishes. *Copeia* 4(4):773–802.

Miller, R.R., and J. Van Conner. 1997. Peces de Catemaco. Pp. 451–456 *In* E. González-Soriano, R. Dirzo, and R.C. Vogt (eds.). *Historia natural de Los Tuxtlas.* Universidad Nacional Autónoma de México, Instituto de Biología: Instituto de Ecología, México. xiv+647 pp.

Miller, R.R., and M.L. Smith. 1986. Chapter 14. Origin and geography of the fishes of central Mexico. Pp. 487–517 *In* C.H. Hocutt, and E.O. Wiley (eds.). *The Zoogeography of North American Freshwater Fishes.* John Wiley & Sons, New York, New York. 866 pp.

Miller, R.R., W.L. Minckley, and S.M. Norris. 2005. *Freshwater Fishes of México.* University of Chicago Press, Chicago, Illinois. xxv+490 pp.

Minckley, W.L., and M.E. Douglas. 1991. Discovery and extinction of western fishes: A blink of eye in geologic time. Pp 7–17 *In* W.L. Minckley, and J.E. Deacon (eds.). *Battle Against Extinction: Native Fish Management in the American West.* University of Arizona Press, Tucson, Arizona. 538 pp.

Minckley, W.L., R.R. Miller, C.D. Barbour, J.J. Schmitter Soto, and S.M. Norris. 2005. Chapter 3. Historical Ichthyology. Pp. 24–47 *In* R.R. Miller, W.L. Minckley, and S.M. Norris (eds.). *Freshwater Fishes of México.* University of Chicago Press, Chicago, Illinois. xxv+490 pp.

Ministério de Medio Ambiente y Recursos Naturales. 2023. Listado Oficial de Especies de Vida Silvestre Amenazadas o en Peligro de Extinctión. *Diario Oficial* 441 (Acuerdo 257):36–58.

Mock, K.E., J.C. Brim Box, J.P. Chong, J. Furnish, and J.K. Howard. 2013. Comparison of population genetic patterns in two widespread freshwater mussels with contrasting life histories in western North America. *Molecular Ecology* 22(24):6060–6073.
Modell, H. 1942. Das natürliche system der najaden. *Archiv für Molluskenkunde* 74(5–6):161–191.
Modell, H. 1945. Die Anodontinae, Ortm. emend. (Najad., Mollusca). Eine Studie über die Zusammenhänge von Klimazonen und Entwicklungsgeschichte (Klimazonentheorie). *Jenaische Zeitschrift für Medizin und Naturwissenschaft* 78:58–100.
Modell, H. 1964. Das natürliche system der najaden. 3. *Archiv für Molluskenkunde* 93:71–126.
Montenegro-Guillén, S. 2003. Lake Cocibolca/Nicaragua. Lake Basin Management Initiative: Experience and Lessons Learned Brief. LBMI Regional Workshop for Europe, Central Asia and the Americas. Saint Michael's College, Colchester, Vermont. 26 pp.
Montero-Martínez, M.J., and O.F. Ibáñez-Hernández. 2017. *La cuenca del Río Conchos: una mirada desde las ciencias ante el cambio climático*. Instituto Mexicano de Tecnología del Agua, Morelos, Mexico. 267 pp.
Morelet, A. 1849. Testacea novissima insulae Cubanae, et Amer. Centralis. *Librairie de L'Académie Nationale de Medecine, Paris Part* 1:1–31.
Morelet, A. 1851. Testacea novissima insulae Cubanae, et Amer. Centralis. *Librairie de L'Académie Nationale de Medecine, Paris Part* 2:1–30.
Morelet, A. 1857a. *Voyage dans l'Amérique Centrale, l'Ile de Cuba et le Yucatan*. Tome Premier. Gide et J. Baudry, Paris. Vol. 1. 337 pp.
Morelet, A. 1857b. *Voyage dans l'Amérique Centrale, l'Ile de Cuba et le Yucatan*. Tome Deuxiéme. Gide et J. Baudry, Paris. Vol. 2. 323 pp.
de Morretes, F.L. 1949. Ensaio de catálogo dos moluscos do Brasil. *Arquivos do Museu Paranaense Curitiba* 7:5–216.
Morrison, J.P.E. 1968. Collecting Mexican freshwater mussels. *American Malacological Union, Inc. Annual Reports* 1967(34):50–51.
Morrison, J.P.E. 1969. The earliest names for North American naiads. *American Malacological Union, Inc. Annual Reports* 1969(36):22–24.
Morrison, J.P.E. 1976. Species of the genus *Uniomerus*. *Bulletin of the American Malacological Union, Inc* 1976:10–11.
Myers, T.L. 2009. Pre-historical, historical, and recent distribution of freshwater mussels (Unionidae: *Anodonta*) in the Colorado River and Río Yaqui Basins (with notes on Guzmán Basin, Río Sonoyta, Río Asunción/Magdalena, and Rio Grande). Final Report to the Arizona Game and Fish Department 83 pp + appendices A–E.
Nagel, K.-O. 2025. Type catalog of the freshwater mussels (Mollusca: Bivalvia: Unionida) in the collection of the Senckenberg Research Institute and Natural History Museum in Frankfurt am Main, Germany. *Archiv für Molluskenkunde* 154(1):79–247.
Naranjo-García, E. 2003a. Historia de la malacología en México con énfasis en la malacología continental. *Revista de Biología Tropical* 51(Supplement 3):463–471.
Naranjo-García, E. 2003b. Moluscos continentales de México: Dulceacuícolas. *Revista de Biología Tropical* 51(Supplement 3):495–505.
NatureServe. 2024. *NatureServe Network Biodiversity Location Data Accessed through NatureServe Explorer*. NatureServe, Arlington, Virginia. Available https://explorer.natureserve.org/. (Accessed: February 14, 2024).
Neck, R.W. 1984. Restricted and declining non-marine molluscs of Texas. *Technical Series* 34:1–17. Texas Parks and Wildlife Department, Austin, Texas.
Neck, R.W. 1987. Notes on the R.D. Camp Mollusk Collection in the Corpus Christi Museum. *Malacology Data Net* 1(6):145–160.
Neck, R.W., and A.L. Metcalf. 1988. Freshwater bivalves of the Lower Rio Grande, Texas. *Texas Journal of Science* 40(3):259–268.
Nico, L.G., J.D. Williams, and H.L. Jelks. 2005. *Black Carp: Biological Synopsis and Risk Assessment of an Introduced Fish*. American Fisheries Society, Special Publication 32, Bethesda, Maryland. 337 pp.
Nördlinger, C. 1899. Obituary notice of the late Joseph Charles Hippolyte Crosse. *Journal of Conchology* 9(7):212–215.
Obregón-Barboza, H., S. Contreras-Balderas, and M.L. Lozano-Vilano. 1994. The fishes of northern and central Veracruz, Mexico. *Hydrobiologia* 286:79–95.
O'Brien, C., A. Maine, D. Nez, and J. Brim Box. 2019. A comparison of glochidial shells of the freshwater mussels *Anodonta californiensis*, *Anodonta kennerlyi*, *Anodonta nuttalliana*, and *Anodonta oregonensis*. *Freshwater Mollusk Biology and Conservation* 22(1):20–24.

O'Dee, S.H., and G.T. Watters. 2000. New or confirmed host identifications for ten freshwater mussels. Pp. 77–82 *In* R.A. Tankersley, D.I. Warmoltz, G.T. Watters, B.J. Armitage, P.D. Johnson, and R.S. Butler (eds.). *Freshwater Mollusk Symposia Proceedings.* Ohio Biological Survey, Columbus, Ohio. xxi +274 pp.

Olson, D., E. Dinerstein, P. Canevari, I. Davidson, G. Castro, V. Morisset, R. Abell, and E. Toledo (eds.). 1998. *Freshwater Biodiversity of Latin America and the Caribbean: A Conservation Assessment.* Biodiversity Support Program, Washington D.C. xvii+61 pp.

Ortmann, A.E. 1910. The systematic position of the unionid-genus *Parreysia. The Nautilus* 23(11):139–142.

Ortmann, A.E. 1912. Notes upon the families and genera of the najades. *Annals of the Carnegie Museum* 8(2):222–365+3 plates.

Ortmann, A.E. 1915. Studies in najades. (cont.). *The Nautilus* 28(12):141–143.

Ortmann, A.E. 1921a. Some Central American species of naiades, belonging or allied to the genus *Elliptio. The Nautilus* 35(1):24–27.

Ortmann, A.E. 1921b. South American naiades; a contribution to the knowledge of the fresh-water mussels of South America. *Memoirs of the Carnegie Museum* 8(3):451–670+plates 34–48.

Ortmann, A.E. 1925. The naiad-fauna of the Tennessee River system below Walden Gorge. *American Midland Naturalist* 9(8):321–372.

Ortmann, A.E., and B. Walker. 1922. On the nomenclature of certain North American naiades. Occasional Papers of the Museum of Zoology, University of Michigan No. 112. 75 pp.

Paetel, F. 1890. *Catalog der Conchylien-Sammlung, mit Hinzufügung der bis jetzt publicirten recenten Arten, sowie der ermittelten Synonyma. Dritte Abtheilung: Die Acephalen und die Branchiopoden.* Verlag von Gebrüder Paetel, Berlin. Vol. 3. 256 pp.+index.

Parodiz, J.J. 1967. Types of North American Unionidae in the collection of the Carnegie Museum. *Sterkiana* 28:21–30.

Parodiz, J.J., and A.A. Bonetto. 1963. Taxonomy and zoogeographic relationships of the South American naiades (Pelecypoda: Unionacea and Mutelacea). *Malacologia* 1(2):179–213.

Parodiz, J.J., and J.J. Tripp. 1988. Types of Mollusca in the collection of the Carnegie Museum of Natural History. Part I. Bivalvia and Gastropoda (Prosobranchia and Opistobranchia). *Annals of the Carnegie Museum* 57(5):111–154.

Pease, A.A., K.A. Capps, M.M. Castillo, D.A. Hendrickson, M. Mendoza-Carranza, R. Rodiles-Hernández, C. Avila, and T. Contreras-MacBeath. 2023. Chapter 12. Rivers of Mexico. Pp. 974–1024 *In* M.D. Delong, T.D. Jardine, A.C. Benke, and C.E. Cushing (eds.). Rivers of North America. Second Edition. Academic Press, Elsevier Inc., Amsterdam. i–iv+1088 pp.

Pereira, D., M.C.D. Mansur, L.D.S. Duarte, A.S. de Oliveira, D.M. Pimpão, C.T. Callil, C. Ituarte, E. Parada, S. Peredo, G. Darrigran, F. Scarabino, C. Clavijo, G. Lara, I.C. Miyahira, M.T.R. Rodriguez, and C. Lasso. 2014. Bivalve distribution in hydrographic regions in South America: Historical overview and conservation. *Hydrobiologia* 735:15–44.

Pérez-Miranda, F., O. Mejia, B. López, and O. Rican. 2020. Molecular clocks, biogeography and species diversity in *Herichthys* with evaluation of the role of Punta del Morro as a vicariant brake along the Mexican Transition Zone in the context of local and global time frame of cichlid diversification. *PeerJ* 8:e8818. 27 pp.

Perrilliat, M.C., F.J. Vega, B. Espinosa, and E. Naranjo-García. 2008. Late Cretaceous-Early Tertiary freshwater gastropods from northeastern Mexico. *Journal of Paleontology* 82(2):255–266.

Pfeiffer, J.M., A.E. Sharpe, N.A. Johnson, K.F. Emery, and L.M. Page. 2018. Molecular phylogeny of the Nearctic and Mesoamerican freshwater mussel genus *Megalonaias. Hydrobiologia* 811(1):139–151.

Pfeiffer, J.M., C.L. Atkinson, A.E. Sharpe, K.A. Capps, K.F. Emery, and L.M. Page. 2019a. Phylogeny of Mesoamerican freshwater mussels and a revised tribe-level classification of the Ambleminae. *Zoologica Scripta* 48(1):106–117.

Pfeiffer, J.M., III, and D.L. Graf. 2013. Re-analysis confirms the polyphyly of *Lamprotula* Simpson, 1900 (Bivalvia: Unionidae). *Journal of Molluscan Studies* 79(3):249–256.

Pfeiffer, J.M., III, and D.L. Graf. 2015 Evolution of bilaterally asymmetrical larvae in freshwater mussels (Bivalvia: Unionoida: Unionidae). *Zoological Journal of the Linnean Society* 175(2):307–318.

Pfeiffer, J.M., J.W. Breinholt, and L.M. Page. 2019b. Unioverse: A phylogenomic resource for reconstructing the evolution of freshwater mussels (Bivalvia, Unionoida). *Molecular Phylogenetics and Evolution* 137:114–126.

Pfeiffer, J.M., III, N.A. Johnson, C.R. Randklev, R.G. Howells, and J.D. Williams. 2016. Generic reclassification and species boundaries in the rediscovered freshwater mussel '*Quadrula*' *mitchelli* (Simpson in Dall, 1896). *Conservation Genetics* 17(2):279–292.

Philippi, R.A. 1848. *Unio* Tab V. Abbildungen und Beschriebungen neuer oder wenig gekannter Conchylien. Cassel. *Theodor Fischer* 3:79–81, plate 5.

Philippi, R.A. 1849 *Unio* Tab VI. Abbildungen und Beschriebungen neuer oder wenig gekannter Conchylien. Cassel. *Theodor Fischer* 3:27–28, plate 6.

Pieri, A.M., K. Inoue, N.A. Johnson, C.H. Smith, J.L. Harris, C. Robertson, and C.R. Randklev. 2018. Molecular and morphometric analyses reveal cryptic diversity within freshwater mussels (Bivalvia: Unionidae) of the western Gulf coastal drainages of the USA. *Biological Journal of the Linnean Society* 124(2):261–277.

Pilsbry, H.A. 1891. Land and fresh water mollusks collected in Yucatan and Mexico. *Proceedings of the Academy of Natural Sciences of Philadelphia* 43:310–334+2 plates.

Pilsbry, H.A. 1900. Notes on some southern Mexican shells. *The Nautilus* 13(12):139–141.

Pilsbry, H.A. 1904. Mexican land and freshwater mollusks. *Proceedings of the Academy of Natural Sciences of Philadelphia* 55:761–789+8 plates.

Pilsbry, H.A. 1909. Mollusks from northeastern Mexico. *The Nautilus* 23(4):45–49+1 plate.

Pilsbry, H.A. 1910. Unionidae of the Panuco River system, Mexico. *Proceedings of the Academy of Natural Sciences of Philadelphia* 61(3):532–539+3 plates.

Pilsbry, H.A. 1920a. Mollusks from Lake Chapala, State of Jalisco and vicinity. *Proceedings of the Academy of Natural Sciences of Philadelphia* 72(2):192–194.

Pilsbry, H.A. 1920b. Costa Rican land and freshwater mollusks. *Proceedings of the Academy of Natural Sciences of Philadelphia* 72(1):2–10.

Pilsbry, H.A., and A.A. Hinkley. 1910. Melaniidae of the Panuco River system, Mexico. *Proceedings of the Academy of Natural Sciences of Philadelphia* 61(3):519–531+2 plates.

Pilsbry, H.A., and L.S. Frierson. 1908. Description of a Mexican *Lampsilis*. *The Nautilus* 22(8):81–82+plate 12.

Plummer, M.V., and J.W. Goy. 1997. *Sternotherus odoratus* (common musk turtle) mortality. *Herpetological Review* 28:88.

Pointier, J.-P. 2015. *Freshwater Molluscs of Venezuela and their Medical and Veterinary Importance*. ConchBooks, Harxheim, Germany. 228 pp.

Poulson, C.A. 1832. *A Monograph of the Fluviatile Bivalve Shells of the River Ohio Containing Twelve Genera and Sixty-Eight Species*. [Translation of Rafinesque, 1820]. J. Dobson, Philadelphia, Pennsylvania. 72 pp.

Powis, T. 2004. Ancient lowland Maya utilization of freshwater pearly mussels (*Nephronaias* spp.). Pp. 125–141 *In* K.F. Emery (ed.). *Maya Zooarchaeology: New Directions in Method and Theory*. Cotsen Institute of Archaeology: University of California, Los Angeles. Monograph 51. xv+310 pp.

Poyard, C. 1899. Hippolyte Crosse, 1826–1898. *Journal de Conchyliologie* 47(1):4–10+portrait.

Preece, E.P., M. Bryan, S.M. Mapes, C. Wademan, and R. Dorazio. 2021. Monitoring from freshwater mussel presence in rivers using environmental DNA. *Environmental DNA* 3(3):591–604.

Quintana, Y., and K. Inoue. 2024. Freshwater mussel project in Central America. *Tentacle* 32:17–18.

Rafinesque, C.S. 1818. Discoveries in natural history, made during a journey through the Western region of the United States. *American Monthly Magazine and Critical Review* 3:354–356.

Ramírez, C., R. Barba, J.M. Caspeta, F. Córdova, H. Espinosa, S. Larre, E.F. Mendoza-Franco, N. Mercado-Silva, E. Moreno, E. Naranjo-García, N. Rosales-Quintero, G. Salgado-Maldonado, J.L. Villalobos, X.G. Vital, and F. Álvarez. 2022. Biota acuática de la cuenca media del río Lacantún, Chiapas y la importancia del monitoreo de largo plazo. *Aquatic biota of the middle basin of the Lacantún River, Chiapas and the importance of long-term monitoring. Revista Mexicana de Biodiversidad* 93(2022):e934844. 35 pp.

Ramírez, R., C. Paredes, and J. Arenas. 2003. Moluscos del Perú. *Revista de Biología Tropical* 51:25–284.

Randklev, C.R., E.T. Tsakiris, M.S. Johnson, J. Skorupski, L.E. Burlakova, J. Groce, and N. Wilkins. 2013. Is False Spike, *Quadrula mitchelli* (Bivalvia: Unionidae), extinct? First account of a very recently deceased individual in over thirty years. *Southwestern Naturalist* 58(2):247–249.

Randklev, C.R., M.S. Johnson, E.T. Tsakiris, S. Rogers-Oetker, K.J. Roe, J.L. Harris, S.E. McMurray, C. Robertson, J. Groce, and N. Wilkins. 2012. False spike, *Quadrula mitchelli* (Bivalvia: Unionidae), is not extinct: First account of a live population in over 30 years. *American Malacological Bulletin* 30(2):327–328.

Randklev, C.R., T. Miller, M. Hart, J. Morton, N.A. Johnson, K. Skow, K. Inoue, E.T. Tsakiris, S. Oetker, R. Smith, C. Robertson, and R. Lopez. 2018. A semi-arid river in distress: Contributing factors and recovery solutions for three imperiled freshwater mussels (family Unionidae) endemic to the Rio Grande basin in North America. *Science of the Total Environment* 631–632:733–744.

Rangaswami, X.L., A.M. Goldsmith, J.M. Khan, C.R. Robertson, R.R. Lopez, and C.R. Randklev. 2023. Thermal tolerances of *Popenaias popeii* (Texas hornshell) and its host fish from the Rio Grande Basin, Texas. *Scientific Reports* 13(4603):1–11.

Reeve, L. 1842. *Conchologia Systematica, or Complete System of Conchology*. Longman, Brown, Green and Longmans, London, Paternoster Row 1(4):105–144+plates 81–104.

Reeve, L.A. 1864. Monograph of the genus *Unio*. *In* L. Reeve, and G.B. Sowerby (eds.). *Conchologia Iconica* 16:[unpaginated]+plates 1–18.

Reeve, L.A. 1865. Monograph of the genus *Unio*. *In* L. Reeve, and G.B. Sowerby (eds.). *Conchologia Iconica* 16:[unpaginated]+plates 19–30.

Regan, C.T. 1906–1908. Pisces. *In* F.D. Godman, and O. Salvin (eds.). *Biologia Centrali-Americana*. Vol. 8, xxxiii+203 pp.+26 plates and 2 maps.

Richards, H.G., and H.J. Boekelman. 1937. Shells from Maya excavations in British Honduras. *American Antiquity* 3(2):166–169.

Rodiles-Hernández, R. 2004. Diversidad de peces continentales en Chiapas. Pp. 141–160 *In* M. González-Espinosa, N. Ramírez-Marcial, and L. Ruíz-Montoya (eds.). *Diversidad Biológica en Chiapas*. Plaza y Valdés, ECOSUR, COCYTECH, Distrito Federal, Mexico. 484 pp.

Rodiles-Hernández, R., D.A. Hendrickson, J.G. Lundberg, and J.M. Humphries. 2005. *Lacantunia enigmatica* (Teleostei: Siluriformes) a new and phylogenetically puzzling freshwater fish from Mesoamerica. *Zootaxa* 1000:1–24.

Roe, K.J., and C. Lydeard. 1998. Species delineation and the identification of evolutionary significant units: Lessons from the freshwater mussel genus *Potamilus* (Bivalvia: Unionidae). *Journal of Shellfish Research* 17(5):1359–1363.

Roe, K.J., and W.R. Hoeh. 2003. Systematics of freshwater mussels (Bivalvia: Unionoida). Pp. 91–122 *In* C. Lydeard, and D.R. Lindberg (eds.). *Molecular Systematics and Phylogeography of Mollusks*. Smithsonian Books, Washington and London. xv+312 pp.

Rumi, A., D.E. Gutiérrez Gregoric, V. Núñez, and G.A. Darrigran. 2008. Malacología Latinoamericana. Moluscos de agua dulce de Argentina. Latin American Malacology. Freshwater mollusks from Argentina. *Revista de Biología Tropical* 56(1):77–111.

Rypel, A.L., W.R. Haag, and R.H. Findlay. 2008. Validation of annual growth rings in freshwater mussel shells using cross dating. *Canadian Journal of Fisheries and Aquatic Sciences* 65(10):2224–2232.

Sabaj, M.H. 2020. Codes for natural history collections in ichthyology and herpetology. *Copeia* 108(3):593–669.

San Miguel, E., S. Monserrat, C. Fernandez, R. Amaro, M. Hermida, P. Ondina, and C.R. Altaba. 2004. Growth models and longevity of freshwater pearl mussels (*Margaritifera margaritifera*) in Spain. *Canadian Journal of Zoology* 82:1370–1379.

Saxon, R., C.R Randklev, K.J Roe, C.R Robertson, and C.H Smith. 2023. A phylogeographic assessment redefines the distributions of *Truncilla* species (Bivalvia: Unionidae) in Texas. *Journal of Molluscan Studies* 89(3):1–12.

Say, T. 1840. *Descriptions of Some New Terrestrial and Fluviatile Shells of North America. 1829, 1830, 1831*. New Harmony, Indiana. 26 pp.

van der Schalie, H. 1940. Notes on Mollusca from Alta Vera Paz, Guatemala. Occasional Papers of the Museum of Zoology, University of Michigan No. 413. 11 pp.

Schönhuth, S., and I. Doadrio. 2003. Phylogenetic relationships of Mexican minnows of the genus *Notropis* (Actinopterygii, Cyprinidae). *Biological Journal of the Linnean Society* 80(2):323–337.

Semarnat (Secretaría del Medio Ambiente y Recursos Naturales). 2019. Modificación del Anexo Normativo III de la Norma Oficial Mexicana NOM-059-SEMARNAT-2010. Protección ambiental-Especies nativas de México de flora y fauna silvestres-Categorías de riesgo y especificaciones para su inclusión, exclusión o cambio-Lista de especies en riesgo. Diario Oficial de la Federación 30 de diciembre de 2010, Segunda Sección, México.

Serrand, N., and K.S. Cummings. 2014. Occurrences of exogenous freshwater mussel shells (Bivalvia: Unionida) during the Precolumbian ceramic age of the Lesser Antilles. Pp. 65–76 *In* K. Szabó, C. Dupont, V. Dimitrijević, L. Gómez Gastélum, and N. Serrand (eds.). *Archaeomalacology: Shells in the Archaeological Record*. BAR International Series 2666. Archaeopress, Oxford.

Sharpe, A.E., T. Inomata, D. Triadan, M. Burham, J. MacLellan, J. Munson, and F. Pinzón. 2020. The Maya Preclassic to Classic transition observed through faunal trends from Ceibal, Guatemala. *PLoS ONE* 15(4):e0230892. 42 pp.

Shively, S.H., and M.F. Vidrine. 1984. Fresh-water mollusks in the alimentary tract of a Mississippi map turtle. *Proceedings of the Louisiana Academy of Sciences* 47:27–29.

Sierwald, P., E.K. Shea, R. Bieler, and G. Rosenberg. 2018. Mobilizing mollusks: Status update on mollusk collections in the U.S.A. and Canada. *American Malacological Bulletin* 36(2):177–214.

Silva-Souza, A.T., and J.C. Eiras. 2002. The histopathology of the infection of *Tilapia rendalli* and *Hypostomus regani* (Osteichthyes) by lasidium larvae of *Anodontites trapesialis* (Mollusca, Bivalvia). *Memórias do Instituto Oswaldo Cruz, Rio de Janeiro* 97(3):431–433.

Simone, L.R.L. 1994. Anatomical characters and systematics of *Anodontites trapesialis* (Lamarck, 1819) from South America (Mollusca, Bivalvia, Unionoida, Muteloidea). *Studies on Neotropical Fauna and Environment* 29(3):169–185.

Simone, L.R.L. 2006. *Land and Freshwater Molluscs of Brazil: An Illustrated Inventory of the Brazilian Malacofauna, Including Neighboring Regions of South America, Respect to the Terrestrial and Freshwater Ecosystems*. EGB, Fapesp., Sao Paulo. 390 pp.

Simpson, C.T. 1894. Types of *Anodonta dejecta* rediscovered. *The Nautilus* 8(5):52.

Simpson, C.T. 1896. The classification and geographical distribution of the pearly fresh-water mussels. *Proceedings of the United States National Museum* 18(1068):295–343+1 map.

Simpson, C.T. 1900. Synopsis of the naiades, or pearly fresh-water mussels. *Proceedings of the United States National Museum* 22(1205):501–1044.

Simpson, C.T. 1914. *A Descriptive Catalogue of the Naiades, or Pearly Fresh-Water Mussels. Parts I–III.* Bryant Walker, Detroit, Michigan xii+1540 pp.

Singley, J.A. 1893. Contributions to the natural history of Texas. Part I, Texas Mollusca. Geological Survey of Texas, Fourth Annual Report 1892:299–343.

Slapcinsky, J. 2017. In Memorium–Fred G. Thompson. *Tentacle* 25:2–3.

Smith, C.H., J.M. Pfeiffer, and N.A. Johnson. 2020. Comparative phylogenomics reveal complex evolution of life history strategies in a clade of bivalves with parasitic larvae (Bivalvia: Unionoida: Ambleminae). *Cladistics* 36(5):505–520.

Smith, C.H., N.A. Johnson, J.M. Pfeiffer, and M.M. Gangloff. 2018. Molecular and morphological data reveal non-monophyly and speciation in imperiled freshwater mussels (*Anodontoides* and *Strophitus*). *Molecular Phylogenetics and Evolution* 119:50–62.

Smith, C.H., N.A. Johnson, K. Havlik, R.D. Doyle, and C.R. Randklev. 2021. Resolving species boundaries in the critically imperiled freshwater mussel species, *Fusconaia mitchelli* (Bivalvia: Unionidae). *Journal of Zoological Systematics and Evolutionary Research* 59(1):60–77.

Smith, C.H., N.A. Johnson, K. Inoue, R.D. Doyle, and C.R. Randklev. 2019. Integrative taxonomy reveals a new species of freshwater mussel, *Potamilus streckersoni* sp. nov. (Bivalvia: Unionidae): Implications for conservation and management. *Systematics and Biodiversity* 17(4):331–348.

Smith, D.G., B.K. Lang, and M.E. Gordon. 2003. Gametogenetic cycle, reproductive anatomy, and larval morphology of *Popenaias popeii* (Unionoida) from the Black River, New Mexico. *Southwestern Naturalist* 48(3):333–340.

Smith, E.A. 1882. On the freshwater shells of Australia. *Journal of the Linnean Society London. Zoology* 16(92):255–317+3 plates.

Smith, E.A., and H.W. England. 1937. Martini and Chemnitz (Kuester's Edition) Systematisches Conchylien-Cabinet, 1837–1918. *Journal of the Society for the Bibliography of Natural History* 1(4):89–99.

Smith, H.M. 1899. The mussel fishery and pearl-button industry of the Mississippi River. *Bulletin of the U.S. Fish Commission* 18(1898):289–314+9 plates.

Smith, J.M., C. Bogan, and A.E. Bogan. 2007. Museum of Fluviatile Mollusks collection donated to North Carolina State Museum of Natural Sciences, Raleigh. *Ellipsaria* 9(3):12–13.

Smith, M.L., and R.R. Miller. 1986. Chapter 13. The evolution of the Rio Grande basin as inferred from its fish fauna. Pp. 457–485 *In* C.H. Hocutt, and E.O. Wiley (eds.). *The Zoogeography of North America Freshwater Fishes*. John Wiley & Sons, New York. 866 pp.

Solem, A. 1967a. New molluscan taxa and scientific writings of Fritz Haas. *Fieldiana Zoology* 53(2):71–144.

Solem, A. 1967b. The two careers of Fritz Haas. *Bulletin Field Museum of Natural History* 38(11):2–5.

Sowerby, G.B., II. 1839. *A Conchological Manual*. Mr. G.B. Sowerby, Bookseller, Bloomsbury, London. 130 pp.+488 figures.

Sowerby, G.B., II. 1866. Monograph of the genus *Unio*. *In* L. Reeve, and G.B. Sowerby (eds.). *Conchologia Iconica* 16:[unpaginated]+plates 31–54.

Sowerby, G.B., II. 1867. Monograph of the genus *Anodon*. *In* L. Reeve, and G.B. Sowerby (eds.). *Conchologia Iconica* 17:[unpaginated]+plates 2–19.

Sowerby, G.B., II. 1868. Monograph of the genus *Unio*. *In* L. Reeve, and G.B. Sowerby (eds.). *Conchologia Iconica* 16:[unpaginated]+plates 1, 61–96.

Sowerby, G.B., II. 1869. Monograph of the genus *Anodon*. *In* L. Reeve, and G.B. Sowerby (eds.). *Conchologia Iconica* 17:[unpaginated]+plates 21–24.

Sowerby, G.B., II. 1870. Monograph of the genus *Anodon*. *In* L. Reeve, and G.B. Sowerby (eds.). *Conchologia Iconica* 17:[unpaginated]+plates 1, 25–37.

Starobogatov, Y.I. 1970. *Fauna Mollyuskov i Zoogeograficheskoe Raionirovanie Kontinental'nykh Vodoemov Zemnogo Shara*. Akademiya Nauk SSSR, Leningrad. 371 pp.

Stern, E.M., and D.L. Felder. 1978. Identification of host fishes for four species of freshwater mussels (Bivalvia: Unionidae). *American Midland Naturalist* 100(1):233–236.

Strayer, D.L., and D.R. Smith. 2003. *A Guide to Sampling Freshwater Mussel Populations*. American Fisheries Society, Monograph 8. Bethesda, Maryland. 103 pp.

Strayer, D.L., J.A. Downing, W.R. Haag, T.L. King, J.B. Layzer, T.J. Newton, and S.J. Nichols. 2004. Changing perspectives on pearly mussels, North America's most imperiled animals. *BioScience* 54(5):429–439.

Strecker, J.K., Jr. 1929. Lorraine Screven Frierson, a southern conchologist. *Contribution of the Baylor University Museum* 21:1–12+portrait.

Strecker, J.K., Jr. 1931. The distribution of the naiades or pearly fresh-water mussels of Texas. *Baylor University Museum Special Bulletin* 2:1–71.

Strenth, N.E., R.G. Howells, and A.C. Sandoval. 2004. New records of the Texas hornshell *Popenaias popeii* (Bivalvia: Unionidae) from Texas and northern Mexico. *Texas Journal of Science* 56(3):223–230.

Surber, T. 1912. *Identification of the Glochidia of Freshwater Mussels*. Report and Special Papers of the U.S. Bureau of Fisheries. [Issued separately as U.S. Bureau of Fisheries Document 771]. U.S. Government Printing Office. 1–10 pp.+3 plates.

Swainson, W.A. 1840. *A Treatise on Malacology: Or the Natural Classification of Shells and Shell-fish*. Printed for Longman, Orme, Brown, Green, & Longman, Peternoster Row, London. 419 pp.

Swainson, W.A. 1841. *Exotic conchology: Or figures and Descriptions of Rare, Beautiful, or Undescribed Shells*. Second Edition. Edited by Sylvanus Hanley. Henry G. Bohn, London. 39 pp. + 48 plates.

Tate, R. 1870. On the land and fresh-water Mollusca of Nicaragua. *American Journal of Conchology* 5(3):151–162+plate 16.

Taylor, D.W. 1966. A remarkable snail fauna from Coahuila, México. *The Veliger* 9(2):152–228.

Taylor, D.W. 1967. Freshwater mollusks collected by the United States and Mexican Boundary Surveys. *The Veliger* 10(2):152–158.

Taylor, D.W. 1981. Freshwater mollusks of California: A distributional checklist. *California Fish and Game* 67(3):140–163.

Taylor, D.W. 1993. Moluscos dulceacuícolas de Costa Rica: Introducción y lista preliminar. *Revista de Biología Tropical* 41(3):653–655.

Teiga-Teixeira, J., E. Froufe, A. Gomes-dos-Santos, A.E. Bogan, A.Y. Karatayev, L.E. Burlakova, D.C. Aldridge, I.N. Bolotov, I.V. Vikhrev, A. Teixeira, S. Varandas, D.T. Zanatta, and M. Lopes-Lima. 2020. Complete mitochondrial genomes of the freshwater mussels *Amblema plicata* (Say, 2817), *Pleurobema oviforme* (Conrad, 1834), and *Popenaias popeii* (Lea, 1857) (Bivalvia: Unionidae: Ambleminae). *Mitochondrial DNA Part B. Resources* 5(3):2977–2979.

Thiele, J. 1934. *Handbuch der systematischen Weichtierkunde*. Teil 3. Scaphopoda; Bivalvia; Cephalopoda. pp. 779–1022.

Thompson, F.G. 1968. *The Aquatic Snails of the Family Hydrobiidae of Peninsular Florida*. University of Florida Press, Gainesville, Florida. xv+268 pp., figs. 1–69.

Thompson, F.G. 2011. An annotated checklist and bibliography of the land and freshwater snails of Mexico and Central America. *Bulletin of the Florida Museum of Natural History* 50:1–299.

Thorp, J.H., and A.P. Covich (eds.). 2010. *Ecology and Classification of North American Freshwater Invertebrates*. Third Edition. Academic Press, Elsevier Inc., Amsterdam. xiv+1021 pp.

Tiemann, J.S., K. Inoue, J.A. Rodríguez–Pineda, M. Hart, K.S. Cummings, E. Naranjo-García, and C.R. Randklev. 2020. Status of freshwater mussels (Unionidae) of the Río Conchos basin, Chihuahua, Mexico. *Southwestern Naturalist* 64(3/4):180–186.

Tomlin, J.R., and B. Le. 1947. Editorial notes. *Journal of Conchology* 22:287–290.

Torres, S., L. Cao, D.E. Gutiérrez Gregoric, M. de Lucía, F. Brea, and G. Darrigran. 2018. Distribution of the Unionida (Bivalvia, Paleoheterodonta) from Argentina and its conservation in the Southern Neotropical Region. *PLoS ONE* 13(9):e0203616. 15 pp.

Turgeon, D.D., A.E. Bogan, E.V. Coan, W.K. Emerson, W.G. Lyons, W.L. Pratt, C.F.E. Roper, A. Scheltema, F.G. Thompson, and J.D. Williams. 1988. *Common and Scientific Names of Aquatic Invertebrates from the United States and Canada: Mollusks*. American Fisheries Society, Special Publication 16. Bethesda, Maryland. viii+277 pp.+12 plates.

Turgeon, D.D., J.F. Quinn, Jr., A.E. Bogan, E.V. Coan, F.G. Hochberg, W.G. Lyons, P.M. Mikkelsen, R.J. Neves, C.F.E. Roper, G. Rosenberg, B. Roth, A. Scheltema, F.G. Thompson, M. Vecchione, and J.D. Williams. 1998. *Common and Scientific Names of Aquatic Invertebrates from the United States and Canada: Mollusks*. Second Edition. American Fisheries Society, Special Publication 26. Bethesda, Maryland. ix+526 pp.

U.S. Fish and Wildlife Service. 2023. Species status assessment report for two Rio Grande mussels: Salina Mucket (*Potamilus metnecktayi*), Mexican Fawnsfoot (*Truncilla cognata*). February 2023. U.S. Fish and Wildlife Service Region 2 Albuquerque, NM. x+99 pp.

Utterback, W.I. 1915. The naiades of Missouri. IV. *American Midland Naturalist* 4(6):244–273.

Utterback, W.I. 1916. The naiades of Missouri. VIII. *American Midland Naturalist* 4(10):432–464.

Valenciennes, A. 1827. Coquilles fluviatiles bivalves du Nouveau-Continent, recueilles pendant le voyage de MM. de Humboldt et Bonpland, et décrites par A. Valenciennes. Pp. 225–237+plates 50, 52–54. *In* A. de Humboldt, and A. Bonpland (eds.). *Recueil d'Observations de Zoologie et d'Anatomie Comparée, faites dans l'Océan Atlantique, dans l'Intérieur du Nouveau Continent et dans la Mer du Sud pendant les années 1799, 1800, 1801, 1802 et 1803*. J. Smith and Gide, Paris. Band Pt. 2, Vol. 2.

Vaughn, C.C., C.M. Taylor, and K.J. Eberhard. 1997. A comparison of the effectiveness of timed searches vs. quadrat sampling in mussel surveys. Pp. 157–162 *In* K.S. Cummings, A.C. Buchanan, C.A. Mayer, and T.J. Naimo (eds.). *Conservation and Management of Freshwater Mussels II: Initiatives for the Future*. Proceedings of a UMRCC Symposium, 16–18 October 1995, St. Louis, Missouri. Upper Mississippi River Conservation Committee, Rock Island, Illinois.

Vega, F.J., E. Naranjo-García, M. Carolina Aguillón, and D. Posada-Martínez. 2019. Additions to continental gastropods from the Upper Cretaceous and Paleocene of NE Mexico. *Boletín de la Sociedad Geológica Mexicana* 71(1):169–191.

Vidrine, M.F., and K.B. Schwartz. 1978. Lorraine Screven Frierson – Naturalist. *Louisiana Renaissance* 1(2):8–9, 35–36+portrait.

Villalobos, C.R., and R.A. Cruz. 1984a. Biología de *Glabaris luteolus* (Mycetopodidae: Bivalvia). I. Distribución de tamaño, crecimiento y mortalidad en Cañas, Guanacaste, Costa Rica. *Revista de Biología Tropical* 32(1):29–34.

Villalobos, C.R., and R.A. Cruz. 1984b. Biología de *Glabaris luteolus* (Mycetopodidae: Bivalvia). III. Distribución de tamaño, crecimiento y mortalidad en 28 Millas, Limón, Costa Rica. *Revista de Biología Tropical* 32(1):57–60.

Wächtler, K., M.C.D. Mansur, and T. Richter. 2001. Larval types and early postlarval biology in naiads (Unionoida). Pp. 93–125 *In* G. Bauer and K. Wächtler (eds.). *Ecology and Evolution of the Freshwater Mussels Unionoida*. Springer-Verlag, Berlin, Ecological Studies Vol. 145. 394 pp.

Walker, J.M., J.P. Curole, D.E. Wade, E.G. Chapman, A.E. Bogan, G.T. Watters, and W.R. Hoeh. 2006. Taxonomic distribution and phylogenetic utility of gender-associated mitochondrial genomes in the Unionoida (Bivalvia). *Malacologia* 48(1–2):265–282.

Watters, G.T. 1994. An annotated bibliography of the reproduction and propagation of the Unionoidea (Primarily of North America). *Ohio Biological Survey Miscellaneous Contributions* 1:1–165.

Watters, G.T., M.A. Hoggarth, and D.H. Stansbery. 2009. *The Freshwater Mussels of Ohio*. Ohio State University Press, Columbus, Ohio. ii+421 pp.

Watters, G.T., and S.H. O'Dee. 1998. Metamorphosis of freshwater mussel glochidia (Bivalvia: Unionidae) on amphibians and exotic fishes. *American Midland Naturalist* 139(1):49–57.

Watters, G.T., T. Menker, S. Thomas, and K. Kuehnl. 2005. Host identifications or confirmations. *Ellipsaria* 7(2):11–12.

Welter-Schultes, F.W. 1999. Systematisches Conchylien-Cabinet von Martini und Chemnitz (1837–1920), bibliography of the volumes in Göttingen. *Archives of Natural History* 26(2):157–203.

West, R.C., N.P. Psuty, and B.G. Thom. 1969. The Tabasco lowlands of southeastern Mexico. Technical Report no. 70., Coastal Studies Institute, Louisiana State University, Baton Rouge. 207 pp.

Whelan, N.V., A.J. Geneva, and D.L. Graf. 2011. Molecular phylogenetic analysis of tropical freshwater mussels (Mollusca: Bivalvia: Unionoida) resolves the position of *Coelatura* and supports a monophyletic Unionidae. *Molecular Phylogenetics and Evolution* 61(2):504–514.

Wiley, E.O. 1976. The phylogeny and biogeography of fossil and recent gars (Actinopterygiii: Lepisosteidae). University of Kansas Museum of Natural History Miscellaneous Publication No. 64, Lawrence, Kansas. 111 pp.

Willey, G.R., W.R. Bullard, Jr., J.B. Glass, and J.C. Gifford. 1965. *Prehistoric Maya settlements in the Belize Valley*. Papers of the Peabody Museum of Archaeology and Ethnology, Harvard University, Cambridge. Vol. 54. 589 pp.

Williams, J.D., A.E. Bogan, and J.T. Garner. 2008. *Freshwater Mussels of Alabama & the Mobile Basin in Georgia, Mississippi & Tennessee*. University of Alabama Press, Tuscaloosa. xv+908 pp.

Williams, J.D., A.E. Bogan, R.S. Butler, K.S. Cummings, J.T. Garner, J.L. Harris, N.A. Johnson, and G.T. Watters. 2017. A revised list of the freshwater mussels (Mollusca: Bivalvia: Unionida) of the United States and Canada. *Freshwater Mollusk Biology and Conservation* 20(2):33–58.

Williams, J.D., M.L. Warren, Jr., K.S. Cummings, J.L. Harris, and R.J. Neves. 1993. Conservation status of freshwater mussels of the United States and Canada. *Fisheries* 18(9):6–22.

Wilson, C.B. 1916. Copepod parasites of fresh-water fishes and their economic relations to mussel glochidia. *Bulletin of the Bureau of Fisheries* 34:333–374+plates 60–74. [Issued separately as U.S. Bureau of Fisheries Document 824].

Winckworth, R. 1936. Journal de Conchyliologie: Dates of publication. *Proceedings of the Malacological Society*, London 22(3):153–156.

Wright, B.H. 1888. *Check List of North American Unionidae and Other Fresh Water Bivalves*. 8 p. (unpaginated). Dore & Cook. Book and Job Printers Portland, Oregon.

Zanatta, D.T., A. Ngo, and J. Lindell. 2007. Reassessment of the phylogenetic relationships among *Anodonta*, *Pyganodon*, and *Utterbackia* (Bivalvia: Unionoida) using mutation coding of allozyme data. *Proceedings of the Academy of Natural Sciences of Philadelphia* 156:211–216.

Zanatta, D.T., and R.W. Murphy. 2006. Evolution of active host-attraction strategies in the freshwater mussel tribe Lampsilini (Bivalvia: Unionidae). *Molecular Phylogenetics and Evolution* 41(1):195–208.

Zepeda Quintero, P.I., and M. Valadez Moreno. 2011. Moluscos en contextos arqueológicos del Noreste de México. pp. 39–70 *In* A. Velázquez Castro, E.R.M. Tísoc and L.G. Gastélum (eds.). *Moluscos Arqueológicos de América*. Universidad de Guadalajara, Guadalajara, México. 366 p.

Zieritz, A., J. Pfeiffer, K.A.A. Rahim, H. Prayogo, M.S. Anwari, F. Diba, E. Froufe, T. Blackwell, H. Hartikainen, and M. Lopes-Lima. 2024. High endemic freshwater mussel (Bivalvia: Unionida) diversity in western Borneo, with description of three new species. *Zoological Journal of the Linnean Society* 2024(3):1–15.

Zilch, A. 1967. Die typen und typoide des Natur-Museums Senckenberg, 39: Mollusca, Unionacea. *Archiv für Molluskenkunde* 97(1/6):45–154.

Zilch, A. 1970. Fritz Haas (1886–1969). *Archiv für Molluskenkunde* 100(1–2):1–17.

Zimmerman, L.L., and R.J. Neves. 2002. Effects of temperature on duration of viability for glochidia of freshwater mussels (Bivalvia: Unionidae). *American Malacological Bulletin* 17:31–35.

Zimmerman, L.L., R.J. Neves, and D.H. Smith. 2003. Control of predacious flatworms *Macrostomum* sp. in culturing juvenile freshwater mussels. *North American Journal of Aquaculture* 65:28–32.

Appendix 1. Glossary

DERIVED FROM THORP AND COVICH (2010)

Actinodont: Descriptive of bivalve shell having a hinge with teeth that radiate outward from the umbos.

Adductor muscle(s): In bivalves, the muscle, either one or two, connecting valves closing the valves on contraction.

Adductor scar(s): The depressed scars of the muscles of a bivalve shell. A differentiated area on the interior of a bivalve shell that marks the attachment point of an adductor and is generally striated, depressed, or outlined. Attachment scars of adductor muscles inside a bivalve shell.

Alate: Winged or wing-like.

Anadromous: Ascending rivers from the sea for spawning.

Anterior end: The shorter end of the shell as measured from the umbo. The foot of the animal is directed toward this end.

Aperture, excurrent: Opening of the mantle at the posterior end of the mussel which expels water, wastes, glochidia, or sperm; may be tube-like and located above the incurrent aperture.

Aperture, incurrent: Opening of the mantle at the posterior end of the mussel which brings water and other resources into the body of the mussel; may be tube-like and located below the excurrent aperture.

Basal: At or pertaining to the base or point of attachment to or nearest the main body.

Beak: The tip of the dorsal margin of the shell generally pointed and located near the hinge. Also called the umbo.

Beak cavity: The depression or pocket on the inside of each valve leading into the beak.

Beak sculpture: The raised loops, ridges, or bumps on the beak or umbo.

Biangulate: Having two angles or corners.

Bivalve: A member of the class Bivalvia of Mollusca, generally bearing a shell of two valves, left and right, and lacking a radula. Commonly called clams, mussels, oysters, scallops, cockles, shipworms, etc.

Bradytictic: Mussel species with females gravid for extended periods, usually becoming gravid in late summer or early fall and remaining gravid through the spring or early summer.

Branchiae: Gills.

Cardinal teeth: Conical projections formed at the center of the shell-hinge plate which interdigitate to form the fulcrum on which the valves open and close.

Chevron: A V-shaped marking.

Compressed: Flattened or pressed together laterally.

Concentric: Having a common center.

Conglutinate: Mass of freshwater mussel larvae bound by mucus; often resembling prey items of glochidial fish hosts.

Ctenidium/ctenidia: The molluscan gill in bivalves, consisting of two demibranchs formed from the reflection of right and left gill filaments upon themselves.

Cusps: The highest elevations on the lateral teeth.

Demibranch(s): Portion of the bivalve ctenidium (gill) formed from a set of filaments reflecting upon themselves; in each ctenidium, the inner lateral filaments form the inner demibranch lying next to the visceral mass and the outer lateral filaments form the outer demibranch lying next to the mantle.

Dentition: Hinge teeth considered collectively.

Depressed: Dorsoventrally flattened.

Distal: Away from the center or origin.

Dorsal: Referring to the top part of the shell where the umbo, hinge, and ligament are located.

Edentate: The condition when teeth are absent from the hinge line.

Elliptical: Having the form of an ellipse.

Elongate: Long or extended.

Endemic: Restricted to a particular geographic region.

Epidermis: See periostracum.

Estivation: General term for a physiologic process whereby animals reduce metabolism and activity levels to survive harsh conditions, such as pond drying for freshwater bivalves.

Exhalent: Mantle opening or siphon where water is expelled from the mantle cavity.

Extant: A population or species present as live individuals in a particular area.

Extirpated: A population or species no longer present as live individuals in a particular area.

Fish host: Species of fish parasitized by the glochidia or lasidium of a freshwater mussel.

Foot: The muscular part of the body used for movement on or within the substrate; located anteriorly.

Gill filaments: In bivalve molluscs, long, thin, fused lateral extensions from the gill axis which extend ventrally and then reflect dorsally to form the inner and outer demibranchs.

Globbose: Globe-like or spherical in shape.

Glochidia, hooked: Glochidia larvae in bivalve molluscs with spiny hooks on their ventral shell margin.

Glochidia/glochidium: The parasitic bivalved larval stage of a freshwater mussel.

Gonochoric: Sexually reproducing species in which individuals have one of at least two distinct sexes.

Gravid: Condition of a female when glochidia or young mussels are present in the gills.

Growth lines: Darkened lines on the surface of the shell indicate periods of rest during growth.

Hermaphroditic: An individual that has both male and female sexual organs.

Hinge ligament: Elastic proteinaceous portion of the shell of bivalve molluscs secreted by the mantle isthmus which connects the shell valves; it is flexed or compressed when the valves are closed and when adductor muscles relax and function to open the valves by returning to an unflexed state.

Hinge: Collective term for structures of the dorsal region that function during opening and closing of the valves.

Hinge teeth: Pseudocardinal and lateral teeth that hold the valves in position and serve as the fulcrum on which they open and close.

Incurrent: Mantle forming a passage for water drawn into the mantle cavity from outside.

Inflated: Swollen or expanded.

Interdentum: A flattened area between the pseudocardinal and lateral teeth.

Interlamellar space: A cavity formed between ascending and descending gill filaments. Water flowing through the ostia of gill filaments enters the interlamellar space to be carried dorsally to the epibranchial cavity, also called as the water tube or water channel.

Iridescent: Exhibiting rainbow colors.

Labial palps: A pair of thin, flattened structures extending from either side of the mouth region of bivalves that lie against the outer surface of the outer demibranch and the inner surface of the inner demibranch. They function to receive filtered particles from the ctenidia and sort them into those accepted for ingestion and those rejected as pseudofeces.

Lanceolate: A narrow oval shape tapering to a point.

Lasidia: Small uncalcified, univalved larval shell with a trilobed body.

Lateral teeth: The elongated lamellar teeth along the hinge line of the shell situated posterior to the pseudocardinal teeth in freshwater mussels.

Left valve: The shell half on the left side when the hinge is facing up and the anterior end directed forward.
Ligamental Sinus: A brown triangular-shaped blotch at the posterior end of the hinge.
Mantle: The folds or lobes of the body wall of a mollusk that lines the shell and bears shell-secreting glands.
Mantle cavity: A space formed between the body and surrounding mantle tissues completely surrounding the body in bivalves.
Margin: The edge of the shell.
Marsupium: A brood pouch for eggs and developing glochidia or lasidia, formed in the outer gill or part of the outer gill in the Lampsilini and Popenaiadini.
Muscle scar: Impression on the interior of the shell marking the former place of attachment of a muscle.
Nacre: The interior layer of the shell, usually white, pink, salmon, or purple and iridescent.
Obliqual: Oblique in shape.
Oblong: Oval.
Obovate: Ovate with the narrower end basal.
Ovate: Egg-shaped.
Pallial line: A line of muscle scars on the outer perimeter of the inside of the shell, marking points at which pallial muscles attach the mantle to the shell. The line of attachment of the mantle.
Papillae: Small finger-like projections associated with the outer margins of the apertures, and in some species, associated with the outer margins of the mantle.
Papillose: Covered with, resembling, or bearing papillae.
Pedal elevator muscle: A thin bundle of muscle fibers attached to the shell in the umbonal cavity and serving to raise the foot.
Perforate: To make a hole through.
Periostracum: The proteinaceous outer layer of mollusk shells secreted by the mantle edge.
Plications: Parallel ridges or folds.
Posterior end: The longer end of the shell as measured from the umbo. The end with the siphons, usually sticking above the stream bottom.
Posterior ridge: The ridge of the external surface of the valve running from the umbo to the posterior-ventral margin.
Posterior slope: The surface of valve running posteroventrally from the umbo along the dorsal part of the shell between the posterior ridges.
Posteroventrally: Situated posteriorly and ventrally.
Pseudocardinal teeth: The triangular often serrated teeth located on the anterior-dorsal part of the shell.
Pustule: A protuberance, bump, or raised knob on the surface of a bivalve mollusk shell.
Quadrate: Square or rectangular in shape.
Rays: Lines of color on the external surface of the shell.
Rhomboid: A quadrilateral in which only the opposite sides and angles are equal.
Right valve: The shell half on the right side when facing up and the anterior end directed forward.
Schizodont: A hinge with reversed V-shaped teeth and often an elongated lateral tooth. This arrangement is found in most Unionidae freshwater mussels.
Serrated: Notched or grooved.
Sexually dimorphic: A condition in which males and females of the same species are morphologically different, usually indicated in mussels by an expanded posterior marsupial area in the female in contrast to a more pointed or bluntly rounded area in the male.
Shell: The calcareous exoskeleton made up of calcium carbonate, such as aragonite or calcite, deposited within an organic matrix, which encloses, supports, and protects the soft parts of an animal.

Siphons: The tubular structure at the posterior end formed by fusion of opposite mantle margins directing inhalant (inhalant siphon) and exhalant (exhalant siphon) respiratory and feeding water currents into and out of the mantle cavity, respectively.
Solid: Thick or heavy.
Species richness: The number of species present in a place or habitat.
Striate: Containing parallel lines or grooves (striations or striae).
Subcentral: Not quite central; off-center.
Subelliptical: Somewhat elliptical in shape; approaching an ellipse.
Subovate: Less than ovate.
Sulcate: Having long, narrow grooves or channels.
Sulcus: A shallow depression or furrow on the surface of the shell.
Synapomorphy: A character shared by two or more taxa and is hypothesized to have evolved in their most recent common ancestor. In cladistics, synapomorphy implies homology.
Trapezoidal: A quadrilateral that has at least one pair of parallel sides.
Truncate: Having the end shortened or squared off.
Tubercle: A pointed, rounded, or knob-like projection on a shell.
Umbo: The inflated dorsal part of the shell. Also called the beak.
Unequilateral: Not having all sides the same length.
Univalved: Having one valve.
Valve: One of the two halves of the shell.
Ventral margin: Referring to the bottom part of the shell, opposite the hinge line.
Vestigial: An organ or part of the body, degenerate, rudimentary, or atrophied, having become functionless in the course of evolution.
Wing: A dorsal structure, more or less elongate, triangular, distally acute, terminal part of a shell.

Index

Note: **Bold** page numbers refer to tables and *italic* page numbers refer to figures.